Principles and Applications of Room Acoustics

Volume 1

Principles and Applications of Room Acoustics

Volume 1

I: Geometrical Room Acoustics

II: Statistical Room Acoustics

III: Psychological Room Acoustics

Volume 2

IV: Wave-Theoretical Room Acoustics

Principles and Applications of Room Acoustics

Volume 1

LOTHAR CREMER

Emeritus Professor of Acoustics
Technical University of Berlin, FRG

and

HELMUT A. MÜLLER

Director of Müller BBM, Munich, FRG

Translated
By

THEODORE J. SHULTZ

Principal Scientist, Architectural Acoustics,
Bolt, Baranek and Newman Inc., Cambridge, USA

PENINSULA
PUBLISHING

Principles and Applications of Room Acoustics

This book is the translation of
Cremer/Müller,
Die Wissenschaftlichen Grundlagen
Der Raumakustick, Vol. 1.

Original Edition
British library Cataloguing in Publishing Data
Cremer, Lothar
Principles and applications of room acoustics.
Vol. 1
1. Architectural acoustics
I. Title II. Müller, Helmut A. III. Die
Wissenschaftlichen Grundlagen der Raumakustik
English
729'.29 NA2800

Original ISBN: 0-85334-114-1

Reprint ISBN: 978-0932146-29-8

Library of Congress Control Number: 2016937048

WITH 4 TABLES AND 109 ILLUSTRATIONS

ENGLISH LANGUAGE EDITION

Published by arrangement with Taylor & Francis Books LTD,
a member of the Taylor and Francis Group.

Peninsula Publishing, Westport, Connecticut USA
Telephone: 203-292-5621 E-mail: sales@peninsulapublishing.com

Preface to the Reprint Edition

Interest in room acoustics, at least the written specification thereof, can be said to begin in biblical times, with the acoustical specifications found in Exodus XXVI: 7—"And thou shalt make curtains of goats' hair for a tent over the tabernacle. . . The length of each curtain shall be thirty cubits, and the breadth of each curtain shall be four cubits. . ."[1] Since that early citation, there has been a large collection of work citing acoustics by many authors including Aristotle, Galileo, Mersenne, Hooke, Taylor, Bernoulli, D'Alembert, Euler, and Saveur. It was not until Galileo's time that the theory of pitch, originally conceived by the Greeks, was associated with the motion of a vibrating body. And Saveur is probably best known for his suggestion of the name *acoustics* for the science of sound. Acoustics is derived from the French word acoustique, which was derived from the Greek word *akoustikos*, meaning of or for hearing, which is from the word *akouein*, to hear. The other authors are better remembered for their contributions to mathematics.[2]

Until the mid-nineteenth century though, the works about acoustics were mainly qualitative, rather than quantitative, in their descriptions. John Henry's[3] work in the mid-nineteenth century was a study of auditorium acoustics, which were mainly qualitative in nature. Despite Henry's work, architects either ignored, or were ignorant of, his studies, and attempts were often made to correct gross acoustical defects by the use of such absurd devices as strung wires and ash-filled glass bottles, among others. The later part of the nineteenth century saw the publication of the works of Hermann von Helmholtz[4] and Lord Rayleigh[5] (John William Strutt). Through research using modern

---

[1] J. H. Hertz (editor), *The Pentateuch and Haftorahs*, Soncino Press, London, 1987

[2] "Books and Acoustics, Especially Wallace Clement Sabine's Collected Papers on Acoustics," N. A. Shaw, J. Klapholz, and M. R. Gander, paper 1aAAb2, *Proceedings Wallace Clement Sabine Centennial Symposium*, Acoustical Society of America, Woodbury, NY, 1994

[3] "Acoustics Applied to Public Buildings," Joseph Henry, Annual Report of the Smithsonian Institution, 1856

[4] Hermann L. F. Helmholtz (Alexander J. Ellis, translator, Third Edition), *On the Sensations of Tone*, Fourth Edition, Longmans, Green, and Co., London, 1895 (reprint by Dover)

[5] John William Strutt, Lord Rayleigh, *The Theory of Sound*, Second Edition, Macmillan, London, 1894 (reprinted by Dover)

scientific experimentation and mathematical descriptions of the phenomena these two giants of science proved, disproved, and further developed many theories, providing the basis for all contemporary work in acoustics.

Modern room acoustics analysis can be said to have begun in the waning years of the nineteenth century with Dr. Wallace Clement Sabine's[6] investigations of the acoustical problems of the Fogg Museum Lecture Hall at Harvard University. With the far-reaching changes in cosmopolitan society of that period, the city soundscape was significantly changing due to the introduction of motor vehicles, high-rise buildings with interior ventilation systems, radio, and electronic sound systems[7], and the acoustics of indoor spaces needed to be taken into account as this had a profound effect on the indoor environment. This new built environment also manifested itself in the increase in the number of spaces for performance. Indeed, the time between the world wars was fruitful for architectural acoustics and saw the publication of two classic books: F. R. Watson's[8] *Acoustics of Buildings* in 1923 and Vern O. Knudsen's[9] *Architectural Acoustics* in 1932.

By this time it was known that in a closed space we perceive the original sound with the addition of many reflections which either improve the original's character and intelligibility or, alternatively, degrade the original sound. Room acoustics is the study of this combination and its total perception from seat to seat in a room. Rather than attempt to analyze such a complex phenomenon in one view, several different idealized models can be used: geometrical acoustics, statistical acoustics, and wave acoustics.

Geometrical acoustics, the employment of optic techniques in acoustical analysis, defines modeling of room acoustics geometrically in a rudimentary way. Basically, one graphically ray-traces a sound from a source, and includes reflections from multiple surfaces in a space via multiple paths to their final observation at a specific location. This technique has a few caveats, similar to geometrical optics. For a specular reflection to take place, a reflecting surface must be comparable in size to the wavelength of a sound, and smooth compared to this wavelength. More importantly, these numerous momentary sound paths do not ordinarily reach their destination simultaneously. Therefore, modeling of the sound path must contain both a spatial and temporal component.

While geometrical acoustics plots the direct sound and the first few reflections, statistical acoustics describes the reverberant sound resulting from an almost infinite number of room reflections. These multitudes of reflections are simplified statistically to correlate the average reverberant sound level in the room and its decay rate to the

---

[6] Wallace Clement Sabine, *Collected Papers on Acoustics*, Harvard University Press, Cambridge, 1922 (reprinted 1992 by Peninsula Publishing for the Acoustical Society of America)

[7] Emily Thompson, *The Soundscape of Modernity - Architectural Acoustics and the Culture of Listening in America*, 1900-1933, MIT Press, Cambridge, 2004

[8] Floyd Rowe Watson, *Acoustics of Buildings*, John Wiley & Sons, New York, 1923

[9] Vern O. Knudsen, *Architectural Acoustics*, John Wiley & Sons, New York, 1932 (reprinted by the Acoustical Society of America)

various room surfaces' absorption characteristics. The first such scientific relation incorporating a statistical vantage point was the Sabine formula:

$$T = 0.161\ V/A$$

where T is the reverberation time in seconds, 0.161 is a constant with units of seconds/meter, V is the room volume in cubic meters and A is the room absorption expressed in Sabins, which in this case, has dimensions of square meters. The Sabine formula, however, breaks down at the extreme when the room becomes fairly absorbent. This dichotomy between the theory and analysis became a topic of much debate in the 1960s and into the 1970s (see, for example, F. V. Hunt[10] and W. B. Joyce[11]) and still continues today.

Wave acoustics is the application of wave theory to acoustics that dates back to Lord Rayleigh, who published wave equations and the expressions for normal modes in rooms and applied them to room resonance control methods. In the mid-1920s, E. T. Paris[12], who invented the stationary-wave tube for measuring normal incidence absorption and impedance of materials, further developed the concept of acoustic impedance and its relation to sound absorption. *Sound Waves in Rooms*, a 1944 review paper by Philip M. Morse and Richard H. Bolt[13], looks at the application of wave acoustics in rooms.

In the years following the Second World War there was another burst of activity in architectural acoustics research and analysis. By the middle of the twentieth century, progress in the development of the quantitative analysis of room acoustics was such that in 1948 Dr. Lothar Cremer, then at the University of Munich, under the imprint of S. Hirzel Verlag, Leipzig, published the first of three planned volumes, *Die wissenschaftlichen Grundlagen der Raumakustik, Band I, Geometrischen Raumakustik*, which presented room acoustics from the geometrical perspective. This was followed by volume three, *Die wissenschaftlichen Grundlagen der Raumakustik, Band III, Wellentheoretische Raumakustik*, in 1950, which presented the wave-mechanical description. The middle volume, *Die wissenschaftlichen Grundlagen der Raumakustik, Band II, Statistische Raumakustik*, did not appear until 1961, and dealt with the statistical treatment of sound rays. Indeed, with the publication of the second and the last of the three original volumes, the three primary analytical models for describing room acoustics were developed and described in a unified and ordered manner.

The years following the Second World War saw the proliferation of concert halls and venues for performance, with the attendant increase in interest by the public,

---

[10] "Remarks on the Mean Free Path Problem," F. V. Hunt, Journal of the Acoustical Society of America, **36**, 556-564, 1964

[11] "Sabine's Reverberation Time and Ergodic Auditoriums," W. B. Joyce, Journal of the Acoustical Society of America, **58**(3), 643-655, 1975

[12] "On the Stationary-Wave Method of Measuring Sound Absorption at Normal Incidence," E. T. Paris, Proceedings of the Physical Society of London, **39**, 269-295, 1927

[13] "Sound Waves in Rooms," Philip M. Morse and Richard H. Bolt, Reviews of Modern Physics, **16**(2), 69-150, April 1944.

architects, and acoustic designers in the acoustics of spaces for listening. The 1950s and 1960s saw a profusion of new titles concerned with room acoustics and architectural acoustics. They presented information not only on the theoretical analysis of rooms, the practical design formulas for the design of rooms, but also the means of measuring the final product. Leo. L. Beranek's[14, 15, 16] several works from this time are well known examples.

In the late 1960s, at the request of the publisher S. Hirzel, Dr. Cremer, now working with Dr. Helmut A. Müller, Director of Müller BBM, revised and updated the three volumes, issuing a second German edition of *Die wissenschaftlichen Grundlagen der Raumakustik*, now in two volumes. Volume II, published in 1976, was now devoted solely to what the authors call *wave-theoretical room acoustics*. Volume I, published in 1978, now had three parts: Geometrical Room Acoustics, Statistical Room Acoustics, and a new section, Psychological Room Acoustics. These two new volumes were well received and many positive reviews of the books appeared in professional and trade journals.[17, 18, 19] The addition of the psychological room acoustics section by Dr. Cremer in the second edition expanded on the physical design tools available to a designer to include the human response to sound in a room. Today, we take the effects of perception of sound for granted in our designs, witness the numerous metrics that are meant to provide an objective description of various perceived or measured phenomena. These include, but are not limited to, metrics that describe loudness, envelopment, and warmth as well as many other time and energy based measures.[20]

In 1982, an English translation of Cremer's and Muller's seminal two-volume work by Dr. Theodore J. Shultz, of the consulting firm Bolt, Beranek and Newman in Boston, appeared, entitled *Principles and Applications of Room Acoustics*. It became an instant—for a technical book—best seller. Copies of the book were soon hard to come by after the initial publication. In fact, copies were known to mysteriously disappear. Attempts to find used copies of either or both of the two volumes were usually unsuccessful. Even with the advent of the Internet, attempts to secure a copy of the volumes were fruitless as were continued efforts to purchase a copy from those who were known to (luckily) still have their copies.

---

[14] Leo L. Beranek, *Acoustical Measurements*, John Wiley & Sons, New York, 1949 (reprinted 1988, as *Acoustical Measurements*, by the Acoustical Society of America)

[15] Leo L. Beranek, *Acoustics*, McGraw Hill Book Company, New York, 1954 (reprinted 1986 by the Acoustical Society of America)

[16] Leo L. Beranek, *Noise Reduction*, McGraw Hill Book Company, New York, 1954 (new edition issued as *Noise and Vibration Control* in 1971; a revised edition by the Institute of Noise Control Engineering was published in 1988; a reprint edition by Peninsula Publishing was published in 1991; and then a new edition called *Noise and Vibration Control Engineering* was published by John Wiley & Sons in 1992)

[17] M. S. Schroder, Journal of the Acoustical Society of America **62**(6), 1977

[18] P. H. Werner, *Technishe Mitteilungen* **248**(6), 1978

[19] M. S. Schroeder, Journal of the Acoustical Society of America **65**(1), 1979

[20] "Architectural Acoustics Tutorial Lecture," Neil A. Shaw, Richard Telaske, and Silvio Bistafa, First Pan-American/Iberian Meeting on Acoustics, Cancun, Mexico, 2002

Dr. Shultz's translation is clearly written. The development of the book flows in a smooth manner, with the authors' taking you to unexpected places. For example in Volume II, in the discussion of thin partition walls, we learn why the crack of a whip sounds so sharp under a circus tent. Similarly, in Part I of the first volume, we are taken from the amphitheaters of antiquity to a simple explanation for the raked seating in auditoria and how the excess absorption characteristic depends, in part, on that rake.

During the years that *Principles and Applications of Room Acoustics* has not been readily available to the researcher, theoretician, and practitioner, there has been a surfeit of alternate texts to turn to and to refer during its absence. Books directed to architects and others allied to the building trades include those by David Egan[21], James Cowan[22], and Madan Mehta, Jim Johnson, and Jorge Rocafort[23]. More advanced treatments of the subject can be found in Yoichi Ando's[24] and Heinrich Kutruff's[25] texts. Books that deal with acoustics in general, and are not limited to architectural acoustics, include those by Lawrence Kinsler and Austin Frey[26], Phillip Morse and Uno Ingard[27], and Allan Pierce[28]. Marshall Long[29] provides a complete contemporary resource for all things acoustical in the built environment. Michael Barrons's[30] text and Leo L. Beranek's[31, 32, 33] three survey books on performance spaces provide a introduction and tutorial to what works, and how well. And for electroacoustics there are the classic texts by Frederick Hunt[34], and Harry Olson[35], and

---

[21] M. David Egan, *Architectural Acoustics*, McGraw-Hill, Inc., New York, 1988

[22] James Cowan, *Architectural Acoustics and Design Guide*, McGraw-Hill Companies, New York, 2000

[23] Madan Mehta, Jim Johnson, and Jorge Rocafort, *Architectural Acoustics—Principles and Design*, Prentice-Hall, Inc., Upper Saddle River, NJ, 1999

[24] Yoichi Ando, *Concert Hall Acoustics*, Springer Verlag, New York, 1985

[25] Heinrich Kutruff, *Room Acoustics*, Fourth Edition, Spon Press, London, 1999

[26] Lawrence E. Kinsler, Austin R. Frey, Allan R. Coppens, and James V. Sanders, *Fundamentals of Acoustics*, John Wiley & Sons, New York, 2000.

[27] Philip Morse and Uno Ingard, *Theoretical Acoustics*, McGraw-Hill, Inc., 1968 (reprinted by Princeton University Press in 1986 with an errata)

[28] Allan D. Pierce, *Acoustics: An Introduction to Its Physical Properties and Applications*, McGraw-Hill, Inc, New York, 1981 (reprinted 1989 by the Acoustical Society of America)

[29] Marshall Long, *Architectural Acoustics*, Second Edition, Academic Press, Oxford, 2014

[30] Michael Barron, *Auditorium Acoustics and Architectural Design*, E & FN Spon, London, 1993 (reprinted 1998, 2000)

[31] Leo L. Beranek, *Music, Acoustics & Architecture*, John Wiley & Sons, New York, 1962

[32] Leo L. Beranek, *Concert and Opera Halls—How They Sound*, Acoustical Society of America, Melville, NY, 1996

[33] Leo L. Beranek, *Concert Halls and Opera Houses—Music, Acoustics & Architecture*, Springer, New York, 2003

[34] Frederick V. Hunt, *Electroacoustics*, Harvard University Press, Cambridge, 1954 (reprinted by the Acoustical Society of America)

[35] Harry F. Olson, *Acoustical Engineering*, D. Van Nostrand Company, Princeton, 1957 (reprint edition in 1991 by Professional Audio Journals)

Mario Rossi's[36] comprehensive treatment, Mendel Kleiner's[37] recent text, as well as John Eargle's loudspeaker[38] and microphone[39] books.

In the very early years of the twenty-first century, Damian Doria of Artec Consultants, New York, made a request to the Books+ Committee of the Acoustical Society of America that a reprint edition of *Principles of Applications of Room Acoustics* be issued. In August 2005 Commander William Murphy, U. S. Public Health Service, suggested to Peninsula Publishing that reprinting *Room Acoustics* would be a valuable contribution to the acoustics community. The publisher researched the project, obtained the rights to republish from Taylor & Francis Books Ltd, Abington, Oxford UK in January 2006, and requested that I write a preface to this edition. Dr. Leo Beranek provided his copy of the original translation edition for review in the preparation of this preface. And, with copies of the book almost as scarce as unicorns, Dr. Lily Wang of the University of Nebraska graciously offered her copy for use in preparing the reprint.

NEIL A. SHAW
MENLO SCIENTIFIC ACOUSTICS, INC.
TOPANGA, CALIFORNIA

---

[36] Mario Rossi, *Acoustics and Electroacoustics*, Artech House, Norwood, MA, 1988

[37] Mendel Kleiner, *Electroacoustics*, CRC Press, Boca Raton, 2013

[38] John Eargle, *Loudspeaker Handbook*, Kluwer Academic Publishers, Norwell, MA, 2003

[39] John Eargle, *The Microphone Book*, Focal Press, Woburn, MA, 2004

Contents

Notation

The following symbols, used in Parts I, II and III, are grouped in alphabetical order according to type of letter. If the same letter is used for different purposes, the Part and Section in which it is first used in each case is given. Letters that appear in only one equation, where the meaning is immediately clear, are omitted here, as are mathematical symbols. Subscripts are mentioned in only a few cases.

Italic Capital Letters

A Equivalent absorption area
B Irradiation strength (II.2.4);
  Slit separation (II.6.4)
C Clearness index
D Diameter of a sphere
E Sound energy density
F Frequency index (III.1.3);
  Force (III.1.4);
  Factor matrix (III.2.8);
  Factor axis (III.2.9)
G Strength index
H Reverberation distance
J Sound intensity
K Coupling coefficient
L Level;
  Sound pressure level (also L_p, III.1.2);
  Loudness level (II.1.2), (also L_s, III.1.2)
L_A A-weighted sound pressure level (III.3.4)
N Number of reflections (II.2.1);

Number of objects (III.2.8);
Loudness (III.1.2)

P Sound power (II.1.3);
Projection surface (II.2.1)

R Radius (I.1.5);
Reduction index (transmission loss) (II.3.2);
Flow resistance (II.6.3);
Reverberation index (II.7.4);
Correlation matrix (III.2.8)

S Surface (II.1.4);
Loudness (III.1.2)

T Duration of period (I.1.1);
Reverberation time (II.1.1)

V Volume

W Weighting matrix (factor load)

X Data matrix

Z Normalized data matrix

Italic Lower Case Letters

a Radius of a circular hole (II.6.4)
Radius of a sphere (II.2.1)

b Slit width

c Sound propagation speed

d Diameter of a circle (I.5.3);
Layer thickness (II.6.3);
Air-layer thickness (II.6.4)

e Receiver distance (I.3.2);
Separation between perforations (II.6.6)

f Frequency (I.1.1);
Factor (-score) (III.2.3)

g Acoustical quality of a concert hall, or of a seat in the hall

h Height (for example, of a dome above the floor) (I.3.4)

k Coupling factor

l Length, width, height (I.2.2);
Plate thickness (II.6.4)

l_m Mean free path length

m Mass per unit area (II.6.4);
Mass (III.1.4);
Number of factors (III.2.8)

n Integral number (I.1.5);
Number of attributes (III.2.8)
p Sound pressure
p_{ik}, p'_{ik} Transition probability
r Radius of a curve (I.3.2);
Radius in polar coordinates (I.6.4)
r_{H} Reverberation radius
s Source distance (I.3.2);
Spring stiffness (III.2.8)
s_{ii} Variance of attributes (III.2.8)
s_{ik} Covariance of attributes i and k (III.2.8)
t Time (I.2.1);
Duration of reverberation (II.4.1)
v Particle velocity
w Weighting (factor load)
x, y, z Position coordinates

Gothic Letters

$\mathfrak{r}$ Location vector
$\mathfrak{v}$ Percent syllable articulation
$\mathfrak{w}$ Row or column in the weighting matrix W

Greek Capital Letters

Δ Prefix indicating a small change (for example, Δl = orifice correction (II.6.4))
Λ Diagonal matrix of eigen-values
Ξ Specific flow resistance
Ω Solid angle

Greek Lower Case Letters

α Absorption coefficient
α_{S} Sabine absorption coefficient
α' Absorption exponent
α'' 'Effective' exponent
γ Strength coefficient
δ Dissipation coefficient (I.6.1);
Damping constant (II.2.3)
ε Glancing angle;

	Viewing angle (I.6.3);
	Echo coefficient (II.7.5)
η	'Signal-to-noise ratio'
ϑ	Incidence angle (I.2.2);
	Distinctness coefficient (II.7.4)
κ	Coupling coefficient
$\kappa_{(\tau)}$	Correlation function
λ	Wavelength (I.1.1);
	Eigen-value (II.2.4);
	'Liveness' (II.7.3).
μ	Scale of model (I.7.3);
	Dissipation coefficient (II.1.6)
ρ	Reflection coefficient (I.6.1);
	Volume density of air (II.6.3)
σ	Porosity factor (II.6.3);
	Standard deviation (II.6.3),
τ	Transmission coefficient (I.6.1);
	Energy decay time (II.1.1);
	Time shift (II.7.5);
	Relaxation time (III.1.4)
ϕ	Angle in polar coordinates (I.6.3);
	Relative humidity (II.1.6);
	Autocorrelation function (II.7.5);
	Azimuth angle (III.1.5)
χ	Structure factor
ω	Radian (angular) frequency

Part I:
Geometrical Room Acoustics

Chapter I.1

Sound Waves and Sound Rays

I.1.1 General Physical Properties of Sound

Sound is wave motion within matter, be it gaseous, liquid or solid. It exhibits, in many respects, behavior similar to other wave motions that we encounter in nature. In particular, there are two kinds of wave phenomena whose propagation laws we can readily observe with our eyes: namely, water waves (which usually come to mind when we think of 'waves') and light waves; in the latter case, special procedures are required for us actually to 'see' the light wave propagation.

Since water waves share many of the physical characteristics of sound waves (because an oscillation of matter is involved in both cases) we will often find it useful, in discussing the properties of sound waves, to refer to the corresponding property in water waves. With water waves the phenomenon is visible, with sound waves (usually) not.

Accordingly, we begin by introducing a fundamental property of *all* wave motion, and illustrate it with water waves. Namely, the matter itself does not travel with the wave; rather, the wave represents a disturbance or modification of the matter, and it is this disturbance that propagates away from the source.

When we see waves on the surface of the sea, the water does not travel. Instead, the particles of water move up and down, and back and forth, describing an approximately circular local motion. The succession of neighboring particles of water, all executing their circular motions 'out of step', gives the typical wave shape to the surface of the water. It is the *shape* of the surface, namely, the wave, that appears to propagate in one direction, not the water itself.

At the seashore, one sees the waves always advancing toward the beach; between waves, one sees the ebbflow washing back out to sea: the waves come in, the flow returns. A swimmer carried out to sea by the undertow would be dangerously mistaken to hope that the incoming waves will bear him safely to shore!

The appearance of wave propagation results, as we have noted, from the sequential behavior of neighboring particles of matter; they all execute the same motion, but with a time delay. The ratio between (1) the distance separating two oscillating particles and (2) this time delay defines the propagation velocity c of the wave.

If the wave in question comprises more than a single, short, up-and-down movement, and, instead, is periodically repeated with the period T, always in the same way, then there will be a waveform that is periodic both in space and in time. Moreover, the distance between points having the same phase on adjacent waves is constant, whatever the phase reference selected for comparison. For water waves, this constant distance is most evident between the crests of successive waves.

The constant distance separating points of equal phase on adjacent waves, corresponding to the time period T, is called the wavelength λ. The speed of propagation of the wave is given by the quotient of λ, the distance traveled by the wave between two instants of equal phase, and T, the time required for this travel:

$$c=\lambda/T \qquad (1.1)^1$$

The quantity which is the reciprocal of the time period indicates how many wave periods occur in a unit of time. This frequency of periods is always meant when we speak simply of frequency.

If, as usual, we choose the second as the unit of time, its reciprocal (which is thus the unit of frequency of periods) has been designated the hertz (Hz), according to international standards, in honor of Heinrich Hertz, the explorer of electromagnetic waves. Thus:

$$f=1/T \qquad (1.2)$$

and:

$$c=f\lambda \qquad (1.3)$$

When we *hear* sound, we perceive the frequency of the wave (an objective quantity) as the pitch of the tone (a subjective experience). The

[1] The numbering of the equations begins anew in each chapter; the first figure in each equation number corresponds to the chapter number. The same is true for the figure numbers. Footnotes are numbered beginning with 1 in each section.

lowest frequency we can hear and still consider a tone is 20 Hz; the highest is about 16 000 Hz.

Below 20 Hz, the individual oscillations tend to be perceived separately, rather than continuously as a tone. Above 20 Hz, the increasingly rapid individual oscillations coalesce and evoke a quite different kind of perception: tonal hearing. Thus, a fast to-and-fro movement of the air is required if we are to hear sound as such.

But fast motion is also a necessary condition for waves (of all kinds) to be propagated away from the point of excitation. If we very slowly dip the point of a cone into a basin of water, the water particles can easily move aside to make way for the intrusion. At some distance away, therefore, no disturbance is perceptible, except that the water level is slightly higher.

But if the cone is inserted quickly into the water, then, because the inertia of the water resists lateral movement, the water finds it easier to overcome gravity and to rise in the vicinity of the cone rather than to move aside. This situation cannot continue: the risen water must flow down again. But, again because of its inertia, the downward accelerating water continues past its equilibrium level, and reverses itself to form another 'water mountain', which propagates away from the point of disturbance as a wave, always composed of new water particles.

The outward-moving wave cannot be prevented (or cancelled) by rapidly withdrawing the cone from the water, because that would only generate a second wave with changed sign (i.e. with crests and troughs reversed). Because the withdrawal of the cone comes *after* the original insertion, the second wave follows the first.

This experiment makes it clear that wave propagation implies not only a kinematic sequence of motion, but also a propagation of mechanical energy. Since, in the outward propagation of water waves, the waves have the form of ever-growing circles, the 'intensity' of the wave (i.e. the power per square centimeter) must decrease as r^{-1}, where r is the distance from the center of excitation. We call this lateral expansion of the wave 'divergence'.

As we move from the two-dimensional world of water waves to the world of airborne sound, we note some important differences. The corresponding process for sound waves in air is more difficult to understand, since the wave motion, instead of moving outward in a plane, expands into a volume, and thus is three-dimensional. The intensity of the wave is distributed over an ever-expanding sphere, and thus decreases as r^{-2}.

We can no longer use the trick of introducing the third dimension (the wave height) to help us visualize the wave *shape*, as in two-dimensional water wave propagation. Moreover, we have seen that the motion of the water particles in a surface wave entails both up-and-down, as well as back-and-forth, movement. In a sound wave, by contrast, the air particles move only back-and-forth along the direction in which the wave is propagating. We call these *longitudinal* waves.

In order to distinguish between (1) the local velocity of the air particles and (2) the velocity with which the waveshape propagates forward, we refer to the first as the 'particle velocity' and to the second as the 'speed of sound'.

Since an air particle at one location along a sound path moves in a direction opposite to that of an air particle one-half wavelength away, it follows that along the sound path there must be periodic changes of the local air density and pressure, the latter being called 'sound pressure'.

The square of the amplitude of any of these time-varying quantities can serve as a measure for the sound intensity. (In the case of water waves, the wave height is the quantity analogous to sound pressure in airborne sound.)

Sound pressures are usually extremely small in comparison with the mean atmospheric pressure. For example, for sounds of moderate strength, such as normal speech at one meter distance, the sound pressure is about 1 microbar ($=1\ \mu b = 0{\cdot}1\ N\,m^{-2} = 0{\cdot}1$ Pa). For sounds that are so loud as to be barely tolerable, the sound pressure is one thousand times greater; the mean atmospheric pressure ($P_0 = 1\ b \approx 100\,000\ N\,m^{-2} = 10^5$ Pa) exceeds normal speech pressures by a millionfold.

Similarly, the particle displacements that occur in airborne sound waves lie far outside our experience with gross mechanical motion. They not only change their direction very frequently with time, but they are so small as to be invisible to the naked eye.

Since the sound intensity depends on the square of the particle velocity, then, for equal sound intensity, the displacement in a sound wave must be greater the lower the frequency. But even at the lowest frequency of interest and the highest intensity of interest, the air particle displacements do not exceed 1–2 mm.

On the other hand, at a frequency of 2000 Hz (the frequency range where our ears are most sensitive to sound), we can detect sounds for which the amplitude of particle displacement is less than 10^{-7} mm, comparable to atomic dimensions.

Thus, with the ear, as well as with the eye, the limits of human perception challenge the limits of measurement.

Electromagnetic light waves propagate so extremely fast that we cannot perceive in daily life that their speed is, in fact, finite; instead, we regard all optical events as simultaneous with our perception of them.

We are, however, quite aware of the finite propagation time for sound waves. It is well known, for example, that one can estimate the distance to a thunderbolt by the time elapsed between the lightning and the thunder: 3 s corresponds to about 1 km.

A more precise accounting of the speed of sound must take into account the temperature of the air (see Section IV. 1.4, Volume 2). The speed of sound at 0°C is about $331\,\mathrm{m\,s^{-1}}$; at 18°C it is $342\,\mathrm{m\,s^{-1}}$. For practical purposes, it is sufficient to use the rounded value $c=340\,\mathrm{m\,s^{-1}}$.

The speed of water surface waves is so slow that we can easily follow their progress by eye, which could never be done with airborne sound. As a further contrast, there is the complication, with water surface waves, that the speed of propagation depends on the wavelength; fortunately, this is not the case with airborne sound.

Not only does airborne sound propagate with a speed that is independent of wavelength (or frequency), it is also, to a first approximation, independent of the sound intensity. To be sure, the sound of extremely intense events, such as explosions or sonic booms, travels faster than ordinary sound; but in room acoustics we do not usually encounter such high sound pressures. Accordingly, we adopt the constant speed recommended above.

For water surface waves, it can be demonstrated that the speed of wave propagation is independent of the wave height; this leads to an observation that is equally important for optics and acoustics. If we cast a stone into calm water, the resulting wave appears as a growing circle. If we cast a second stone some distance away from the first, we observe a second growing circle. The important point is that neither circle is influenced by the other, and what prevails finally is only the additive effect of the two waves.

In the optical regime, this 'superimposability without interaction' exhibits itself in the fact that, although all the objects in a room scatter light in all directions, the eye has no difficulty in analysing the resulting mix of light waves, and can localize and identify each object.

In the acoustic regime the same holds true in principle, but our ears are less capable of discriminating among the various simultaneously sounding sources; it is especially difficult if we close one ear. Also, sonic localization is made more difficult by the fact that sound waves are reflected back and forth within a room to a much greater extent than light waves.

This independence of the individual waves makes it possible to pick out, from the total sound field established by a radiating source, only that part containing the waves that arrive at the observation point.

I.1.2 Tone Analysis and Synthesis

The mathematician Fourier has given his name to a powerful analytical tool based on the principle, which he established, that every function of time that appears in nature can be *analysed* in terms of simple sinusoidal functions of time (in acoustical language, 'pure tones'). As a consequence, when, by the application of Fourier analysis, the 'recipe' for a complicated time function has been determined, that same function can later be *synthesized* identically by superimposing a series of sinusoidal signals according to the recipe.

The tones of a violin and a clarinet are examples of complicated functions of time; if they are playing the same note, they will agree with respect to the period T but otherwise they are very different.

This difference is best described by an analysis in terms of 'partial tones'. Physically, one can think of the sound of one instrument as a combination of pure tones, the partial tones, in much the same way as one thinks of the sound texture of an instrumental ensemble as a combination of the characteristic sounds of the individual instruments.

For single notes only, the frequencies of the partial tones are related to the lowest frequency (the fundamental) in ratios of small whole numbers (integers). We call these 'harmonic ratios'; for this reason, the process of sorting out the frequencies and amplitudes of the tonal components of a sound is called Harmonic Analysis. Musicians express a corresponding qualitative analysis of the tones of their instruments in terms of the fundamental, the octave, the twelfth, etc., that is, the 'partials'.

Even unperiodic functions can be analysed in terms of pure-tone components. In this case, however, instead of dealing with harmonically related partial tones, the frequencies of the components are, in general, infinitely dense on the frequency scale.

This difference in objective composition of the sound may correspond to a transition from a tonal (musical) sound to a 'noise'.[1] (We recall that, if we play together several neighboring keys on the piano, the resulting sound is more noise-like than tonal.)

With respect to noises, we must distinguish between single, impulsive sounds, like sonic booms or explosions, and continuous sounds, like a running brook or the rustling of leaves.

Fourier analysis becomes especially useful in acoustical problems because of the law of independent superposition of different sound waves. With this tool, we may analyse the propagation of *any* sound signal, however complex, provided that we know how all the component pure tones behave at their various frequencies.

We have mentioned that, in contrast to water surface waves, the speed of propagation for airborne sound is the same at all frequencies. It would be a tremendous complication for acoustical analysis (and a musical disaster!) if the high-frequency and low-frequency components of speech and music traveled at different speeds and arrived at different times at the listener's ear. Moreover, the sound would be, in principle, different at different distances from the source. Fortunately, this problem does not arise.

But even so, we must account for the differences in behavior between sounds of different frequencies when they interact with reflecting and absorbing wall surfaces, and also for frequency-dependent differences in the losses sustained during propagation between boundary surfaces.

The chief implication of such frequency-dependent processes is that, as sound propagates within a room, there may be significant changes in the composition of the partial tones that govern the timbre of the sound.

We will encounter these problems in Part IV (Volume 2) and consider them quantitatively there. In the present chapters, dealing with Geometrical Room Acoustics, we emphasize the various ways in which sound can propagate.

I.1.3 The Geometrical Laws of Sound Propagation

A general principle, first established by Fermat, states that every wave propagates from the source to the receiver by way of the fastest path: not the shortest, but the fastest.

[1] The expression 'noise' is also sometimes used to refer to unwanted sound; that is, sound that is generated without purpose or that cannot be avoided.

If the speed of sound is uniform throughout the space of interest, as it usually is for air, then the fastest path *is* the shortest path: that is, for free propagation, a straight line.

We can observe this straight-line propagation clearly when sunlight penetrates through a break in the clouds and illuminates dust or water vapor in its path. Then the straight paths of light can be seen against a darker background; we call these 'light rays'. Figure 1.1 shows light rays coming from above, together with water waves below.

Fig. 1.1. Water waves and light rays: the two wave propagation processes appear entirely different. (Photo: G. Schulze.)

Even when no aperture restricts the lateral motion of the waves, and when the path is not made visible by scattering from intervening particles, we can still retain the concept of the 'ray' so long as a significant portion of the total energy travels in the neighborhood of a narrowly defined line.

Thus, we can divide the total power of a source into portions that propagate along rays in different directions, though these rays will not necessarily carry equal amounts of power. We know, for example, that the sound of a trumpet is strongest along its axis; and we recall, as well, the directivity of the human voice.

The simple, straight-line law of wave propagation is upset, however, when the wave encounters changes in the medium through which it passes.

Let us put in the path of the wave an obstacle which does not participate at all in the wave motion; this is usually the case with rigid bodies in air. The motion excited in the obstacle by the incident sound pressure is small enough to be entirely negligible. We can demonstrate with water waves that, when the waves impinge upon the obstacle, new waves are generated. Some part of the incident energy is turned about and travels back toward the direction from which it came; this is called *reflection.*

If the obstacle is not large, however, there will be another wave propagating beyond it in the original direction (see Fig. 1.2(a)).[1] Here, we see that the law of straight-line propagation fails, because the rigid obstacle (a region in which no wave motion exists) is interposed directly in the line of straight propagation between the source and receiver. If any waves reach the region behind the obstacle, they must do so by going around it, departing from straight-line propagation. We call this bending mode of propagation *diffraction.*

The diffraction phenomenon is especially evident if the obstacle is very small: it hardly hinders the passage of sound and the incident wave seems to ignore it. (See Fig. 1.2(b). Here, instead of steady excitation a short impulse is used, in order to make the faint wave scattered backward by the obstacle visible at all.) This phenomenon is often seen at the seashore: large waves sweep past the pilings that support the pier with no interruption.

On the other hand, if the obstacle is quite large, we do not expect to find any sound waves behind it; they are completely shielded by the obstacle. This behavior is well known from light studies as 'shadow'. In the example of the pier, given above, one can see, on calmer days, that small ripplets are reflected back out to sea by the pilings; for these small waves, the diffraction is negligible and reflection is dominant. The limits of the shadow are determined by straight, tangential rays from the source to the periphery of the obstacle.

The uncertainty (or fuzziness) sometimes observed at the shadow boundary occurs because of the finite size of the source; a true *point* source always produces a sharply defined shadow. The limits of the 'gray

[1] Fig. 1.2 is taken from the famous lectures of R. W. Pohl (*Einführung in die Physik*, Vol. 1, Springer Verlag, Berlin, 1955).

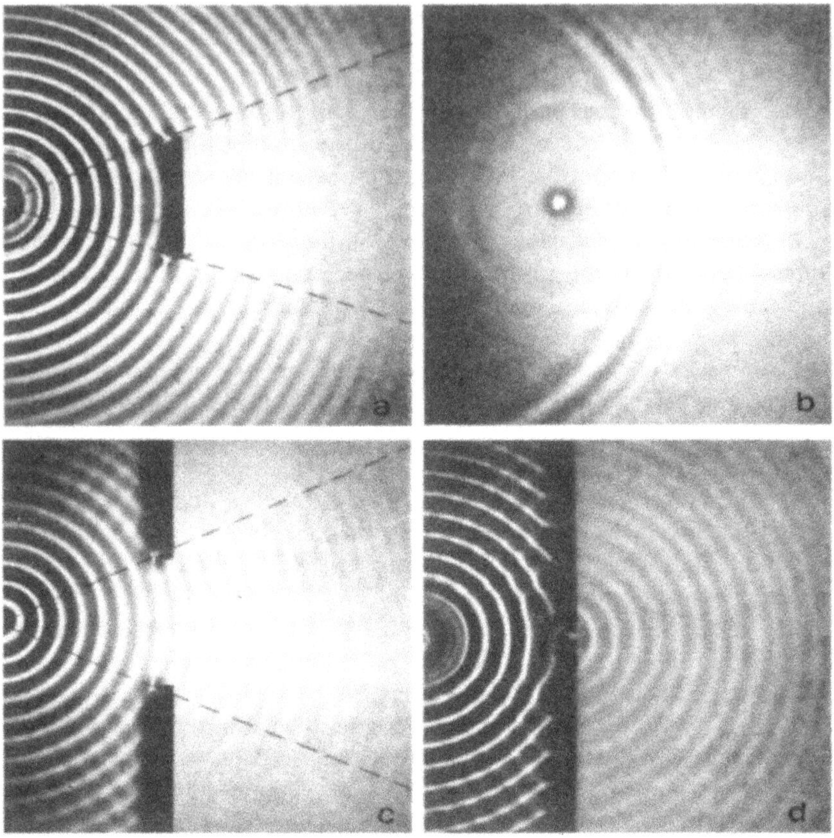

Fig. 1.2. Photographs of the propagation of water waves under various conditions. Top: wave propagation past obstacles, large on the left, small on the right. Bottom: wave propagation through an aperture, broad on the left, narrow on the right. The 'sizes' of the obstacles and the apertures are reckoned in relation to the wavelength.

zone' for a finite source can be constructed in terms of light rays, emanating from the edges of the source.

No sharp distinction can be drawn between reflected and diffracted sound, because, in both cases, energy is scattered in all directions.

In general, ray constructions are more suitable for light waves than for water waves, but that is not a special consequence of the electromagnetic nature of light. Indeed, in water wave studies with large obstacles, we can often neglect the 'bent' (diffracted) portion of the energy in comparison with the portions that go straight through or are reflected. Even in Fig. 1.2(c), where an aperture in a large screen replaces the obstacle of similar size in Fig. 1.2(a), one can see a shadow zone on either side of the waves that have passed through the opening. No waves have bent around to fill these shadow zones. The shadow boundaries are approximately marked by the dashed straight lines, which also delimit the portion of the wave field occupied by diverging straight-line water wave rays.

By contrast, the diffraction of the water waves is clearly evident when the aperture is quite small, as in Fig. 1.2(d).

The criterion for what is 'large' or 'small' in wave fields is always the wavelength. Since the wavelengths for visible light are smaller than 1/1000 mm, all of the objects that we encounter in our daily life, which naturally are adapted to the human scale, are very large indeed! And not only are the objects themselves large, but this is also true for the roughness of their surfaces: to a lightwave, even a rather shiny surface would look as though it were made up of ridges and crevices. From such a textured surface (its irregularities being measured in wavelengths of light), the secondary (reflected) waves scatter in all directions, even when the incident light comes from only one direction. This is called *diffuse reflection.*

The phenomenon of diffuse reflection of light waves explains why we can see *from any direction* an object that produces no light itself, but is illuminated by another light source.

It requires an exceptionally smooth surface to reflect an incident ray of light in a single direction. Such a surface is called a 'mirror' and such reflection is called a 'specular' or geometric reflection. In this case, the following rules apply for the incident and reflected rays, with respect to a perpendicular to the mirror surface:

1. The incident ray, the reflected ray and the perpendicular to the mirror at the point of incidence lie in the same plane.
2. The angle between the incident ray and the perpendicular, called the angle of incidence, is equal to the angle between the reflected ray and the perpendicular, called the angle of reflection.

These rules can be derived from Fermat's principle: we need only ask 'for what point on the mirror is it true that the distance to both the source and the receiver are shortest, compared with all neighboring paths?'.

We start with an arbitrarily chosen point on the mirror. The distance from this point to the source is the same as the distance to the source-image behind the mirror; the image lies on the perpendicular to the mirror through the source at a distance behind the mirror equal to that of the source. Now, in general, the path from the source image, through the chosen point of incidence on the mirror, and thence to the receiver will have a bend in it where it crosses the mirror surface; therefore, it will be longer than the path *directly* from the source image to the receiver. According to Fermat's principle, then, this cannot be the path traveled by the reflected ray. The true reflection point must be the point where the direct line between source image and receiver intersects the mirror surface, since this is the 'fastest path'.

In geometric optics, the redirection of light rays by lenses (i.e. by changes in the properties of the transmitting medium) is much more important than reflections in plane mirrors. But it is also true that the direction of a sound ray can be changed as it passes through a region occupied by two adjacent media with different propagation speeds. This is called *refraction*.

The rule for refraction can also be derived from Fermat's principle (here, since the transmitting medium is not uniform, we must be careful to seek the fastest, not the shortest, path). If the wave spends most of its time in the medium with the higher propagation speed, the time for propagation from source to receiver will be less than if it followed a straight line between source and receiver. This implies a bend in the preferred ray path, because the straight line path is not the fastest.

These changes of direction are important in room acoustics, particularly for sound waves incident on a room boundary at oblique incidence. We shall treat the problems of refraction very carefully in Part IV (Volume 2).

For propagation *within* a room, however, the changes of direction are not important, because, in general, the properties of the medium are practically uniform throughout the space: there are no discrete changes of medium to cause sharp refraction nor any gradual transitions to cause curved rays. If one tried to introduce such changes artificially, the individual media would soon intermix into a single homogeneous medium; or there might be drafts that would have unhealthy consequences for the occupants.

The only exceptions to this general observation are the hot air columns rising from large heated bodies, which may form acoustic lenses. In optics, we think of convex lenses as 'converging' lenses, because the speed of light in glass is less than in air: the light passing through the (thick) center of such a lens is delayed more than the light passing through the thinner outer edge; therefore, the rays converge after passing through the lens. With respect to the air column, we recall from Section I. 1.1, that the sound speed increases with increasing temperature; thus, sound passing through the midst of a hot air column travels faster than in the ordinary air at the boundary of the column, and thus the sound rays spread farther apart as they leave the column; the effect is similar to that of a 'diverging' lens in optics. For a listener behind the air column, the sound source seems to be more distant and less loud.

There are apparently no quantitative data on this phenomenon, but there are reports of subjective judgments from people who have sat with a hot air column between themselves and a lecturer; they complain of degraded speech intelligibility.[2]

We may conclude, then, that any concentrated heating apparatus between a speaker and a listener should be avoided. Fortunately, contemporary ventilating systems in buildings distribute the heat at moderate temperature from numerous locations on the side walls and ceiling, so that noticeable refraction phenomena do not occur.

In the open air, however, where large distances must be taken into account, and where, moreover, there are no sound reflections from the side walls or ceiling, the sound ray curvature due to inhomogeneities in the air may play an important role.

If the temperature increases with increasing height above the ground (e.g. if the sun heats the air above a frozen lake or a cold ground surface), the sound waves will be curved downward. This, too, follows from Fermat's principle: the sound can travel faster, from one location to another, by staying in the high-speed layers of air farther above the ground surface than by following a direct-line path; sound in the lower layers travels more slowly and thus the rays bend down. Hearing conditions are improved in such cases, especially in comparison with the more frequent occurrence where air temperature decreases with increasing height. In that case, the rays bend upward, carrying the sound over the heads of the listeners on the ground.

Such an upward curvature is especially interesting for the grazing sound ray that touches the ground tangentially at the location of the listener. This entire ray constitutes an 'acoustical horizon', in the sense that any sound source that lies above it is audible, but any sound source

---

[2] Sabine, W. C., *Collected Papers on Acoustics*, No. 4, Harvard University Press, Cambridge, 1923; also, Michel, E., *Handb. d. Arch.*, IV Teil, 1 Abt., Kap 4d, Leipzig, 1926.

below it is not heard. This phenomenon can have a startling effect in conjunction with the normal (visual) horizon. For example, when an airplane appears above the horizon and approaches an observer, he expects to hear the sound of the engines, first rising above the background noise, and then gradually increasing in loudness as the plane grows near. If, however, there is an 'acoustical horizon', due to a negative thermal gradient in the atmosphere, he will hear nothing at all until the plane crosses that invisible barrier, and then the noise is suddenly audible.

Even more important in the open air are the influences of wind; they, too, follow from Fermat's principle. If the air is moving with velocity u, then the sound will propagate with absolute speed $(c+u)$ in the direction of the wind, and with speed $(c-u)$ in the direction against the wind.

The sound is typically heard to be louder in the downwind direction, but this is not because it reaches the listener sooner with the help of the wind. The important factor is that the propagation speed $(c+u)$ changes with height above the ground. At ground level the wind speed is reduced by ground friction and by encounters with various obstacles; consequently, both u and $(c+u)$ increase with increasing height above the ground. As a result, numerous downward-curving rays in the higher regions can reach the listener at the same time as the direct rays; they significantly reinforce the sound, concentrating it upon the listener.

An opposite effect occurs for sound travelling upwind. In this case $(c-u)$ decreases with increasing height, and the rays curve upward, seriously weakening the sound at the listener's location.

It may, therefore, be advantageous, in designing open air theatres, to choose the orientation of the audience with respect to the stage in such a way as to take advantage of downward-curving sound waves from the prevailing wind direction. A possible countervailing principle must also be taken into account, however: the direction of the sun. The spectators naturally prefer to see the stage with the sun behind them.

It may be supposed that, if the wind direction is so important in outdoor theatres, one should therefore apply the same principles in indoor theatres: that is, make the supply air from the air-conditioning system flow from the stage toward the audience, with increased flow speed higher in the hall. But the possible variation of air mass flow in a hall is so small, and the beneficial influence of properly designed reflective ceiling and walls is so great, that architects and mechanical system engineers are advised not to waste time on such matters.

Instead, we may conclude that, in the realm of geometric room acoustics, we can base all our ray diagramming on straight-line wave propagation and on specular reflection from the boundary surfaces, when such reflections are of interest.

I.1.4 The Range of Wavelengths in Geometrical Room Acoustics

How, we may ask, do the wavelengths of sound compare with the dimensions of the obstacles, the surfaces and the roughnesses of frequently encountered objects?

In contrast to optics (where all wavelengths are very short with respect to life-size objects), one must face in acoustics the difficult and sometimes intriguing necessity to deal both with long waves and with short waves. Long waves can bend around most obstacles and proceed onward, but (paradoxically) they are reflected specularly from a corrugated surface as though it were shiny. Short waves, on the other hand, can produce a shadow (because they *do not* bend) and are diffusely reflected from rough surfaces in all directions.

We can observe this contrasting behavior of long and short sound waves (corresponding to low and high frequencies) in the shadowing effect of obstacles. If we walk beside a rushing mountain brook, and happen to pass behind a boulder, so that we can no longer see the brook, we still hear it even if there are no sound reflections from the opposite slope or from other surrounding surfaces. But the sound has a different quality behind the boulder: not only is it less loud, but we hear a 'murmur' instead of a 'rushing' sound, because only the low-frequency components bend easily around the obstacle to reach us; we are in a shadow zone for the high frequencies. In a similar manner, the timbre of an opera singer's voice changes as he steps from behind the scenery, or as he turns to face upstage. The effect is less noticeable for female singers because the frequency range of their partial tones is smaller.

For a speed of sound of 340 $m s^{-1}$, we can calculate the wavelength corresponding to the musical tone Contra-C (c_{-2}), which has a frequency of about 32 Hz.[1] From eqn. (1.3), $\lambda = c/f = 340/32$, or about 10 m. Therefore, in small rooms the boundary surfaces are nowhere near large

[1] The frequencies used in the following examples are not those corresponding to musical scales; instead, we use the closest approximations from the series of preferred frequencies internationally standardized by ISO–266–1975(E), International Standards Organization, Geneva.

enough to be regarded as 'large with respect to the wavelength'; consequently, the rules of geometrical reflections do not apply: such 'small' surfaces cannot be mirrors. Even two octaves higher, at c_0 (125 Hz, the baritone range), where the wavelength is reduced to 2·5 m, similar conditions hold.

It is evident that architectural columns are so small that the sound bends around without giving rise to diffuse waves. On the other hand, ceilings with coffers and walls with stucco-decorations will reflect as ideal mirrors.

But another two octaves higher, at c_2 (500 Hz, the middle of the speech frequency range and thus of special interest in room acoustics), the wavelength is only 68 cm. In this important frequency range, the practical relationships become very complicated, because most life-size objects and wall-surface modulations are neither 'large' nor 'small' with respect to the wavelength. They tend to generate diffuse reflections, which are difficult to calculate.

Another two octaves higher, at c_4 (2000 Hz), with $\lambda = 17$ cm, we begin to expect shadowing by life-size objects; and certainly at c_6 (8000 Hz), with $\lambda = 4$ cm (a frequency which is so high that it is no longer of interest as a fundamental, but is still very important as a partial tone), the individual steps and facets of the ceiling coffers reflect independently as separate mirrors.

Accordingly, a zig-zag wall will reflect at low frequencies like a large flat mirror, at middle frequencies it will generate more or less diffuse reflections, and at high frequencies it will behave like a series of small perfect mirrors inclined to one another.

We have seen in Fig. 1.2(b) that an obstacle in a wavefield can act like a new source of sound, reflecting waves in all directions; the radiation pattern for an actual sound source also depends on the size of the source in comparison with the wavelength. An oscillating plate—for example, the diaphragm of a loudspeaker—at low frequencies (where its size is small with respect to the wavelength) radiates sound power equally in all directions, as did the small obstacle of Fig. 1.2(b). At high frequencies (where the diaphragm is large with respect to the wavelength), if the diaphragm oscillates 'in phase' over the entire surface, the radiation will be concentrated in a direction along its axis, as though it were a reflected wave excited by a wave incident from the axial direction.

Naturally, in view of these relationships between obstacle size and wavelength, wires or strings stretched across a room are far too thin to have any influence on the sound wave propagation, even at the highest frequencies of interest. Similarly, oscillating strings, of themselves, cannot be effective sound radiators unless they are coupled to large plates, which, when set into vibration by the motion of the string, can effectively radiate the 'string sound'.

For this reason, it is pointless to attempt to cure acoustical problems by introducing wires into an auditorium, because the sound field will simply ignore them. The popularity of this old-time remedy rested on the hope of improving the sound of an existing hall with an inexpensive, easily installed device that has minimum impact on the architectural appearance.

The concept of the sound ray leads to simple propagation rules only if we can restrict our attention to specular (geometrical) reflections, and do not have to take into account diffraction (ray-bending) or diffusion (ray-scattering). According to our definition of a ray as a well-defined path in whose neighborhood a large portion of the sound energy travels, we should have to break up each ray into parts every time it encounters a scattering object, and keep account of each of the resulting new rays.

Although in principle it is possible by means of ray-tracing procedures to follow the course of sound rays diffusely scattered from trees in the forest or objects in a room, it is not possible to predict how the energy incident on each object will be distributed among the different directions each time it is reflected.

The proper application of geometrical room acoustics is, therefore, restricted to straight-line propagation and to specular reflections, as in optics. The rules are applicable so long as the reflecting surfaces are large and the dimensions of surface roughness are small with respect to the wavelength.

Strictly speaking, it would appear that these conditions are seldom fulfilled in real rooms, and certainly not for all frequencies. It is surprising, therefore, how useful the principles of geometrical room acoustics can be, even in situations where their validity is undoubtedly questionable.

Accordingly, the concept of geometrical room acoustics is applied, in practice, in a much broader sense, including sometimes more-or-less diffuse reflections. Primarily, it covers problems where we are interested only in the paths along which sound is propagated.

But since, in a closed room, those paths become extremely numerous and (practically) unpredictable, we add the further restriction that geometrical room acoustics finds its greatest usefulness when the number of sound rays of interest is few, either because only a few rays exist, or because we are interested only in the first reflections.

I.1.5 Fresnel Zones

In cases where the requirements for specular reflection are only approximately fulfilled, it is nevertheless possible, by relatively simple means, to estimate the deviations from ideal specular reflection.

We start, for expediency, with a special case involving rather large departures from ideal geometrical behavior. We have seen (Fig. 1.2(b)) that sound bends around an obstacle if the obstacle is small in comparison with the wavelength. If the obstacle is large (re λ), we get more or less geometrical reflection.

Thus, we might expect that the amount of sound reflected from an object would decrease monotonically with a reduction in its size. And so it does in most practical cases.

However, for particular shapes of obstacle, and for particular source locations and observation points, and for pure-tone excitation, some striking deviations are possible.

The diagram at the left of Fig. 1.3 shows an arrangement devised by Spandöck,[1] in which a loudspeaker L radiates sinusoidal impulses

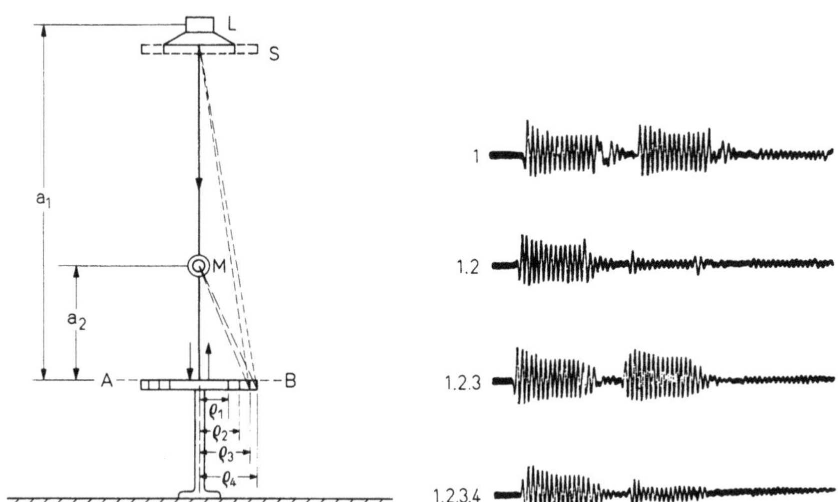

Fig. 1.3. Demonstration of Fresnel zones. Left: arrangement of reflecting zonal plates. Right: short tone-impulses of incident and reflected sound pressure. (After Spandöck.[1])

[1] Spandöck, F., *Ann. d. Phys. (V)*, **20** (1934) 328.

toward a circular plate AB, which reflects the sound back along the axis. On the way to and from the plate, the sound passes the microphone M.

Actually, the plate is a composite of four separate elements: a circular center portion and three successively larger annular sections, which can fit tightly together in various configurations: the central circle only, and (with the addition of successive annular rings) three more circles of increasing size.

The diagrams at the right in Fig. 1.3 show the experimental results, in which the sound pressure registered at M is plotted as a function of time for the four plate configurations. The first pulse (shown at the left of each trace) represents the incident signal, the second, the reflected signal.

Instead of a reflection whose strength increases monotonically with the size of the plate, the reflections come and go: with the central plate only (top trace) the reflection is strong; in fact, if one takes the normal divergence of the sound wave into account, this reflection is twice as strong as it should be, based on geometrical acoustics. Astonishingly, when the first annulus is added to increase the size of the reflector (second trace), the reflection disappears! The addition of the next annulus (trace three) brings back the reflection, and the final annulus nearly cancels it again.

This extraordinary behavior cannot be explained by the rules concerning the propagation of sound *rays*, in the framework of geometric acoustics, alone. We have to take into account the wave nature of sound radiation, as well. For this purpose, we make use of Huygen's principle, according to which the waves propagated beyond a certain position—in this case, the location of the reflecting plate AB—can be constructed by regarding each point on the wavefront passing AB as a point source of further waves. It is necessary, in the application of this principle, that the relative phases of these elementary sources be taken into account. It may happen, that, at some observation point, the contributions from the elementary sources must be added together, while at other locations they must be subtracted. At certain locations, the combined contributions of all the elementary sources may completely cancel each other, leaving a null pressure. This process is called *interference*. If two wave components are in phase with one another, their amplitudes will reinforce one another when the waves are added, and we speak of 'constructive interference'. If the components are out of phase, their amplitudes will tend to cancel one another when the waves are added, and we speak of 'destructive interference'.

The experiment shown in Fig. 1.3 has been designed so that all of the wave elements reflected from a single annular zone have approximately the same phase; also the reflections from neighboring annular zones are opposite in phase. Note that the reflections from the outer rings have farther to travel from L to M than reflections from the central zone.

The radii ρ_1, ρ_2, ρ_3, and ρ_4 are so chosen that the paths from L to M, via the center of the plate and via each successive annulus boundary, differ by one-half wavelength, according to the following formula:

$$\sqrt{a_1^2+\rho_n^2}+\sqrt{a_2^2+\rho_n^2}-a_1-a_2=n\frac{\lambda}{2}, \qquad (n=1,\ 2,\ 3,\ 4)$$

where n refers to the circular or annular element in question, four elements altogether in this case.

Under the assumption that the square of the outer radius is small compared to the squares of the source and receiver distance, we can expand in a series both of the radicals in the formula given above, to derive the relation:

$$\rho_n^2=\frac{n\lambda}{\dfrac{1}{a_1}+\dfrac{1}{a_2}}=n\ (\text{const.}) \tag{1.4}$$

which implies that the areas of the central plate and all of the annular elements must be equal.

These zones, which (according to the Huygen's principle) reflect nearly equal amplitudes with opposite phases, are called Fresnel zones, in honor of the natural philosopher of that name.

Thus, the reflections from zones 1 and 2 cancel each other (trace 2 in Fig. 1.3(b)), the addition of zone 3 restores the reflected impulse (trace 3), and with the addition of zone 4, the impulse is again almost cancelled (trace 4). But this time the cancellation is not quite complete, because the path length for the fourth annulus is greater than for the third annulus, and the out-of-phase reflected components are thus not quite equal in amplitude.

In fact, for this reason, each successively larger annulus cannot quite completely cancel the reflection from the next smaller one. But since the changes in intensity and also the deviations from eqn. (1.4) occur gradually with increasing ρ, we may still conclude that one zone effectively cancels the *nearest* half of each of its neighboring zones.

For a large circular reflecting plate, then, this means that only the inner half of the central circular Fresnel zone and the outer half of the outermost annulus contribute effectively to the reflected energy: the reflections from all the remaining zones cancel each other! If the latter contribution is negligible, as it may be for large plates, then *all* of the reflected sound from the plate comes from one-half of the central circular zone only, and is independent of the total area of the plate! Incidentally, this observation also determines the strength of each of the (equal) zone contributions.

In the concentric arrangement of Fig. 1.3 we may expect this asymptotic behavior only for a very large outer radius, and only if both the source and receiver lie on the axis of the circular reflector.

The cancellation of reflections from neighboring Fresnel zones on a rectangular reflector is almost as impressive as for a circular reflector; and this is true no matter where the source and receiver are located. Figure 1.4 shows the zone system for an eccentric rectangular plate ('eccentric' meaning that the line joining the source and receiver does not pass through the center of the reflector). Here the reflective contributions of the effective parts of the half-zones decrease more rapidly with increasing radius, because the zones themselves fall increasingly short of full circular Fresnel zones. Nevertheless, it remains generally true that (particularly at high frequencies) each zone cancels approximately half of

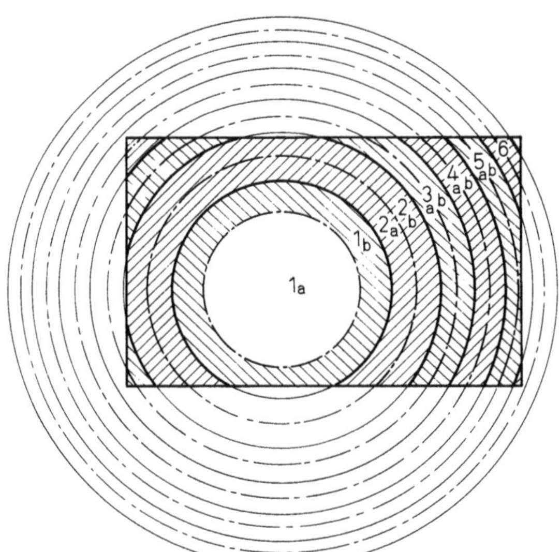

Fig. 1.4. Fresnel zones on a rectangular plate.

each adjacent zone. Even in Fig. 1.4, where the total area contains the parts of only the first six zones, the difference $2a+2b-1b-3a$ amounts to only 7% of $1a$.

Thus, we see, in general, that a reflector containing only a few zones nevertheless can approximate geometric reflection very closely. Furthermore, the interference exhibited in the experiment of Fig. 1.3 is smoothed out for the 'broad-band' mixture of frequencies in speech and music. The larger the number n of zones on the reflector, the smaller the relative frequency variation for a change in the interference pattern; the result is that, instead of the contribution from the outermost zone adding to that of the central zone, it may detract from it.

It also follows that ideal geometrical reflection is more closely approached, the more Fresnel zones there are on the reflecting surface. Equation (1.4) shows that the number of zones increases, the larger the reflecting area, the shorter the wavelength, and the closer the reflector lies to the source and receiver. This result will be of great importance for the reflectors we shall consider in Section I.5.3, for it means that if the wavelength is small with respect to the reflector dimension, there always exists a distance beyond which the reflector loses its effectiveness. In particular, for reflection of a plane wave ($1/a_1=0$) from a large plate with radius R, eqn. (1.4) states that at the limiting distance

$$a_{\text{lim}}=2R^2/\lambda \qquad (1.5)$$

only the first half-zone occurs on the entire plate. For larger distances, the radius R would have to increase to maintain a full half-zone of reflection; so the energy actually reflected decreases in comparison with the expected geometrical reflection. The sound pressure decreases proportionally to the ratio of actually reflecting area to the area of the first Fresnel half-zone: that is, it is inversely proportional to distance.

Huygen's principle, together with the Fresnel zones, not only establishes the frequency limits for geometrical reflection, but, outside these limits, allows us to estimate the deviations from the expected geometrical behavior.

Chapter I.2

Sound Reflections from Plane Surfaces

I.2.1 The Single Source Image: First-Order Reflections

The concept of sound rays and the rules of specular reflection allow us to analyse, using pencil and paper, the course of sound as it travels around a room. The procedure is called 'ray-diagramming'.

Such geometric constructions are the oldest tools of room-acoustical studies; even today they are frequently used, and are particularly employed by architects, who create their world on the drawing board.

We shall see, indeed, how many practical problems can be solved by this means. On the other hand, we must be careful not to overestimate the validity of the ray-diagramming results; since the procedure is based on geometrical reflection, it is subject to the same limitations as discussed at the end of Section I.1.4, above. Thus, to attempt to analyse by this means the precise scattering of sound rays from each tiny crevice or bump in a plaster wall decoration would not only be exceedingly tedious, it would even be misleading for most wavelengths of interest.

A further restriction results from the fact that the ray-diagram construction is simple only if the wall (or other boundary) is perpendicular to the drawing plane; if this is the case, the incident and reflected sound rays all lie in that plane. Thus, we can study horizontal rays in the plane containing the sound source by using the plan drawing; and rays with vertical components can be studied from the sectional drawings. Furthermore, in cases of circular symmetry (such as a dome), the analysis of a sectional drawing applies to a large set of rays: for a dome, 'any section speaks for the entire volume'.

This restriction relating to space may also be accompanied by a restriction relating to time. We shall assume that the sound signal under study is a brief impulse, such that, while it lasts, it travels only a distance which is short with respect to the room dimensions. This does not restrict the general validity of the results, because any desired signal can be broken up into short time-increments. Because of the rule of 'superimposability without interaction' for different wave fields, this synthesis is possible in the time domain, analogous to the Fourier synthesis in the frequency domain, which we encountered in Section I.1.2.

By using impulsive sound as the source of excitation in our paper analysis, we can think of the sound wave as 'reaching' a certain point of interest, just as in the case of water waves we can follow the outgoing wave from a pebble cast into the water because the disturbance at the circular periphery is small compared with the increasing diameter: in the ray diagram, the line-width of the impulse on the drawing is small compared with its diameter.

Even if the sound signal were somewhat longer, as in the case of a spoken word, we can still focus on a particular brief time-segment and follow its instantaneous progress. If we choose, for example, to follow the onset of the signal in the undisturbed medium, then the paper analysis should trace the progress of the wavefront.

We can construct a wavefront by marking off the same path-length from the source on each of the sound rays emanating from the source (including reflections); then we connect these points together. If the path length so chosen is less than the distance from the source to the nearest boundary (that is, if no reflection has occurred), then the wavefront so constructed will be a sphere; in a plan or sectional drawing it will be represented as a circle.

When the sound strikes a plane boundary, we know, from the rules of specular reflection, the direction of the reflected ray: the angle of reflection equals the angle of incidence. If, for a number of rays reflected from different points of incidence on the boundary, we prolong these rays backwards, *behind* the surface, they will meet at a common point S_1, called the image point. It lies on the line perpendicular to the surface through the source, and at the same distance from the surface as the source-point S_0 (see Fig. 2.1). For observers in the room, all of the reflections of the source in the wall seem to come from the image point.

In order to construct the wavefront, we get the unreflected part by drawing a circle around the source point S_0 with the path-length ct as radius; we get the reflected part by drawing a circle with the same radius around the image point S_1. Thus, the total wavefront comprises a growing crescent, made up of two expanding arcs, one around the source,

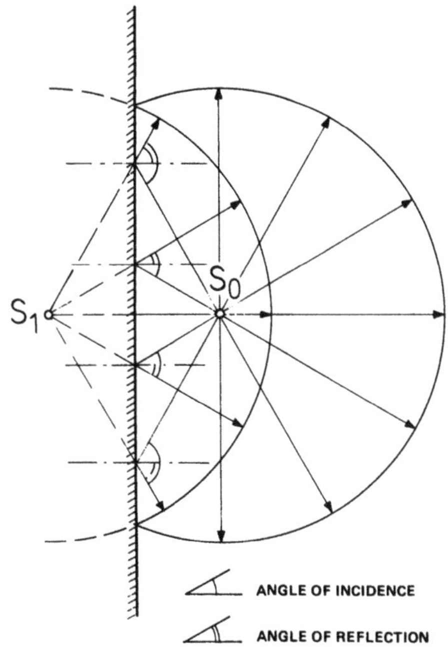

Fig. 2.1. Wave reflection at a plane surface.

the other around the image; they cross each other at the reflecting surface, as shown in Fig. 2.1.

This replacement of a plane reflector by an image source is not merely a mental experiment; both the source field and the image field have real identities, as can be observed with public address systems having two equal loudspeakers driven by identical signals.

When we are near one loudspeaker, we localize upon it as the source of sound; this is not surprising, because at this location its signal is louder. As we move from the first loudspeaker toward the second, the impression that the first loudspeaker is the only source persists until almost the midpoint between the speakers. Suddenly, as we pass the midpoint, the localization jumps to the second loudspeaker, which is now the nearer. This change cannot be based on the difference in sound pressures produced by the individual loudspeakers, because at the location where the sudden change of localization occurs these two components are virtually equal. It must depend on the difference in time delay for the two signals. (See Section III.1.6.)

Our mirror principle, taken in reverse, allows us to replace the more distant loudspeaker with an image in a plane reflecting surface, perpendicular to the line connecting the two loudspeakers and midway between them. Now, the image of the first loudspeaker in the reflecting surface occupies, in the virtual space behind the reflector, the same position as did the second loudspeaker in real space.

So long as we stay on the source side of this wall, there is never any doubt about localizing upon the true source; its wavefront always arrives at our ears before that of the image, even though their strengths are nearly the same. If we could magically pass through this reflecting surface (equivalent to passing an 'invisible' wall when we moved from the first to the second loudspeaker in real space), we would enter an image room, where now the image source becomes the original and the original source becomes the image. We now localize unmistakably on the image source, because its wavefront reaches us first.

I.2.2 Source Images of Higher Order in a Rectangular Room

We can now extend the principle of image sources to a rectangular room with smooth, ideally reflecting walls. Such a room is not only the simplest, but is also the most frequently encountered; from the standpoint of building construction, it is the most natural one.

In practice, of course, the rooms we live in are never ideally rectangular. The walls are seldom plane surfaces but are interrupted by niches for doors and windows; and their geometric simplicity is disturbed by the furniture. Nevertheless, by studying the ideal rectangular room, we can learn how the propagation of sound in such a space differs from that in, say, a triangular room, or even in a room with curved walls and vaulted ceilings.

As a first step, we consider a sound source located at S_0, between two infinite parallel walls, as shown in Fig. 2.2. It is evident that we get source images on both sides when the original outgoing circular wavefront has reached both walls. There are corresponding reflected wavefronts, centered on the images S_a and S_b, as shown at the top of the figure. The resulting first-order reflected wavefront is made up of three arcs.

When the reflected wavefront from one wall reaches the opposite wall, as shown in the bottom of the figure, we get a source image of second order S_{ab}, for which the first-order source image S_b may be regarded as the 'original' (S_b and S_{ab} are equidistant from the left-hand wall). This

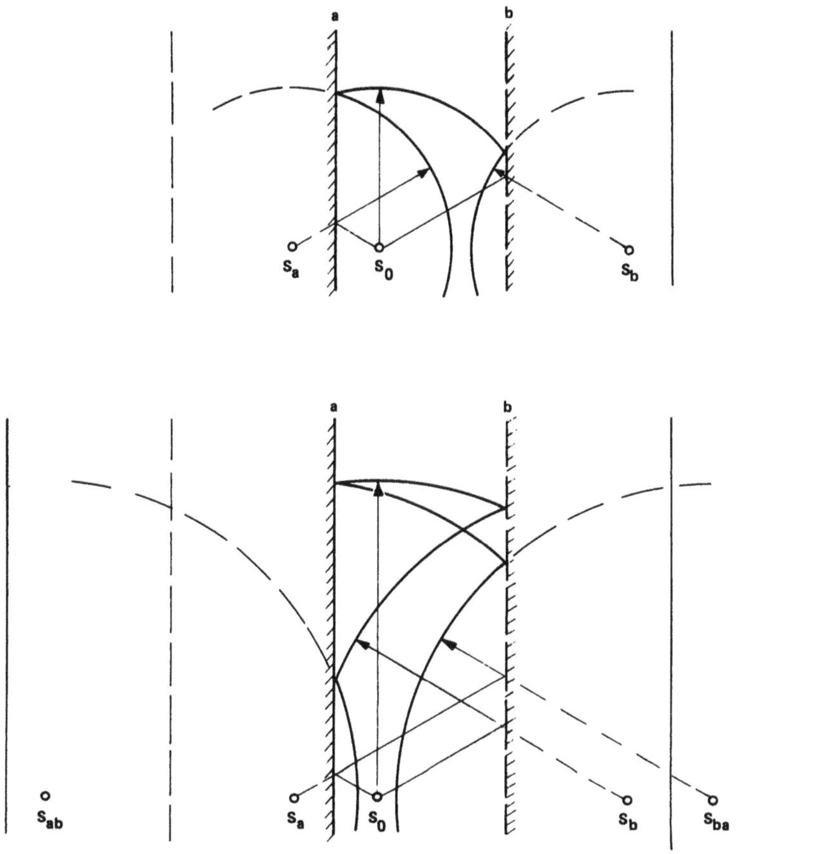

Fig. 2.2. Source images in two plane-parallel reflecting surfaces.

process is repeated again and again, so that we get two infinite series of source images, one on each side, all of which (together with the original source) lie on a line perpendicular to the walls. Again, the wavefronts consist of circles drawn around each of the images with the radius *ct*.

As the second step, we consider two walls perpendicular to each other in the neighborhood of the source. In this case, also, we get source images, but never of order higher than the second. Indeed, as Fig. 2.3(a) shows, we find only a single location for the second order images, where two second-order images coincide: one of these images may be imagined as generated by reflection first in the *x*-wall and then in the *y*-wall; the other corresponds to two reflections in the opposite sequence.

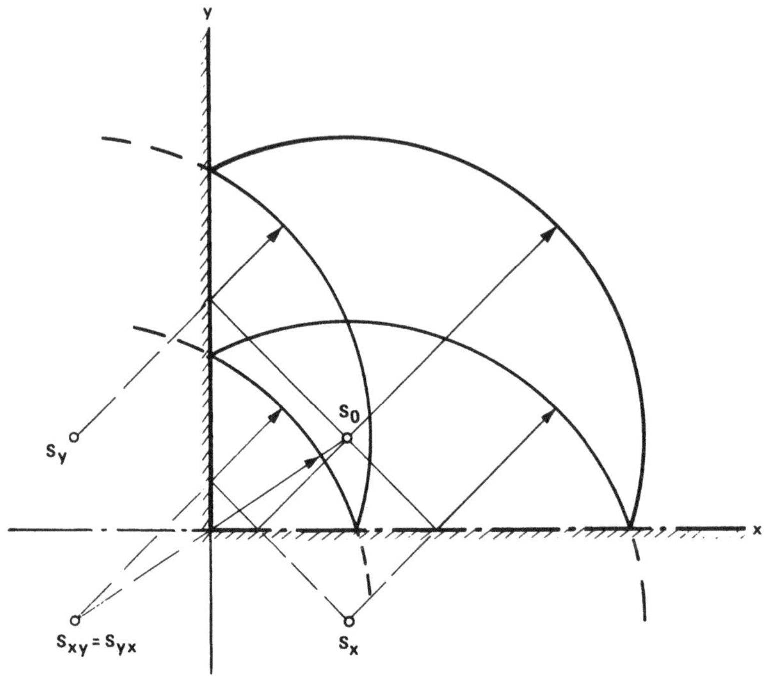

Fig. 2.3(a). Source images, wavefronts and reflected rays at a 90° angle.

Another important property of the 90° corner is that each incident ray, after two reflections, returns in the direction from which it came (a 'cue-ball' reflection). In the limiting case, where the incident ray strikes the corner exactly, it returns upon itself, just as if the two intersection reflectors were replaced by a single reflector perpendicular to the incident ray. In the ray construction, this shows up in the fact that the source, the corner and the second-order image all lie on a straight line.

To emphasize the exceptional behavior of the 90° corner, we compare it with the situation shown in Fig. 2.3(b), where the x-wall and the y-wall form an obtuse angle.[1] In this case, we have two images corresponding to the twice-reflected waves; they are *not* opposite the source through the corner, but are displaced from each other by an angle with respect to the corner that is double the amount by which the obtuse angle of the walls

[1] Weisse, K., *Leitfaden der Raumakustik für Architekten*, Druckhaus Tempelhof, Berlin, 1949, p. 15.

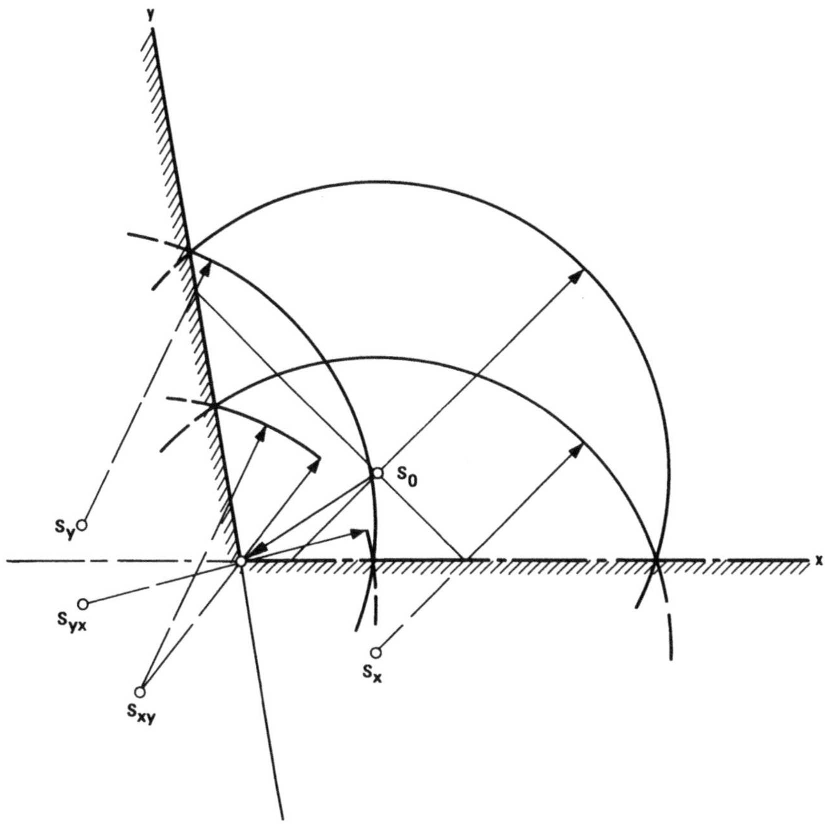

Fig. 2.3(b). Source images, wavefronts and reflected rays at an obtuse angle.

deviates from a right angle. The rays from these second-order images through the corner into the source room represent reflected rays, corresponding to an incident ray going from the source directly into the corner. This incident ray is split into two rays upon reflection, and there are two corresponding second-order wavefronts with radius *ct* that do not coincide and that are interrupted by a gap. These are the arcs shown nearest the corner in Fig. 2.3(b).

We get similar results when the two walls form an acute angle, as shown in Fig. 2.3(c). Instead of a gap between the second-order rays through the corner (as in Fig. 2.3(b)), we find an overlapping as the second-order rays cross each other (see the arcs and rays nearest the

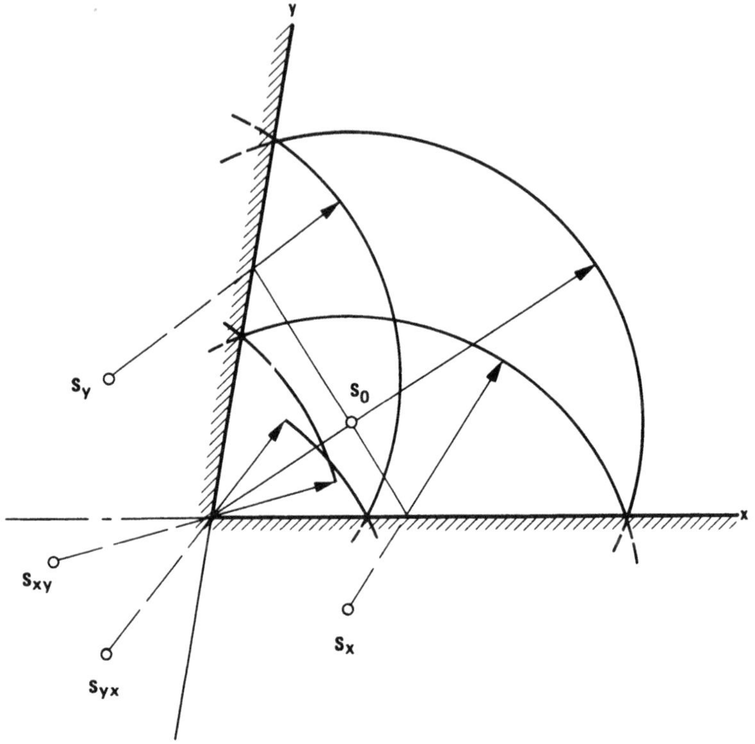

Fig. 2.3(c). Source images, wavefronts and reflected rays at an acute angle.

corner in Fig. 2.3(c)). Again, the second-order 'wavefront' is made up of two non-coincident parts.

Thus, when the angle between the reflecting walls differs from a right angle, whether obtuse or acute, the second-order wavefronts are not closed curves (surfaces), as we found for single walls, for parallel walls, and for 90° walls. (Those earlier examples represent special cases, although they are the most frequently encountered ones.) The general rule is that the second-order wavefronts form closed curves only when the angle at which the two reflecting walls meet is an integral fraction of 180°.

In reality, the sound intensity can never change abruptly, as implied by the gaps in the ray analysis of Fig. 2.3(b) and 2.3(c), above. Instead, we must expect diffraction of sound into the shadow zone; and the extent to which the sound is diffracted increases with the wavelength.

As a reasonable approximation, we may fair across the gaps in the wavefronts (in Figs. 2.3(b) and 2.3(c)) with a circle, drawn with its center at the corner, whose radius is less than that corresponding to the two arcs that it connects. This approximation, in fact, agrees with experimental photographs of such wavefronts, as shown in Fig. 7.6. But these parts have smaller intensities and thus less importance (see again Fig. 7.6).

It is clear that reflection directly back to the source, as from a right-angled corner, is avoided if the angle at which the reflecting surfaces meet is either obtuse or acute.

Now, we consider a source in a rectangular room, or rather, in a rectangular area on the drawing board. We now find source images in all directions, completely covering the plane (see Fig. 2.4). In this case, the mirror images of the walls form parallel straight lines. The correspondence of the image walls to the original walls is indicated by different

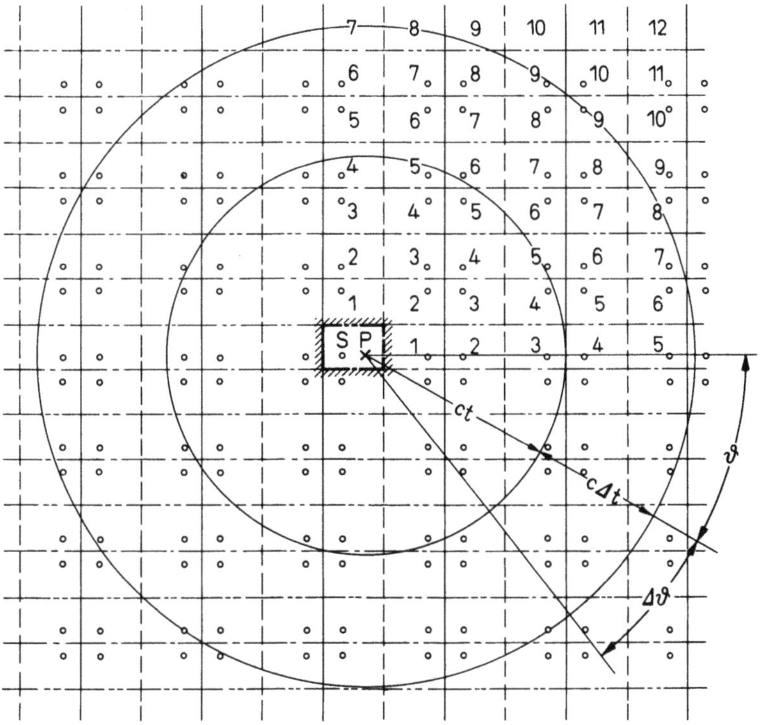

Fig. 2.4. Images of a rectangular room with source images.

kinds of straight lines. Furthermore, each image room contains an image of the source, which is placed a bit eccentrically on purpose.

In order to construct the wavefront that exists after time t, we must draw circles with the radius ct around every source image and then account for only those arcs that occur in the original room.

Figure 2.5 shows the wavefronts constructed by this procedure for eight instants equally spaced in time, until the appearance of reflections of the sixth order.[2] In the first two frames one can see the wavefront (the 'direct sound') moving from the source to the opposite corner; in this stage the sound intensity varies greatly from point to

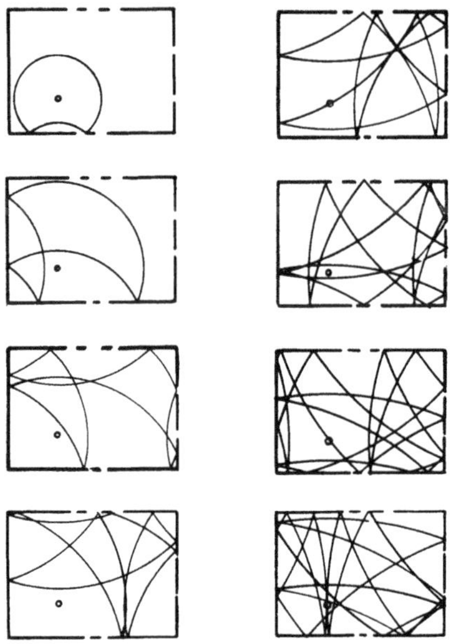

Fig. 2.5. Propagation of wavefronts in a rectangular room.

point in the room. Subsequent frames show the successive arrival of wavefronts from source images of higher order and from greater distances. As more and more wavefronts of higher order arrive, they overwhelm the few wavefronts of lower order; and because they all come from distances much greater than the original room dimensions, they tend to be of

[2] Anyone who wishes to see this radiation of wavefronts in a rectangular room as a continuous process in time may borrow the educational film, *Schallschutz im Wohnungsbau*, (FT II48) from the Institut für Film und Bild in Wissenschaft und Unterricht, in Munich.

comparable strength and to fill the room uniformly; we call this a homogeneous distribution of sound energy. Also, the radius of curvature of the wavefronts increases with increasing order of reflection, so that the waves approach more and more closely to plane waves.

In this stage, the original source and its 'direct sound' contribute negligibly to the total energy, which is dominated by the numerous contributions from distant images of high order; the result is a very nearly uniform energy distribution in the original room, a characteristic typical of rooms with large plane boundaries. (We emphasize 'large' here, because *any* room boundary can be approximated by small plane surfaces.)

In the rectangular room, there are certain preferences for direction of propagation at the beginning; but here, also, there is an equalizing tendency. In Fig. 2.4, we have drawn two circles with radii ct and $c(t+\Delta t)$ around the observation point P. (The location of P was chosen arbitrarily, but this choice has no influence on the conclusions.) Wavefronts from all the source images lying in the annular zone between these circles reach the observation point P during the time interval Δt. If we assume a duration of 1/5 s for Δt, which is a rather short interval in our daily lives, the value of $c\Delta t$ is 68 m, which is likely to be 'large' with respect to the dimensions of the room in question. The total number Δn of impulses arriving at P during Δt is equal to the ratio of the annular area to the area of the original rectangular room (this is simply a means of counting the number of images inside the annulus):

$$\Delta n = \frac{2\pi c^2 t \Delta t}{l_1 l_2} \tag{2.1}$$

The number of impulses arriving during Δt from the directions between ϑ and $\vartheta + \Delta\vartheta$ are:

$$\Delta(\Delta n) = \Delta n \frac{\Delta\vartheta}{2\pi} = \frac{(c\Delta t)(ct\Delta\vartheta)}{l_1 l_2} \tag{2.2}$$

This means that, once t has become large enough, there will be the same number of reflections per second from all directions, since $\Delta(\Delta n)$ is not a function of ϑ.

We might expect that, for a long, oblong room, more waves would be directed laterally; and, in fact, this is the case at the beginning. But with source images of higher order this tendency disappears, provided that the area $(c\Delta t)$ $(ct\Delta\vartheta)$ is large enough to contain several image rooms. Thus, the result is, for large enough Δt, independent of the choice of observation point P.

The results, derived here from purely geometrical considerations, do not take into account any differences in intensity of the individual waves. In practice, of course, differences in intensity in different directions are of great importance. The equal distribution of sound in all directions, suggested by the example above, can be expected only if the original source and all the image sources radiate the same intensity in all directions. For a highly directive source, which radiates significant sound energy in only one direction, it would be sufficient to study the fate of the sound ray in that direction only.

In order to account for directivity of the source, we reverse the foregoing analysis of Fig. 2.4 and follow a particular ray direction, starting outward from the original source room and passing through many image rooms, one after another. We note from Fig. 2.4 that the image rooms occur in only four different configurations, which also means that our ray can pass through any image room in one of only four directions. Since all of the walls appear as parallel lines, we see, furthermore, that in a rectangular room each wall is encountered by the ray again and again at the same angle of incidence. If, now, the walls absorb sound preferentially at particular angles of incidence—which does happen in practice—then the intensities of the rays encountering the walls at these angles diminish more rapidly than others, and the distribution of intensity with direction becomes uneven, although the directional distribution may have been equal at the beginning.

An impressive example is the case in which two parallel walls are highly absorptive (or even only one of them) and the other walls are reflective. Then those rays that propagate parallel to the absorptive walls will persist longer than the rays that encounter the absorptive walls, and eventually they will dominate the sound field. We shall encounter this situation again and again (see, for example, Fig. 4.7). It sometimes leads to the simple periodic reflections that we call 'flutter echo'.

I.2.3 Triangular Room and Wedge-shaped Room

In order to preserve the favorable effect of equal distribution of sound energy by large plane walls but to avoid significant dependence on direction (particularly the flutter echo), we may turn two opposite, strongly reflecting walls slightly away from their parallel orientation, in such a way that their prolongations would meet in an acute angle. Such a symmetrical trapezoidal plan has been frequently used in auditoriums.

Usually it is only the long walls in a hall that are turned in this manner, since the short walls are occupied by the stage or the balconies at the rear. In either case, the wall is rather absorptive or its surface is well broken up.

A room in which none of the walls are parallel is seldom found. Such a plan may be useful for acoustical measurements, however; so sometimes one finds reverberation chambers designed to have no walls parallel.

For trapezoidal-shaped rooms, the principles of specular reflection lead to general geometric properties that are less simple than for the rectangular room. We therefore consider several triangular examples, admitting from the start that these shapes are almost never encountered in practice, not even in laboratory measurement facilities. But the study of these triangular-shaped rooms is easy, and it gives us an insight into the behavior of non-rectangular polygons. Specifically, we shall study the right equilateral triangle, the isosceles triangle and the half-isosceles triangle.

In Fig. 2.6, the entire plane is covered with right equilateral triangles that are images of the original room shown at the center. (In this case, we forego the construction of wavefronts.) Here, again, the distant source images fill the original room uniformly with sound from all directions, as in the rectangular room. This results, as before, from the large plane

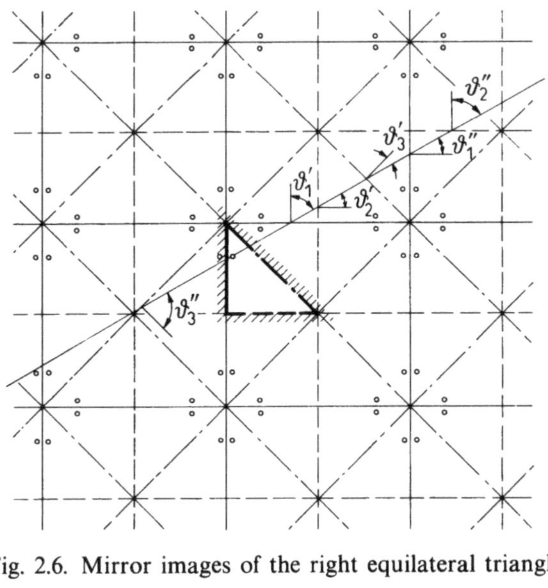

Fig. 2.6. Mirror images of the right equilateral triangle.

walls. which do nothing to hinder the divergence of the sound waves.

In this case, however, the directional distribution presents no opportunity for flutter echo. The image rooms in the shape of right equilateral triangles in this plan can occur in eight different configurations, and thus a ray can cross an image room in eight different directions. The images of all the walls form closed squares, and it is impossible for a ray to travel in such a way as to avoid hitting one of the walls. In general, each wall is encountered at two different angles of incidence. Even if one wall is sound absorptive, it must be encountered by all the rays, so no flutter echo can exist.

The same characteristics are presented by the other two types of triangular room, depicted in Fig. 2.7. The system of neighboring triangles is considerably more complicated in these cases, because the images of the walls of the original room change back and forth along the same straight line in the image network. However, it is easy to see, from Fig. 2.7, that the images of the isosceles triangle appear in six configurations, while those of the half-isosceles triangle appear in twelve configurations, corresponding to the same numbers of possible directions for the image ray crossing the original room. This means that, for both cases, each wall is encountered by a ray at three different angles of incidence.

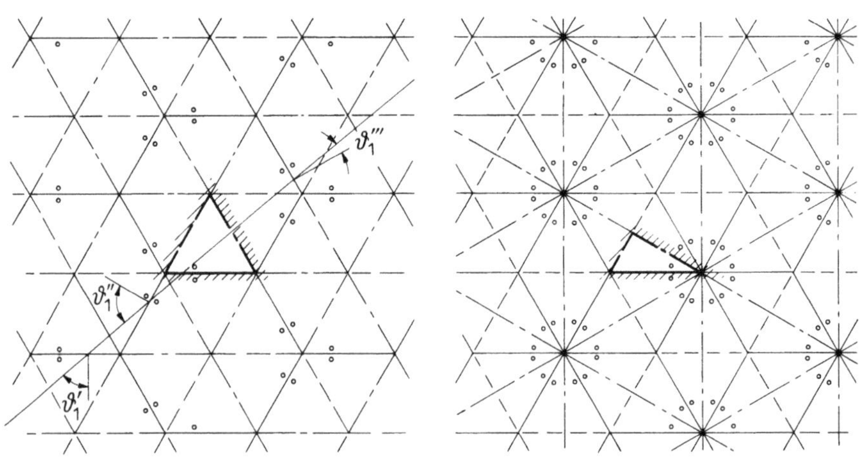

Fig. 2.7. Mirror images of an isosceles triangle and a half-isosceles triangle.

(It should be mentioned that the rectangular room plan and the three triangular room plans studied here are the *only* plans for which the drawing plane is completely filled by mirror images, without gaps.)

For walls intersecting at angles that are not integral fractions of 180°, the reflected rays never return along their original paths (see Fig. 2.8). Therefore, the walls are encountered again and again at new angles of incidence. We will find this fact important in the theory of reverberation (to be discussed in Part II), which assumes the ideal case for which the sound energy in a room is not only distributed equally over all positions, but is also flowing equally in all directions. We call this an isotropic homogeneous distribution of energy.

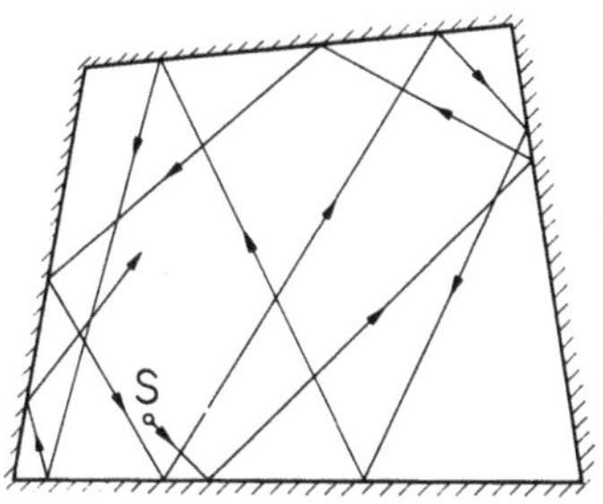

Fig. 2.8. Irregular course of sound rays in a room with a general quadrilateral ground plan.

Instead of the awkward designation 'isotropic homogeneous distribution', the condition above is sometimes referred to as a 'diffuse sound field'. But, as we shall see later (Section IV. 13.5, Volume 2), in a diffuse sound field we must assume, in addition, that the phase relations among the waves following the various rays are incoherent, so that the intensities of the partial waves may be simply added together, without accounting for phase (a condition that we assume in this and Part II, anyhow).

We return now to the acute angle already discussed in relation to Fig. 2.3(c); but this time we consider an extremely acute angle, such as is common with wedges. The wedge-shaped room—again discussed here as a two-dimensional space only—is assumed to be closed on the short side by a totally absorptive wall or by a non-reflecting opening.

If a sound ray enters the wedge through this short wall, and encounters one of the 'wedge-walls' with an angle of incidence $(90° - \beta)$, we can determine by geometrical construction that, as this ray approaches the vertex, it is reflected more and more frequently between the wedge-walls, but it never reaches the vertex. Instead it is forced to return to the short wall (see Fig. 2.9).

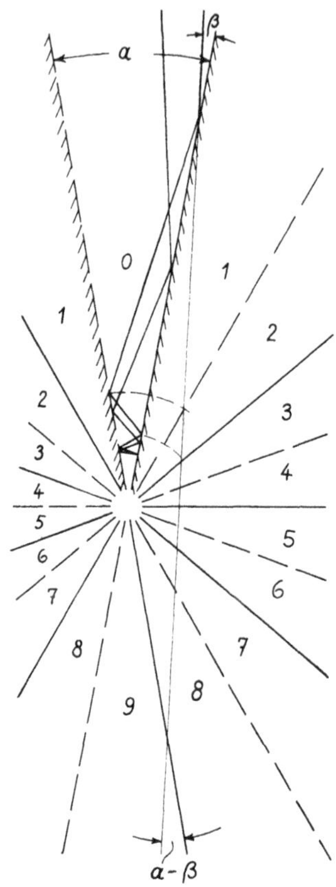

Fig. 2.9. Ray path in a wedge room (0), and the 17 wedge images, for a wedge angle $\alpha = 360/18 = 20°$.

If we assume that the wedge angle is an integral fraction of 360° (in Fig. 2.9, $\alpha = 20°$), we can again apply the mirror principle and follow the ray through the image rooms. We find that the ray starts its return toward the short wall at a distance from the vertex of ($d \sin\beta$), where d is the distance from the vertex of the ray's first encounter with a wedge-wall.

It would be wrong, however, to conclude from this example that the vertex is unreachable by incoming rays from the open end of the wedge. Here, we encounter one of the weaknesses of ray-diagramming as an analytical tool: it must always start with a particular ray (which may or may not be typical) and follow it through to its end. Instead, we need to take into account all of the rays starting from a source and to investigate the distribution of all the wavefronts throughout the room.

We can thus achieve a better overview if, as in Figs. 2.4 and 2.5, we take into account all the source images in the image rooms, and investigate what happens in the original wedge when all the image sources radiate identically.

The mirror principle makes it clear that all image sources radiate to every observation point. But, when we take the time structure of the sound field into account (which we shall do carefully in Chapters I.4 and I.5), we find that the vertex point is exceptional. At the vertex, *all* of the wavefronts from the image sources arrive together at the same time. This makes the signal at the vertex very distinct. On the other hand, the ray paths from the image sources to any other observation point differ more and more the farther that point is from the vertex. At such points the sound is more reverberant and unarticulated.

A famous example of a vertical wedge room, with hard wedge-walls and a rather absorptive floor, is the cave near Syracuse in Sicily called 'The Ear of Dionysius'. This room has an S-shaped ground plan which is of no interest in the present context; but the vertical section approximates the shape of a wedge. The height is 22 m; the width at the ground is about 10 m, and the opening at the top is about 0·2 to 0·3 m.

One must conclude today, from the smooth surface finish of the walls, that this extraordinary shape was not an accident of nature, for it seems to have been made on purpose. It is said that the tyrant Dionysius used the cave as a prison, and that his guards, listening at the small opening at the top, could hear the conversations of the prisoners, even when they spoke in whispers.

For this reason, Sabine included the 'Ear of Dionysius' in his examples of 'Whispering Galleries', although it is not a gallery at all.[1]

This unique phenomenon would be a great tourist attraction today, if access to the top were permitted, so that the tourists could listen, as did the guards of Dionysius. But people are not allowed there nowadays for fear of an accident. Even so, the effect could be demonstrated almost as effectively with a microphone near the top opening. Instead, tourists are restricted to the ground level, where they are given a not-very-interesting demonstration of the reverberation time of the cave (about 2·5 s) and a simple echo at the entrance.

[1] Sabine, W. C., 'Whispering Galleries', in: *Collected Papers on Acoustics*, Harvard University Press, Cambridge, 1927.

Returning now to the model with the wedge angle of 20°, we can estimate the amount of 'reinforcement' of the source sound, in comparison with the strength of a single wavefront. If the observation position is a number of wavelengths distant from the vertex, then we may add together the intensities of the source and its images (assuming broadband signals); the resulting intensity would be increased in the ratio of 18/1. But if the distance from the vertex is small with respect to the wavelength, we must take the phases of the component waves into account and add together the sound pressures. The resulting pressure would be 18 times as great as a single pressure component, but the intensity would be increased in the ratio of 324/1!

Certainly, we could not expect such a great reinforcement in the cave at Syracuse: first, because it is not a perfect wedge shape, and second, because the walls surely are somewhat absorptive, which would weaken the sound propagation.

I.2.4 Extension to Three Dimensions

It is an easy matter to extend the two-dimensional problems, discussed above, to the corresponding three-dimensional problems.

Just as the mirror images of the rectangular area entirely cover the drawing plane in the two-dimensional case, the mirror images of the rectangular room completely fill all of space. Therefore, everything that we have said about the tendency to equal sound wave distribution and even homogeneous isotropic distribution of energy in two-dimensional rectangles holds true for the rectangular room.

The number of image sources that become effective between the instants t and $(t+\Delta t)$ is now given by the ratio of the volume of the spherical shell $4\pi c^3 t^2 \Delta t$ and the room volume $l_x l_y l_z = V$:

$$\Delta n = \frac{4\pi c^3}{V} t^2 \Delta t \tag{2.3}$$

The number of image sources in the increment of solid angle $\Delta\Omega$ in the time increment Δt is given by

$$\Delta(\Delta n) = \frac{\Delta\Omega}{4\pi} \Delta n = \frac{c^3 t^2}{V} \Delta t \Delta\Omega \tag{2.4}$$

The images of the original rectangular room appear in eight different configurations, and therefore an oblique ray can take eight different directions in the original room. But each boundary plane is always met by a specific ray at the same angle of incidence.

The two-dimensional triangles may also be easily extended into prismatic rooms, if we regard the triangles as the ground plans of rooms with vertical walls and with horizontal floors and ceilings. Then the networks given in Figs. 2.6 and 2.7 should be thought of as repeated in an infinite number of storeys.

Similar extensions hold for each prismatic room, whatever the ground plan may be. If we know the behavior of a horizontal ray in the ground plan, then we know the behavior of a ray whose projection starts horizontally in the same direction but which is inclined away from the horizontal, so that it meets the floor or ceiling with the angle of incidence ϑ_z. The projection in a vertical plane then moves with a speed $c \cos\vartheta_z$ and the vertical path component is $ct \cos\vartheta_z$. Thus, it follows that if a reflected ray bundle concentrates at time t_1 in a point (or nearly so) in the horizontal plane, all the corresponding rays which meet the ceiling and floor under the angle of incidence ϑ_z will also concentrate in one point (or nearly so), which lies above or below the horizontal; but this will happen later, at time $t_2 = t_1/\cos\vartheta_z$. (The analysis above is true only with strictly vertical walls everywhere.)

In the case of the triangles, the extension to three-dimensional prisms doubles the number of possible configurations of the image rooms to 16, 12 or 24, respectively, for right equilateral triangles, isosceles triangles and half-isosceles triangles. These numbers also correspond to the number of different directions a ray can take in passing through the original room in each case. But the number of possible angles of incidence for a given ray on a side wall remains unchanged, because it does not matter whether the ray meets a vertical wall at $(90° - \vartheta_z)$ from above or from below.

For the ceilings and floors, we have only one angle of incidence ϑ_z, and there remains the danger of a flutter echo between them.

It is tempting, therefore, to try to find a room with oblique walls in all three dimensions that still is amenable to the application of the mirror principle. We can, indeed, find such a room. If we imagine a cube aligned with the Cartesian coordinates x, y, and z, and partition it with the three planes defined by $x = y$, $y = z$, and $z = x$, we get six pyramids that form image rooms to each other. Since the images of the cube fill all of space, with eight different possible configurations, the possible number of configurations of the pyramids is $8 \times 6 = 48$. Furthermore, the images of the boundaries always form closed surfaces, so that it is not possible for *any* ray to avoid encounters with *all* the wall surfaces and thus to create a flutter echo.

Finally, it can be shown that all 'walls' are encountered at eight different angles of incidence. (We shall come back to this pyramidal room in Part IV (Volume 2), Fig. 11.4, as a suitable room for the measurement of sound absorption.)

Just as we proceeded from the triangle room, extending the acute angle of Fig. 2.3(c) into a two-dimensional wedge room with an absorbing short wall, we now extend the three-dimensional pyramid room, discussed above, into a 'funnel room', of which one of the small triangular surfaces is non-reflecting; the tip of the funnel is at the opposite end.

If we introduce an impulse into the non-reflecting opening, its wavefront will reach the tip of the funnel simultaneously (in phase) with the wavefronts of the 47 image sources. Therefore, we observe at the tip a sound pressure 48 times greater than at the same distance in open air. This example may help us understand the extraordinary amplifying effect of all horns, whether associated with a receiver (microphone or ear) or with a sound source (loudspeaker or brass instrument).

We may summarize the results of this chapter as follows: all rooms with large plane boundaries lead, after the first few reflections, to a homogeneous distribution of wavefronts throughout the room. In addition, if parallel walls with similar absorptive characteristics are avoided, there will be a homogeneous isotropic distribution of energy.

Chapter I.3

Curved Surfaces

I.3.1 The Circle

The tendency to equal sound energy distribution in a room is immediately destroyed if some of the room boundary surfaces are concave. This becomes most evident if we imagine a circular ground plan with a sound source at the center. All of the radiated rays meet the wall at the same time and at perpendicular incidence, and are thus reflected back upon themselves. Therefore, the reflected waves all converge simultaneously upon the original source position.

If the circular plan represents a section through a cylinder, the reflected sound energy concentrates along the axis, with different time delays for different heights. But if it is a section through a sphere, the reflected energy from the entire volume is focused at a single point: an extreme concentration of sound energy!

In this geometrical argument, we have tacitly extended the rules of specular reflection so that they apply not only to plane surfaces but also to curved surfaces: in this case, the tangent surface at the point of incidence on the curve replaces the plane reflecting surface.

Furthermore, from our discussion in Section I. 1.3, we know that the wavelength must be small with respect to the radius of curvature. Otherwise, the curvature of the boundary would have to be considered as merely a roughness or modulation of the reflecting surface, which would produce a diffuse, rather than concentrated, reflection. Certainly, with a fully circular surface it is evident that only a radius large with respect to the wavelength would be of interest.

With these extensions and assumptions, we can proceed to construct reflected ray paths at curved boundary surfaces. But with arbitrarily curved surfaces, it may be difficult to determine exactly the tangent to the curved surface at the point of incidence, and therefore also the perpendicular of interest. The matter is relatively serious, because any angular error in determining the perpendicular is doubled in determining the direction of the reflected ray.

With a circle, however, the perpendicular is always the straight line from the center to the periphery; so it is easy to construct the reflected rays (angle of reflection equals angle of incidence), even if the source is off-center, as shown in Fig. 3.1. Furthermore, we can again mark off equal path lengths from the source along each ray and connect these points to determine a wavefront. In this case, we do not expect the wavefront to be an easily constructed circle with radius *ct* around an image source, as in the case of plane reflectors; it must be drawn by the more tedious procedure described above.

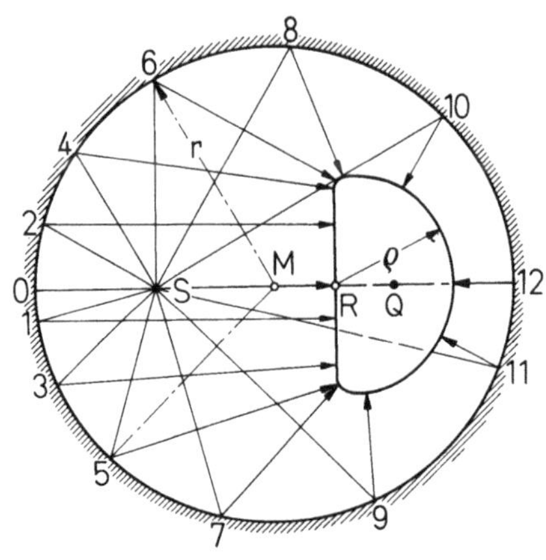

Fig. 3.1. Construction of ray paths and a wavefront in a circular room.

Figure 3.1 shows the result for the case in which the source point is located at a distance $r/2$ from the center (r being the radius of the circle). Although there is not a strong concentration of rays at a single receiving point (as with the source at the center), it is nevertheless clear that the sound shows no tendency toward equal distribution throughout the circle. Instead, there is a tendency to concentrate in the vicinity of the point Q, symmetrically opposite the source point S.

We may express the effect of this energy concentration by the ratio of the perimeter of a circle with radius *ct* (characteristic of the wavefront for unhindered propagation or for plane-surface reflections) to the perimeter of the wavefront in Fig. 3.1. The concentration will be greatest when the latter perimeter is shortest; but it is not always easy to estimate where that occurs.

Figure 3.2 shows the wavefront at several successive arbitrarily chosen instants;[1] we may compare these with the wavefronts in a rectangular room, shown in Fig. 2.5. Although the wavefronts depicted in Fig. 3.2 do not correspond to equally spaced instants in time, at least the total elapsed time corresponds approximately to doubled room length in both cases. We see (sixth frame down, left-hand column) a strong concentration near Q, as expected from Fig. 3.1, followed, after a comparable time delay, by a similar (but not so extreme) concentration at the

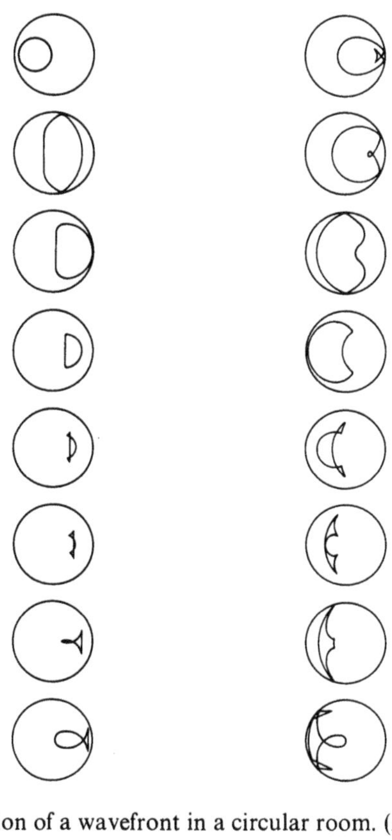

Fig. 3.2. Propagation of a wavefront in a circular room. (After R. W. Wood.[1])

[1] The constructions of Fig. 3.2 are due to R. W. Wood (*Phil. Mag., Ser. X*, **50**, 148); he has also made a film showing these frames, as well as intermediate instants.

source position. But even after these alternating concentrations at Q and S have been repeated several times, each time with less pronounced concentration, there is still no tendency toward equal energy distribution in the room.

I.3.2 Reflection Laws for the Concave Mirror

It would be very tedious indeed, if in order to analyse the sound reflections from every dome, vault or apse, we had to construct the ray paths and wavefronts according to the procedure outlined for Figs. 3.1 and 3.2.

For an estimate of the effects of these concave surfaces on the room acoustics, it is sufficient to make use of the simple laws for rays reflected at concave mirrors, well known in geometrical optics. The laws apply strictly only to small segments of cylindrical or spherical surfaces; but the concave surfaces that appear in buildings are also frequently only small segments. Even if we must deal with a larger part of a circle, it is possible to split this surface up into smaller segments to which the concave mirror laws do apply.

For the derivation of these laws, we now consider the practical case of a lecture room, in which the architect wishes to place an apse-like cylindrical enclosure behind the speaker. The distance between the lectern S (the source position) and the center of the cylindrical surface P is chosen on aesthetic (architectural) grounds. But we may choose on the drawing board different values for the radius of the cylinder.

We start again with the simplest case, where the center of the cylinder M and the source point S coincide (see Fig. 3.3(a)). Then again the total reflection will be concentrated on the speaker. From the listener's standpoint, this is the same (neglecting differences in time delays) as though the lecturer produced a greater sound pressure. For the speaker, however, it is as though he is 'shouting in his own ear'.

We next make the radius of curvature $MP = r$ larger than the distance between the source and the cylinder $SP = s$, in such a manner that $s < r < 2s$.

In this case, the construction of the reflected rays shows that they converge approximately on the receiver (*E̲mpfänger*) point E; and the construction of 'equal paths' indicates a very dense concentration of the wavefronts there.

We would get a precise focus at E if the shape of the curved reflector

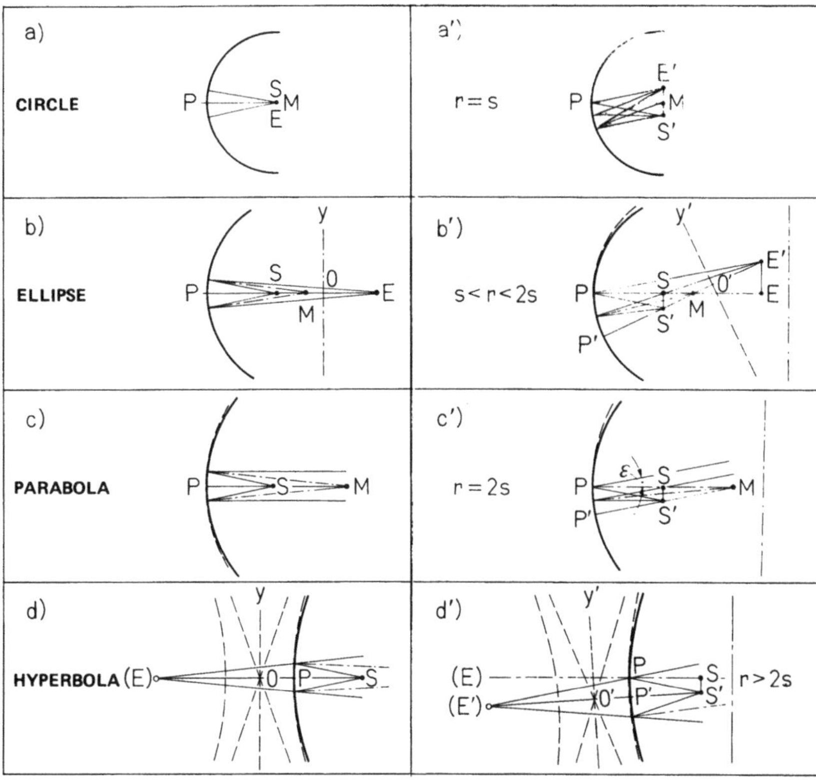

Fig. 3.3. Illustrations of the concave mirror laws. The sketches on the left pertain to a source on the axis of the curved reflector; those on the right are for a source off-axis. PS$=s$ is the distance of the source from the reflector; this distance is regarded as given. PM$=r$ is the radius of curvature of the reflector; this value varies. The designations (a) circle, (b) ellipse, (c) parabola, and (d) hyperbola in the sketches indicate that *any* curved reflector may exhibit the reflecting characteristics of any of these conic sections with the source at a focus, depending on the value of the ratio s/r.

were that of an ellipse with S and E as the foci. From the 'string construction' of an ellipse, it is well known that the sum of the distances from the two foci to any point on the ellipse is constant. But the ellipse has the further peculiarity that the rays from the two foci form the same angle with the tangent at the perimeter. Therefore, the rays radiated from one focus, no matter what the direction, are all collected at the other focus by reflection. These peculiarities follow from Fermat's principle, stating that sound will be reflected from such a point on a surface that the travel time from source to receiver is shortest. If all points on an ellipse obey this rule, then all the source–receiver reflected paths must be shortest, and therefore all are equally long.

We may conclude generally that if all *neighboring* rays reflected from a surface tend to concentrate nearly in one point, then they reach that point at nearly the same time.

It does not follow from this that *every* point of convergence must be reached by the intersecting rays at the same time. We need only consider the counter-example of two different ellipses that share the same foci, such that the reflecting surface in question is formed partially from one and partially from the other ellipse. The rays emanating from one focus all converge at the other focus, due to reflection from one or the other of the two ellipses, but the time of arrival will depend on which ellipse was involved in each reflection.

The reflecting areas for those two kinds of rays are separated by surfaces that are so inclined as to send the reflected waves in other directions. Fermat's principle is valid only for small variations and cannot be applied over regions with totally different behavior.

It is now possible to replace the circle assumed in Fig. 3.3(a) by an ellipse that has S as one focus and PS as the axis, and that best matches the curvature of the circle at the center of the enclosure. Here we introduce certain considerations of the geometry of curves.

First, we introduce the procedure for determining the tangent to an arbitrary curve. If we draw a straight line between two neighboring points P_0 and P_1 on an arbitrary curve, the line will cut off the curve at those two points. Now we let point P_1 move toward P_0 along the curve. The tangent to the curve at P_0 is the limit of the straight line when P_1 coincides with P_0.

Second, we introduce the procedure for matching the curvature of a circle to the curvature of an arbitrary curve. If we choose three neighboring points on an arbitrary curve, P_0 the point of interest and P_1 and P_2 on either side, these three points uniquely define a circle. Now we let points P_1 and P_2 approach point P_0. The circle defined by these points in the limit as they coincide with P_0 has the same radius of curvature as the original curve at P_0.

In Fig. 3.3(b), the circle with center M coincides so closely with the ellipse through P with foci S and E that it is not possible to distinguish them in the drawing.

It is now of interest to know where E lies, when S, P and M are given.

For this purpose, we define E by its distance from the center of the reflector, $e = PE$. For the ellipse, then, we have the simple law of concave mirrors:

$$\frac{1}{s}+\frac{1}{e}=\frac{2}{r} \tag{3.1}$$

Those who recall the analytic geometry of ellipses will remember the following relations, expressed in terms of the half-axes a and b of the ellipse: $s=a-\sqrt{a^2-b^2}$; $e=a+\sqrt{a^2-b^2}$; and $r=b^2/a$. Substituting these relations into eqn. (3.1), we find it confirmed.

If S coincides with M (i.e. if $s=r$), then eqn. (3.1) states that $e=r$. If we make $r<s$, then it must also be true that $e<r$. This means that a second source appears behind the speaker. This case is also included in Fig. 3.3(b): we have only to interchange the roles of S and E in that figure. This case would not be particularly objectionable from the room-acoustics viewpoint. But the situation actually illustrated in Fig. 3.3(b) implies that there is a focusing point in the audience where the sound is concentrated, and that other audience locations are deprived of sound. Both the concentration and the deprivation of sound should be avoided.

For $s=r/2$, E is shifted to an infinite distance as the ellipse approaches in the limit a parabola (see Fig. 3.3(c)). Sound rays emanating from the focus S leave the parabolic mirror as rays parallel to one another, aiming at the infinitely distant receiving point E. This principle is the basis for the design of searchlights. The parabola differs so little from the circle having the same curvature at P that, in practice, we can regard every vault with radius $r=2s$ as being parabolic. This case would be favorable if the circular opening embraced the entire width of the audience, since all seats would be equally covered with reflected sound.[1]

If we make $r>2s$, then we meet the third kind of conic section: the hyperbola. In this case the reflected rays seem to come from a point E behind the wall, just as we found previously for plane wave reflections. But the second focus of the hyperbola behind the wall is farther from the wall than is the source in front of the wall. (See Fig. 3.3(d).)

Whereas the parabolic reflector of Fig. 3.3c takes the divergent bundle of rays from the source and reflects it with no divergence at all (in

[1] Paul Sabine describes an American hall with parabolic boundary surfaces (Hill Auditorium, Ann Arbor, Michigan) and contrasts it with The Auditorium Theater, Chicago and Carnegie Hall, New York. (*Acoustics and Architecture*, McGraw-Hill, New York, 1932, pp. 182–7.)

a beam of parallel rays), the divergent tendency of rays from behind a hyperbolic reflector, though it is less than that of the rays from the original source, is greater than those from a parabola. Thus, the hyperbola may be more suitable as a back reflector than the parabola if the width of the audience is greater than that of the curved reflecting surface.

As Fig. 3.3(d) shows, the exact hyperbola corresponding to the same values of r and s differs from the circle only in regions far from the axis. Formally, the hyperbolic case differs from the elliptic case by changing the sign of e, when $s < r/2$, thus indicating that the point E (which, in this case, we may designate as (E)) lies behind the wall.

If (E) and S change their roles, the hyperbolic case demonstrates what happens at a *convex* reflecting surface. In this case, the apparent source behind the wall is closer to the wall than the real source in front of it, and the divergent tendency of the reflected waves is increased. Thus, convex boundaries increase the tendency toward homogeneous sound distribution to an even greater extent than do plane boundaries.

The transition from concave to convex surfaces, passing through the limit of a plane surface, corresponds to a transition from positive r to negative r, passing through the limit of $r \to \infty$. For $r < 0$, according to eqn. (3.1), we always have $e < 0$, whatever we assume for s. For the same reason, we can find, for each convex surface and source distance, a hyperbola with the same curvature as the reflecting wall at P.

So far, we have assumed that the sound source lies on the central axis of the curved reflector, corresponding to the illustrations at the left in Fig. 3.3. The illustrations on the right-hand side show what happens when the source moves to one side or the other of the axis. If this shift is small, then we may take over all of the 'on-axis' results for the line MS′. The vertices move to P′ and the opposite ends of the circular reflector now deviate further from the shapes of ideal conic sections. We can set $P'S' \approx PS$ and $P'E' \approx PE$; the latter means that E′ is approximately lateral to E, as assumed in geometrical optics.

In the hyperbolic case, the apparent source point (E)′ moves to the same side as the original source S. In the elliptic case, on the contrary, the second focus E′ moves to the opposite side. In both cases, the amounts of those lateral shifts are given by

$$\frac{EE'}{SS'} = \frac{e}{s} \tag{3.2}$$

In the parabolic case, the eccentric source S′ produces parallel reflected rays in a direction that deviates by

$$\varepsilon = \frac{SS'}{s} \tag{3.3}$$

from the center axis.

Although in the case of the full circle, which is not included in the concave mirror laws, each eccentric shift of the original source position produces, according to Fig. 3.2, a focusing region symmetrical about the center, we can develop the details more accurately by applying eqn. (3.1), piecewise, to small sections of the periphery. For instance, the portion of the circle to the left of the source in Fig. 3.1 acts approximately parabolically, since $s=r/2$; it thus produces reflected rays that are nearly parallel to one another and a corresponding straight-line wavefront. On the other side of the figure, where $s=3r/2$, we must expect a concentration at $s=3r/4$, that is, at point R; and we can see how the right-hand (nearly semicircular) portion of the wavefront (rays 10, 11 and 12) is, indeed, converging toward that point.

In the development of Fig. 3.3, we accepted the value of s as given and the value of r as variable. Four different cases can also occur where r is given and s is variable—meaning that the source may change its position in front of the curved reflecting surface. It is always the ratio s/r that determines whether the (actually) circular surface behaves like a circular, elliptic, parabolic or hyperbolic reflector, excited at its focus.

Finally, it is not even necessary that the surface be truly circular (cylindrical or spherical), although for architectural reasons this is most likely to be the case. Each curved surface may be approximated by a circular curvature in the neighborhood of a mean reflection point; and this curvature, with radius r, will exhibit the behavior of the various conic sections described above, according to the value of the ratio s/r. For example, a true parabola reflects as a parabola only if $s=r/2$; for $s<r/2$, it reflects like a hyperbola; for $s>r/2$, it reflects like an ellipse, including the circular case $s=r$.

Thus we may summarize:

1. *Every* concave curved boundary surface is to be treated as an ellipse, a parabola, or a hyperbola with the source at its focus, according to whether the distance from the source is larger than, equal to, or smaller than the radius of its curvature.
2. In the elliptic case the reflected rays produce a second focus, in the parabolic case they become parallel, and in the hyperbolic case

they appear to emanate from a source behind the reflecting wall. In all these cases the divergence of the reflected rays is less than that of the rays from the source.

3. Every convex surface may be approximated by a convex branch of a hyperbola; it always results in an increase of divergence of the reflected rays, compared to those from the source.

I.3.3 Whispering Galleries

In the preceding paragraphs, we have concerned ourselves with the reflecting behavior of curved walls, making use of the laws of geometric optics. However, for certain purposes we do not need to adhere strictly to those laws, since in acoustics we are usually not interested in a precise correspondence between object-points and image-points; it is enough if we can account for a higher concentration of sound energy in the vicinity of the ears due to some particular reflection, for this will help explain the listener's observations. The listener compares this increase of sound pressure with what he would expect under normal conditions, either in a rectangular room or in open air. Also, it may puzzle him that this strong sound seems to come, not directly from the speaker, but from a vault or niche.

In this section, we are concerned with a particular kind of curved reflecting surface called the 'whispering gallery'.

An unusual acoustical condition becomes especially striking when the speaker whispers, for our ears have the peculiarity that we can hear sound only when it exceeds a certain threshold, which (in absolute quiet) is determined by the limits of human hearing, but most often is determined by the background murmur of noise from the surrounding environment.

If the sound of interest is very weak (for example, whispering) then the listener may not hear it at all—unless he can stand in a focusing region.

Whispering is especially suitable for observing focusing phenomena because it contains only high frequencies; such sounds are readily directable, because the wavelengths are always small in comparison with the size of the reflecting surfaces and the radius of curvature.

Finally, the very concept of concentration of sound 'at a focus point' in principle requires short wavelengths.

Therefore, architectural arrangements that exhibit extraordinary concentrations of sound are called 'whispering galleries'. They are often regarded as acoustical miracles and are demonstrated as such, particularly in buildings that are visited primarily for historical or artistic reasons.

Such effects are to be expected to a greater or lesser extent wherever we find extended concave surfaces. But it is unusual to find such surfaces that are smooth enough, that are *not* interrupted by doors, windows, columns or decorative plaster relief which spoil the focusing, and that are so placed that the focus locations for both speaker and listener are readily accessible.

We mention here some examples that are already regarded as classics in room acoustics. Wallace Clement Sabine, the great pioneer of room acoustics, has given in his *Collected Papers on Acoustics* a detailed description of a number of such curiosities.[1] In that work he defined a whispering gallery in general as a room of such form that soft sounds can be heard at unusually great distances.

An example of a sphere excited and observed at or near its center is the Hall of Statues in the Capitol Building, Washington DC. (This corresponds to the example discussed above in Figs. 3.3(a) and 3.3(a').) Its dome is indeed spherical, and (which seldom happens!) its center is at the visitors' head height. (Mostly the centers of architectural domes lie so high that they are not accessible for observation.) Of utmost acoustical importance is that the surface was originally completely smooth; the visible coffers on this dome were painted there.

After a fire, however, the dome was restored; on this occasion the painted coffers were replaced with plaster coffers, having real cavities and surface modulation, in accordance with the taste and construction techniques of the era. To everyone's dismay, the famous focusing effect was remarkably reduced, since the earlier precise geometric reflection had been supplanted by a fuzzier diffuse reflection.

Such acoustical disappointments, following changes in a building, seem only to exalt the nimbus of the acoustical marvel! The occurrence even seems to support the popular notion that such calamities cannot be foreseen at all—whereas, in fact, the unhappy result was the direct and predictable consequence of grossly flouting the physical laws of sound reflection.

Sometimes the whispering gallery is comprised of several concave mirrors in succession, lying in the ray path, just as in geometrical optics a

[1] Sabine, W. C., 'Whispering Galleries', in: *Collected Papers on Acoustics*, Harvard University Press, Cambridge, 1927.

number of lenses are combined to make a compound lens. Thus, we can get a whispering gallery of second order when two parabolic mirrors are placed opposite each other, with the speaker's mouth at the focus of one and the listener's ear at the focus of the other. This case—also well known in its electromagnetic wave application—is a convenient arrangement for lecture demonstrations.

If, instead of facing the two mirrors directly toward one another, we direct the rays reflected from the source parabola toward a plane reflecting surface, such as a flat ceiling, and collect the reflected rays from the ceiling in the receiver parabola, we get a whispering gallery of third order. An example of this type is found in The Louvre, in Paris. The roles of the parabolic mirrors are taken over, there, by two antique marble bowls; the speaker and listener must bend over the rims of the bowls to place their heads in the proper positions. The bowls are located at such a great distance apart that the partners cannot understand each other by normal speech from these positions; in addition, another large art object blocks the direct line of sight. The shallow vaulted ceiling that forms the intermediate reflector between the two bowls is slightly concave in the lateral direction. (This is probably an indication that the two bowls do not work like true parabolas, sending their reflected rays in parallel lines, but more likely as hyperbolas, whose divergent beams must be collected by the concave ceiling in order to preserve the channel of communication.)

It is interesting that this rather complicated and sizable geometrical arrangement has been moved. Sabine's paper contains a photograph of the two bowls set up in the Salle des Caryatides; it was later installed in quite another room, again with a smooth and slightly vaulted ceiling.

A whispering gallery need not always be made up of a sequence of concave mirrors as described above. And when it is not, it is usually not possible to carry out a systematic analysis of sound focusing according to the order of reflections. This is especially true if a real circular gallery is involved, where the sound rays are reflected again and again at the wall.

The classic example is the famous gallery under the dome of St. Paul's Cathedral in London. Below the spherical dome, there is a narrow annular platform with a nearly cylindrical outer wall, completely uninterrupted by niches or pillars, where people can walk around. If someone whispers against the outer wall, the focus is not limited to a single location, but listeners all around the platform can hear him if they put their ears near the wall.

Lord Rayleigh[2] offered the following explanation for this phenomenon.

If the speaker turns his head more or less parallel to the wall, then all the sound rays from his mouth within the angle α remain within the annular region limited by the two radii R and $R \cos\alpha$ (as shown in Fig. 3.4), that is, within a strip of width $R(1-\cos\alpha) \approx R\alpha^2/2$. This means that the mean intensity in the annular strip increases with $1/\alpha$ as we move toward the outer wall. (Incidentally, the ray construction in Fig. 3.4 supplements our discussion in Section I. 3.1 by showing that even sources far off to the side do not lead to equal filling of the room with wavefronts. Sources near the center of a circle lead to a concentration of reflected sound there; sources near the perimeter lead to a concentration of reflected sound near the perimeter. Even the last frame (bottom right) of Fig. 3.2—drawn for a source-to-center distance $r/2$—shows the eventual dissolution of a narrowly defined focus region, replaced by a mere tendency for sound to accumulate in a broad annular region near the periphery.)

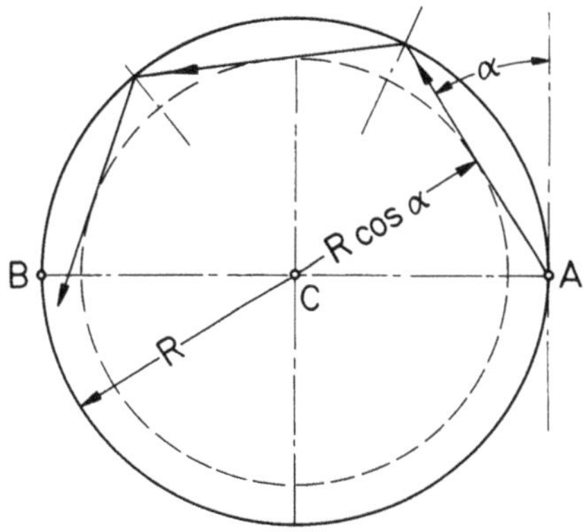

Fig. 3.4. Ray paths around a circular gallery.

Sabine has pointed out, in the course of a discussion of other galleries in which this effect is not so prominent, that Rayleigh's explanation, although elegant, does not tell the whole story, because one observes, in St. Paul's gallery, an additional increase of sound power at point B, opposite to point A in Fig. 3.4. It is conjectured that this may be due to a

[2] Lord Rayleigh, *Theory of Sound*, Vol. II, 2nd. edn., Macmillan, London, 1929, Section 287, p. 126.

collecting of sound in the vertical direction, perhaps by an inward slant of the outer wall, or by a contribution from the dome high above.

For example, if that gallery consisted not only of the cylindrical wall, but were roofed over with a hemispherical dome of the same radius R, then the ray construction of Fig. 3.4 would hold for every section through the perimeter (such as AB), and the sound concentration at B, for a source at A, would be enormous.

As an extreme example of focused sound, we mention the completely spherical room, dedicated to a geo-political map of the world, in the Christian Science Publishing House in Boston. In contrast to the usual maps of the globe, which are viewed from outside, this one is viewed from the inside. The sphere, two storeys high, is crossed by a catwalk, on which the visitor can stand to study the map of the world, displayed all around, above and below him. His head is approximately on a diameter of the sphere, so that when he stands at the middle of the catwalk, his head is at the center of the sphere. He can thereby place his ears so that, with a slight movement of the head, he can whisper into either his own left or right ear.

Although, today, whispering galleries may be regarded simply as curiosities, we should not forget that the problems of communication between people had practical importance long before the era of electroacoustics.

The reader who is interested in an early view of the field should refer to the *Phonurgia* of Athanasius Kircher.[3] Among other suggestions, he recommends providing elliptical rooms in buildings as a means of telephonic contact between prominent persons; and elliptical 'hearing tubes' for the hard-of-hearing. We know today, of course, that we can speak over extraordinary distances through ordinary tubes, and that a horn shape is much better than an ellipse for an ear trumpet.

But the book is interesting for its demonstration that the geometrical analysis of sound rays is the oldest method of room-acoustical investigation. It is even more fascinating to watch how Kircher applies this method to cases for which it is inappropriate, because the wavelengths are *not* small with respect to the wall dimensions or the radius of curvature—and thereby gets himself into trouble! For example, he tries to explain the increase of sound pressure at the small end of a horn by ray diagramming (as we did in the wedge example of Fig. 2.9, above), and

[3] Kircher, A., *Phonurgia*; translated into German with the title *Neue Hall- und Tonkunst*, Nördlingen, 1684.

he explains the reinforcement by the frequent reflections of the rays near the top.

I.3.4 The Avoidance of Focusing

It is seldom, nowadays, that the occasion arises for the deliberate construction of a whispering gallery. Such occasions are likely to become even rarer in the future, as people begin to understand how such galleries work and their miraculous acoustical properties take their place among other ordinary and familiar acoustical events.

In normal room-acoustics practice we are much more interested in *avoiding* such curiosities, so we shun the use of large, smooth, concave, curved surfaces. Domes especially, which are so widely admired by architects, can present serious acoustical hazards.

But if the center of a dome lies nearer to its apex than to the floor, then the reflection from the curved surface, for a source near the floor, is even weaker than it would be from a plane ceiling at the same height, as illustrated by Fig. 3.5. Here, the center of the spherical dome M lies closer to the dome than to the floor. A source in the source–receiver plane at S, at a distance h below the center of the dome, directs a ray-bundle with a certain included angle toward the ceiling. If the ceiling

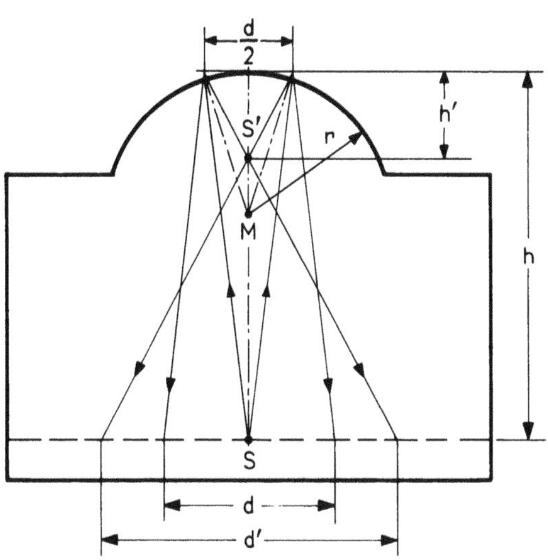

Fig. 3.5. Ray paths for a dome or a vault with a high center.

were a flat plane, this ray-bundle would return to the receiver plane confined within a circle of diameter d. But if the ray-bundle is reflected from the curved surface of the dome, this bounding circle will be larger, with diameter d'. Since the reflected sound energy in the initial ray-bundle is spread over a greater area in the second case, the dome reflection is weaker than that from the flat ceiling. Although the focus point S′ for this ray-bundle, at a distance h' from the apex of the dome, lies closer to the floor than the image source of a plane ceiling, the rays reflected from the dome are more divergent than those in the original ray-bundle.

Since the original ray-bundle covers a circle of diameter $d/2$ at the ceiling plane, we get:

$$d' = \frac{d}{2}\left(\frac{h-h'}{h'}\right) \tag{3.4}$$

The condition for the dome reflection to be weaker than the flat-ceiling reflection in the receiver plane is $d'/d > 1$. Thus, this requirement

$$\frac{d'}{d} = \tfrac{1}{2}\left(\frac{h}{h'} - 1\right) > 1 \tag{3.5}$$

implies:

$$h > 3h' \tag{3.5a}$$

which can, with the aid of the concave mirror formula,

$$\frac{1}{h} + \frac{1}{h'} = \frac{2}{r} \tag{3.6}$$

or alternatively:

$$\frac{h}{h'} = \left(\frac{2h}{r} - 1\right) \tag{3.6a}$$

be transformed into

$$\left(\frac{2h}{r} - 1\right) > 3 \tag{3.7}$$

that is

$$r < \frac{h}{2} \tag{3.7a}$$

Only if the center of the dome lies higher above the floor than halfway to the ceiling is the dome reflection weaker than the flat-ceiling reflection.[1]

The same holds for cylindrical concave mirrors, which often appear as vaults in the transverse section or as apses in the ground plan. The concave mirror formula is equally valid in the two-dimensional problem; therefore, we may interpret Fig. 3.5 as representing either a section through a spherical dome or a section through a barrel vault. In the latter case, the requirement that the reflection at the floor from the curved ceiling be weaker than a reflection from a flat ceiling is also given by eqn. (3.5).

If we are interested in comparing the deviations from the limiting case for two- and three-dimensional situations, we must note that the intensities vary as $(d'/d)^2$ for the dome and as (d'/d) for the vault. If $r<h/2$, then the dome reflection is more divergent (weaker) than that of the barrel vault; if $r>h/2$, the concentration of the dome is greater.

In both cases the simple rule holds: in order to determine whether or not a circular focusing surface will cause problems, we have only to complete the partial circle to make a full circle. If within this full circle there is neither a source nor a receiver point, then no problem with focusing is to be expected. (If the curved surface in question is not circular, the curvature-matching procedure described in Section I.3.2 can be used to find the equivalent circle, after which the above rule may be applied.)

Even the largest domes built during the Renaissance necessarily respected this rule because of the limits on the possible length of span imposed by the structural technology of the time. But modern concrete technology makes it possible to span entire auditoriums with a thin and shallow domed shell. In these cases, if the radius of curvature exceeds $2h$ (in the sense of Fig. 3.5), then we also exceed the parabolic limit whereby the reflected rays are distributed equally over the audience.

The elegance of form and the simplicity of construction of shallow domes, with a radius of curvature between $h/2$ and $2h$, recommend them to architects for application in large rooms, even when those rooms are intended to serve as auditoriums; in these cases, one may confidently expect that there will be serious acoustical problems of focusing. But

[1] The observance of this simple rule has also been recommended by E. Michel (*Akustik und Schallschutz im Hochbau*, Vieweg, Leipzig, 1938, pp. 43–4). Paul Sabine discusses acoustical defects due to curved shapes in concert halls, as well as allowable curved shapes (*Acoustics and Architecture*, McGraw-Hill, New York, 1932, pp. 171–81).

architects are accustomed to the need for compromise in building design, and they rely on their acoustical consultants to find architecturally acceptable means of avoiding the troublesome effects of focusing.

There exist three different methods for the avoidance of focusing, as illustrated in Fig. 3.6.

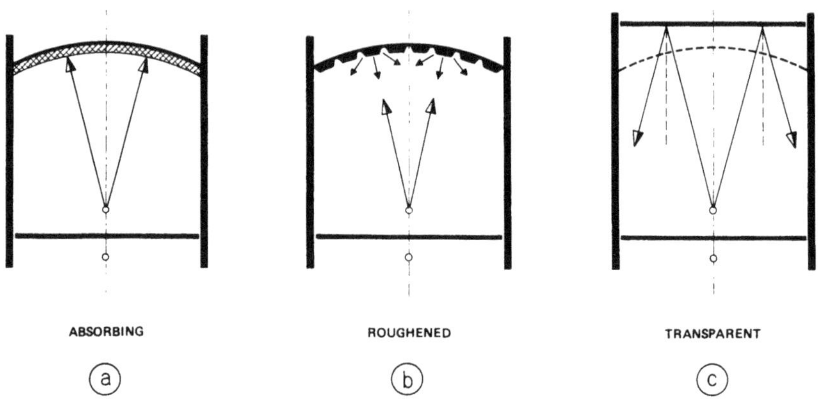

Fig. 3.6. Methods for avoiding focusing: (a) by absorption; (b) by breaking up the focusing surface; (c) by making the visual domed surface acoustically transparent.

An obvious and simple method is to cover the concave curved ceiling (or wall) with sound absorptive material (Fig. 3.6(a)). But this method is simple only in appearance. If the absorption is to be effective down to very low frequencies, and equally effective at high frequencies, then the required treatment involves multiple absorbing layers of great thickness (see Chapter IV.8, Volume 2). If this approach is followed, it allows the architect to preserve the appearance of a shallow dome if he wishes.

An example of such use of absorptive surfaces to cure focused reflections is the treatment of the Festival Hall of the Farbwerke Hoechst, near Frankfurt/Main.[2] Figure 3.7 shows the longitudinal section, the plan, and the reflected ceiling plan, as designed by F. W. Kraemer, of Braunschweig. Notice the spherical ceiling and the cylindrical wall at the floor level.

[2] Meyer, E. and Kuttruff, H., *Acustica*, **14** (1964) 138.

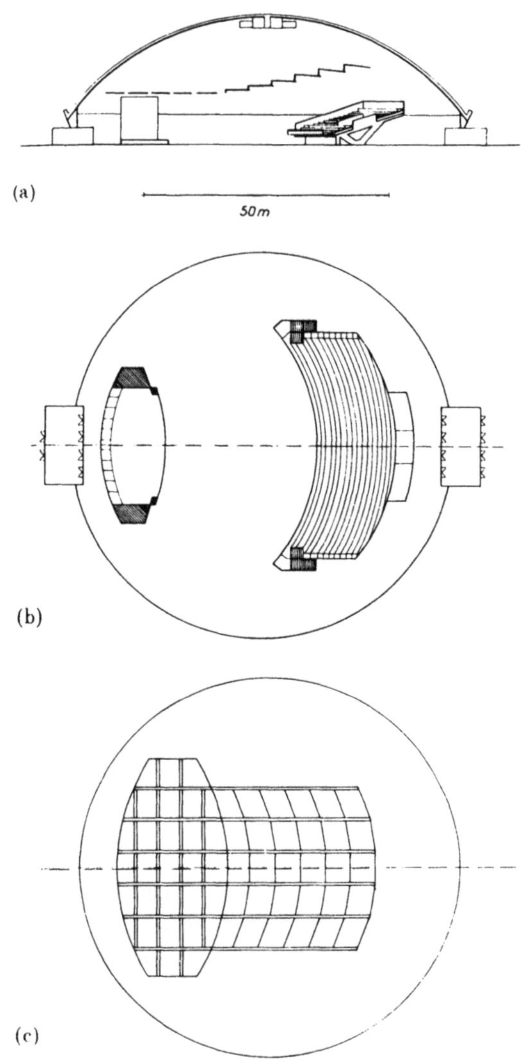

Fig. 3.7. Festival Hall of the Farbwerke Hoechst, Frankfurt/Main. (a) Longitudinal section; (b) ground plan; (c) ceiling plan. (After Meyer and Kuttruff.[2])

The ceiling is covered with a 15 cm layer of mineral wool over a 20 cm airspace; the walls are draped with heavy curtains at a distance of 70 cm from the wall.

This elaborate sound absorptive treatment is effective in avoiding the room-acoustics problems due to focusing from the curved boundary surfaces; the difficulty is that it does so by eliminating the room acoustics altogether! Reflections from the walls and ceiling are not intrinsically detrimental; in fact, as we shall see in Chapter I.5, they may be quite essential. The complete lack of wall and ceiling reflections is a very serious defect, at least for all musical performances.

Therefore, the acoustical consultants had to go two steps further to replace the useful reflections that were lost in the heavy sound absorptive treatment. (1) A flying, stepped ceiling was provided over the stage and most of the audience, as shown in Figs. 3.7(a) and 3.7(c), in order to provide early sound reflections. (2) In addition, in order to replace the reverberation normally provided by multiple reflections from the room boundaries, an electroacoustical artificial reverberation system was installed.

As admirable as this pioneering work was, it showed how difficult it is to produce in this way a room-acoustical impression that satisfies musicians' ears.

This is not to criticize the use of electroacoustical systems in auditoria; in fact, for rooms as large as the Festival Hall, which are often used for 'multi-purposes', it is quite essential to provide reliable, high-quality sound systems, for the reinforcement of speech and certain kinds of musical performance.

The second method of avoiding focusing is to break up the concave curved surface into smaller scattering surfaces (see Fig. 3.6(b)), and thus to change the geometrically focused reflection into a diffuse reflection. According to what we have said in this respect in Section I.1.4, the dimensions of these scattering elements must be comparable with the wavelengths, and this means shapes of considerable depth (see Section IV. 13.3, Volume 2). It is evident that a compromise must be found, to provide adequate depth for the scattering elements without stealing too much volume from the room. It is also an architectural challenge to incorporate such 'strong' visual elements into an aesthetic design scheme.

An impressive example of cooperation between the architect (Henselmann) and the acoustician (Reichardt) is presented in Fig. 3.8. The architectural/acoustical design was developed with the aid of acoustical model tests[3] at a scale of 1:20. In this way it was possible to

[3] Reichardt, W., Budach, P. and Winkler, M., *Acustica*, **20** (1968) 149.

test a number of new proposals for breaking up the focuses from the cylindrical side-walls, until both the acoustician and the architect were satisfied.

Figure 3.8 shows radial vertical glass panels, around the seating level, that prevent the gallery-effect illustrated above in Fig. 3.4. In addition, one can see that the wall panels between them lean inward, so that they reflect sound toward the floor before it can come to a focus farther out in the room. The horizontal stripes on these panels indicate a series of diffusing surfaces. Above these panels are the numerous interpreters' booths surrounding the seating area, and above them we see an ornament like a bracelet arranged about the

Fig. 3.8. Breaking up of a cylindrical wall and a spherical ceiling ($r = 22$ m, $h = 13$ m) by diffuse reflecting elements in the 'Haus des Lehrers'. (East) Berlin (architect, Henselmann; acoustician, Reichardt).

periphery of the upper wall. Below the ceiling are hung four 'wreaths' of plexiglas panels at various inclinations, to inhibit focusing from the ceiling. The architect intended the ceiling to be visible as a smooth, spherical surface, symbolizing infinite space.

A more famous example of the use of suspended acoustical reflectors to minimize focusing from a concave ceiling is the Aula Magna in

Caracas (architects, Villanueva, Briceño-Ecker, Ellenberg), see Fig. 3.9. Here, a world-renowned artist and sculptor, Alexander Calder, was engaged to design the shapes and colors of the 'convex downward' diffusing reflectors. When the aesthetic demands of such a prominent artist are in competition with the scientifically based requirements of the acoustician, it may sometimes happen that unfortunate compromises are made. Fortunately, in this case, the recommendations of the acousticians (Bolt Beranek and Newman Inc.) were respected, and it appears that the collaboration in Caracas was successful.

Fig. 3.9. Calder's mobiles, used as diffusers in the Aula Magna, Caracas. (After L. L. Beranek, *Music, Acoustics and Architecture*, John Wiley, New York, 1962 (reprinted Krieger Publishing Co., Huntington, New York, 1979).)

Since in all the foregoing cases, the problem was to get an intrinsically unfavorable room shape to work, the acoustician sometimes wonders if it would not be better to avoid resolutely all focusing elements, from the very beginning of the project!

When Hans Scharoun conceived, for the New Berlin Philharmonie, a stage that would be surrounded on all sides by the audience, it was a natural extension of his concept to cover the hall with a dome. But when he was told by his acoustical consultant that such concave ceilings always lead to unfavorable focusing effects, whereas a convex shape would help to make the sound distribution more nearly equal, he decided to provide a ceiling in the shape of a tent[4] (see Fig. 3.10).

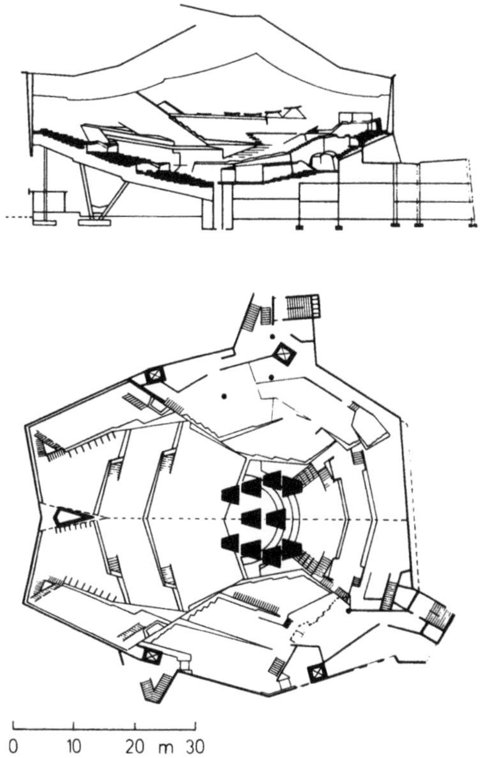

Fig. 3.10. Section and ground plan of the Berlin Philharmonie.

There is one type of room in which an exactly spherical, smooth ceiling is unavoidable, namely the planetarium. The strict requirement for the ceiling shape arises because the ceiling is the projection screen on which the stars and constellations are shown. But here it is possible to

[4] Cremer, L., *Deutsche Bauzeitung*, Deutsche Verlagsanstalt, Stuttgart, 1965, Part 10.

make use of the third approach to the avoidance of focusing, as shown in Fig. 3.6(c); namely, let one surface be the acoustical boundary of the room, while another surface (the visual boundary, on which the pictures of the heavens are projected) is made of finely perforated metal, which forms a suitable screen for light waves but permits sound waves to pass through freely.

Recalling the discussion of diffraction in Section I.1.3, the bending of sound waves around small obstacles does not only occur with single objects; even a large number of small obstacles in one plane may be quite transparent to sound if they are separated by small airgaps (see Section IV.9.1, Volume 2). In general, it is easy to make perforated sheet metal highly transparent to sound, even if the open area represents only 10 to 12% of the total area. Such a surface is still suitable for optical projection, because, for the much shorter light waves, approximately 90% of the screen is still diffusely reflecting. Behind the perforated metal, any convenient acoustical treatment can be installed to control the room acoustics. (Focusing is no problem because there is no longer a curving acoustically reflective surface.)

The example shown in Fig. 3.11 relies on a rectangular, stepped system of coffers[5] as acoustical scatterers. In addition, absorbing mats of rockwool were installed between the perforated metal screen and the coffer system.

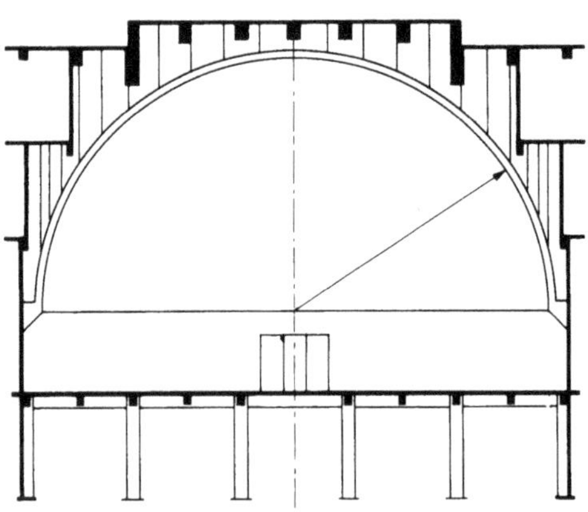

Fig. 3.11. Section of an anechoic (echo-less) planetarium in Philadelphia. (After P. E. Sabine.[5])

[5] Sabine, P. E., *Journal of the Franklin Institute*, **217** (1934) 452.

This approach is rendered more difficult if the architect wants the building to proclaim its purpose by means of a dome that is visible from the outside. In this case, there must be an inner dome on which to project the stars and an outer architectural dome as a symbol and a weather barrier. The inner dome must be perforated to avoid severe acoustical focusing in the audience, and the means for control of focusing from the inner side of the outer dome must be installed between the two domes.

In the example of Fig. 3.12, the planetarium in Nürnberg (architect, Schlegtendahl), a highly diffuse series of reflecting elements was used. The

Fig. 3.12. The dome of the planetarium in Nürnberg, showing the space between the inner (sound-transparent) and outer (concrete) domes filled with sound diffusing elements.

treatment is so successful in breaking up the focus that this planetarium is even used for chamber music, although the sound-transparent screen (made in this case of a perforated plastic foil manufactured for use in sound motion picture screens) has such tiny perforations that the friction of the air in the holes may add some unwanted sound absorption to the room (see Section IV.9.2, Volume 2). Such perforated screens are optically opaque when lit only from the front (as with projection); but when the space behind it is illuminated, the screen becomes optically transparent and one can easily see what lies behind.

Acoustically transparent walls and ceilings lend themselves to being painted or decorated without losing their sound transparency (though naturally, care must be taken not to clog the holes with paint); thus, the architectural function of such surfaces can be made independent of the acoustical function.

Not all architects and designers are happy with the thought that visual and acoustical functions can in some cases be separated. A strong school of design contends that 'form follows function'; its adherents insist that every element in a building should 'look like what it does'. To these people, the strategems described above for planetarium design smack of fakery. In the present case, however, no delusion is intended at all: the visitors to buildings where acoustically transparent boundaries have been used may simply know that they are in the presence of an optical limitation that differs from the acoustical one. They are already very familiar with the opposite case, which has so far raised no objections of the form-and-function kind: windows!

Chapter I.4

Echo Problems

I.4.1 Conditions for the Existence of Echoes

In Section I.2.3 we have considered the importance of the transit time in the subjective evaluation of a sound reflection. We become especially aware of the time delay in relation to the most conspicuous of room-acoustical reflections, the echo. We may define an echo as a repeated signal that gives us the impression of coming from somewhere other than the position of the true source. Thus, the definition is based on a subjective perception, or a so-called 'hearing event'.[1] To this end, several objective features, or so-called 'events of sound', must coincide.

First of all, the echo reflection must come with such a delay that the direct sound has passed the listener's ear before the echo sound arrives. For short impulses, even a very short delay may be sufficient to fulfil this condition. But if the sound in question is a monosyllable (a typical case), then a time delay of 100 ms is required, corresponding to a path-length difference of about 34 m. In the case of a first-order reflection returning to the speaker, the reflecting surface must be about 17 m away from him.

Since we speak at an average rate of about five syllables per second (i.e. one syllable every 200 ms), this means that an echo will appear as quite distinct if it returns in the interval between syllables as in the case above; the speaker will feel as if he is continually interrupted by another speaker.

If the time delay is significantly shorter, on the other hand, then the reflection is regarded, not as a disturbance, but as a useful reinforcement of the original signal. The limit between 'useful' and 'harmful' sound was demonstrated and given the name 'limit of perceptibility' by Henry as

[1] Blauert, J., *Räumliches Hören*, Hirzel Verlag, Stuttgart, 1974, p. 2.

early as 1854;[2] it is not a sharp limit, but depends on several circumstances, which we shall discuss in Section III.1.4.

But in this and the next chapter, it will be sufficient—at least for speech—if we take this limit to be

$$\Delta t = 50\,\text{ms} \tag{4.1}$$

corresponding to the path-length difference

$$\Delta l = 17\,\text{m} \tag{4.2}$$

In this case, the 17 m limit includes the path out and back, so we can regard sound reflections from a wall or ceiling as being useful if these surfaces are within 8·5 m.

A second condition for the appearance of an echo, in addition to an appropriate time delay, is that it must be strong enough to compete effectively with the other sound impulses arriving at about the same time. Since the sound pressure decreases during propagation, either by divergence or by losses sustained in the medium, most audible echoes occur with time delays just above the echo limit (>100 ms). Reflections from distant sources are lost in the background noise, and in a quiet landscape the more distant reflections usually do not exceed the threshold of hearing. On the other hand, with sharp sounds of high intensity, such as pistol shots or explosions, echoes are nearly always heard. Even in a noisy harbor, each blast of the foghorn produces strong echoes.

The noise level that masks an echo may be produced by the same signal that generates the echo, because this signal can excite so many closely spaced reflections (reverberation) that the echo does not stand out among them. This, in fact, is usually the case in closed rooms. We observe echoes more readily in the open air, because the competing reflections are fewer; if there is a reflecting surface that fulfils the conditions for an echo, the echo has a better chance of being heard.

For an echo to be heard indoors, the room surfaces other than the echoing surface must be rather sound-absorptive, or else the echoing surface must be concave so that it returns an especially strong focused reflection to the observer, as we have discussed in Chapter I.3.

---

[2] See Shankland, R. S., *J. Acoust. Soc. Am.*, **61** (1977) 250. In Germany this limit is usually called *Verwischungsschwelle* (blurring threshold), following a proposal of E. Petzold (*Elementare Raumakustik*, Bauwelt-Verlag, Berlin, 1927, p. 8). A number of examples cited in Knudsen's classic work, in his chapter on the effect of the hall shape on auditorium acoustics, were drawn from German sources, such as Petzold: see Knudsen, V. O., *Architectural Acoustics*, John Wiley, New York, 1932, Chapter XVI.

Evidently, much more than the time delay and the intensity of a reflected impulse must be known before we can determine whether or not it will be perceived as the 'hearing event' called an echo.

The impression of an echo does not necessarily depend on a reflected sound at all. Since the invention of the electroacoustic public address system, the delayed signal ('echo') may come from a distant loudspeaker; it arrives much later than the same sound from nearby loudspeakers on the same circuit. We call this an artificial echo.

Although such echoes can create as much disturbance as natural echoes, they can be avoided by proper installation of the loudspeaker system; thus, they do not belong strictly to room acoustics. We shall return to this problem in Section I.5.2, when we discuss electroacoustical means for adding useful sound reinforcement to the direct sound. Here we only call attention to an important rule: if—as is usual—the loudspeakers are synchronized with the (live) speaker, then each loudspeaker creates the same effect as if there were a reflecting wall, located half-way between it and the original speaker (see Section I.2.1). Therefore, the loudspeaker may be placed at a distance of 17 m from the speaker without risk of exceeding the limit of perceptibility in the intervening area, even in the neighborhood of the speaker (provided, of course, that the loudspeaker signal is not too greatly amplified). The impression of an echo would not be expected until the distance between speaker and loudspeaker exceeded 34 m.

If the loudspeakers are fed with time-delayed signals (a procedure which has advantages that will be discussed in Section I.5.2), the time delay of the loudspeaker near the original source may be even longer than that of a natural reflection from a wall placed at the loudspeaker location. It is always easy to decide whether a natural or artificial echo is present by switching off the loudspeaker.

I.4.2 The Single Echo

A famous example in which a long-delayed echo was reinforced by focusing occurred in the Royal Albert Hall, London (see Fig. 4.1). The listeners in the rear part of the main floor (stalls) received a concentrated reflection from the vaulted ceiling, which is curved in both the longitudinal and transverse directions. The focal point for the transverse curve is about at floor level, while that for the longitudinal curve is a bit below floor level. The path difference between the ceiling-reflected sound and

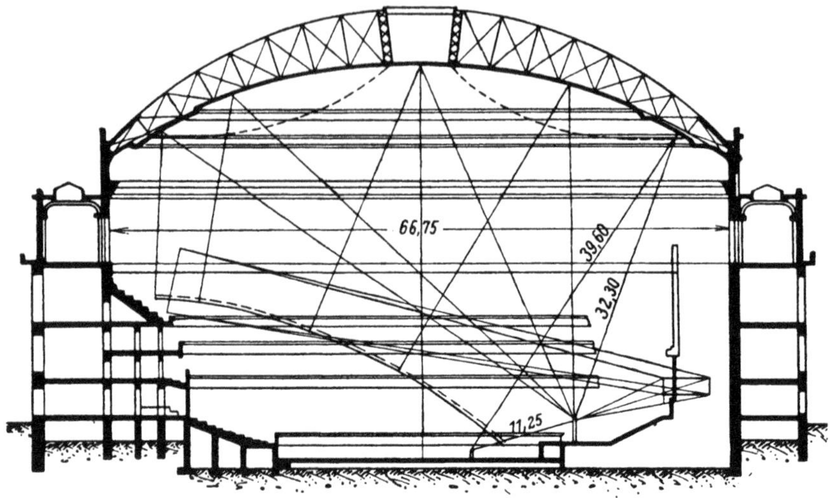

Fig. 4.1. Longitudinal section through the Royal Albert Hall, London. (After Bagenal and Wood, *Planning for Good Acoustics*, London, 1931.)

the direct sound is about 60 m, corresponding to a time delay of about 1/6 s.

The first attempted remedy was to install large draperies, spanning from the top of the dome to the periphery, as shown by the dashed lines in Fig. 4.1. Sounds radiating from the stage toward these draperies first encountered their convex lower surface, which reflected a small portion of the sound, but with such strong divergence that it was greatly weakened. The sound that passed through the draperies struck the concave ceiling and was focused back toward the usual location; but, since it had passed twice through the draperies, it was somewhat attenuated and the disturbing echo was much reduced. The presence of the draperies, however, had the additional unwanted effect of reducing the reverberation time of the hall, which had been judged insufficient in the first place. Accordingly, the draperies have since been replaced with hanging reflecting panels, as discussed in Section I.3.4.

The time delays for reflections from hall ceilings are generally not so long that they give rise to echoes. It is quite different, however, with reflections from the rear wall, and particularly from the upper right-angled corner between the rear wall and ceiling (see Fig. 2.3(a)). This long-delayed reflection is of second order (a so-called 'cueball' reflection), and is especially troublesome if the rear wall follows the typical curve of the seating rows, and thus focuses the echo in a particular section of the audience.

An example of such design is shown in Fig. 4.2, where the rear wall, following the amphitheater seating plan, is circular, with its center on the stage.[1] This hall was not merely a drawing board exercise, but was actually built! The ray-diagram shows the concentration of long-delayed reflections from the rear wall to the stage.

As an inexpensive, yet effective, remedy, sound absorptive mats of rockwool were hung in the ray path around the rear wall, as shown by the dashed line in Fig. 4.3. The concave shape of the mats was chosen on

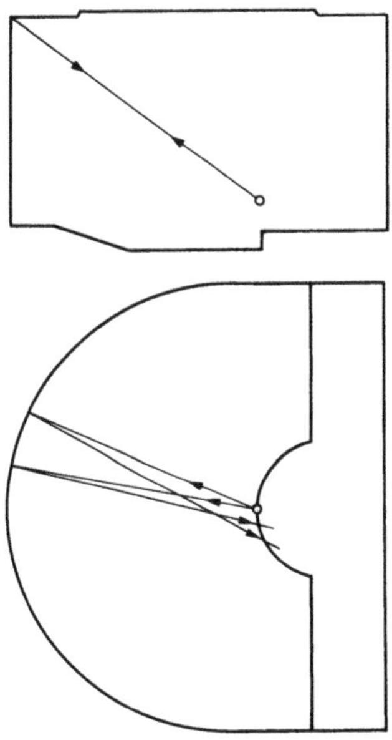

Fig. 4.2. Longitudinal section and plan for a multi-purpose hall with semicircular ground plan and a focusing reflection from the upper rear corner.

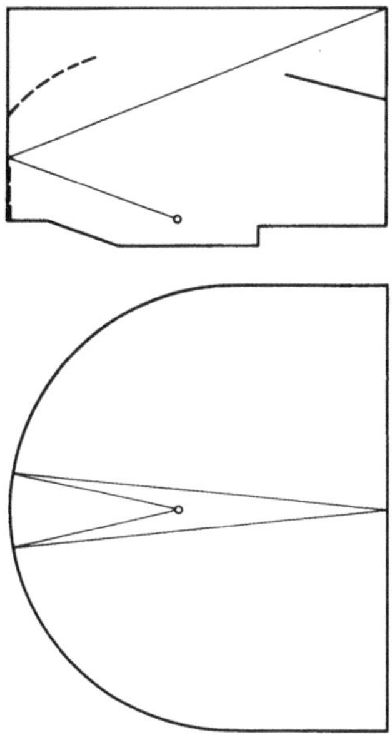

Fig. 4.3. Echo of fourth order, in the hall shown in Fig. 4.2. It involves the rear wall as a concave mirror and a cueball reflection from the upper front corner.

[1] This theater was built in one of the houses of the 'Party' in Munich. The changes described in the text were made when the house was used as an Amerikahaus at the end of World War II. Today, the room serves, after further extensive changes recommended by the authors, as a concert hall for the Musikakademie that occupies the building.

architectural grounds; but since the mats were highly absorptive, no ill effects occurred. In addition, a sound-reflective surface was installed above the stage, also shown in Fig. 4.3.

However, once the mats and the over-stage reflector were installed, another echo problem became evident. But since it involves a source-location in the audience, where performers never go, it is primarily of academic interest because it involves a fourth-order echo; Fig. 4.3 shows the ray paths. Sounds originating at the indicated point in the audience, radiate back to the curved rear wall, which focuses them at a point in the upper front corner of the hall; since this is a 90° corner, the rays return upon themselves to the same location on the rear wall, which then refocuses the rays back to the source location. Neither the absorptive mats around the rear of the hall nor the reflecting panel over the stage interferes with this multiple focused reflection.

Another example of the same kind is the Prinzregententheater in Munich (see Fig. 4.4). It also has a ground plan in the form of an amphitheater, but with a smaller 'open angle'. All of the seating rows, as well as the upper corner of the hall between rear wall and ceiling, are concentric circles, having as their common center a point at the middle of the stage. The focused reflection of second order follows the same path as shown in Fig. 4.2. Sounds from the performers on stage are focused back to them with a delay of 72 ms, causing them very serious disturbance.

This defect is all the more disappointing since the Prinzregententheater was deliberately intended as a copy of a room that is highly praised for

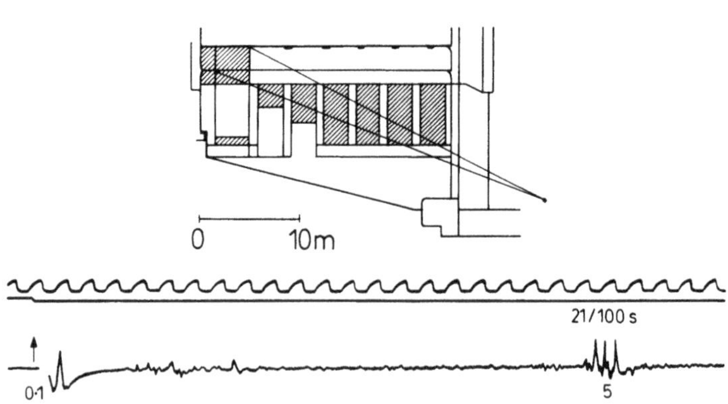

Fig. 4.4. Echogram from the Prinzregententheater in Munich. (After Crone *et al.*[2])

its acoustics, the Festival Hall in Bayreuth (see Section II. 3.2, Fig. 3.2). But besides differences both in ground plan and in the choice of building materials, there is a major difference between the two halls: in Bayreuth, the rear wall is filled with boxes, nearly up to the ceiling, so the double reflection does not occur.

Although these significant architectural alterations were made in 'copying' the Prinzregententheater from Bayreuth, the acoustical problems that developed in the Munich hall have been taken as proof that Bayreuth must possess some mystical acoustic secret that has not been found to this day!

There is no secret. It has been clearly demonstrated that, of all the reflected rays that can occur in the Prinzregententheater, only the one shown in Fig. 4.4 is responsible for the disturbance of the artists.[2] Zenneck and his assistants, Crone and Seiberth, fired pistol shots from the stage toward the auditorium and recorded as a function of time the corresponding signals from a microphone placed on the stage by means of an electrodynamic oscilloscope.[3] The impulses from a tuning fork were simultaneously recorded to serve as a time reference.

The trace at the bottom of Fig. 4.4 shows a strong sound pressure peak immediately after the pistol shot, which corresponds to the direct sound wave. The only other outstanding signal is the reflection at 210 ms, corresponding to the ray path drawn in the upper part of the figure.

Following this experiment, the offending reflecting surfaces were covered with sound-absorptive materials, as indicated by the shaded areas in the sketch of Fig. 4.4. As a result, the disturbing echoes were much reduced and the artists were satisfied. But since these coverings represented a considerable change in the architectural appearance, they were not retained when the building was restored after World War II. In fact, no attempt was even made to replace them with an alternative treatment that would be compatible with the architecture. Once again, it is disappointing for the acoustician to see what the audience and performers are expected to suffer in the name of an architectural goal. (See also Section III. 3.3.)

The method of echo diagnosis with the aid of recorded sound

---

[2] Crone, W., Seiberth, H. and Zenneck, J., *Ann. Phys.* (*Leipzig*) *V*, **19** (1934) 300.

[3] With respect to the electroacoustical means used for room-acoustical measurements, the reader is referred to specialized textbooks, such as Reichardt, W. *Grundlagen der Technischen Akustik*, Academische Verlagsgesellschaft Geast und Portig, Leipzig, 1968; or Beranek, L. L., *Acoustic Measurements*, John Wiley, New York, 1949.

pressures following a pistol shot, was named '*Stossprüfung*', or 'shock-testing' by Scharstein and Schindelin,[4] who were also assistants of Zenneck. The word 'shock' in this context, however, may be misleading, since, strictly speaking, shock refers to a 'one-sided' impulse, whereas a pistol shot is characterized by a sound pressure signature in the far field that shoots first above and then below the mean (rest) pressure (see Part II, Fig. 7.1).

In room acoustics nowadays, it has become customary to call such traces 'echograms', even when they are made for purposes other than echo diagnosis. They provide records of *all* the significant reflections, whether or not they qualify as echoes according to the definition given above.

It is not always necessary to use strong impulses such as a pistol shot to evaluate sound reflections: even hand-clapping may be very useful. Indeed, it is one of the trademarks of an acoustician to clap his hands and listen upon visiting a hall, because this is an informative qualitative acoustical evaluation of the hall for which the required tools are always 'at hand'.

I.4.3 Multiple and Repetitive Echoes

Quantitative echograms are essential if a room presents several echoes whose correspondence with particular ray paths is not evident. Figure 4.5 shows an example of another shock-test from Zenneck's school. Figure 4.5, top, gives the longitudinal section and plan of the Aula of the University of Freiburg (Baden-Württemberg), with some selected ray paths; Fig. 4.5, bottom, shows the echogram recorded at point *g*, in the vicinity of an impulsive source, at point *a*. (It had been observed that listeners at point *g*, fairly near the lecturer at point *a*, had great difficulty in understanding him.)

This echogram exhibits several reflections with long time delays. The elliptical ground plan, near one of whose foci the lecturer stands, was first suspected as the cause of the trouble: it was thought that the sound of the lecturer, leaving one focus in the horizontal plane, might collect at the second focus and then return to the first focus, to disturb the listeners there. But this explanation fails, because the entrance hall

[4] Scharstein, E., and Schindelin, W., *Ann. Phys.* (*Leipzig*) *V*, **2** (1929) 194.

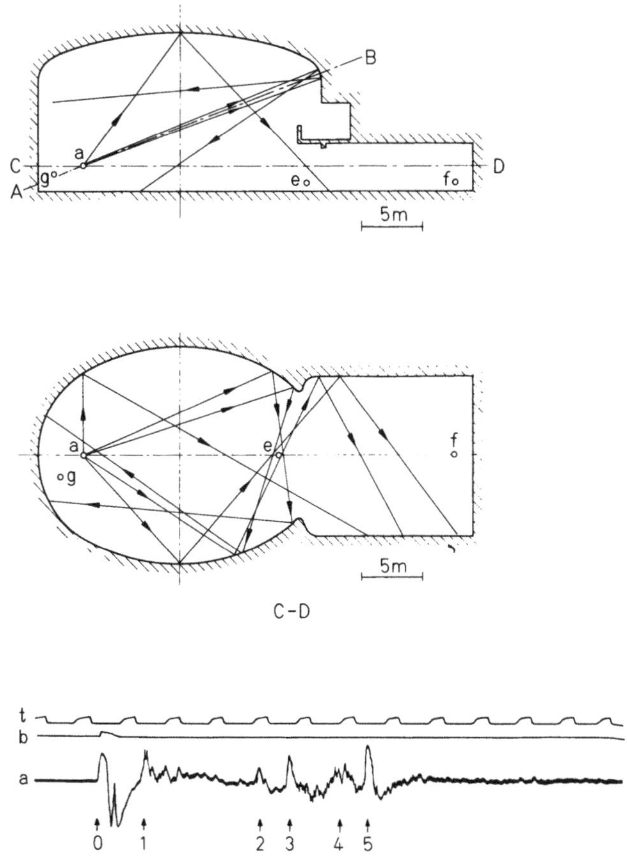

Fig. 4.5. Top: section and ground plan of the Aula of the University of Freiburg (Baden-Württemberg). Bottom: echogram: source (pistol shot) at *a*, receiver at *g*. (After Scharstein and Schindelin.[4])

interrupts the path at the end of the hall opposite the lecturer and the side walls are broken up by diffusing decorative elements.

There exists, however, an oblique plane through the source point, passing through the sharp curve in the ceiling above the balcony (see line AB in the section); it, too, has an elliptical perimeter. Comparison with the echogram shows that peak 3 in the series of reflections corresponds to a reflection of first order at B; peak 4 corresponds to a second-order reflection involving the elliptical perimeter; peak 2 comes from a first-order reflection at the balcony railing; and, finally, peak 5 comes from the rear wall of the entrance hall. (The latter is especially strong since the pistol was aimed in that direction.)

The occurrence of multiple echoes sounds especially peculiar if they follow one another in an ordered sequence. If their separations are great enough, a shouted cry may be repeated several times. Since, in general, the later echoes are weaker, these repeated cries gradually dwindle into silence. One can hear such echoes in the mountains.

If, however, the recurring echoes follow one another rapidly, at short and equal intervals, the individual impulses cannot be distinguished by the ear and the impulse-sequence is perceived as a tone, called a 'picket-fence' echo or tonal echo. If the common distance between the reflecting elements, as seen from the observer's location, is $\Delta l/2$, the period of the tone is $\Delta l/c$ and the frequency of the tone is $c/\Delta l$. In this case, the fundamental frequency is rather high. We find a similar repetitive, tonal echo in front of large expanses of steps outdoors (entrances to large libraries, public buildings, etc.). The effect is especially impressive under focusing conditions, as with impulse excitation from the center of the concentric seating in amphitheaters (unoccupied, of course, since the absorptive audience would spoil the echoes); see the discussion of Fig. 4.6, below.

Figure 4.6 shows an echogram recorded by Plenge and Kürer[1] in the

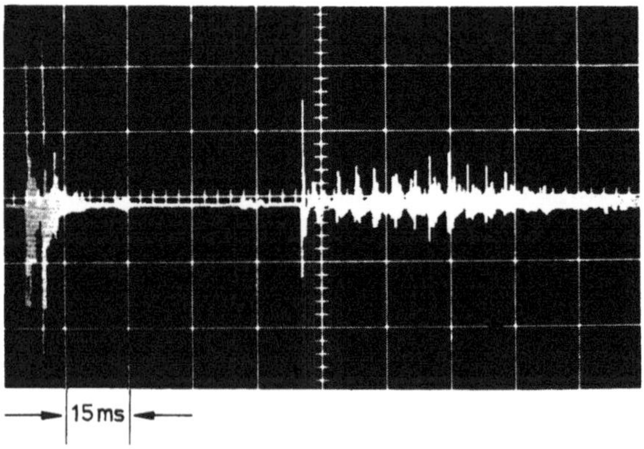

Fig. 4.6. Echogram of a 'picket fence' echo, recorded in the empty theater of Epidauros. (After Plenge and Kürer.[1])

[1] Kürer, R., Techn. Ber. No. 96 of the Heinrich Hertz Institute, Berlin-Charlottenburg.

famous theater of Epidauros. The receiver was placed at about the middle of the 'orchestra' (the circular part of the stage) and the source at the higher part of the stage (see Chapter I.6, Fig. 6.7). (A comparison between this echogram and those in Figs. 4.4 and 4.5 demonstrates the technical progress in echogram recording in the course of 20 years.) The first reflection, immediately after the direct sound, corresponds to a wall behind the stage, the second reflection is from the first row of the audience. At the right of the trace in Fig. 4.6, however, we see an interesting sequence of reflections, equally spaced in time; these correspond to the succession of reflections from the regularly spaced seating banks of the amphitheater. If the row-to-row spacing is about 1 m, the increment of path-length Δl between reflections is 2 m and the frequency of the tone is $340/2=170$ Hz, rather low.

So far, in Figs. 4.4, 4.5 and 4.6, we have seen a succession of individual reflections or echoes, which arrive at instants corresponding to specific, identifiable ray paths. These are called *multiple* reflections; they may or may not be periodic.

In contrast to multiple reflections (for which the paths are all different), we can speak of repetitive reflections if the same ray path is retraced again and again and the wavefront repeatedly passes the listener's position. It is sometimes difficult for the listener to distinguish between multiple and repetitive reflections, because repetitive reflections also die away gradually, giving the impression of coming from a greater and greater distance.

We have already considered the appearance of repetitive reflections in rectangular rooms (Section I. 2.2), where sound rays travelling parallel to two absorbent walls persist longer than the other rays; if the distance between the reflecting surfaces is not too large, the repetitive reflections sound like the fluttering of a bird, and are therefore called 'flutter echoes'. At a distance of 7 m the period frequency is about 25 Hz, in this case the flutter echo may sound like snoring.

Similar effects can be observed in living rooms or offices with the furnishings concentrated at one side; they can be especially prominent in entrance halls with hard floors and ceilings (the latter sometimes vaulted), where the side reflections are suppressed by adjacent stairways, corridors, etc. Often one gets the impression that something is vibrating.

The flutter echo demonstrates clearly that both a positive and a negative condition must be fulfilled if an echo is to be heard: the positive condition is the existence of a strong reflection, the negative condition is the absence of significant competing reflections. An echo, to be heard as an echo, must stand alone. Therefore, in order to eliminate a disturbing echo, we may either eliminate (with diffusing or absorptive treatment) the reflection that causes it, or supplement this reflection with numerous other reflections that hide it.

Figure 4.7 shows an echogram recorded in a rectangular room, using an electric spark as the source of impulsive sound. For trace 1, the floor and one side wall were covered with sound absorptive material; in this case, we get a clear flutter echo: the succession of *double* pressure peaks occurs because of the off-center position of the receiver. If we remove the absorptive material from the side wall (trace 2), the periodic structure disappears from the echogram, and the flutter echo is no longer perceived. Finally, if we also remove the absorptive material from the floor (trace 3), we get a long, continuous reverberant sound, without discernible peaks of pressure and without audible echoes.[2]

Even more remarkable are the repetitive echoes based on higher order reflections in rooms. It would be tedious (and would contribute little to the understanding of practical room acoustics) to exhaust all of the possibilities, so we content ourselves with the consideration of one example, which occurs rather frequently and is easily understood. Namely, a vaulted ceiling above a plane reflecting floor, on which the source and receiver stand. The surrounding geometry is such that side-wall reflections are negligible.

If the radius of curvature of the ceiling equals twice its height ($r = 2h$), the ceiling acts like a parabola. (We neglect for simplicity, in contrast to Fig. 3.5, the difference in height between the plane of the floor and the plane containing the source and receiver.)

If we produce a hand clap under the center of the vault, the first reflected sound from the ceiling reaches the floor as a set of parallel rays, or a broad plane wave; after reflecting specularly at the floor, this wave returns to the curved ceiling. This time the ceiling focuses the incident parallel rays back to the original source-point at the floor. Thereafter, the process repeats itself. In fact, the exchange repeats itself again and again, so that the echoes heard at the receiver position alternate between weak (the parallel rays) and strong (the rays focused back to the source-point). If the observer (acting both as source and receiver) moves to a point off-

[2] The authors are indebted to F. Thele, Institute of Technical Acoustics, Berlin, for these photographic records.

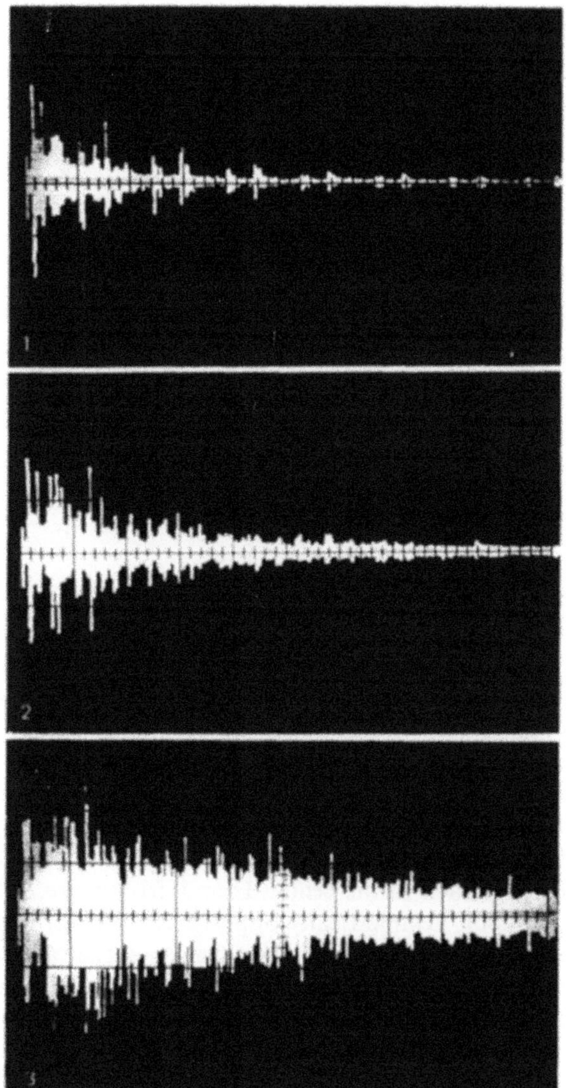

Fig. 4.7. Echogram in a rectangular room: 1, floor and one side wall absorbent; 2, floor only absorbent; 3, no absorptive treatment.

center (see Fig. 4.8(a)), the weak reflections may become inaudible and only the strong focused reflections are heard, at double the period and double the effective path-length.

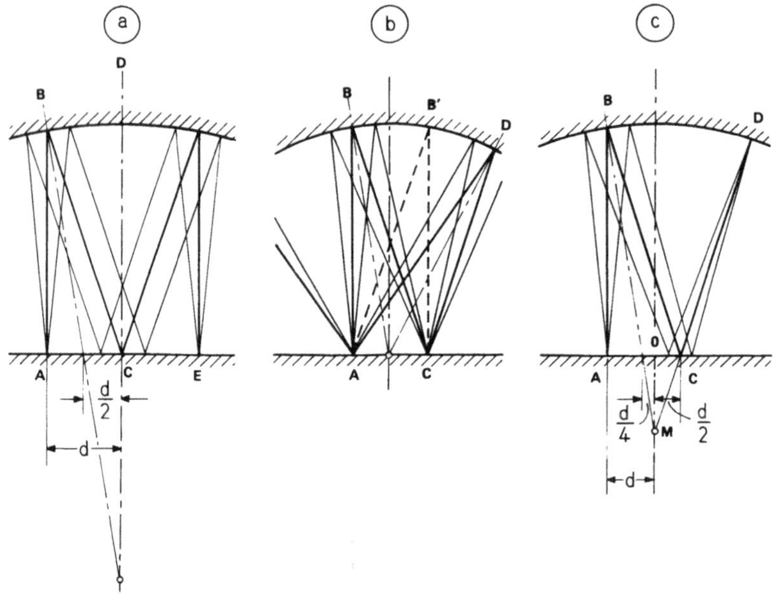

Fig. 4.8. Ray paths between plane floor and vaulted ceiling: r = radius of curvature of ceiling; h = height of vertex. (a) $r = 2h$; (b) $r = h$; (c) $r = 4/3h$.

If, as in Fig. 4.8(a), the ray-bundle is directed from an eccentric source position A toward the ceiling, it leaves the ceiling B obliquely as a set of parallel rays and strikes the center of the floor C; it is then reflected as a set of parallel rays back to the ceiling D and, from there, is focused at a point E on the floor, symmetrically opposite the source. From here, the process reverses itself so that, after eight traverses, the rays return to the original source position. This particular set of reflections is defined by the fact that the central ray of the bundle must encounter and then leave the floor at E in a perpendicular direction.

In this case of $r = 2h$, the round trip of the ray requires eight reflections; for each return of the initial ray to the original source position, the room-height must be travelled eight times. This is an astonishing effect, since one instinctively expects, for such a long delay, that the ceiling must be four times as high, and obviously the room dimensions are much too small for that. Moreover, it is surprising that the ray could make eight trips back and forth without encountering the receiver.

We now compare with the preceding example ($r=2h$) the case in which the radius of curvature of the ceiling is *equal* to the ceiling height above the floor ($r=h$). Upon the first reflection from the ceiling B, this configuration leads to a concentration of rays at a point C on the floor symmetrical to the source position, as shown in Fig. 4.8(b). Here, only the ray (shown dotted) that reaches the floor C perpendicularly is reflected back upon itself and returns to the source position A. All of the other rays from this first ceiling reflection are reflected from the floor C to one side or another. Moreover, even the ray that is returned to the source position tends in subsequent reflections to be lost to the sides, and finally disappears. Thus, as we have seen from Fig. 3.2, the tendency to focus at the point of symmetry becomes less and less. In such a case, we get a true flutter echo only with the source and receiver at the center.

Not only does the case with parabolically reflecting ceiling ($r=2h$) lead to a flutter echo, but so, rather surprisingly, does the case for which $r=4/3h$; in this case the ceiling acts approximately elliptically (see Fig. 4.8(c)). Since, in the present context, the source distance s in the concave mirror equation, eqn. (3.1), corresponds to the ceiling height h, we can find the distance e from the ceiling vertex to the focus by setting $s=3/4r$ in the formula:

$$\frac{1}{e}=\frac{2}{r}-\frac{4}{3r}=\frac{2}{3r} \tag{4.2}$$

That is,

$$e=\frac{3}{2}r=2h$$

This means that the distance from the ceiling vertex to the focus is twice the height of the ceiling, and thus the focal point for the first reflection lies *at* the ceiling itself, at D. Now, regarding this focal point as the source for the next reflection, the source distance s is zero, and therefore, from eqn. (3.1) also $e=0$ and the curved ceiling at D acts as a plane mirror. The rays returning from there are reflected next at the floor C, a third time at the ceiling B, and are finally focused once more on the floor at the observer's location A.

For eccentric positions Fig. 4.8(c) shows similar peculiarities; the focus at the ceiling lies on the opposite side. We can also show that the ray leaving the floor (at A) perpendicularly eventually returns upon itself.

This perpendicular meets the ceiling at B; the line perpendicular to the ceiling at B (i.e. the line between B and the center of curvature M, $r = 4/3h$) meets the floor at a distance of $d/4$ from the central axis O, or at $3d/4$ from A. Therefore, the ray reflected from B must strike the floor at twice this distance from the source point A, that is, at distance $d/2$ on the far side of the axis ($AC = 2 \times 3d/4$). The angles of incidence and reflection there are equal to ∢ ABC, the tangent of which is $3d/2h$. But the angle ∢ OMC has the same tangent; and this means that the ray reflected at the floor at C must meet the ceiling perpendicularly at D. The ray reflected from D, then, must return back upon itself to C and to B, and the process is repeated when the ray strikes the floor perpendicularly at A.

So far, we have considered only the plane containing the source, the vertex and the center of curvature. But it may also be shown that rays radiated outside this plane can return to the source point. From symmetry of the problem, we may conclude that the projections of the ray path in the ground plane form, in the case of eight reflections, a square, and, in the case of six reflections, a triangle. These polygons are shown in Fig. 4.9, bottom.

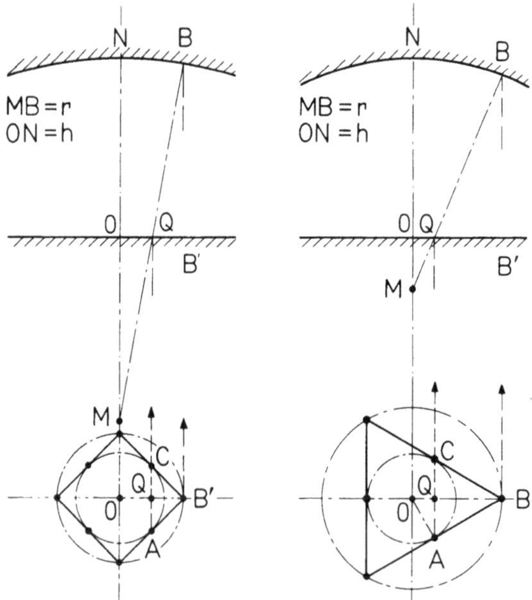

Fig. 4.9 Three-dimensional extension of Fig. 4.8. Left corresponds to Fig. 4.8(a), $r = 2h$; right corresponds to Fig. 4.8(c), $r = 4/3h$.

Here A represents the source location, B is the point of first reflection at the ceiling, and C is the point of the next reflection at the floor. AB and BC lie in one plane, which contains the perpendicular to the ceiling at B and, therefore, also the center of curvature M. This plane is perpendicular to the sections in Fig. 4.9, top. The perpendicular to B meets the floor at point Q. Furthermore, we designate the center on the floor as O, the vertex as N, and the projection of B on the floor as B′.

The assumed ray requires that:

$$\frac{h}{r}=\frac{\mathrm{ON}}{\mathrm{MB}}\approx\frac{\mathrm{QB}}{\mathrm{MB}}=\frac{\mathrm{QB'}}{\mathrm{OB'}} \tag{4.3}$$

In the case of the square, it is easy to see that QB′/OB′ is 1:2; in the case of the triangle, we may combine the relations:

$$\mathrm{QB'}=\mathrm{AB'}\cos 30^\circ$$

and

$$\mathrm{OB'}=\mathrm{AB'}/\cos 30^\circ$$

to get

$$\mathrm{QB'}/\mathrm{OB'}=3/4$$

In both cases, we have the same relation of h to r as before.

For repetitive reflections, it is not necessary that the geometric conditions derived here be fulfilled exactly; in fact, certain deviations can be rather interesting.

In the first edition of this book, we described with echograms, as an example of such deviations, the ballroom of the Kaiser-Wilhelm Palace in Berlin, which was unfortunately destroyed during World War II. Michel[3] had investigated this room earlier by drawing sound rays in a section, as in Fig. 4.8, except that, in his case, the ratio r/h was 2/3, thus lying between the ratios of Figs. 4.8(a) ($r/h=2$) and 4.8(c) ($r/h=4/3$) which we found would lead to perfectly repetitive reflections.

This difference in r/h did not hinder the production of a nearly repetitive reflection, which apparently moved about in a circle while changing its strong periodicity. It alternated between loud and soft reflections, which seemed to come sometimes from the left, sometimes from the right, depending on the instant of arrival. We should also note that the wandering of these repetitive reflections can occur simultaneously in both directions—a complication that makes the phenomenon very intriguing. The first edition of the present book extended Michel's sectional ray

[3] Michel, E., *Zentralblatt der Bauverwaltungen*, **48** (1928) 486.

diagram with an analysis in three dimensions (as in Fig. 4.9) and a recorded echogram.

Repetitive reflections, regarded as curiosities, can be just as much tourist attractions as whispering galleries: the greater the number of audible echoes, the greater the fame of the attraction. But, as we have seen, this depends not only on the geometric arrangement of the reflective surfaces, but also on the strength of the sound source and on the level of the background noise. If we regard both the source strength and background noise as given, then we may rate the echoes according to the number of reflections that can be heard.

Equally interesting is the time delay for the reflections: namely, whether there is time for one or two syllables to be spoken before the return of the reflection, for this delay determines whether the echo can 'speak' one word or two.

Finally, if we compare the room dimensions with the path length required for a reflection with a given time delay, then the *order* of the reflection may hold a special interest.

In general, we may assume that echoes have not been built into rooms on purpose, but that they have been an accidental result of architectural choices made on other grounds.

On the other hand, it would not be difficult to design interesting echoes deliberately. If the reader elects to do so, he should not neglect beforehand, to read Mark Twain's cautionary tale[4] about the man who collected echoes!

[4] Mark Twain, 'The canvasser's tale', *The Complete Short Stories of Mark Twain*, Doubleday and Co., Garden City, NJ, 1957.

Chapter I.5

The Use of Geometrical Reflections to Guide Useful Sound

I.5.1 Single Sound Source

The last two chapters were concerned with avoiding the harmful aspects of sound reflections, namely focusing effects and echoes. We turn now to consider the positive aspects: how to use geometrical reflections to provide useful sound at the desired locations with suitably short time delays.

At the beginning of the last chapter, we noted that the addition of reflected sound to the direct sound can be regarded as 'useful', if the time delay for the reflected sound is less than the 'limit of perceptibility'. We chose for this limit the round figure of 50 ms for speech, corresponding to a path-length difference of 17 m. For music this limit is about 50% higher. We shall discuss the technical background and the details for this rather loosely defined limit in Section III. 1.4.

As a general rule, we can regard the first reflections from the ceiling of a room as useful if the ceiling height is no greater than 8·5 m; the first reflections from the side walls of a room are useful if the width of the room is no more than 17 m.

But we must also take care that the time delay Δt and the path-length difference Δl are not too small. If Δl is so small as to be *comparable* with the wavelength, then we must account for the relative phases of the direct and reflected waves and allow for the possibility of interference between the two waves, as discussed in Section I. 1.5. And if the separation between a source and its image is *very small* with respect to the wavelength, we must add the sound pressures, not the intensities, of the direct and reflected waves in the direction perpendicular to the wall. If the path difference is half a wavelength, the sound pressures cancel each other; if it is a full wavelength, the pressures must be added together.

As a consequence, the amplitude of the resulting pressure wave is quite dependent on the wavelength (and thus, also, the frequency); therefore, the timbre of the sound of the combined direct and reflected waves depends on the distance of the source from the reflecting surface. This frequency-dependence, moreover, is different for different propagation directions of the resulting wave.

With increasing separation between the source and reflector, the peaks and valleys of sound pressure amplitude at different frequencies and directions of propagation tend to smooth out for the broad-band (wide-frequency-range) signals of interest in room acoustics. Thus, we may add together the intensities of the various sound components, without regard to their phases, and ignore the problems of change of timbre associated with phase-sensitive interference phenomena.

The lower limit of Δl depends on the type of sound signal and on the room-acoustic situation.

In providing useful sound by means of geometrical reflections, we begin with measures that can be applied in the neighborhood of the source; then we discuss the effect of sound reflections at the ceiling and walls of the room, and finally we discuss what can be done in the neighborhood of the listener. In the latter case, however, the question of audience seating arrangement must be postponed to the next chapter, because it goes beyond the simple rules of geometrical room acoustics.

With respect to reflectors near the sound source, it matters greatly whether there will be only one sound source, such as a lecturer, or several, as with a musical ensemble. We start with the simple case, a single sound source.

If the room is high and wide, as in many large churches, the speaker must be provided with a sound reflector.

The canopy over the pulpit (called a 'baldachino' in the language of church architecture) is one of the oldest and most effective of such devices: it reflects the words of the priest to the congregation and prevents the sound of his voice from reaching the upper parts of the building where it would generate a late-returning echo or unwanted reverberant sound, either of which would degrade the speech intelligibility (see Fig. 5.1). In other words, a canopy can enhance the useful sound and suppress the harmful sound.

Most baldachinos are so small, however, that they serve only the second purpose of helping to suppress late echoes and reverberant sound; and even this they do imperfectly, particularly in high churches. The usual horizontal arrangement of the reflector is not helpful, because the

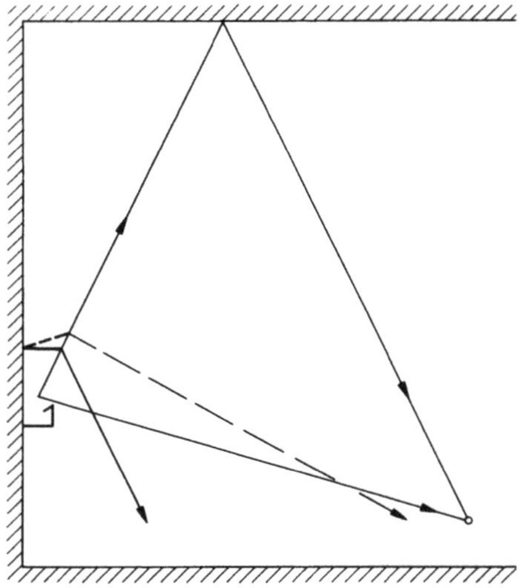

Figure. 5.1. Construction of limiting rays for both a horizontal and an upward slanting reflector over the pulpit.

useful reflections toward the congregation extend only a short distance into the seating area, where the congregation already profits from strong direct sound. A slight upward tilt of the reflector is helpful, since it guides the sound to the more distant members of the congregation, as shown in Fig. 5.1.

The underside of the pulpit reflector must not have any significant roughness or surface modulation, lest the sound reflection be diffuse and 'uncontrolled': it will be scattered rather than directed. If the architect wishes to provide decorative elements on the reflector, they must be confined mostly to the side and rear edges and to the upper surface. Decoration of the lower surface is permissible if it consists of paint or gold-leaf only, or if the surface relief is very small compared to the wavelengths of interest—in this case, speech. But this means frequencies up to 4000 Hz, or wavelengths as small as 8 cm; therefore, the surface relief must be no more than 8–10 mm deep.

Finally, most pulpit reflectors are too small to be effective at low frequencies; acoustically speaking, the pulpit reflector is better the larger it is.

Pulpit canopies designed from a purely acoustical standpoint first appeared in the 1920s, in The Netherlands. The most effective shape depends, in each case, on the distribution of the congregation, both in plan and in section. In the simplest situation the listeners are seated on only one floor, as in many churches and lecture halls. Here, the bundle of rays to be reflected from the pulpit canopy needs to cover only a rather small angle in the vertical section. If we approach this design problem in terms of sound propagation in a single plane, we may conclude, from the discussion of Section I. 3.2, that the reflector over the pulpit should be parabolic. Figure 5.2 shows how such a reflector can be installed at a side wall; moreover, the mirror image of this reflector in the side wall effectively doubles its size, as the mirror image of the speaker doubles the strength of his voice. On account of the *cylindrical* extension in the cross-direction, the lateral divergence of the sound in the horizontal plane is not hindered.

In most cases the distant listeners, who have the greatest need of an increase of useful sound, are confined to a rather narrow angle in the ground plan. Indeed, considering the directivity of the human voice, such an arrangement is desirable. In extreme cases, one might install a reflector in the shape of a

Fig. 5.2. Parabolic-cylinder pulpit canopy in the St.-Johannes-Kirche in Oberhausen. (After Gabler W., *Die Schalltechnik*, **13** (1953) No. 5.)

parabola of rotation, instead of the parabolic cylinder. The spherical wave, radiating from the speaker, would be transformed upon reflection from the pulpit canopy into an outgoing *plane* wave (parallel rays), with minimal divergence, designed to cover the congregation and nothing but the congregation.

This limit, however useful it may be for signalling, is usually too extreme for churches and auditoriums, where, after all, some divergence is always desirable. The ideal reflector might then be one that transforms the direct, outgoing spherical wave from the speaker into a weakly diverging cylindrical wave, for which the axis would lie rather far behind the reflector. From Section I. 3.2, we know that in such a case the horizontal section must become a hyperbola, while the vertical section remains a parabola. As for the configuration of the overall surface, we can again make use of Fermat's principle.

If a sound wave leaves the reflector in the form of a cylindrical wavefront, then all the paths from the source to this wavefront must be 'shortest paths', and therefore have the same length. On the other hand, all points on the cylindrical wavefront must also have the same distance from the axis of the cylinder. The difference between these distances must be constant, equal to b. Therefore, the path from the reflector to the cylindrical wavefront drops out of consideration, and we get, with the coordinates indicated in the sketch of Fig. 5.3:

$$\sqrt{(x+a)^2+y^2}-\sqrt{x^2+y^2+z^2}=b \qquad (5.1)$$

where a is the distance between the source-point and the cylindrical axis.

Figure 5.3 presents, in the two sketches, the principles of the configuration, and, in the photograph at the upper left, the appearance of the finished reflector, made of plywood. (For the construction in Fig. 5.3, two other parameters are needed: the distance s between the source point and the vertex, and, depending on the distribution of the audience, the angle α which the outermost rays form in the x–y plane.)

The fabrication of this reflector, curved in both directions, was difficult. Fokker and his collaborators[1] managed it by dividing the surface with horizontal cuts (all hyperbolic) and vertical cuts (all parabolic) through the cylinder axis, making use of eqn. (5.1).

The speaker seen in the photograph of Fig. 5.3 gives an idea of the

[1] Nuyens, M. and Philippi, G. Th., *Physica (Utrecht)*, **X** (1930) 19.

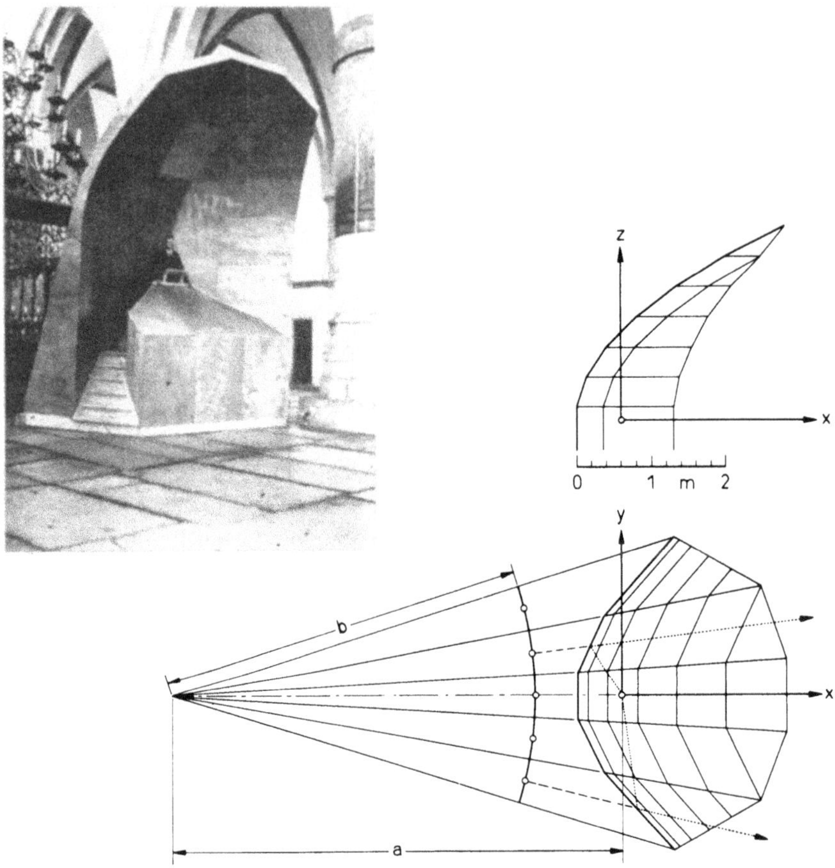

Fig. 5.3 The parabolic–hyperbolic reflector. The two sketches show the principles of construction; the photograph shows the appearance of the finished reflector.

large size of this reflector, and helps us understand why only one reflector of this design was ever built! It had, however, the advantage that the reflector extended in front of and even below the speaker; the directivity of the mouth makes lateral extensions of the reflector unnecessary, and (even more so) reflections from behind the head contribute no enhancement of the loudness and clarity of speech.

Even before the development of eqn. (5.1), however, Mulder[2] had built

[2] Mulder, A. J. M., *Physica (Utrecht)* **XII** (1932) 311; also *Zentralbl. d. Bauverw.*, **55** (1935) 208.

reflectors empirically whose form changed from a hyperbola in the x–y plane to a parabola in the x–z plane. For this purpose, he subdivided the surface into small sectors of 10°, extending radially from the crown of the reflector ($x=s$, $y=0$, $z=0$), where they were nailed to the common circular foundation board, shown around the origin in Fig. 5.4. Later, he

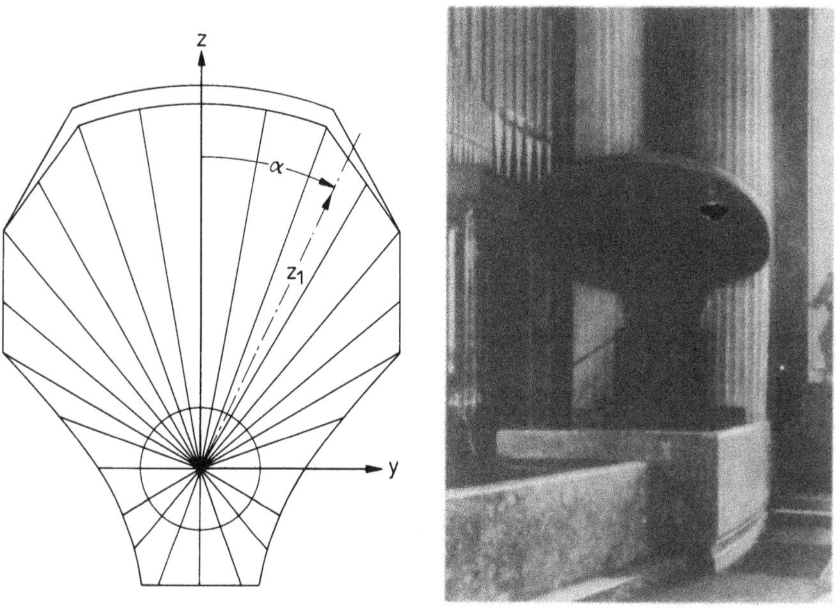

Fig. 5.4. Parabolic-hyperbolic reflector. Left: Principle of the construction. Right: Photograph of a finished example in St. Hedwig's Church in Berlin, before World War II. (After Mulder.[2])

presented a formula for the exact shape of the center of each sector, in terms of $x_{(z_1,\alpha)}$. Figure 5.4 shows, in the sketch at the left, Mulder's principle, defining z_1 and α; and, at the right, a photograph of the corresponding pulpit canopy in St. Hedwig's Church in Berlin, before its destruction in World War II.

Mulder's construction sketch can give only a general idea, but his approach has the advantage that the sectors of the reflector are small where the curvatures are large, and vice versa. Thus, it is possible to construct a practically crack-free surface that can be terminated anywhere without aesthetic disadvantage, a feature which permits the architect considerable freedom in design.

Some architects refuse, on aesthetic principles, to accept design elements which have been developed on purely acoustical grounds. But, there are always a number of opportunities to achieve the desired architectural effect without diminishing the desired acoustical effect. For example, one would hesitate to install either of the pulpit canopies of Figs. 5.3 and 5.4, in their 'naked' form, in a Gothic or Baroque church. But these are only the technical skeletons of the acoustical concept, which may be freely adapted to a specific building style, provided that the cautions, mentioned above, with respect to their size and decoration, are respected.

Furthermore, in most practical cases it is not necessary to adhere strictly to the ideal shape. Between the 'ideal' and the pulpit canopies usually found in churches, there is such an enormous difference that almost any step toward the ideal, in the form of properly inclined planes, may be regarded as significant progress, and will result in better hearing conditions for the congregation.

Such a composition of plane surfaces would avoid some of the focusing difficulties to be expected with smoothly curved surfaces. As we have explained in Section I.3.2, the ultimate effect depends on the source location. The reflector works optimally only if the speaker's mouth is at the focus. If he moves his head backward, the divergence increases; even worse, if he moves forward, he creates a focal point in the audience. A shift of his head by a distance equal to half the radius of curvature would move the focus from a location very far in front of him to his immediate neighborhood. Such a shift must therefore be prevented.

Finally, it may be disturbing to the speaker himself to stand in a focal point for sounds from the audience: their noise will be concentrated upon him, to his considerable distraction.

I.5.2 Artificial Sound Amplification

The concluding comments in the last section imply a reciprocity between the sending and receiving position: a natural reflector can not only direct sound from the speaker to the listeners, but it can also concentrate the noises from the audience back to the speaker. This reciprocity fails, however, when electroacoustical means are employed to provide 'artificial' amplification for the speaker's voice; the reason is that such systems always involve elements that can transmit signals in only one direction.

The existence of electroacoustical means of amplification, however, raises a new question: does it make sense to spend time looking for optimal 'natural' means of amplification, using properly designed reflecting surfaces, when the same effect can be achieved with simple loudspeaker arrangements? (A lecturer in an auditorium can make himself heard only by a limited audience, even under the best room-acoustical conditions; with the electroacoustic help of radio or television, he can be clearly heard anywhere in the world.)

In fact, nowadays, the pulpit canopy in churches and the raised pulpit itself are becoming unnecessary; even the smallest churches and lecture rooms have provided themselves with sound systems, which are usually not necessary, and by no means always work to the advantage of the listeners.

As we have mentioned in connection with the electronic equipment needed for recording impulse responses, it would go well beyond the scope of this book to survey all the available elements of electroacoustical sound-reinforcement systems; but it is important to recognize that a wide choice of equipment exists, together with the necessary technology, so that it is readily possible to reproduce and amplify speech and music in the frequency range of interest, without distortion and with great realism.

Nevertheless, even if the room-acoustics consultant is not responsible for the design of the sound-reinforcement system, he must have the cooperation of the system designer. He must be informed about the placement of the loudspeakers with respect to room reflecting surfaces, the frequency and directional characteristics of the loudspeakers, and whether time delays will be provided in the system.

All of the care that goes into a room-acoustics design can come to nothing if the electroacoustical system is installed without consideration of the room acoustics. (If there are problems in the finished auditorium, the listeners will know only that they have difficulty in understanding; they will not know whether the room or the sound system is to blame.) The same holds true even for presentations where the singers 'swallow the microphone' and where musical instruments are amplified far beyond their normal levels: that is, events where the sound system dominates. (We omit, here, any consideration of heavy rock music, where the instruments are amplified to ear-damaging loudness.)

Two quite different principles can govern the placement of loudspeakers in a sound-reinforcement system: one provides either a distributed loudspeaker arrangement, with many small loudspeakers throughout the audience area, each loudspeaker radiating a small amount of sound power; or one provides, near the live sound source, a central loudspeaker system, radiating a large amount of sound power to all of the audience.

The advantage of the distributed system is that all the listeners hear the reinforced sound at approximately the same level; this holds true especially for events in the open air. It also avoids disturbing the people living in the neighborhood of the facility, who may not wish to hear the announcements from sports stadiums, swimming pools, racetracks, etc. Moreover, the 'artificial echoes' that we discussed in Section I. 4.1 will be avoided if the loudspeakers are installed at distances less than 17 m apart and—as is usually the case—all the speakers are fed from the same circuit, without time delay.

On the other hand, the use of a distributed system of multiple, synchronized loudspeakers has the disadvantage, for performances with live sound sources, that the listener always hears as the 'original source' the sound coming from the nearest loudspeaker; there is no chance to preserve 'directional realism' with a synchronized distributed system. We have mentioned this effect in Section I. 2.1, explaining it in terms of the equivalence of an image source (the loudspeaker) and a reflecting wall at half the distance; we will discuss this 'law of the first wavefront' in greater detail in Section III. 1.6.

In addition, there always exists a small zone of 'localization confusion', about midway between two synchronized sound sources, in which the listener feels uncertain as to which source he should regard as the original. In the choice of loudspeaker placement, therefore, every attempt should be made to locate those confusion zones in the aisles, on staircases or in other areas outside the seated audience.

Two important rules for the placement of synchronized loudspeakers can be deduced from the equivalence between reflecting walls and image sources:

1. If, especially outdoors, there are no useful reflecting surfaces in the neighborhood of the source, equivalent 'reflections' can be supplied by loudspeakers. If, for example, we install three loudspeakers around the speaker, like the images S_x, S_y, and S_{xy} in Fig. 2.3(a), this will create the same acoustical effect as if two reflecting walls were to meet at right angles behind the speaker. Thus, it is sometimes possible to replace room-acoustical methods with electroacoustical methods.

2. A loudspeaker installation can have the same effect as natural amplification, provided that each loudspeaker is located so that it can be regarded as the mirror image of the live source in a normal room boundary. Among other things, this implies that the loudspeaker must not be operated at too high a sound level, though some increase in sound power, compared with the original source, is permitted (see Section III. 1.6).

According to this second rule, it would sound unnatural if loudspeakers, synchronized with the live speaker, were placed at the rear of the audience. In this case, the 'equivalent reflecting walls' corresponding to these loudspeakers cut across the audience halfway back from the stage. The rear half of the audience thus finds itself in an 'impossible' position, *behind* the wall with respect to the stage.

By contrast, a central loudspeaker arrangement, installed above the original source, can simulate a reflecting ceiling at half the height; but care must be taken that no listeners occupy positions 'above' this imaginary ceiling.

If the loudspeakers can be placed very high, they may be installed over the middle of the audience without risk that the amplified sound reaches the listener before the sound from the live speaker. Fig. 5.5 shows such an arrangement, used temporarily during the restoration of the Frauenkirche in Munich.[1] Not only was speech from the pulpit perceived as coming from that location, when loudspeakers II and III were switched on, but the congregation localized the prayers and the singers in the vicinity of the altar *a*, when loudspeaker I was in use.

The risk of a concentrated cluster of loudspeakers above the stage is that sound from this source may be reflected as a disturbing echo from surfaces at the rear of the hall, which would not be excited by the natural source. Natural sound striking vertical rear-wall surfaces from the stage will be reflected above the audience up to a diffusing ceiling, where it causes no harm. But sound from the loudspeakers may strike the same surface and be reflected back down into the front rows of the audience with an annoying delay. The solution depends on collaboration between the designers responsible for the room acoustics and for the sound system.

[1] This loudspeaker installation was in use during a part of the restoration when a temporary scaffolded ceiling was installed, as shown in Fig. 5.5; the reverberation time of the building during this period was abnormally low. When the restoration was completed, the temporary ceiling was removed and the reverberation time increased; not enough absorptive treatment was allowed because of the concern for the preservation of historical monuments. Thus, the system shown in Fig. 5.5 was replaced with a distributed system.

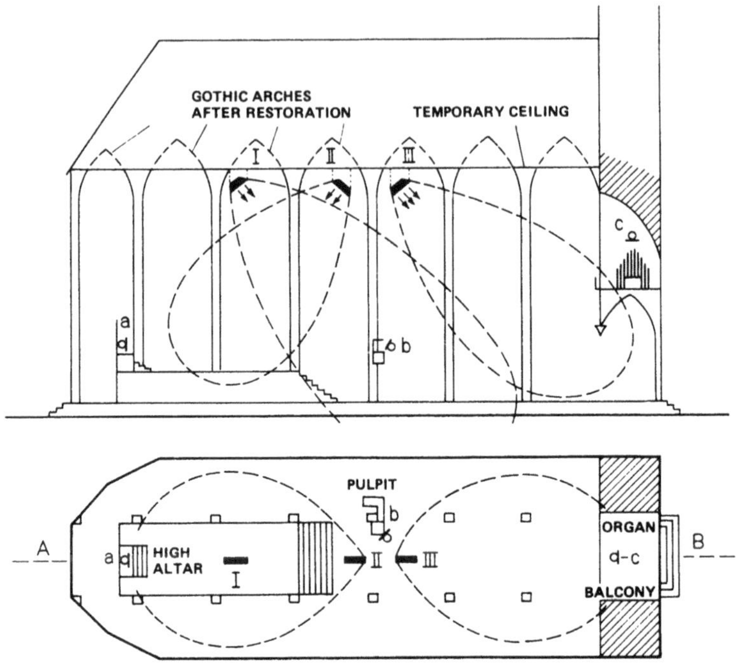

Fig. 5.5. Loudspeaker installation, with natural time delays in the seating areas, in the Frauenkirche in Munich.

It is essential, in the design of the loudspeakers for a sound reinforcement system, to use loudspeakers of very high directivity and to control absolutely the areas of the auditorium covered by the loudspeakers.

The directional loudspeakers most frequently used in Europe[2] consist of a line-array of single loudspeakers in a row; they are installed with the long dimension vertical or tilted slightly down toward the audience. Such an array beams the sound forward in any plane that includes the line of sources; it confines the sound to a rather narrow beam in the vertical section, but distributes the sound freely in front of the array in the plane of the audience. As usual, the beamwidth of the radiated sound depends

[2] Loudspeaker system design practice in North America typically uses, instead of directional line arrays, a cluster of horn-type loudspeakers, which (for typical horn dimensions) can control the beamwidth in both the vertical and horizontal planes for frequencies above about 300–500 Hz. Low-frequency sound is provided by relatively non-directional 'woofers' (large folded horns).

on the size of the radiator, in comparison with the wavelength. The line array is effectively directive only in the frequency range for which the length of the array is equal to or greater than the wavelength (from 340 Hz up, for a one-meter array), and only out to a certain distance, which is shorter the lower the frequency. (This result may be derived from the principle of Fresnel zones, as discussed in Section I. 1.5. According to eqn. (1.5), directivity begins to fail at distances greater than approximately $L^2/2\lambda$, where L is the length of the line array.)

Fortunately, for the reinforcement of speech sounds, the frequencies below about 300 Hz are not necessary for satisfactory intelligibility. But it is a quite different matter when music of high quality must be reinforced (live performers) or reproduced (recordings); in that case, such a frequency limitation is unacceptable. The solution can again be found only in a timely collaboration between the room-acoustics designer and the electroacoustics designer, proper consideration being given to the requirements of both disciplines.

The best results are usually achieved when no greater power and no greater directivity are demanded of the loudspeakers than are necessary for the purpose, for in this way one gets a more natural sound.

Directional loudspeakers are required in sound-reinforcement installations, not only for room-acoustics reasons (to achieve precise coverage of the audience without exciting unnecessary reverberation), but also for electroacoustical reasons: if the amplified sound from the loudspeakers strikes the microphone, it will be picked up and reamplified, over and over. This process may, in fact, be unstable, if the gain of the amplifiers is set too high, and a threshold of oscillation may be exceeded, usually first at a single frequency, with the result that the loudspeaker system begins to 'howl'.

We shall return to this phenomenon, called 'acoustic feedback', in Section II. 3.5. There, we discuss the possibility of restricting the gain of the system so that, while it remains below the threshold of instability, it nevertheless achieves an effective prolonged reverberation time.

Evidently it is easier to avoid acoustic feedback when the loudspeakers are highly directional, because then the microphone can be placed in a null-position in the radiation pattern of the loudspeakers. For the line array, this can occur on a prolongation of the line containing the sources.

It is possible, in addition, to combine the advantages of a central loudspeaker system with those of a distributed system of line arrays if the latter system is fed by signals 'suitably delayed' with respect to the sound from the central cluster. The suitable delay, in this case, is slightly greater than the natural delay for sound reaching the secondary loudspeaker from the live source on stage. Under these circumstances, the first sound heard anywhere in the audience is the live sound from the stage, followed very quickly by the amplified sound from the system. Thus, the listener benefits from the amplification without losing the impression that *all* of the sound that he hears originates on the stage. Parkin and Taylor made use of this principle in St. Paul's Cathedral in London, in 1952.[3]

Figure 5.6 shows an application of the same principle in the Herkules Hall in Munich. The time delays are achieved by temporarily recording the microphone signal on a rotating magnetic disk and picking up the same signal slightly later, at various angular locations around the disk, corresponding to the desired time delays. The recorded signal is erased after one rotation, so the disk will be ready to receive the next microphone signals: the process is continuous.

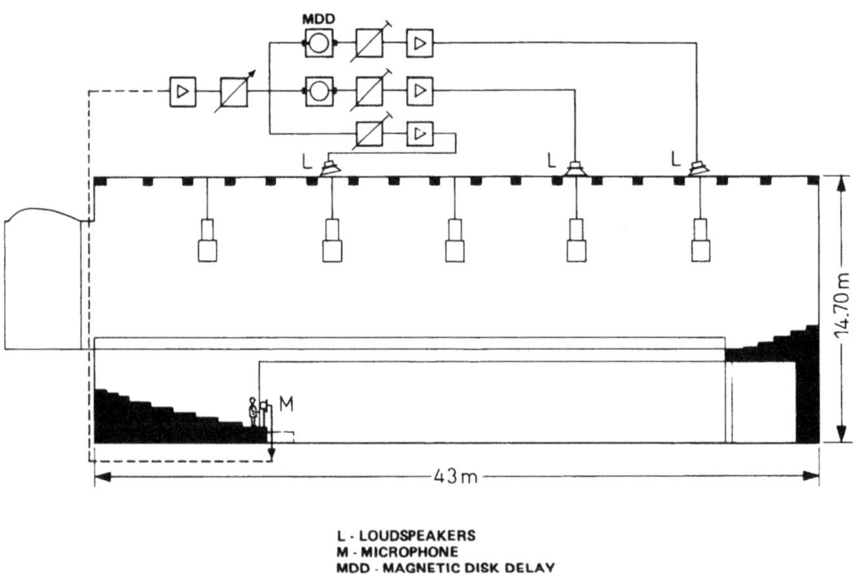

Fig. 5.6. Scheme for a distributed loudspeaker system with an artificial time delay.

[3] Parkin, P., and Taylor, H. H., *Wireless World*, **58** (1952) 54 and 109.

Note that the time-delay principle works in one direction only, the more distant loudspeakers being fed with greater time delays. In the opposite direction all the delays will be double their natural values, and the audible result is chaotic. For this reason, loudspeaker systems involving time delays must always utilize highly directional loudspeakers, separated by no more than 8·5 m.

In the early days, magnetic time delay devices were expensive and noisy, and they required continual maintenance; thus, the advantages of time-delayed systems were not widely exploited. Nowadays, with the development of inexpensive digital delay systems that are highly reliable and very quiet, the use of time-delayed sound reinforcement is becoming more common; in the United States such applications have become routine, even in rather modest installations.

This is as it should be; it is no longer excusable that audiences must endure serious discrepancies between the directions of the visible and the audible sound sources, or even zones of uncertainty as to where the source lies. It is especially confusing at a congress, when one of the speakers uses a microphone in his neighborhood but the sound of his voice comes from a central cluster above the stage. In such installations, it is possible to preserve the desired directional realism with a distributed system of loudspeakers, hanging like 'lamps' from the ceiling; each loudspeaker (or small group) is connected only to the microphones below.

The problem becomes more difficult if the sources to be distinguished are located close together. For such applications, we need, in addition, directional microphones. As early as the 1930s, attempts were made, particularly in open air theaters, to amplify speech and singing in operas and dramas in such a way as to preserve directional realism (i.e. the visual and audible locations of the source should be the same). The problem is still difficult to solve, even to this day; at any rate the success of such methods, so far, has not been so spectacular as to replace the methods of room acoustics—with the exception of the pulpit canopy!

I.5.3 Multiple Sound Sources

We have already seen that it may be difficult to optimize a sound reflector, such as a pulpit canopy, for a single sound source at a fixed source position. It is even more evident that concave reflecting enclosures for multiple sound sources (for example, a symphony orchestra) present much greater difficulties. The chief problems are: (1) they reflect the different instruments preferentially into different parts of the audience; (2) they provide only 'patchy' communication between the musicians on the stage; and (3) they provide a distorted impression of orchestral balance at the conductor's position.

Nevertheless, most orchestra shells in public gardens have been built with at least concave cylindrical boundary surfaces, and sometimes even as a concave quarter-sphere. It is probable that these forms were chosen on architectural grounds or on the basis of structural stability.

In order to avoid unequal sound distribution from an orchestra enclosure, it is better to surround the orchestra by a few, properly directed, plane and stepped surfaces.

Figure 5.7 shows such an orchestra shell developed by G. Izenour for the ancient theater in Caesarea (Israel). The overall enclosure is basically made of large plane surfaces; but, on a smaller scale, these are folded so that they provide a certain amount of diffusion in the reflected sound, to improve the communication among the musicians. (Flat surfaces that throw *all* of the sound to the audience are not satisfactory in this respect.) The construction of this enclosure, because of an additional requirement of the archaeologists for the site, is such that it can be moved laterally out of view of the visitors when it is not needed

Fig. 5.7. Portable orchestra shell in the theater at Caesarea. (After G. C. Izenour, *Theater Design*, New York, 1977, p. 329.)

for a performance. (The problem with such portability is that, once the enclosure has been moved to one side, out of sight, it may never be returned to enclose the stage when it is needed.)

Orchestra shells are required not only outdoors, but also in theaters and opera houses when an orchestra is to give a concert on stage. Otherwise, the sound would be lost into the sound-absorbing wings and stagehouse, and the music would reach the audience very much weakened. It is also essential that the necessary side walls, rear wall and ceiling can be taken away in the shortest possible time and with the least manual effort, in order to return the theater to a stage condition suitable for plays and opera. Fig. 5.8 shows an example of a portable stage enclosure developed for the theater in Karlsruhe.

Even multi-purpose rooms, which have no stage tower, wings or the usual complement of equipment for staged productions, need an enclosure for orchestras, particularly if, as is often true, they are very wide and may be fan-shaped in plan.

Figure 5.9 shows a solution for this situation, developed for the Congress Center in Hamburg. As seen in vertical section, the configuration resembles an enormous chest of drawers, with the upper drawers pulled out farther than the lower ones. The horizontal under-surfaces of these 'drawers', working together with the adjacent vertical elements, provide cueball (second-order 90°) reflections back to the musicians, in a more or less diffuse manner. The stepping outward of the successively higher elements of these shelves heightens this effect.

Even in pure concert halls, the installation of sound reflectors near the orchestra has advantages. Especially, if the ceiling of the hall is very high, a canopy above the orchestra greatly improves the contact between the musicians, and may also reflect useful sound to the often-neglected front rows of the audience. (See, for example, the canopy divided into three convex elements above the stage in the Royal Festival Hall, London,[1] Part II, Fig. 7.3.)

When the Herkules Hall in Munich (named after the valuable tapestries depicting the deeds of Hercules that decorate the walls) was being restored in classic style, the acoustical consultant wished to include an array of sound reflectors over the stage, because the ceiling of the hall was very high and the audience floor was flat. The architect (Rudolf Esterer), although he was willing to go along with this acoustical proposal, wanted, for aesthetic reasons, to make these panels as incon-

[1] Parkin, P. H., Allen, W. A., Purkis, H. J., and Scholes, W. E., *Acustica*, **3** (1953) 1.

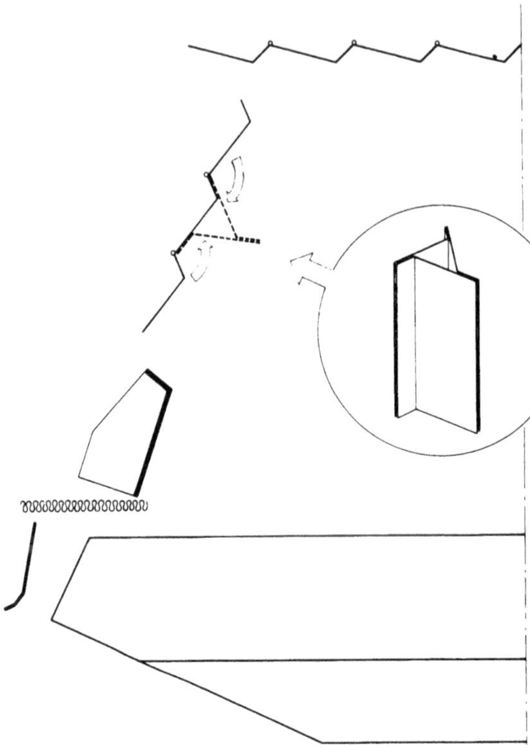

Fig. 5.8. Portable orchestra shell for the stage of the theater in Karlsruhe (architect, H. Bätzner).

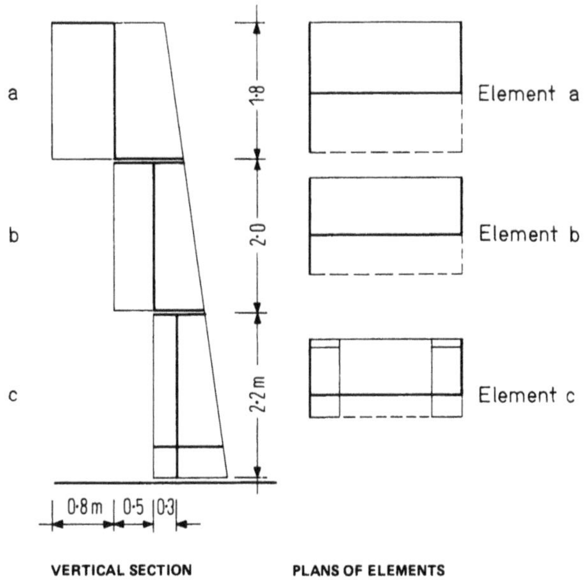

Fig. 5.9. 'Drawer-like' side-wall elements for the stage of the Congress Center in Hamburg. (The dashed lines indicate the upper surface, the solid lines the lower surface, of each drawer element.) (After Cremer, L. and Zemke, H. J., *Technik am Bau*, **3** (1973) 253.)

spicuous as possible. Therefore, the plexiglas reflectors to be seen in Fig. 5.10 were developed.[2] This photograph has been retouched to make the steel frames of the panels visible; in the photograph of the Herkules Hall in Beranek's *Music, Acoustics and Architecture* (John Wiley, New York, 1962), the panels are almost invisible.

The purpose of the reflectors in the Herkules Hall was to reflect parts of the orchestra sound, particularly the strings, to the rear part of the audience on the flat main floor.

Simultaneously and independently, similar plexiglas sound reflectors were installed, in order to provide better contact between the musicians, in a studio of the South German Radio Broadcasting Corporation in Karlsruhe.[3]

[2] Cremer, L., *Die Schalltechnik*, **13** (1953) No. 5.
[3] Keidel, L., *Die Schalltechnik*, **13** (1953) No. 5.

Fig. 5.10. Plexiglas sound reflectors in the Herkules Hall of the Munich Residenz. (After Cremer.[2])

In both the Munich and the Karlsruhe examples, it was understood that the intended reflection of sound could be expected only at high frequencies, and that the interruption of the reflecting area by the gaps between the reflectors would allow diffraction of sound around the panels at low frequencies.

On the other hand, this unavoidable transparency to sound at low frequencies made it reasonable to restrict the weight of the reflector material to about $5\,\mathrm{kg\,m^{-2}}$. (There is no point in making the reflectors *heavy* enough to reflect very low frequencies if they are not *large* enough to reflect those frequencies.)

The lower frequency limit for *ideal* geometric reflection from these panels can be derived from the construction of Fresnel zones. The procedure of Section I.1.5 has only to be extended to include oblique incidence of the sound rays.[2] That is, we must ask: at what frequency do the path lengths of the outer rays, in Fig. 5.11, exceed the path length of the central ray by one-quarter wavelength? Using the definitions given in Fig. 5.11, we get:

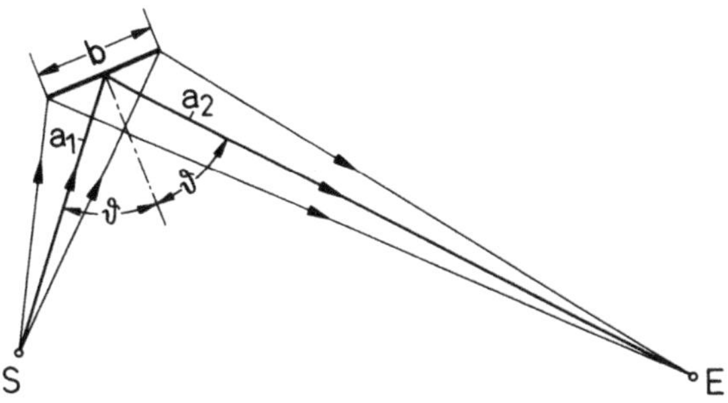

Fig. 5.11. Ray construction to find the first half-Fresnel-zone for oblique incidence.

$$\sqrt{a_1^2+\frac{b^2}{4}-a_1 b \sin\vartheta}+\sqrt{a_2^2+\frac{b^2}{4}+a_2 b \sin\vartheta}-a_1-a_2=\frac{\lambda_{\max}}{4} \qquad (5.2a)$$

$$\sqrt{a_1^2+\frac{b^2}{4}+a_1 b \sin\vartheta}+\sqrt{a_2^2+\frac{b^2}{4}-a_2 b \sin\vartheta}-a_1-a_2=\frac{\lambda_{\max}}{4} \qquad (5.2b)$$

Assuming, now, that the length b of the reflectors is small in comparison with the distances from the reflectors to the source and receiver, a_1 and a_2, we can develop the radicals of eqn. (5.2) in a series of the form

$$a_{1,2}\left[1+c_1\left(\frac{b}{a_{1,2}}\right)+c_2\left(\frac{b}{a_{1,2}}\right)^2+\cdots\right]$$

and, retaining only the first three terms, we get (from both eqns. (5.2a) and (5.2b)):

$$\frac{1}{8}(b\cos\vartheta)^2\left(\frac{1}{a_1}+\frac{1}{a_2}\right)=\frac{\lambda_{\max}}{4} \qquad (5.2c)$$

Comparing this with eqn. (1.4), in which we must set $n=1/2$ and $\rho_n=b/2$, we find that we have only to replace b, the reflector length in the ray plane, with its projection $b\cos\vartheta$, as seen from the source S and receiver E.

This analysis, incidentally, shows that the size of the reflectors can be smaller in the cross-hall direction (since in this direction we can set $\vartheta=0$) than for the longitudinal direction. Choosing, as mean values, $a_1=10\,\text{m}$,

$a_2 = 20$ m, $b = 2{\cdot}5$ m and $\vartheta = 45°$, we get a value of $\lambda_{max} = 23$ cm, corresponding to $f_{min} = 1460$ Hz. We can, of course, expect the reflectors to provide diffuse, but still useful, reflections at frequencies below this limit of ideal geometric reflection.

On the other hand, it is the high frequencies that determine the brilliance and the clarity of the sound, and for this purpose, the short time delays provided by the reflecting panels take high priority.

The architect for the Herkules Hall refused to install these sound reflectors for the opening ceremony; but they were installed for the first public concert. Therefore, there was a good opportunity to compare the sound of the hall, both with and without the reflectors, in quick succession. There was a noticeable improvement when the reflectors were used, particularly in the timbre of the violin sound at the rear of the hall.

Although these first plexiglas sound reflectors had certain technical deficiencies, it is regrettable that they were later removed for non-acoustical reasons. For recordings and live broadcasts from the Herkules Hall, the reflectors are not essential, because the recording engineers can place their microphones wherever they wish, to achieve the balance and tone that they desire. But the string timbre for concerts, in the middle and rear seats on the main floor, has lost its brilliance and the definition has become a bit poor since the reflecting panels were removed.

One of the operational difficulties with the reflectors at the Herkules Hall was that, on each occasion when their position had to be changed—for example, for concerts with and without chorus—this had to be done manually with a hand-winch system. By contrast, the copies of the Herkules Hall plexiglas panels that were later installed in the Concert Hall in Stockholm[4] are operated by motor-winches, and their heights can be quickly adjusted by push-button.

In the recent restoration of the Stockholm hall,[5] the sound-reflector concept was retained; but this time, instead of plexiglas panels, new opaque panels were designed, with small-scale decorative relief to match the other decorative elements in the hall. This relief on the panel surfaces adds the advantage of a slight diffusion in the panel reflections.

The sound reflectors in the Berlin Philharmonie (see Figs. 3.10 and 6.12) were regarded by the architect (Hans Scharoun) as essential architectural elements in the hall design; there they were called 'sails'. Later, these reflectors were replaced with larger panels, and some of them, over the opposition of the acoustical consultant, were hung somewhat higher.

---

[4] Sundblad, G., *Gravesaner Blätter* (1956) No. 2/3, 34.

[5] Architect, A. Tengbom; acoustical consultant, S. Dahlstedt.

It may be feasible, in principle, to develop the original acoustical design for a hall in such a way as to make it impossible (or at least very difficult) for other people to come in later and make changes that spoil the initial concept. But it is extremely difficult, in practice, to preserve the original plan. For example, in de Doelen Concert Hall,[6] Rotterdam, it was hoped that combining the sound-reflecting panels above the stage with the lighting for the orchestra and welding the entire structure solidly in place would discourage any tampering with this acoustically critical element in the hall. Today those reflectors are altogether gone! The recording engineers feared that the panels were creating acoustical interference with the recording microphones, so they removed the panels.

Arrays of sound-reflecting panels above the stage have been applied more and more regularly, particularly in North America. It must be acknowledged, however, that, following the 1962 opening of Philharmonic Hall in New York, serious doubts were raised, on fundamental physical grounds, about the advisability of using sound-reflecting panel arrays for the provision of early sound. The concern is particularly acute when too large a portion of the ceiling is covered with reflectors. In the case of the Shed at Tanglewood, the summer home of the Boston Symphony Orchestra, a 50% coverage of the ceiling with suspended reflecting panels evidently produces an excellent effect.[7]

When a panel array with similar ceiling coverage was installed in Philharmonic Hall,[8] however, the critics complained of a lack of low-frequency sound on the main floor; they attributed this deficiency to excessive coverage of the ceiling with the suspended sound reflectors.

Subsequent model tests[9] showed that the frequency at which the sound energy transmitted *through* such a reflector system equals the diffuse reflected sound lies around 250 Hz; that is, rather low, compared with the lower frequency limit for ideal geometrical reflection. But 250 Hz is still an important frequency, especially for the lower string tones. Although the sound transmitted between the reflectors may be reflected back into the hall at the ceiling, we must remember that some of this sound is absorbed in the space above the reflectors; thus, if they cover the ceiling too densely, the reflectors may behave as a 'frequency switch'.

---

[6] Kosten, C. W. and de Lange P. A., 'The New Rotterdam Concert Hall', *Proc. 5th. ICA, Liège, 7–14 Sept., 1965*, Paper G-43.

[7] Johnson, F. R., Beranek, L. L., Newman, R. B., Bolt, R. H. and Klepper, D. L., *J. Acoust. Soc. Am.*, **33** (1961) 475.

[8] Beranek, L. L., *Music, Acoustics and Architecture*, John Wiley, New York, 1962 (reprinted Krieger Publishing Co., Huntington, New York, 1979).

[9] Meyer, E. and Kuttruff, H., *Acustica*, **13** (1968) 183.

Model tests[10] have shown, in addition, that changes of reverberation time may occur, in a room where most of the sound absorption is on the floor, if 50% of the ceiling is covered with reflectors. Therefore, it may be wise to restrict the ceiling coverage by reflectors to about 30%, at most.

Watters *et al.*[11] found, in model tests, that the steps in the array, and the second layer of reflectors added after the tuning week tests in Philharmonic Hall, produced interference that weakened the reflected sound in the frequency region around 200 Hz. Such interference patterns are to be expected (and, thus, to be avoided!) whenever equally spaced steps occur in the ceiling or side walls of a hall. It must be admitted, however, that interference problems from these geometrically reasonable steps are not observed in practice.

As a general rule, we may say that modular building elements of similar size, which are often preferred for aesthetic reasons or because of construction economy, are usually disadvantageous from the acoustical viewpoint; this is particularly true when the elements in question occur in locations that provide early sound reflections to the performers or the audience. For the acoustician, the proposal to use many identical building elements makes as much sense as designing a pipe organ with all the pipes of equal length!

It is possible to diminish slightly the high-frequency components in the sound reflections, and thereby to remove the 'sharp edge' of the sound, by covering the undersurface of the reflector with fabric. If the fabric hangs free of the reflector, as is the case in a concert hall in Biel,[12] the absorbing effect extends down to 1000 Hz. Furthermore, in Biel this white linen material, when floodlit, becomes a major architectural element in the hall. (Architect, Max Schlup.)

Sound reflectors are usually hung from the ceiling, because it is easy to do so; besides, the lateral surfaces of the hall are often needed for other purposes. But it is always an advantage to provide, in addition, lateral reflectors, because our ears are particularly sensitive to reflections from the sides of the hall.

In the first edition of this book, we mentioned that Berg and Holtsmark[13] installed concave sound reflectors in a small lecture room, at both sides of the speaker, to allow him some mobility.

---

[10] Zemke, H. J., unpublished model studies, carried out under the auspices of ITA.

[11] Watters, B. G., Beranek, L. L., Johnson, F. R. and Dyer, I., *Sound*, **2** (1963) 26.

[12] Cremer, L. and Furrer, W., *Schweizerische Bauzeitung*, **85** (1967) 783.

[13] Berg, R. and Holtsmark, J., *D K N V S Fordhandlinger*, **XIII**, No. 16; ref. *Akust. Zh.*, **7** (1942) 119.

Meyer and Kuhl[14] later placed similar large reflectors, at both sides of the proscenium, in the provisional restoration of the Opera House in Hamburg, and also in the Landestheater in Hannover. They stated that the sound source seemed to be enlarged laterally, without losing its localization, the latter effect depending on the law of the first wavefront.

It is appropriate, however, to mention here a negative experience of the authors in connection with lateral reflections. In the Landestheater in Hannover, the lateral reflectors were so placed that the artist could make use of them only if he stood well forward, at the front of the stage. If he changed position on stage, large changes in the amount of early reflection occurred. Here, a natural means of sound reinforcement caused as much inequality of sound distribution as a change in distance from a microphone. In this case, Meyer and Kuhl had been asked to hide the reflectors from sight; instead of choosing an optically transparent reflector material, such as plexiglas, they hid the reflectors behind an acoustically transparent screen of fine wire mesh.

In contrast to this approach, Fig. 5.12 shows lateral reflectors in the Theater of Wolfsburg,[15] which Hans Scharoun included, once again, as architectural elements in his design. Since the reflectors were to be installed rather high, they had to be angled in both the ground plan and the section. It was very difficult to determine their optimum angular orientation by analysis of the plan and section drawings at the drawing board. These reflectors were, therefore, designed to be mounted on ball-joints, so that when the building was completed, they could be adjusted in angle according to tests made in the hall.

Although we expect more or less diffuse (therefore, uncontrolled) reflection from reflectors comparable in size to the wavelength, it is nevertheless always expedient to try to determine their mean inclination and radius of curvature during design. Insofar as this involves rays in the drawings to which the reflectors and their curvatures are perpendicular, we can easily apply the procedures already discussed above. But if the reflector plane is oblique to the plan and sectional drawings, the analysis becomes more difficult.[16]

---

[14] Meyer, E. and Kuhl, W., *Acustica*, **2** (1952) 77.
[15] Cremer, L., *Technik am Bau*, **1** (1976) 45.
[16] The authors are indebted to Thomas Fütterer for his comments on the following discussion.

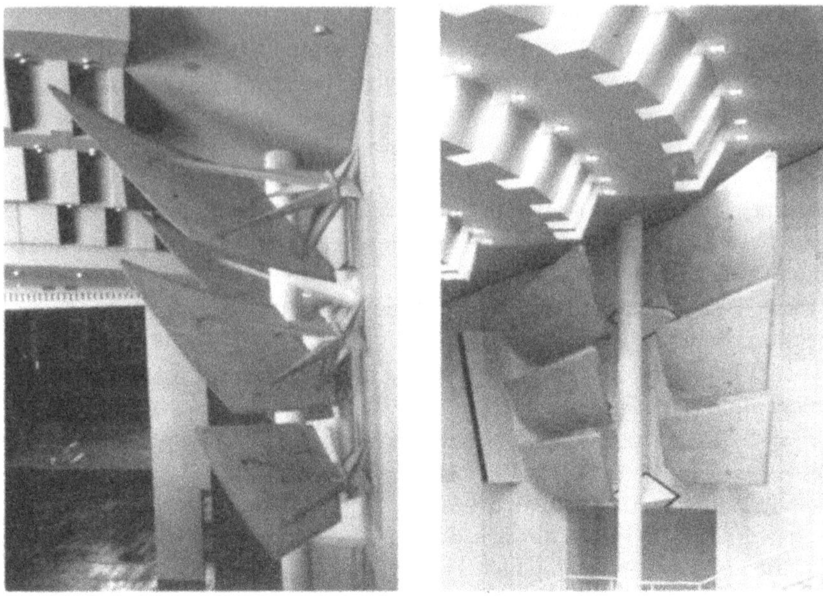

Fig. 5.12. Lateral reflectors in the Theater of Wolfsburg. (Architect, Hans Scharoun.)

Figure 5.13 shows first a simple example for the ray construction. The hall has, as in Wolfsburg, a fan-shaped ground plan, which is often adapted from the antique theaters because of its good sightlines. In this case, the side-wall reflections do not reach the middle of the audience, as is proved by the dashed ray from source S, in the plan. This fault is remedied by adding the reflector ABCD. Its effect is easy to analyse if it appears as a line in (i.e. perpendicular to) the cross-section. Then we find immediately the image source S′ in the cross-section. Drawing the rays from S′ through the edges AD and BC in the section, we find the points A′D′ and B′C′ on the floor. These two points limit the side-to-side extent of the reflection area in the audience for this reflector.

Now, to get the fore-and-aft boundaries A′B′ and D′C′ in the ground plan, we must find the projection of the image source S′ in the ground plan. It lies vertically beneath that in the section and in the same cross-plane as the source (that is, at the same distance from the front wall as the original source S).

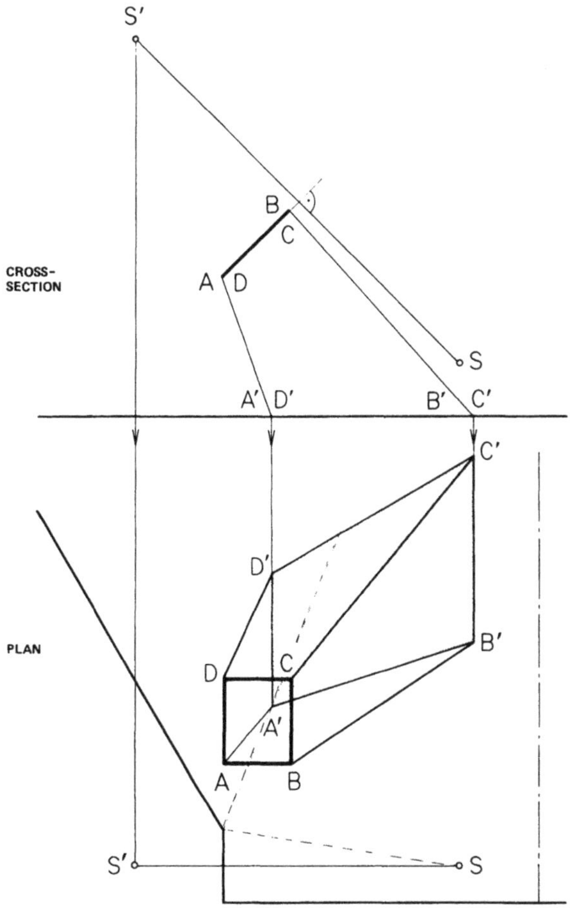

Fig. 5.13. Geometric construction of the receiving area, if the reflector plane is perpendicular to the cross-section. Top: transverse cross-section. Bottom: ground plan.

By drawing, in the plan, the rays from S′ through ABCD, we find the cornerpoints A′D′ and B′C′ on these rays, vertically below the known points in the section.

The intensity of the reflected rays in the receiving area is no less than if the reflector were part of an oblique ceiling, since reflections outside of ABCD would not fall in this area. If we curve the reflector into a convex surface, in order to enlarge the receiving area, we must reckon on a decrease in intensity, because now the same reflected sound power is distributed over a larger area of the audience.

The case in which the reflector is inclined to *both* of the usual projection planes may also be analysed, but with some additional trouble. We may assume that the reflector is rectangular and that two of the parallel edges are horizontal. Then the reflector appears as a rectangular area ABCD in the ground plan. In longitudinal or transverse section, however, the reflector will appear as a parallelogram (see Fig. 5.14).

The source S is given at a similar position, with respect to the reflector, as in Fig. 5.13. Again, we must look for the image source S′, because, as soon as this point is located, the rays from it, in section and plan, again determine the receiving area, as in the example above.

For this purpose, we need the perpendicular l to the reflector in both projection planes. Here, we make use of the rule that the projection of a rectangle is itself a rectangle only if one side is parallel to the projection plane. Thus, through one point of the reflector, say, C, we draw in its plane two straight lines g_1 and g_2, which are parallel to the ground plan and section, and which, therefore, appear in both the section and plan as 'horizontal' lines. Then the projection of l must be perpendicular to g_1 in the ground plan and perpendicular to g_2 in the section. For g_1 we may choose the edge BC; its direction is in the ground plane, and we can construct there the direction of the projection of l.

On the other hand, g_2 is given by C and the point E, where g_2, known in the ground plan, cuts the edge AD. Vertically above, we mark E on AD in the section. So we have there also (defined by points E and C) the direction of g_2, and, perpendicular to it, that of the projection of l.

We now need the point Q where l intersects the plane of the reflector. We may, for instance, mark in the ground plane the projections of the points U and W, where g_1 and g_2 intersect the plane that contains l and is perpendicular to the ground plane. Since g_1 and g_2 lie in the reflector plane, this must hold also for the straight line WU, which is the intersection between the reflector plane and a plane containing l. Therefore, the point Q must also lie on UW. We find it easily in the section as the intersection point of the projections of UW and l. (We could have obtained it also by exchanging the roles of the plan and the section, but with less favorable intersections.)

Having SQ, we have also the equal distance QS′ in the ground plan and section. The rest of the construction follows as in the first example, to find the reflection area A′B′C′D′ in the audience.

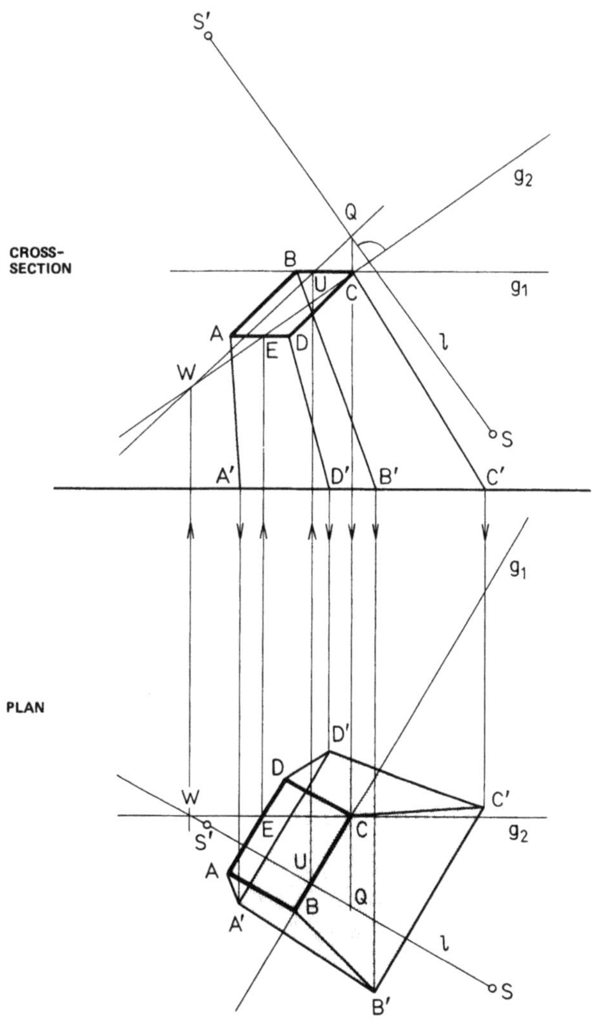

Fig. 5.14. Geometric construction of the receiving area for a reflector inclined toward both projection planes.

We may ask whether such a geometrical construction on the drawing board is worthwhile nowadays. Algebraic solutions to such problems are not only more exact, but they can easily be carried out by computers. With computer assistance, it is even possible to follow a particular ray through a number of reflections at several algebraically defined surfaces.[17] (We shall return to this method in Chapter I.7.)

Despite the possibility of computer calculations, which permit very rapid analysis of the effects of changes at the room boundaries, the drawing-board constructions will always be helpful because the task of going through these geometrical constructions provides valuable insights and suggestions for the solution of room-acoustical problems.

I.5.4 The Shape of the Ceiling

When the large canopy in the Royal Festival Hall, London was under discussion, the director of the London Philharmonic at that time, Thomas Russell, asked whether it would not be more reasonable to design the basic architecture of the room so as to provide the same reflections that the canopy was expected to provide.[1]

We can ask, even more pointedly: is not the necessity for the additional reflectors itself a symptom of an inadequate room design?

Certainly, there is the danger that the use of reflecting panel arrays may become a room-acoustical fad. So it is essential that, for each case, the provision of such an array must be technically justified. For example, in rooms so small that the limit of perceptibility is not exceeded, because the ceiling height is less than 8·5 m and the width is less than 17 m, no supplementary reflectors should be used. And yet it is surprising how often, in such small rooms, permanent sins against sensible room-acoustics design are committed.

It is no doubt true that walls and ceilings have other important functions besides providing sound reflections. The walls are interrupted for doors and windows, and assorted equipment may be mounted there. The ceiling surface is also interrupted, but to a lesser extent, by lighting fixtures and air-conditioning outlets.

With respect to room boundaries, then, we begin with the ceiling, because it is usually the largest reflecting room boundary, and it is the surface most easily accessible to the sound; thus, it is the most effective reflecting area.

It is, unfortunately, a widespread room-acoustical mistake to provide

[17] Krokstad, A., Strøm, S. and Sørsdal, S., *J. Sound Vib.*, **8** (1968) 118.
[1] *Royal Institute of British Architects Journal*, **59** (1951) 47.

sound absorptive treatment on the ceilings of lecture rooms and conference rooms, as a means of controlling reverberation with the hope of increased intelligibility. Where this has been done, it is often necessary later to install a sound system, even in lecture rooms for only 100 listeners and in conference rooms with less than 30 attendees. Such a practice betrays a complete loss of perspective in room-acoustics design: it is the equivalent of amputating healthy limbs and fitting prosthetic devices.

Also, if the ceiling is overloaded with technical equipment or is given a rough or highly modulated surface for architectural reasons, its function as a source of useful early reflections is diminished: the reflections will be scattered to locations where they are not needed. On the other hand, it is not necessary to dedicate the entire ceiling area to providing early reflections: in a conference room, for example, only the ceiling over the conference table and chairs is needed for early reflections; in a lecture room, only the front half of the ceiling needs to be reflective (as shown in Fig. 5.16(a)), and even there only the middle part.

If the ceiling is so high that the reflected sound follows the direct sound with a detour of more than 17 m, the limit of perceptibility may be exceeded and we will get a loss of clarity; but this results only if no lateral reflections have arrived at the listener's position in the meantime. For design purposes it is prudent to assume the worst case: that the ceiling is the only significant reflecting surface.

It is not necessary that the ceiling be lower than 8·5 m, because, for locations very near the speaker, the direct sound always predominates over the first reflection. As we shall see in Part III, the reflected sound may be neglected if its intensity is less than 10% of that of the direct sound.

With a ceiling of height h above the source–receiver plane and a distance r between the source and receiver, this rule of thumb leads to:

$$\frac{(2h)^2 + r^2}{r^2} > 10 \tag{5.3}$$

or

$$h > (3/2)\,r \tag{5.3a}$$

This straight line is plotted as the solid line a in Fig. 5.15. If the ceiling height exceeds this line for a given value of r, then the intensity of the reflected sound is less than 10% of the intensity of the first sound, and therefore no disturbance is to be feared, however great the time delay.

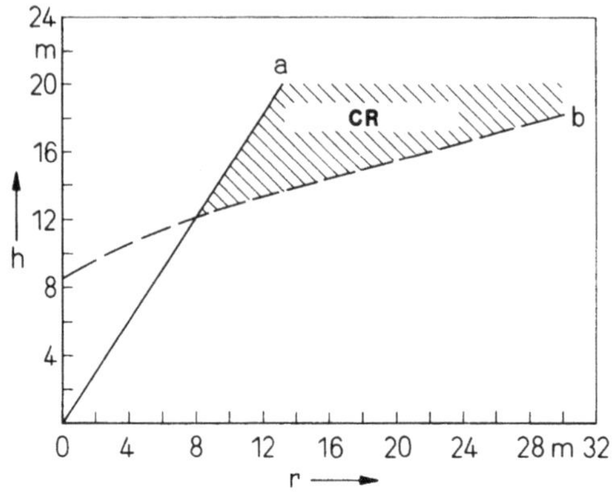

Fig. 5.15. Limitations on the effective ceiling height as a function of source–receiver distance *r*. *a*, Intensity of the ceiling-reflected sound is 10% of the intensity of the direct sound. *b*, The detour of the reflected ray path, with respect to the path of the direct sound, does not exceed 17 m. CR = critical region.

For ceiling heights below this line, the intensity of the reflection exceeds 10% of that of the direct sound, and, therefore, makes a useful contribution if it arrives soon enough.

In the same figure the 'maximum detour' condition is plotted (the dashed line *b*):

$$\sqrt{(2h)^2 + r^2} - r < 17\,\text{m} \tag{5.4}$$

or

$$h < \sqrt{8{\cdot}5\,\text{m}\ (r + 8{\cdot}5\,\text{m})} \tag{5.4a}$$

For ceiling heights below this curve, the detour for the reflected sound ray is less than 17 m; for heights above this curve, the reflected sound arrives too late to be considered useful for speech reinforcement. This area is shaded and is designated a 'critical region'. Ceiling heights that lie below both lines are favorable.

No disturbance is to be expected for any height less than 12 m, the height at which the two curves intersect. The critical zone above this height is bounded, on the left-hand side by negligible intensity in the ceiling reflection, and, on the right-hand side, by admissible time delay. For a ceiling height of 15 m, for example, this critical region extends from $r = 10$ to $r = 18$ m.

Although it would be an exaggeration to call this region a 'dead zone' (it would be more correct to speak of a 'blurring zone'), we can see from this example that the problem regions in a hall are not always the most distant ones. In fact, the famous pin dropped on the stage may be heard better at the rear than in the middle of the hall.

It is possible to avoid the critical zone altogether with a plane ceiling that slopes upward toward the rear of the hall (see Fig. 5.16(b)). In this case, for a given average ceiling height, the image source comes nearer to the audience, but the area of the ceiling available for reflections to the

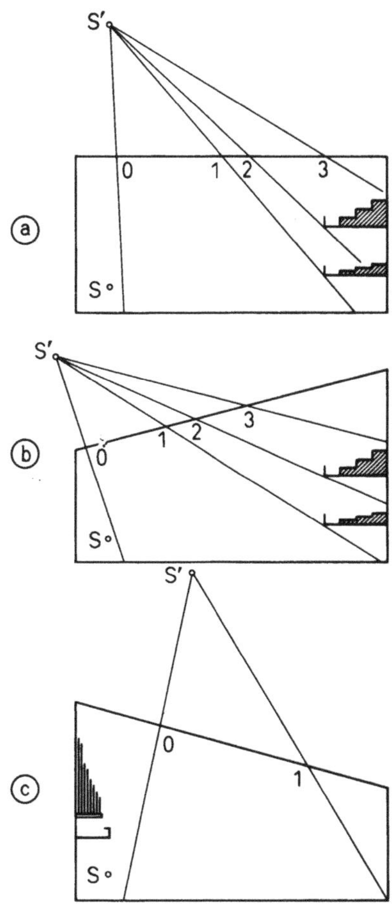

Fig. 5.16. Comparison between ceilings of different configurations. (a) horizontal; (b) sloping upward toward the rear; (c) sloping upward toward the front.

main floor (only 0 to 1) and to the balconies (2 and 3) is smaller. Furthermore, the ceiling reflections do not reach the floor until they are some distance from the source, and thus are not useful for contact between the musicians.

Figure 5.16(c) shows the case of a ceiling that slopes upward toward the front of the hall. This has the advantage of bringing the ceiling closer to distant listeners seated on a horizontal floor, as is usual in churches. Acoustically speaking, the situation is similar to that in a lecture room in which the seating slopes upward toward the rear. In this case, the organ (if any) must be placed on the front wall.[2]

Considering the previous discussion of the dangers of concave boundary surfaces, it would seem reasonable to exclude them altogether from any discussion of ceiling reflections. But to do so would overlook the historical fact that cylindrical–parabolic reflectors behind an orchestra were once conceived as being highly desirable, in the same sense as the pulpit canopy discussed in Section I. 5.1 and illustrated in Fig. 5.2.

Figure 5.17 shows the longitudinal section through the first such parabolic hall, La Salle Pleyel, built in Paris in 1927. The builder, A. Lyon,[3] who had previously made careful investigations of the limit of

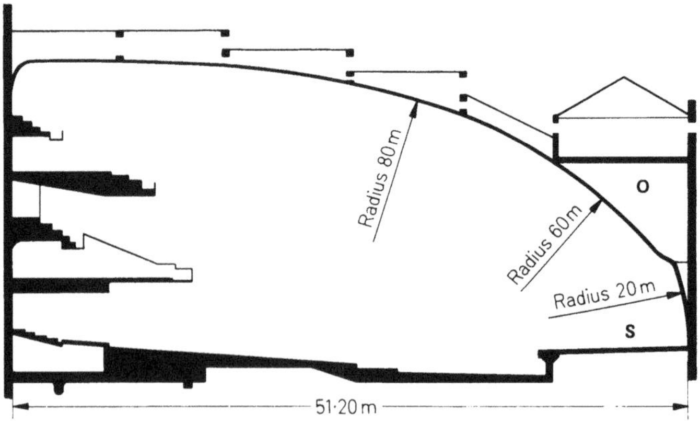

Fig. 5.17. Longitudinal section through La Salle Pleyel in Paris. O = organ, S = stage. (After Trendelenburg, F., *Z. Techn. Phys.*, **13** (1932).

[2] Cremer, L., *Kirchenmusikalisches Jahrbuch*, 1961.
[3] Calfas, P., *Le Genie Civil*, **91** (1927) 421.

perceptibility, replaced the customary flat front-wall and ceiling by three circular cylinders, which, overall, had a more or less parabolic effect. The effective focus was near the front of the stage, and the ceiling beamed the sound to the audience on the main floor and in the balconies. Unfortunately, since the rear wall between the balconies was a reflecting surface, the sound returning from that surface was refocused by the ceiling onto the musicians, with a very great time delay. After the rear wall was covered with absorptive fabric, this strong echo was diminished; but the solution was still not fully acceptable, because the musicians could hear a great deal of the noise from the audience, but very little sound from each other. Their own sound was directed only toward the audience.

In 1928, a fire destroyed the hall, fed by the highly inflammable hanging fabrics. In the course of the subsequent restoration, Osswald[4] placed high zigzag screens around the orchestra, so that, instead of a concentrated geometrical reflection, a highly diffuse reflection was provided. In addition, in place of the early inflammable fabrics at the rear, a fireproofed absorptive treatment was used. With these changes the serious defects of the hall were eliminated, and the original goal of the designer became evident: the extraordinary clarity with which even rapid runs could be heard at all locations.

But, as we shall see in Part III, distinctness and clarity are not the only important criteria of acoustical quality in a concert hall.

The poor communication between the musicians on stage, which was especially evident in La Salle Pleyel, is a problem that occurs not only with concave ceilings; it may also happen with flat ceilings, as we have seen in Fig. 5.16(b). Such ceilings, sloping upwards toward the rear, are sometimes preferred because they simplify the lighting of the stage, particularly for drama or opera. But this important consideration should not be made the excuse for dedicating the entire ceiling to that purpose. (At any one performance, it is seldom that more than 20% of the spotlights are in use!) The acoustical consultant in these cases must depend upon the understanding of the theater consultant and/or lighting engineer in working out acceptable compromises.

The part of the ceiling most important to the acoustical engineer is the central strip along the axis, while the lower side-wall locations are often favored for most of the lighting, because of the more flattering lighting

[4] Osswald, F. M., *Schweizer Bauzeitung*, **95** (1930) 47.

angles possible from such locations. In principle, it may also be agreed that the slots for spotlights that are not needed for lighting can be closed, to provide useful acoustical reflections; but in practice no one has the special responsibility to do this, and it seldom gets done.

In any case, the arrangement shown in Fig. 5.18 should be avoided, with a ceiling sloping steeply upward, just in front of the proscenium, toward a spotlight booth. In this case, no part of the ceiling is effective for sending early reflections to the main floor. At best, the sloping ceiling duplicates the balcony reflections already provided by the flat ceiling.

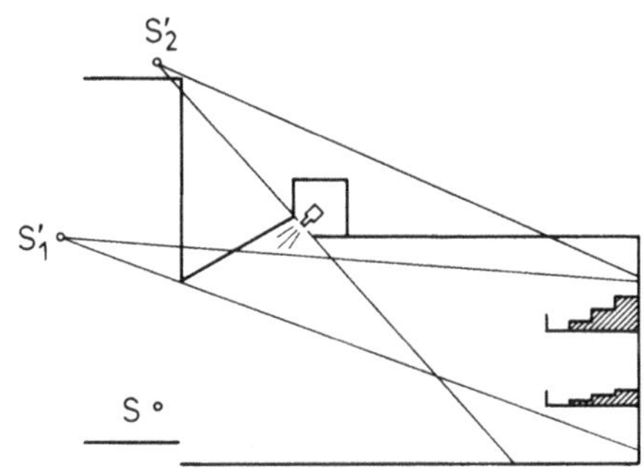

Fig. 5.18. Example of an unfavorable ray distribution due to the steeply angled ceiling in front of a lighting bridge.

Modern theater design tends to minimize the extent of the front wall between the proscenium arch and the high ceiling, the surface that was often decorated, in earlier times, with a clock. Instead, the ceiling now tends to slope down to meet the proscenium arch, and presents an exaggerated version of the situation shown in Fig. 5.18 or even Fig. 5.17. Figure 5.19 (top) shows a proposal for an alternative to the monotonically upward sloping ceiling, namely, suitably inclined convex steps, for the German Opera in West Berlin. It was constructed on the basis of rays that supply early reflections both to the audience and to the orchestra pit.[5] The numbers shown at the reflection locations indicate the reflected-ray path-detours in each case: from 16·5 m near the front of the audience, down to 3·5 m in the balcony. In the pit, it was impossible not to exceed the 17 m limit because of the usual geometry near the front

[5] Cremer, L., Nutsch, J. and Zemke, H. J., *Acustica*, **12** (1962) 428.

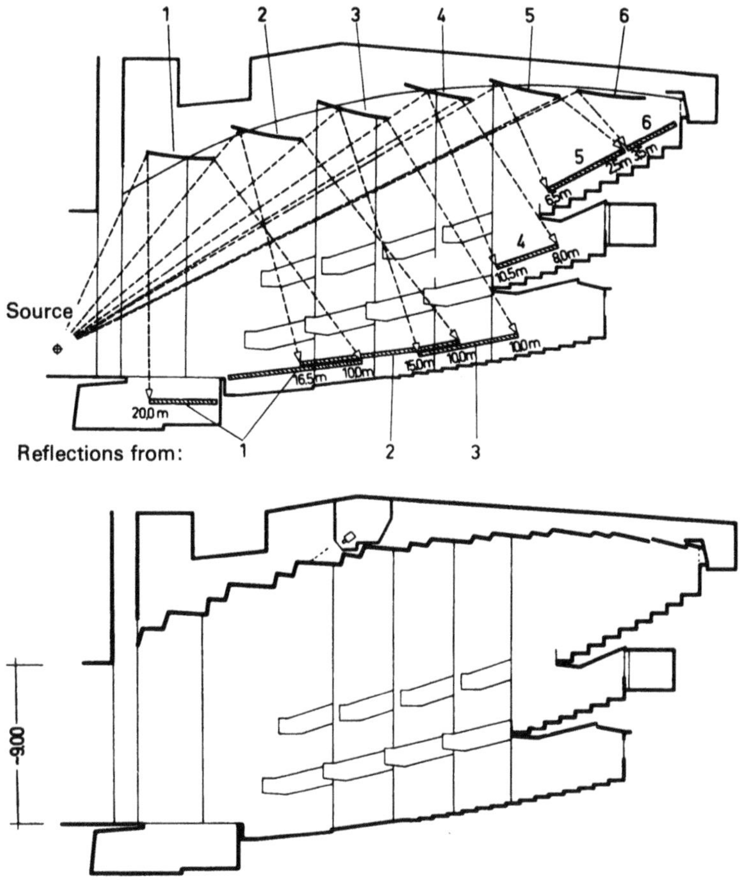

Fig. 5.19 Ceiling profile for the German Opera in West Berlin. Top: proposal based on ray construction (see text). Bottom: ceiling as built (architect, Bornemann; acoustical consultants, Cremer and Gabler).

of the hall—but we recall that, for music, this limit is about 50% higher anyway, say 25 m.

It was clear from the beginning that the proposal of Fig. 5.19 (top) could serve only as a principle to guide the architectural design. The architect, Fritz Bornemann, chose to divide these large steps into smaller ones, alternately large and small (see Fig. 5.19, bottom); it had the effect of diminishing the purely geometrical reflection and enhancing the diffuse reflection. In the audience, one is conscious of sound arriving from above, but not from a particular direction.

This raises another important point in the geometrical construction for ceiling design. Expecially for high or slightly concave ceilings, it is very undesirable that every receiving point should correspond to only a single reflection point on the ceiling. It is much better if, at each receiving point, sound arrives from a number of different ceiling reflection points, thanks

Fig. 5.20. Beethoven Hall, Bonn, showing a ceiling configuration that combines diffuse and geometrical reflections.

modulated ceiling, in the Beethoven Hall in Bonn. The architect conceived of this surface as slightly vaulted, but agreed that it could be completely covered with scattering elements about 30 cm deep: a combination of spherical segments, double pyramids and cylindrical segments. Model studies showed that this configuration provided both a diffuse reflection and the desired geometrical reflection combined.[6]

In low and long rooms, however, one should avoid diffuse ceiling reflections, because, in such cases, they hinder the propagation of sound to the rear of the room, and strongly favor the front seats. On the other

[6] Meyer, E. and Kuttruff, H., *Acustica*, **9** (1959) 465.

hand, if used at the front of the room, they can be very helpful for good contact between musicians.[7]

Finally, the optimal shape for the ceiling depends on the purpose of the room. For lecture rooms and drama theaters a low ceiling is desirable; for concert halls, a high ceiling. For opera the best choice lies in between. Therefore, for rooms which must serve all purposes, and even allow for different audience capacities for the various purposes, mechanically operated ceilings have been developed that can be raised or lowered to serve the needs of various types of events. The ceiling would be raised for musical concerts, but would be lowered for events requiring strong early reflections (such as drama); at the same time it would close off one or more balconies to reduce the audience capacity to a size appropriate to each event.

Figure 5.21 shows such a development by G. C. Izenour, with acoustical consultation by V. O. Knudsen.[8] A special difficulty for the design of this ceiling was created by the amphitheater-like seating of the hall. It necessitated a lateral subdivision of the parts of the ceiling, with sections that overlap in various degrees as the ceiling is raised and lowered.

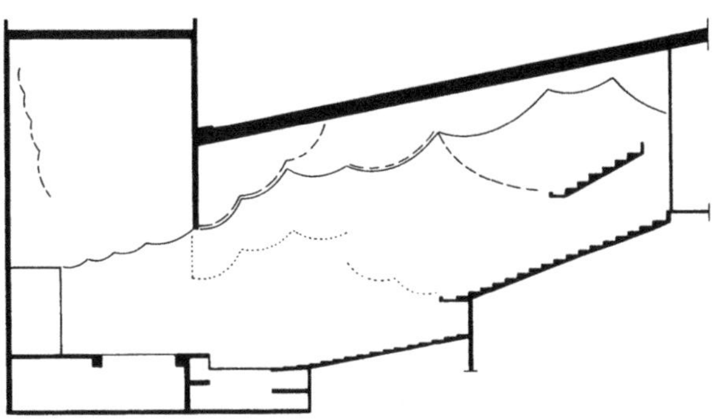

Fig. 5.21. Scheme for a variable ceiling in Edwin J. Thomas Hall, University of Akron, Akron, Ohio. ———Concert, — — —opera, ... drama. (After Izenour and Knudsen.[8])

I.5.5 The Shape of the Side Walls

With respect to the surfaces that bound a hall from above, we are seldom dealing with load-bearing structures. The visible ceiling is often, itself,

[7] The new 'Palais de la Musique' in Strasbourg may be mentioned as an example.

[8] Izenour, G. C., *Architectural Engineering*, March (1974) 143.

supported from a heavy structure, and when it spans the hall it supports only its own weight. Consequently, the details of the ceiling design may be put off until rather late in the building planning. The configuration of the ceiling may even be guided and modified by the results of model tests, after construction of the building has begun.

By contrast, the side walls of an auditorium are often load-bearing structures that must be designed, not only to meet the room-acoustical requirements of the auditorium, but also to carry the structural loads from the upper parts of the building down to the footings and foundations, in conformance with structural and safety codes. The nature of these wall structures will depend on the site contours, the accessibility, the seating capacity, and (last but not least) the ideas of the architect, particularly concerning the external appearance. Even when the acoustician participates in the planning of the facility from the very beginning of design, he seldom has the opportunity to propose basic changes. He may be allowed to modify the shape of the hall somewhat by setting acoustically designed reflecting panels in front of the load-bearing structure; but he must not use up too much space with these changes, otherwise he steals space needed for the audience seating and the circulation areas outside the auditorium.

Beside these restrictions, a geometrical–acoustical limitation must be recognized: while a ceiling can be useful for early reflections over its entire surface, only a small strip of the vertical side-walls is useful; it is defined roughly by the neighborhood of a plane that contains the source and most of the audience. Furthermore, the sound reflected from these walls crosses the audience at grazing incidence, just as does the direct sound. We will learn in the next chapter that, in this case, the sound pressure reaching the ears of the audience seated in the middle or the rear of the main floor is much smaller, particularly at low frequencies, than sound reaching them from above.

On the other hand, we shall see in Section III. 3.1 that laterally incident sound plays an especially important role in determining the room-acoustical quality of a hall.

It is acknowledged that the highly praised acoustical qualities of the famous concert halls of the last century (for example, in Leipzig, Vienna, and Boston) owe their distinction to the parallel orientation of the side walls and to the rather small widths of these halls. The reflections from the side walls of these small halls, even to the center of the hall, have only a small detour to make compared with the direct sound, a point that was strongly emphasized by Beranek (see Section III. 2.3). For a hall width of

20 m, the listener is never farther than 10 m from the nearest side-wall; and we can see, from the limit curves of Fig. 5.15, that the critical zones for early reflections never arise in these halls. Also, the lateral reflections from nearby side-walls arrive with greater intensity.

But the design of narrow halls is out of the question, nowadays, if the performances and the halls themselves cannot be subsidized, as they never are in North America. The seating capacity required to support an unsubsidized hall is about 2400 to 3000, depending on the community. Smaller halls are built in universities, but they are not expected to support themselves fully. Even in Europe, the trend is toward larger audiences.

Moreover, audiences will not accept today the seats in deep boxes and balcony overhangs, with muffled sound and very restricted sightlines typical of the older halls where less than one-third of the stage was visible.

Therefore, today, there is an architectural tendency to design halls with a fan-shaped ground plan, like a section of an antique theater. Unfortunately, these halls have characteristic acoustical problems. (The theater at Bayreuth was based on a similar principle, but its acoustical virtues stem from reasons other than its plan shape.)

Figure 5.22(a) demonstrates that the side wall of a fan-shaped hall, if it is a simple flat plane, does not provide early reflections to the middle of the main floor; the sound from the stage is all thrown to the rear of the hall.

This tendency is accentuated if the side walls have a saw-tooth configuration, as shown in Fig. 5.22(b); the shorter sides are often used for entrance doors. In this case some sound is returned to the stage (which may give confidence to the performers) and the audience enjoys the advantage of entering the hall with a good view of the stage. But no lateral reflections are provided for *any* part of the audience.

Acoustically speaking, a much better solution is provided by the 'reverse' saw-tooth arrangement of Fig. 5.22(c), where alternate elements of the side wall run parallel to the longitudinal axis of the hall. (These surfaces may contain doors if they are kept closed during the performance; but it would be better to put the entrances in the short, rear-facing side-wall elements.)

Even this reverse saw-tooth treatment of the side walls does not solve all the problems of the fan-shaped plan; in between the reflected rays aimed toward the middle of the hall, as shown in Fig. 5.22(c), there will be stripes of shadow across the hall that receive no early reflections from

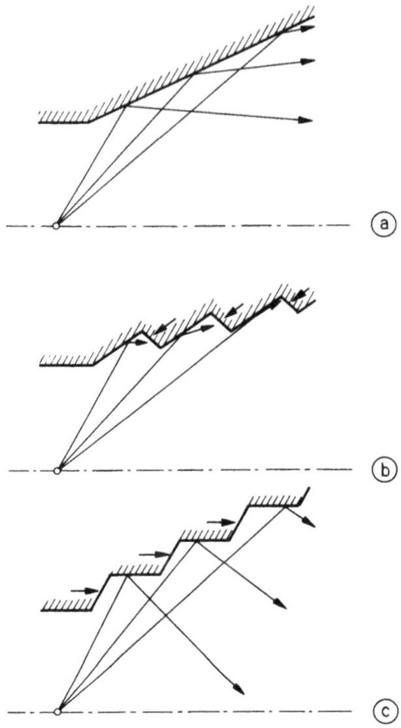

Fig. 5.22. Plan views of side walls in a fan-shaped hall: (a) plane side-wall; (b) saw-tooth shape with entrance doors facing the stage; (c) saw-tooth shape with wall surfaces parallel to the longitudinal axis.

the sides. The only way to remedy this deficiency in a stepped side-wall is to give to the side-wall elements parallel to the centerline a slight convex curvature, which will spread the lateral reflections and fill in the shadows.

A similar problem is created by stepped ceilings, particularly if the scale of the steps is on the order of 1 to 4 m. For any given source location on the stage, a stepped ceiling creates alternating stripes of sound and shadow, running across the audience from side to side in the auditorium.[1]

[1] An example is the Birmingham Jefferson Auditorium in Birmingham, Alabama. With a source of continuous random noise on the stage, one may walk up and down the aisles, and move alternately into and out of regions where the ceiling reflections of this source are audible. The differences are striking. If these regions are identified, one finds, in subsequent listening with an orchestra as the sound source, corresponding stripes of alternating clear and dull orchestra sound in the audience.

The disposition of part of the audience in balconies, which was taken over for 'classical concert halls' from 'classical theaters', has not only the advantage of making the room narrower, but also provides the opportunity to utilize the upper parts of the side walls to bring lateral reflections back to the main floor. We refer, here, not so much to the reflections of first order from the balcony faces (which, if they are to return reflections effectively to the main floor must be inclined to an extreme degree), but to the second order reflections that involve the side walls and the balcony soffits, assuming that these soffits are approximately horizontal (see Fig. 5.23). For this purpose, the reflected rays must pass high enough above the heads of the balcony audience so as not to be absorbed there. In turn, this requires that only one or two rows of seats are provided in these balconies, and that the vertical separation of the balconies is sufficient; this separation must be greater for higher balconies. But a great vertical separation between balconies restricts the number of balconies that can be installed.

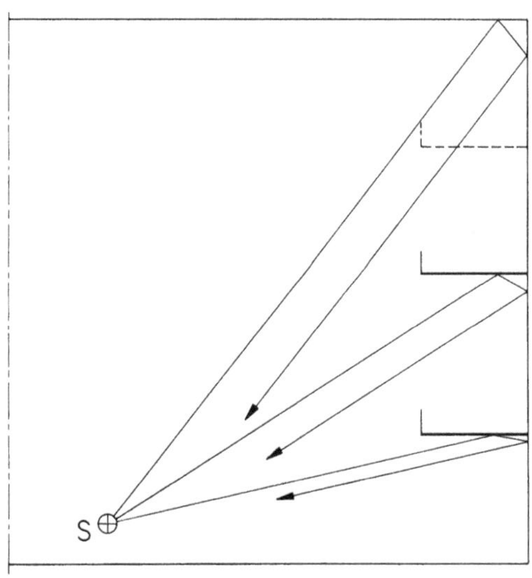

Fig. 5.23. Reflections of second order at the side walls and soffits of the balconies.

For instance, as shown by the dashed lines of Fig. 5.23, the soffit of the third balcony is no longer useful for a second-order reflection to the main floor, because it would only be grounded into the second balcony. Moreover, the second-order reflection at the ceiling is grounded in the third balcony and fails to reach the floor. Thus, at least from the viewpoint of lateral reflections, the addition of the third balcony is no help.

It is necessary also that the corners (defined by the side walls and either the balcony soffits or the ceiling) run parallel to the longitudinal axis of the hall and that they be horizontal. The first requirement marks an abrupt departure from the traditional horseshoe shape of classical opera houses; Harris and Jordan succeeded in this respect at the New Metropolitan Opera House in New York City.[2]

The second condition (horizontal balcony soffits) has tended to fall out of favor in recent years, because the sightlines from the balcony are improved if all the balconies are both raked and oriented toward the stage. The rationale for the improved sightlines with raked balconies can be clearly demonstrated with a simple geometric sketch; and the necessary dimensions for laying out the seats can be quickly derived by computer calculations. The room-acoustical disadvantages of balconies in this form, however, are much more complex and are virtually impossible to quantify. Thus, in case of conflict, the seating designer has at his disposal some apparently 'hard facts', while the acoustician must rely on intuition and experience in judging how much compromise is acceptable.

The consequence, often, is that the second-order reflections from side walls and balcony soffits may be lost altogether, or may provide early sound reflections only to the sides and rear of the main floor, but not to the critical central region.

We must question the requirement of substantially perfect sightlines from the side balcony seats (meaning that a clear view must be provided even to the feet of the performers), since this advantage is gained for relatively infrequent events and for only a small number of seats, and it means that all the seats in the middle of the main floor are deprived of early lateral sound reflections in each and every orchestral performance.

A reasonable compromise may perhaps be struck in step-wise descending balconies, as shown, for example, in Clowes Auditorium, in Indianapolis,[3] see Fig. 5.24.

In general, between the balconies we have only narrow vertical surfaces, so that they reflect very little diffuse sound to the main floor. One exception seems to have been the balconies of the famous Dresden Opera, designed by Semper and destroyed during World War II.

---

[2] Cremer, L., *Applied Acoustics*, **8** (1975) 173.

[3] Beranek, L. L. and Schultz, T. J., *Acustica*, **15** (1965) 307.

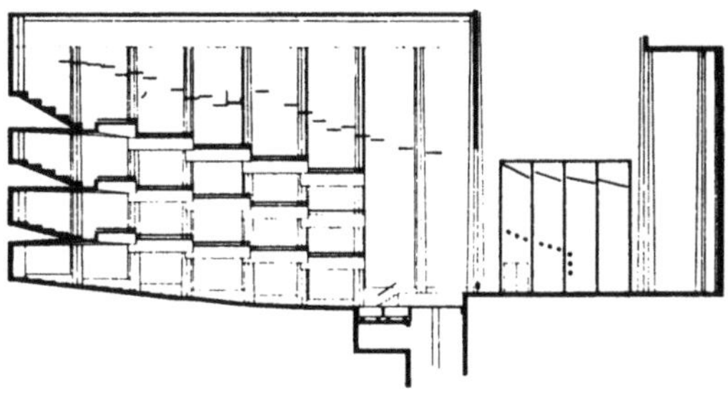

Fig. 5.24. Balconies descending step-wise toward the stage, in Clowes Auditorium in Indianapolis. (After Beranek and Schultz.[3])

According to model tests by Reichardt,[4] the fascia of those balconies were so extended and shaped (shell-vaults) that strong first-order reflections to the main floor could be expected.

In our discussions, so far, concerning the provision of useful sound reflections, we have made the simplifying assumption that the direction of the reflecting surface with respect to the sound source is not important. That is, we have not taken into account the directivity of the individual musical instruments; in fact, these directional characteristics may be quite pronounced in particular cases. But the directivity changes so much from instrument to instrument, and sometimes from tone to tone on the same instrument, that, even if we grossly simplify the directivity of the instruments (by stating the angular range within which the sound intensity exceeds half the maximum intensity[5]), it is still impossible to achieve the same frequency balance for early reflections at all audience seats. Since few complaints are heard about differences in the balance of early sound at different seats, however, it is at least probable that the impression of timbre depends on *all* of the sound reflections arriving at a seat, including the later, statistically distributed, reflections. (See Sections III. 2.9 and III. 3.1)

Certainly, the first reflections are of critical importance for good communication between the musicians and the conductor. But the arrangement of the instruments on the stage is not fixed; it changes according to the number of musicians and the style of the music.[6] Thus, it makes no sense for the acoustician to try to optimize the distribution of early reflections for a particular stage set-up, whether using reflectors

[4] Reichardt, W., *Gute Akustik-aber wie?*, VBE-Verlag Technik, Berlin, 1979, p. 175.

[5] Meyer, J., *Acustica*, **36** (1976) 197.

[6] Meyer, J., *Acoustics and the Performance of Music*, Vol. 33, Technical Handbook Series on Musical Instruments, Verlag das Musikinstrument, Frankfurt am Main, 1978.

in the hall or a special stage enclosure. He must provide a facility that works well in all situations without favoring a single special arrangement.

I.5.6 Acoustical Design Measures in the Neighborhood of the Listeners

We have seen that sound-reflecting surfaces in the neighborhood of the source can add useful reflections. The same is true for reflectors near the listeners. If we concerned ourselves only with a single source and a single listener, the problem would be 'reciprocal' and the acoustical means employed would be the same. Instead, however, we must deal with a 'group of musicians' and a 'group of listeners'; and for the musicians we must also respect their need to hear each other well, a requirement that does not apply to the listeners. Furthermore, it is usually the case that rather small source areas (concert hall stage $=200\,m^2$) must supply sound to rather large listener areas (audience seating $=2000\,m^2$). Therefore, special reflecting surfaces in the neighborhood of the audience are the exception rather than the rule; the important exceptions concern, for example, audience members under deep balconies.

Actually, we have already included some mention of reflectors in the neighborhood of the listeners in our discussion of side-wall reflections. In that case, however, we were primarily interested in guiding early reflections to the center of the main floor.

The situation is quite different when it comes to the low and deep boxes that were usual in the opera houses of the last century. (One frequently hears high praise for the acoustics of La Scala, Milan; but we may be certain that this assessment applies only to those seats that command a full view of the stage and the ceiling—effectively, of the whole room. Listeners seated deep within the boxes see only a fraction of the stage and none of the ceiling; their experience of the performance must be both dramatically and acoustically unfavorable.) Properly shaped sound-reflecting soffits and partition walls in these boxes would be helpful; but in fact, as often as not, these surfaces were upholstered with sound-absorptive materials.

At least, in order to prevent whispering gallery effects,[1] the rear walls

[1] Cremer, L. and Müller, H. A., *Deutsche. Arch. U. Ing. Zeits.*, **12** (1965) 14. The authors regret that they were not able to persuade the architects for the National Theater (Staatsoper) in Munich to respect their proposals in this regard; the architects' primary concern was the restoration of an historical building.

of the balconies (if they are not partitioned into boxes), and also the rear wall on the main floor, should avoid a smooth concave surface (an almost automatic architectural choice in a house with a horseshoe ground plan!).

In nearly all theaters with lateral balconies the heights of the side balconies are the same as those at the rear of the hall; this is by no means favorable from the acoustical standpoint. Whereas the second-order reflections from the side walls and balcony soffits bring useful sound to the middle of the main floor, the corresponding reflections at the rear wall may produce echoes on stage and in the front rows of the audience.

We have already discussed the latter problem in Section I.4.2, where we showed that the echoes may be suppressed with sound-absorptive treatment at the rear. But it is always better to avoid the addition of sound absorption, if possible, because it diminishes the reverberation time that is to some degree desirable in musical performances. The echoes may also be avoided by a diffusing treatment on the rear wall, or, even better, by such a forward tilt to the wall that the rays arriving from the stage are reflected down into the audience nearby. Figure 5.25 shows an example.

Fig. 5.25. Forward-tilted rear wall behind the balcony of the Meistersinger Hall in Nürnberg. (Architects, Löbermann and Puchner.)

If the rear seats extend to the ceiling, as in Fig. 5.26, such means are not needed. Indeed, in all opera houses the seats in 'the gods' (the top of the house) are considered to be best on account of the loudness and distinctness of the sound there. But this is true only if the ceiling extends all the way to the rear wall, more or less as one surface, with only minor 'acoustically rough' decorative elements. If, instead, the ceiling steps upward before the top balcony, the balcony listeners effectively are not in the same room with the performers and the sound they hear may be weak and dull.

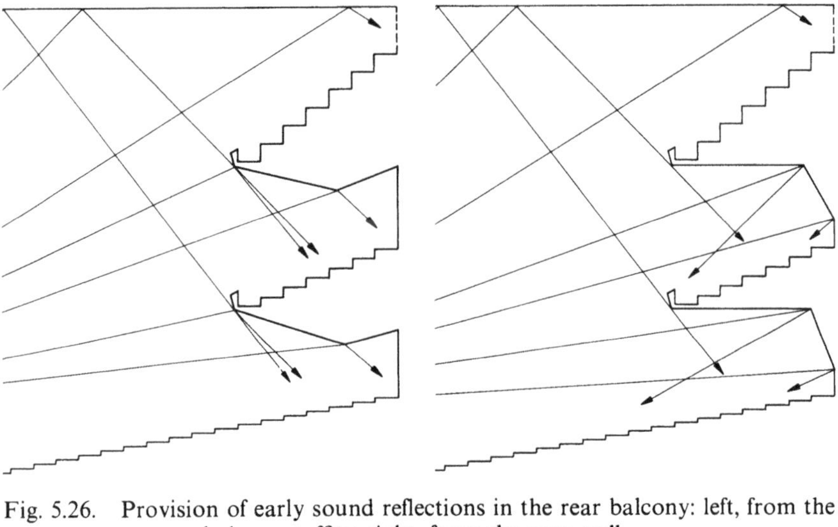

Fig. 5.26. Provision of early sound reflections in the rear balcony: left, from the balcony soffits; right, from the rear wall.

Useful reflections from the ceiling may also be heard in the highest side balconies. In fact, because the sight-lines (and therefore the 'sound-lines') are often poor in these seats, there may be no direct sound from the stage at all; the ceiling reflections provide all the early sound for these seats, and it often sounds marvellous. One of the great advantages of Hans Scharoun's design concept for the central orchestra for the Berlin Philharmonie was that it resulted in many seats near the ceiling.

For seats in the middle of the balcony, the direct sound from the orchestra pit may also be important; but there are not great differences in this respect compared with the sound in the next lower balcony. There, in the rear seats where the soffit hides the ceiling reflections, the sound is usually regarded as less good. In this case, it becomes important to slope the soffit downward toward the rear wall, as shown in Fig. 5.26, left. Alternatively, one may tilt the rear wall forward (as shown in Fig. 5.26 right) to supply the listeners with early reflections, similar to Fig. 5.25.

For the lower balconies, and particularly on the main floor, the ceiling reflections tend to arrive so late that they approach, or even exceed, the limit of perceptibility. This is relatively harmless in balcony theaters, because each balcony soffit provides an intermediate reflection between the direct sound and the ceiling reflection, for the seats below and behind it.

But even on the main floor an alert listener can hear the ceiling reflection (unless he is under a balcony), particularly where the ceiling is concave, as, for instance, in the Paris Opera. (The same effect may be observed in La Salle Pleyel, mentioned earlier.) The less acute listener may feel disturbed by the late ceiling reflection without knowing why!

This disturbance vanishes as soon as the listener moves back under the lowest balcony; these main floor seats are well known for their usually soft but distinct sound. In addition, in these seats one hears the stage sound better than the pit sound; but this is a characteristic common to all seats on the main floor. If at those (rear wall) locations the soffits are horizontal, they will be struck only by a very narrow ray-bundle. Therefore, the soffits should be inclined downwards toward the rear wall if possible. An artist on the stage should see as much reflecting area as possible at the soffits.

On the other hand, if the number of main floor rows under the balcony is large, it may be impossible to avoid sloping the balcony soffit upward toward the rear, to provide adequate head clearance underneath. In the limit, this slope would be determined by the rake of the balcony seats. Such a soffit, of course, is useless for providing early sound reflections for the seats underneath, and cannot even interact with the rear wall to add early sound. Such soffits, in fact, only reinforce the noise of the other nearby audience members; one could even make a case for sound-absorptive treatment of these soffits. This is a situation in which early sound for the underbalcony seats might be supplied with loudspeakers, fed by suitably delayed signals.

In some cases, lateral reflections can also be produced by surfaces that are distant from the side walls. Figure 5.27 shows a hall with the ground plan generally in the form of a trapezoid; the floor is mostly horizontal in order to accommodate television lights, cameras, and other equipment.

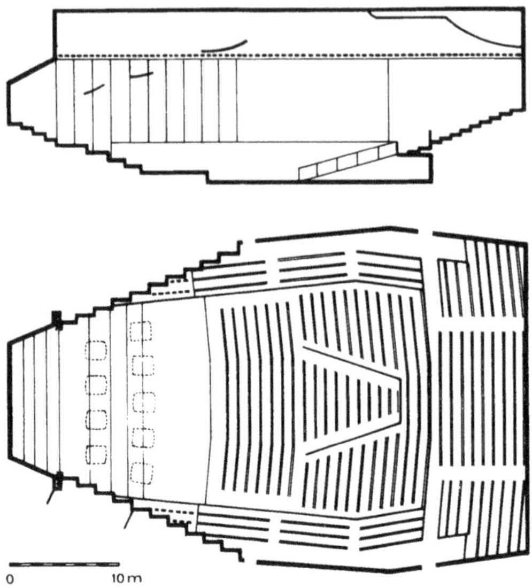

Fig. 5.27. Longitudinal section and ground plan of the large hall of the Senderfreies Berlin, with wedge-shaped side-walls at the middle of the main floor.

The horizontal portion of the floor is surrounded by upward-sloping seats toward the rear, and the side walls actually converge toward the rear in the back of the hall.[2] Without the wedge walls, this would be a critical zone, lacking adequate early reflections. With the wedge walls, the sound is good, and the violins sound especially wonderful there.

Finally, it is also possible to supply early sound reflections, to seats that are otherwise deprived, by means of reflecting panels suspended directly above the audience, similar to those that we have seen in Section I. 5.3, above the orchestra.

Probably the most important acoustical measure that affects the neighborhood of the listeners, however, is sloping the audience seating on the main floor upward from the stage to the rear wall. We shall discuss this matter in the next chapter.

[2] The configuration is shown very clearly in Beranek, L. L., *Music, Acoustics and Architecture*, John Wiley, New York, 1962 (reprinted Krieger Publishing Co., Huntington, New York, 1979) p. 263 ff.

Chapter I.6

Sound Reflections with Losses and Phase Shifts

I.6.1 The Sound Absorption Coefficient

In the preceding chapters, we have assumed that all geometrical sound reflections occur without any loss of energy. Also, when several sound rays were superimposed, we have usually assumed that, on average, their energies may be added together without regard to the phases of the waves. And, when we have been interested in different time delays, as in the case of wedge rooms in Section I. 2.4, we did not take into account any additional time delay (or, for sinusoidal signals, additional phase shift) upon reflection. Instead, we assumed that, immediately at the reflecting boundary, the incident and reflected sound pressure waves are in phase and that, therefore, their amplitudes are to be added together.

Only occasionally have we allowed for the possibility that the incident sound may be absorbed, as, for instance, in the discussion (Section I. 3.4) of the avoidance of focusing effects, or in the explanation of how flutter echoes are possible even in rectangular rooms.

But the problems of boundary shaping, which constitute the chief application of geometrical room acoustics, cannot be entirely separated from the properties of the boundary materials, which govern the energy losses and phase changes upon reflection from the boundaries. Even though those material properties may have a more profound effect upon the 'Statistical Room Acoustics' (Part II), and may be calculated quantitatively only by the methods of 'Wave-Theoretical Room Acoustics' (Part IV, Volume 2), our survey of the problems of 'Geometrical Room Acoustics' (Part I) would be incomplete unless we gave some account of energy loss and phase shifts upon reflection.

Moreover, the quantity that describes the fraction of sound energy lost upon reflection is based on the supposition that the incident and reflected plane waves are extended over an area that is very large with respect to the wavelength: this means that we are dealing with the geometric concepts of incident and reflected 'sound rays'.

W. C. Sabine introduced to the science of room acoustics the definition of the 'sound absorption coefficient' α, namely the ratio between the non-reflected sound intensity and the incident sound intensity. Sometimes it is more convenient to think in terms of the complementary concept, the reflection coefficient ρ, defined as the ratio between the reflected sound intensity and the incident sound intensity. According to their definitions, α and ρ are related by:

$$\alpha+\rho=1 \tag{6.1}$$

Note that the concept of absorption coefficient is not defined exactly this way in all branches of physics. Sometimes it may describe the energy losses incurred during propagation through the medium; indeed, we will find that this phenomenon also shows up in sound fields. Where it occurs, it implies that a part of the sound energy is transformed into thermal energy, or heat; in mechanical engineering this process is called 'dissipation'.

The concept of absorption in acoustics goes somewhat beyond the notion of dissipation of sound energy into heat. For example, if we have an open window with a drapery drawn across it and a sound wave incident upon it, then a part of the 'non-reflected' sound is accounted for by transformation into heat in the pores of the drapery; but a large part is transmitted through the drapery, passes through the window to the outdoors, and is lost to the sound field in the room. So far as the balance of sound energy in the room is concerned, it makes no difference whether the non-reflected sound was partly absorbed and partly transmitted out of the room or whether it was all absorbed in the drapery.

If we divide both parts of the lost sound intensity by the incident intensity, we can distinguish two components of the absorption coefficient, the dissipation coefficient δ and the transmission coefficient τ:

$$\alpha=\delta+\tau \tag{6.2}$$

If the drapery is withdrawn, then all of the incident sound passes outdoors, and

$$\alpha=\tau=1 \tag{6.3a}$$

If the drapery is made of a plastic sheet, the inner losses may be so small that we may write

$$\alpha = \tau < 1 \tag{6.3b}$$

It is customary to neglect the inner losses when calulating τ in connection with the sound insulation of walls. But in that case we are dealing with values of τ so small that their contribution to α is quite negligible.

The other extreme,

$$\alpha = \delta, \qquad \tau = 0 \tag{6.4}$$

occurs when a highly absorptive surface treatment is applied in front of a heavy, rigid, non-porous wall—the case most frequently met in room acoustics. We will survey the behavior of such absorbing arrangements in Chapter II. 6, and will enter into the quantitative treatment and physical details in Chapters IV. 8, IV. 9 and IV. 10 (Volume 2). Nevertheless, some general rules are mentioned here because they are essential in dealing with the problems of boundary shaping; they thus pertain to geometrical room acoustics.

From the visual standpoint, the architect is accustomed to believing that the walls, once plastered, determine the form of the space, and that (for the most part) any further change in appearance is to be accomplished only with paint, tapestries, panels, or other thin elements mounted close against the wall. In these cases, a microscopically thin layer is responsible for the effect upon the incident light rays, and thus for the visual appearance of the walls.

If the surface treatment of the walls is to have a comparable effect on the incident sound waves, however, we must take into account the much larger wavelengths involved. The sound waves must be able to penetrate deeply into the wall. Therefore, not only are the finish materials of special interest but also the basic construction at the room boundary.

In the case of carpets, their thickness and that of the backing pad (if any) are of interest. For wall panels, the thickness of the panel itself, and that of the air cushion behind it, are important, as well as whether or not the airspace is filled with rockwool or other sound-absorptive material.

All these details lead to quite different effects for sounds of different wavelengths (frequencies), see Section I. 1.4. A carpet of 5 mm thickness may be regarded as relatively 'thick' for a wavelength of 4 cm ($f = 8500$ Hz), but not for a wavelength of 4 m ($f = 85$ Hz). It is understandable that such a carpet exhibits significant sound absorption only at high frequencies. Accordingly, if we wish to achieve effective sound absorption at low and middle frequencies, we must provide deeper constructions for the sound-absorbing treatment and must be prepared to sacrifice some of the useful space in the room in exchange. It is possible to reduce the required depth for sound absorption at low frequencies by using a combination of resonant panels and air cushions; but what we save in the depth of the treatment must be made up by the necessity of treating larger wall areas (see Section IV. 9.2, Volume 2).

Fortunately, the greater expense of providing effective sound absorption at low frequencies is offset somewhat by the fact that our ears are less sensitive to low-frequency sound, so that the requirements as to the amount of low-frequency absorption can be diminished.

The definition given above for the sound absorption coefficient assumes geometrical reflection; but the concept may also be extended to include the diffuse reflections expected at rough wall surfaces, ceiling coffers, arrays of resonating wood panels, etc. The only provision is that the dimensions of such structures, and their modular lengths, must be small in comparison with the dimensions of the total area covered by the treatment. Otherwise, it would not make sense to divide the absorbed sound intensity by the area to determine a mean sound absorption coefficient. Under this condition, it is even possible to speak of the sound absorption coefficient of the area occupied by theater seats, either empty or occupied by the audience.

I.6.2 Geometric Methods for Measuring the Absorption Coefficient

In most cases where we make use of the absorption coefficient, we assume geometric reflection. Then we can measure α by comparing the sound pressure amplitude of the incident (*ankommend*) wave p_a with that of the reflected wave p_r, and then computing:

$$\rho = (p_r/p_a)^2 \tag{6.5}$$

or,

$$\alpha = 1 - (p_r/p_a)^2 \tag{6.6}$$

If we measure both the incident and reflected pressures with the same equipment (same microphone, amplifier, voltmeter, etc.), we need no absolute calibration; only a stable proportionality between the sound pressures and the indicated voltages must be guaranteed.

Figure 6.1 shows (at the top) the principle of a method based on this comparison, as developed by Kühl and Meyer[1] in 1932. It permits the measurement of absorption coefficients at frequencies above 1000 Hz and at angles of incidence between 10° and 70°. In this procedure, the incident ray, radiated from the loudspeaker L, is reflected either from the test sample B, being measured, or from the rigid plate A. Since there is no

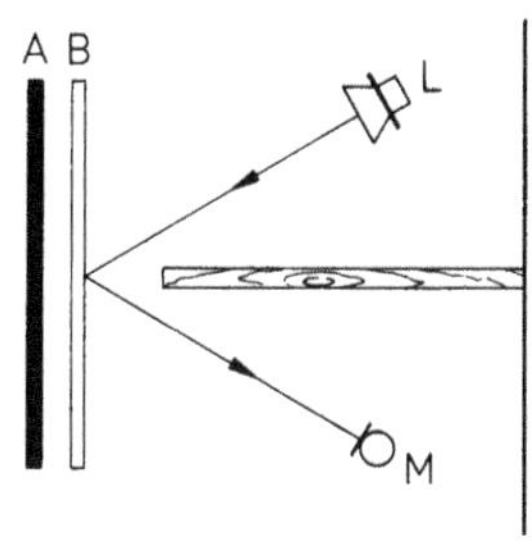

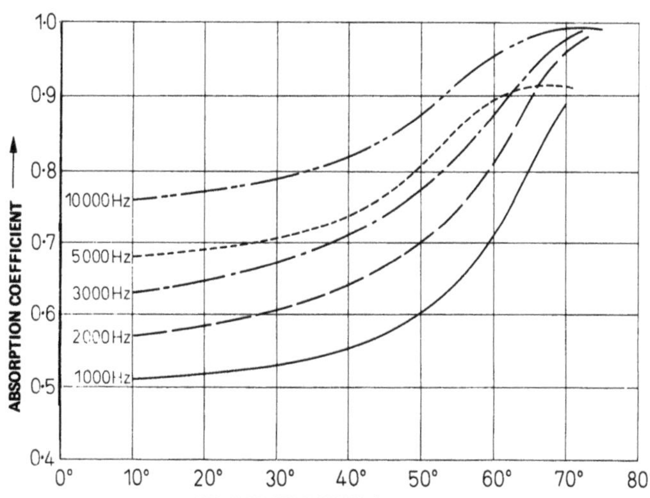

Fig. 6.1. Measurement of the absorption coefficient by exchanging test panels. Top: sketch showing the principle of the test. Bottom: dependence of the absorption coefficient on the angle of incidence, for a panel made of porous wood-fiber. (After Kühl and Meyer.[1])

[1] Kühl, V. and Meyer, E., *Berl. Ber. Phys. Math. Kl*, **XXVI** (1932).

measurable absorption at plate A, the latter reflection corresponds to the sound intensity incident on the test area, including the divergence of the sound rays on the path toward the microphone M. The microphone compares only the reflected wave amplitudes, since it is shielded by a large screen from any direct sound from the loudspeaker. If u_B and u_A are the indicated voltages corresponding to the pressures in the reflections from panels B and A respectively, then:

$$\alpha = 1 - (u_B/u_A)^2 \tag{6.6a}$$

If both plates are removed, we can determine whether (and how much) the absorption measurement is compromised by unwanted sound leaking around the shield, that is, sound reaching the microphone from the loudspeaker without reflecting from the test samples. In order to reduce the effect of this flanking transmission it is helpful to use a loudspeaker and microphone with high directivity. Care must be taken, when the test panels are interchanged, that the reflected ray toward the microphone in each case corresponds to the maximum of the directivity characteristic.

As with all geometric methods for measurement of the absorption coefficient, this procedure yields information about the sound absorption as a function of the angle of incidence.

The lower part of Fig. 6.1 shows this angular dependence for a test sample of a material frequently used in treating over-reverberant rooms, made of compressed wood-fiber. The observed increase in absorption coefficient at high values of the incidence angle is typical of many kinds of sound-absorptive panel, for reasons that we shall discuss in Section IV.7.4 (Volume 2).

Figure 6.1 also shows another tendency that is typical of porous absorbers; for all materials where the sound can penetrate into the pores and be transformed into heat by friction, the absorption coefficient increases with increasing frequency. This is not merely a consequence of the increasing ratio of panel thickness to the wavelength with increasing frequency; there are other significant physical relations that we shall describe in Chapter IV. 8 (Volume 2).

The disadvantages of measuring the sound absorption coefficient by the exchange of test samples lie in the possible failure of the assumptions: (1) that no significant sound leaks from the loudspeaker to the microphone; (2) that the test samples A and B behave as infinitely large reflectors; and (3) that panel A reflects all of the incident sound.

Alternative test methods permit us to measure independently the sound pressures of the incident and reflected waves in the sound field in front of the test sample.

One such method makes use of the so-called 'pressure-gradient' microphone,[2] that is, one where the sound pressure in the field acts on both sides of the microphone diaphragm. It is evident that, under such conditions, the microphone diaphragm will remain at rest (and the microphone signal will be zero) when the direction of sound propagation is parallel to the plane of the diaphragm, because the same pressure exists on both sides. The excitation will be maximum when the sound propagates in a direction perpendicular to the diaphragm. In other words, the microphone is sensitive to the gradient of the sound pressure, not to the pressure itself.

If we now turn the diaphragm, first parallel to the direction of propagation of the incident wave and then parallel to the direction of the reflected wave, we measure in each case only the 'other' wave, and with the same sensitivity in both cases. This measurement can be carried out at several points in the sound field in front of the test specimen, for calibration purposes.

One must take into account that the angle of sound incidence on the microphone, and thus its sensitivity, depend on the measurement location, and also that the reflected sound wave is always intrinsically weaker than the incident wave because, having travelled farther from the source, it has suffered greater attenuation due to divergence. Both effects may be easily corrected if a point source of sound is used.

Willig[3] combined the principle of the exchange of test samples with the principle of nulling the direct sound by the use of a pressure-gradient microphone. In this case, the shielding wall of Fig. 6.1 becomes unnecessary. Since his microphone always occupied the same location in relation to the test samples, no corrections were needed for the divergence or for the directional characteristics of the microphone.

All the measurement methods described above assume that the sound field is composed of only two components: a direct and a reflected sound ray. Such sound fields can be produced only in the open air (where we are at the mercy of the weather) or in large reflection-free (so-called 'anechoic') laboratory rooms.

Spandöck[4] proposed a method for measuring sound absorption coef-

---

[2] Cremer, L., *Elektr. Nachr. Techn.*, **13** (1936) 36.
[3] Willig, F., *J. Acoust. Soc. Am.*, **10** (1939) 293.
[4] Spandöck, F., *Ann. Phys. (Leipzig)* (*V*), **20** (1934) 328.

ficients that can be used even in rooms that are not highly damped, provided that they are of sufficient size. We have encountered this method earlier, in the discussion of his demonstration of Fresnel zones, in Section I. 1.5. The set-up is shown in the upper part of Fig. 6.2. The loudspeaker radiates a tone impulse toward the test specimen P, of such short duration that the direct sound pressure p_1 and the reflected sound pressure p_2 are separated in time, from each other, from the later reflection on wall W and from all other reflections.

The amplitudes of the incident (p_1) and reflected (p_2) pressure waves can be determined from the echogram in the lower part of Fig. 6.2.

In the evaluation of p_1 and p_2 we must again take account of the difference in divergence, and, for this purpose, a point source is preferable. This is also desirable for the possible extension of the method to

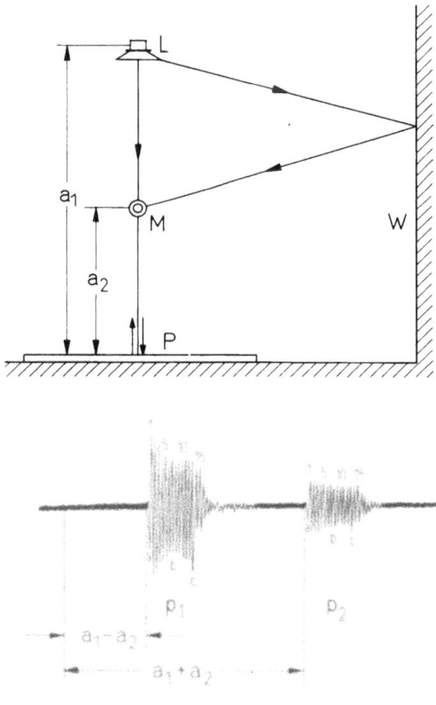

Fig. 6.2. Measurement of the sound absorption coefficient using tone impulses. Top: Sketch showing arrangement of the apparatus. Bottom: Example of a resulting 'echogram' showing sound pressure at the microphone. (After Spandöck.[4])

measurements with the sound rays at oblique incidence on the test sample; furthermore, we would have to take into account the directional characteristics of the source.

After p_1 and p_2 are corrected for divergence, the absorption coefficient is found from the application of eqn. (6.6).

The impulse method was improved by Wittern and Ernsthausen[5] with the use of tone impulses having a bell-shaped envelope, as shown at the top in Fig. 6.3. By this means, they could more accurately compare the

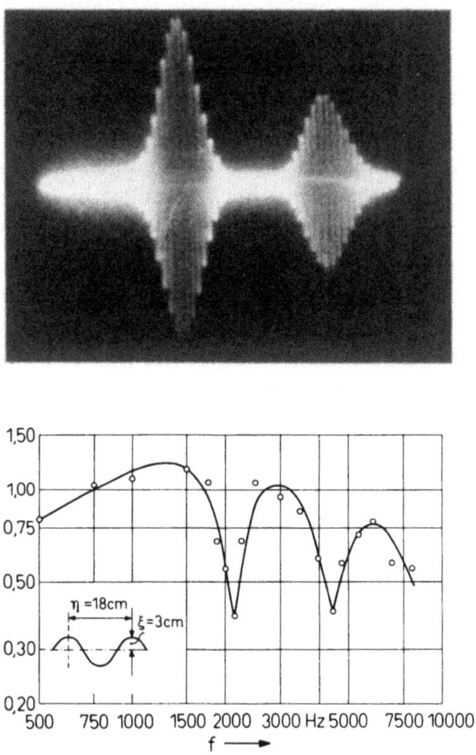

Fig. 6.3. Measurement of the ratio of reflected to incident wave pressures, with periodically repeated tone impulses. Top: echogram before adjustment of the gain of the reflected impulse. Bottom: measurement results for a sinusoidally corrugated surface (see insert for dimensions). (After Wittern and Ernsthausen.[5])

[5] Wittern, W. V. and Ernsthausen, W., *Akust. Z.*, **4** (1939) 353.

peaks of direct and reflected pulses as observed on a cathode ray oscilloscope; the measurement technique was to increase the amplification for the reflected pulse (but not the direct signal) until the two peaks appeared equal on the oscilloscope screen. The increase in gain required to equalize the two impulses was a measure of the ratio of reflected to incident wave pressures. For this test procedure, they needed periodic repetitions of the impulses, and again, the set-up had to be arranged so that the direct and reflected pulses were separated in time from each other and from all other reflections.

The precision of this method made it possible to demonstrate—passing beyond the limits of geometric reflection—the effect of a periodically corrugated surface on the reflected impulse for perpendicular incidence; here the different parts of the reflected pulse interfere with each other. Figure 6.3, bottom, shows the frequency dependence of the ratio (p_r/p_a), on a logarithmic scale. The minima at 2200 and 4500 Hz are the results of destructive interference between reflections from the 'crests' and the 'troughs' of the corrugations. If the profile of the corrugations were rectangular instead of sinusoidal, then the minima would occur at frequencies for which the depth of the relief is one-quarter wavelength, or an odd multiple thereof; under those conditions, the waves reflected from the crests and troughs would have opposite phase. The sinusoidal profile of Fig. 6.3 shifts the frequencies of the minima somewhat,[6] but the interference effect requires relief depths of the same order of magnitude in both cases.

It should be emphasized that the difference in amplitude between the direct and reflected pulses is not due to sound absorption at the test surface; the energy missing from the reflected pulse was scattered in other directions by the corrugated surface. We measure here only the specular part of the reflection (see Section IV. 3.3, Volume 2).

Finally, we include another method of measuring sound absorption that makes use of tone impulses with bell-shaped envelopes:[7] it allows a wave-theoretical evaluation, but it is based on geometric reflection from a right-angled corner, as discussed earlier in Section I. 2.2, Fig. 2.3(a). Figure 6.4, top, shows the ground plan of a flat room whose height (10 cm) is much smaller than its length (8 m) and its width (3 m). The left-

---

[6] Lord Rayleigh, *Theory of Sound*, Vol. II, 2nd. Edn., Macmillan, London, 1929, Section 272.

[7] Schröder, F. K., *Acustica*, **3** (1953) 54.

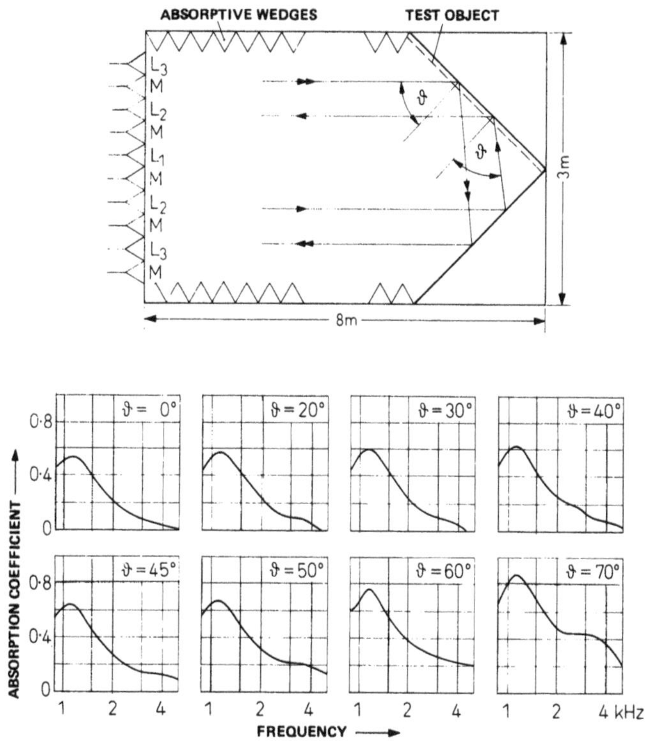

Fig. 6.4. Measurement of sound absorption coefficient with tone impulses and a 90° corner in a flat test chamber. Top: schematic sketch showing arrangement of the equipment. Bottom: examples of the test results for a foil-covered porous sheet. (After Schröder.[7])

hand wall is filled by an array of loudspeakers L_i, alternating with microphones M. When the loudspeakers are driven with the same signal, they generate a plane wave, corresponding to a bundle of parallel rays. The test object (see the dashed line) occupies one side of the 90° corner; the other side is totally reflective. The 90° corner can be turned so that the sound rays strike the test sample at different angles of incidence. It will be noted that if the direct rays strike the test sample with angle of incidence ϑ, then the reflected rays from the totally reflecting surface also strike the test sample at the same angle of incidence, and vice versa. After two reflections (one from the test sample and one from the totally reflecting surface) all the rays return to the plane of the loudspeaker array where they are picked up by the microphones. When the test sample is removed both sides of the 90° corner become reflecting surfaces and the signal picked up by the microphones, according to the principle of 'wall exchange' (see Section I. 6.2), corresponds to the incident wave. The side walls of the test chamber are covered with sound-absorbing wedges, to eliminate stray reflections.

The lower part of Fig. 6.4 shows some measured results. The test sample consisted of a layer of porous material covered with foil and mounted against a rigid wall. The abscissa indicates the frequency, the ordinate the absorption coefficient. We see that this test sample is, in effect, a resonant vibrating system, in which the foil acts as a mass and the porous layer (and included air) acts both as a stiffness and a resistance due to friction. As such, it exhibits a maximum of absorption at the resonance frequency (about 1200 Hz), above which the absorption decreases monotonically. The dependence on angle of incidence (which is the parameter for the different curves) is not pronounced; but if the peak values of absorption in the different curves are compared, it can be seen that there is again a tendency toward greater absorption for greater angles of incidence.

I.6.3 Sound Propagating at Grazing Incidence over the Audience

The chief part of the sound absorption in a concert hall is contributed by the audience and their porous clothes, and, in the unoccupied hall, by the theater seats, particularly if they are upholstered. We shall deal with these facts in detail in Chapter II.6, when we discuss the influence of sound absorption on the reverberation time in the hall. The late sound reflections of higher order that make up the reverberant sound strike *all* of the room boundary surfaces, and at *all* possible angles of incidence; for these reflections it is the total number of audience members (or of seats) that counts, practically independent of their distribution.

But the direct sound and the first-order reflections from the front wall and the side walls travel more or less in the plane roughly defined by the heads of the listeners or by the seat backs. They pass over the audience with an angle of incidence that is scarcely different from 90°; we call this 'grazing incidence'.

This holds especially for an audience seated on a horizontal floor; this arrangement was customary in the famous concert halls of the nineteenth century, because they had to serve for banquets and for dancing as well as for concerts. (For similar reasons flat floors are sometimes required even today; an example is the Beethoven Hall of the Stuttgart Liederhalle.)

According to the sound absorption data shown in the lower part of Fig. 6.1, we might assume that near grazing incidence ($\vartheta \rightarrow 90°$) the sound absorption becomes very high and the reflection more or less vanishes; but this is not the case. As we shall see from the wave-theoretical analysis of the problem, in Section IV. 7.4 (Volume 2) a sharp reversal in behavior occurs in the range of incidence angles near grazing (not shown in Fig. 6.1). Instead of strong absorption we find total reflection, but with a 180° phase shift (i.e. a phase reversal) in the sound pressure wave.

For the analysis of sound reflections in the range near grazing incidence, we will sometimes find it convenient to use the complement of the angle of incidence:

$$\varepsilon = 90° - \vartheta \tag{6.7}$$

ε is called the 'glancing angle'. The sudden shift in behavior from strong absorption to strong reflection with phase reversal occurs as ε tends toward zero.

This phenomenon is very detrimental for the audience, because the sound that they hear is the combination of the direct and reflected waves, that is, the sum of two nearly equal and out-of-phase components; the resultant wave is therefore much weaker than the direct component alone. If there were no absorption to reduce slightly the amplitude of the reflected wave, it would completely cancel the direct wave and the resultant sound pressure would vanish.

But it also follows that the resultant sound pressure increases rapidly with increasing distance above the plane where the out-of-phase reflection occurs; it recovers and becomes nearly equal to the direct pressure at a distance of about a head-height above the 'cancellation plane'.

In closed rooms, this cancellation effect is not so striking, because the listeners receive other reflections, especially from the upper parts of the hall, that have not travelled at grazing incidence over the audience. But it is very evident in outdoor situations.

Békésy[1] was the first to observe quantitatively the anomalous

[1] Békésy, G. v., *Z. Tech. Phys.*, **14** (1933) 6.

attenuation of sound propagating over an audience. The important elements of his experiment are described by him as follows:

> 'In order to determine the reduction in sound pressure by the audience, benches were set up, at 70 cm distance behind one another, outdoors; four persons were seated on each bench. In front of the first bench, a loudspeaker was placed at a height of 40 cm above the heads of the listeners. The observer took his place in the last row, at a distance of about 12 m from the loudspeaker. In order to simulate as nearly as possible the conditions of an actual theater, the seated persons were asked to choose their head positions so as to give them the best view of the loudspeaker and its surroundings. The changes of sound pressure (with height) were evaluated by the observer by means of equal-loudness comparisons: one of his ears was exposed to the sound field of the loudspeaker, the other to the sound from an earphone whose loudness could be adjusted until it matched that of the loudspeaker, as the signal was switched alternately to the loudspeaker and the earphone.'

Békésy found that the sound pressure was weak in the plane of the audience's ears and that it increased with increasing height above this plane till it reached a constant value.

These first observations, made with subjective comparisons and using only a few people, could not establish generally reliable quantitative results; but they demonstrated that the increase of sound pressure with increasing height of the observer above the heads of the audience depends on frequency. At high frequencies, the sound pressure regains the amplitude corresponding to the direct ray at very small elevations. It is obvious that such sound rays, not influenced by the presence of the audience, must be possible at distances of many wavelengths.

Schultz and Watters[2] studied a similar problem in real concert halls, by using impulse responses. They eliminated the later reflections by integrating the square of the sound pressure over only the first 50 ms, beginning with the reception of the direct sound. Furthermore, they eliminated the effect of divergence by dividing the mean measured sound pressure by the pressure to be expected at the same distance from the source in open air. Thus, their results represent anomalous attenuation not explained by the normal divergence of the sound rays.

Figure 6.5 shows this anomalous attenuation in dB, plotted as a

[2] Schultz, T. J. and Watters, B. G., *J. Acoust. Soc. Am.*, **36** (1964) 885.

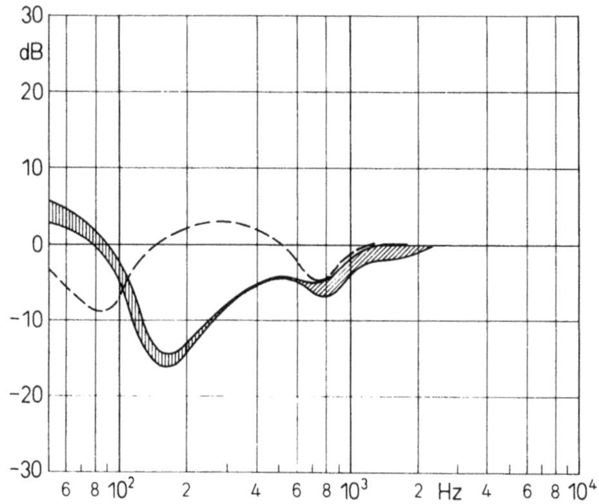

Fig. 6.5. Anomalous sound attenuation (after correction for spherical divergence) for the early sound pressure level (dB) above the audience seating in a concert hall, as a function of frequency (Hz).——— Various seats on the main floor; - - - first row in the first balcony. (After Schultz and Watters.[2])

function of frequency, for different seat locations in an unoccupied concert hall, La Grande Salle in Montreal. The solid curves indicate the range of results obtained for a number of seats on the main floor, at distances of 10 to 30 m from the stage. The fact that these curves are so similar is evidence that we are dealing with an interference phenomenon, and not simply energy absorption.

The dashed line refers to a seat in the first row of the balcony, at 30 m distance; in this case, the direct sound did not propagate at grazing incidence over any part of the audience. The pronounced dip at 150 Hz, observed in all the curves for the main floor, is not seen in the curve for the balcony seat.

The same effect was found in other halls, even in one with very thinly upholstered seats (Boston Symphony Hall); and a further study in a hall with occupied seats gave similar results.

Schultz and Watters found similar results in model tests, where the sound propagated at grazing incidence over an array of open vertical tubes (open honeycomb structure); the maximum attenuation occurred at frequencies for which the tube depth was one-quarter wavelength. They concluded that vertical wave motion can also be excited between the seat rows in a concert hall, and can, by interference, produce a pressure minimum near audience ear height. (This anomalous attenuation at rather low frequencies is the clearest example of the interference effect (see Volume 2, eqn. (7.26a)).

Sessler and West[3] found that this 'seat-dip' phenomenon is also observed (though at a somewhat higher frequency) when the seat backs, as is frequently the case, do not extend all the way to the floor.

The wave theory presented in Section IV. 7.4 (Volume 2) permits a quantitative treatment of this problem.

Schultz and Watters conclude further, from the fact that this interference effect occurs even in very highly esteemed concert halls, that it is not necessarily detrimental, at least so long as other sound reflections follow soon with non-grazing incidence. The pronounced differences between the solid and the dashed curves of Fig. 6.5 (up to 17 dB around 150 Hz) correspond to only very slight differences in the subjective impressions of music heard in the two locations.

Although the out-of-phase reflection is the most conspicuous characteristic of sound at grazing incidence over the audience, the loss of sound energy in the clothes of the listeners and in the seat upholstery also plays a significant role. For one thing, this loss is responsible for the slight reduction in amplitude of the reflected sound wave, which prevents it from completely cancelling the direct wave in the plane of the listeners' heads, and also for the fact that the resultant sound pressure never quite reaches asymptotically the direct wave amplitude at some distance (say 1 m) above the interference plane.

Furthermore, these energy losses cause the direction of sound propagation to bend slightly into the absorbing surface, corresponding to a certain amount of sound energy directed vertically into the seating area (see Section IV. 7.4).

Janowsky and Spandöck[4] have measured this inclination in a test room all of whose surfaces were covered with sound-absorbing layers of cotton batting. The upper part of Fig. 6.6 shows the equipment arrangement, with a loudspeaker at a distance *a* from the absorbing surface, and two identical microphones simultaneously connected into the receiving circuit. The lower microphone is fixed near the absorbing surface; the 30 cm higher microphone is shifted back horizontally until the two microphone signals (connected out of phase) cancel each other, signifying

[3] Sessler, G. H. and West, J. E., *J. Acoust. Soc. Am.*, **36** (1964) 1725.
[4] Janowsky, W. and Spandöck, F., *Akust. Zh.*, **2** (1937) 322.

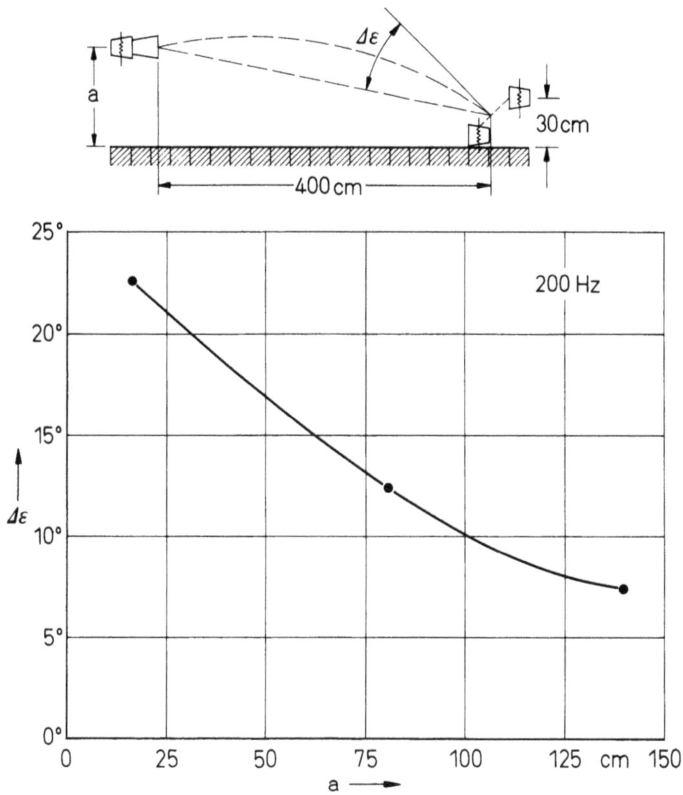

Fig. 6.6. Inclination of the wavefront toward an absorptive surface. Top: the measurement arrangement. Bottom: the experimental data, $\Delta\varepsilon$ plotted as a function of the loudspeaker height a above the absorptive surface. (After Janowsky and Spandöck.[4])

exactly equal phase in the received sound pressures. At that time the locations of the microphones define two points on the same wave front; the perpendicular to the midpoint of the line joining the microphones is, therefore, tangent to the path of energy flow toward the absorptive material. The inclination of this path with respect to the direct path is given by the angle $\Delta\varepsilon$. In the lower part of Fig. 6.6, $\Delta\varepsilon$ is plotted as a function of the source height a. This inclination illustrates how the sound wave must propagate partially toward the absorbing surface to supply it with energy lost from the direct wave.

This diversion of sound energy into the absorptive audience means an additional attenuation of sound for the audience members seated at the rear of a horizontal floor, or in any plane that is met at grazing incidence by the incident sound wave.

I.6.4 Implications for Audience Seating Arrangements

The experimental results of the last section indicate that it is preferable to aim for as high as possible a value of the glancing angle ε, particularly if reflections from the side walls and ceiling are missing, as in open-air theaters.

Evidently, this fact must have been known to the architects of the antique theaters. They stepped the seats at a much steeper rake than was required for good sightlines alone. Figure 6.7 shows the longitudinal section and ground plan for the best preserved of the Greek amphitheaters, at Epidauros in the Peloponnesus. The inner circle, built

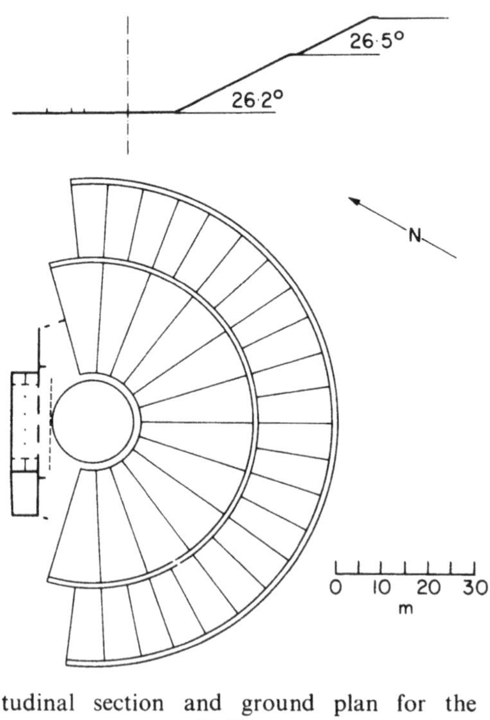

Fig. 6.7. Longitudinal section and ground plan for the amphitheater at Epidauros.

about 300 BC, has a capacity of about 6400 persons. The outer ring, built 100 years later, increased the total capacity to 14 000. The theater, after being restored to its original condition, is used today for performances; and it is astonishing how loudly and clearly each listener, even in the rear rows, can hear the sounds from the stage without any electroacoustic amplification.

This example of good acoustics does not depend only on the arrangement of the audience seating. Two further conditions are essential that are frequently lacking today: a quiet location and (despite the large number) a quiet audience. It may be that the protection of the audience from the wind, afforded by the crater-shape of the amphitheater, and even directing the airflow toward the audience by this shape, is helpful.[1]

Canac, an enthusiastic connoisseur of antique theaters, has found that the slope of the seating was so chosen that, for a speaker on the stage, even the ray reflected from the 'orchestra' floor meets the highest row of the audience with a glancing angle greater than 5°. This result could have been achieved more easily with a higher position of the actor; but the design of the steep seating rake gave, in addition, a greater advantage for the chorus located in the orchestra.[2]

A uniform seating rake, which is given by providing step-widths proportional to the seat height, is detrimental to the sightlines and 'soundlines' for seats in the rear rows, because the glancing angle is always smallest there. A more just acoustical balance can be reached if the slope of the more distant seats is increased, as is usually the case for open-air theaters built today. Figure 6.8 shows the profile of the 'Waldbühne' in West Berlin, with its three slopes of 15°, 23° and 30°.

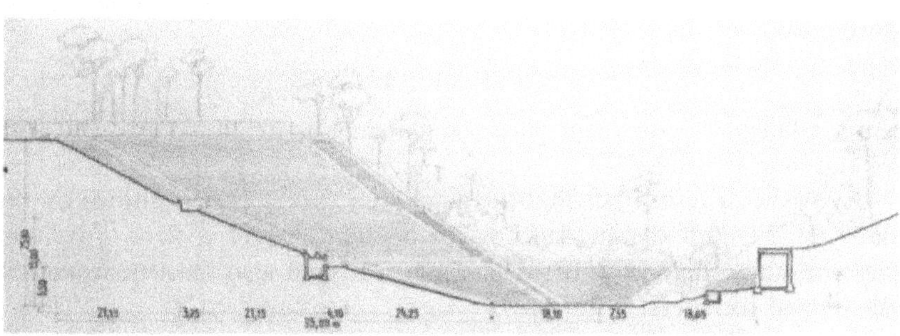

Fig. 6.8. Profile of the Waldbühne in West Berlin (architect, Werner March).

[1] Cremer, L., *Applied Acoustics*, **8** (1975) 173.

[2] Canac, F., *Proc. 5th. ICA, Liège, 7–14 Sept., 1965*, Paper G–13.

The ideal seating rake, if it could be afforded, would increase from step to step, to form a curve such that at all distances the value of the glancing angle ε is constant for a particular (important) source point. This curve is called the logarithmic spiral, and is given, in polar coordinates (r, ϕ) by:

$$r = r_0 e^{\phi \cot \varepsilon} \approx r_0 e^{\phi/\varepsilon} \qquad (6.8)$$

or by:

$$\phi = \varepsilon \ln (r/r_0) \qquad (6.8a)$$

(see Fig. 6.9). Here, r_0 is the distance from the source to the origin O where the spiral starts, and ϕ is the angle between (1) the line from the source to the origin and (2) the ray to the seat row of interest. If we

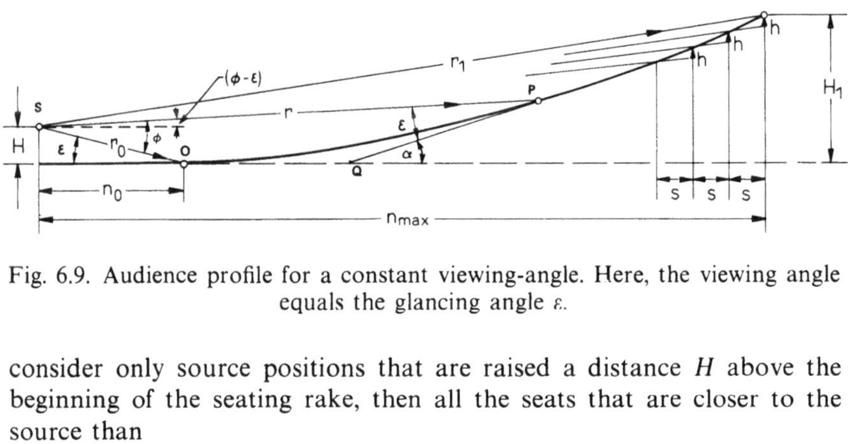

Fig. 6.9. Audience profile for a constant viewing-angle. Here, the viewing angle equals the glancing angle ε.

consider only source positions that are raised a distance H above the beginning of the seating rake, then all the seats that are closer to the source than

$$r_0 = H \cot \varepsilon \approx H/\varepsilon \qquad (6.9)$$

lie on a flat plane; for these seats the glancing angle always exceeds ε, so no stepping is needed.

But we must remember that the actors may take other positions on the stage, at different distances from the apron. Therefore, it is important that the stage slope upward toward the rear, and also that the orchestra be stepped on risers.

Equation (6.8) is not in a form convenient for construction; so, since ε and ϕ are always small, we may replace the ray-length variable r by its projection x on the horizontal plane, the x-axis, setting

$$r \approx x \qquad (6.10)$$

Then we regard the vertical component z (above the beginning of the spiral) as the independent variable of interest:

$$z \approx r(\phi - \varepsilon) + r_0\varepsilon = \varepsilon\left[x\left(\ln\frac{x}{x_0} - 1\right) + x_0\right] \tag{6.11}$$

With this formula, we can calculate the maximum height required:

$$z_{\max} = \varepsilon\left[x_{\max}\left(\ln\frac{x_{\max}}{x_0} - 1\right) + x_0\right] \tag{6.11a}$$

Now we can express the variable x in terms of the row number n, assuming that the row-to-row spacing s is determined by considerations of seating comfort and the need for adequate room to pass; it is thus constant. Then we have

$$n_{\max} = x_{\max}/s \text{ and } n_0 = x_0/s$$

where $n_{\max}$ may be estimated from the required total audience capacity and the mean number of seats per row, and x_0 is the distance from the source to the first row. Then eqn. (6.11a) can be changed to

$$z_{\max} = \varepsilon s\left[n_{\max}\ln\frac{n_{\max}}{n_0} - (n_{\max} - n_0)\right] \tag{6.11b}$$

A more suitable representation, particularly for the drawing board, would give the dependence of the inclination angle α on the number of the seat row (always including the distance from the source to the front row). Since $(\phi - \varepsilon)$, ε and $(180° - \alpha)$ are the three angles of a triangle[3] in Fig. 6.9, their sum must be 180°; from this fact a remarkably simple relation is found:

$$\alpha = \phi \approx \varepsilon\ln\frac{r}{r_0} = \varepsilon\ln\frac{x}{x_0} = \varepsilon\ln\frac{n}{n_0} \tag{6.12}$$

The acoustical condition of constant ε can equally well be regarded as an optical condition, such that the sightline to the speaker always makes the same angle with the tangent to the audience slope at the seat in question; we may thus also call this angle ε the 'viewing angle'.

Between the acoustical and optical requirements, however, there is always an essential difference: if the viewing angle is adequate, no further improvement in sightlines can be achieved by increasing the value of ε.

[3] Two angles of the triangle are at P and Q; the third angle is far off the figure to the left.

But from the acoustical standpoint, where the dimensions are judged in terms of acoustical wavelengths, every reasonable increment in ε results in improved hearing conditions.

Without question, the primary reason for raked seating in theaters and opera houses is to give good sightlines. If, in some concert halls, the seating on the main floor is not raked, this stems, as we have seen, from the requirement for other social uses of the hall (banquets and balls). In such cases, acceptable acoustical conditions on the main floor depend on the provision for sound reflections from the ceiling and other places high in the hall. But even in such halls where the main-floor hearing conditions are judged to be extremely good, this does not prove that the acoustics at the rear would not be better with raked seating. In fact, the discussion of glancing incidence, above, makes this very probable.

The rules given here for open-air theaters may also be applied for the seating in concert halls, but with smaller values of ε; otherwise, too much of the room volume needed for reverberation would have to be sacrificed. Such a rake is desirable because today it is expected that, if possible, all of the instruments of the orchestra should be seen; moreover, since halls intended only for concerts are becoming rare, the raked seating provides an advantage for audiences at events where good sight is as important as good sound.

Again, we must remind ourselves that the visual and acoustical requirements do not always coincide. With respect to the viewing angle, it makes a difference whether the sightlines are perpendicular to the seating rows or oblique; in the latter case the viewing angle becomes significantly smaller. To be sure, the same relation holds for the glancing angle ε of the direct sound. But in the acoustical case, it is also essential to take into account the number of seat rows between the listener and the stage; this is true not only because of the possibility of energy losses but also because of the potential for interference effects when sound is incident at near-grazing angles. The interference effect is independent of the direction of incidence, but it requires a certain number of seat rows for its full formation.

Balcony seats, even in the second row and in the side balconies, are acoustically privileged because they receive the direct sound unimpeded by energy losses or by interference effects. By contrast, the sightlines in the second row of a balcony may suffer, and, particularly in a side balcony, they may be very poor.

The disposition of the audience on a large, fan-shaped surface may yield excellent sightlines to the stage, but it provides an acoustically monotonous solution. No sound reflections are returned toward the stage to give the performers a desirable sense that the room is responsive to their efforts; and no early (lateral) sound reflections are provided for the seats in the critical area in the center of the main floor. All this is changed if the large seating area is interrupted with steps, as in the Berlin Philharmonie, where the steps resemble the steps on a hillside vineyard.

Figure 6.10 shows an application of this principle in the Mozart Hall

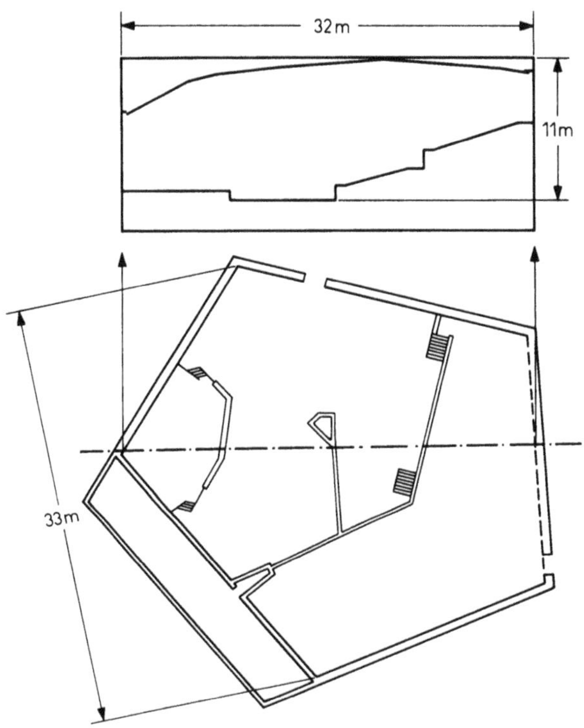

Fig. 6.10. Longitudinal section and ground plan for the Mozart Hall in the Liederhalle, Stuttgart (architects, Abel and Gutbrod).

in the Liederhalle in Stuttgart, designed by the architects Abel and Gutbrod. Figure 6.11 shows the longitudinal section of the Berlin Philharmonie, designed by architects Scharoun and Weber. The rays represented by dashed lines indicate that the first row of each 'block' of seating gets unimpeded direct sound, as in the first row of a balcony; the

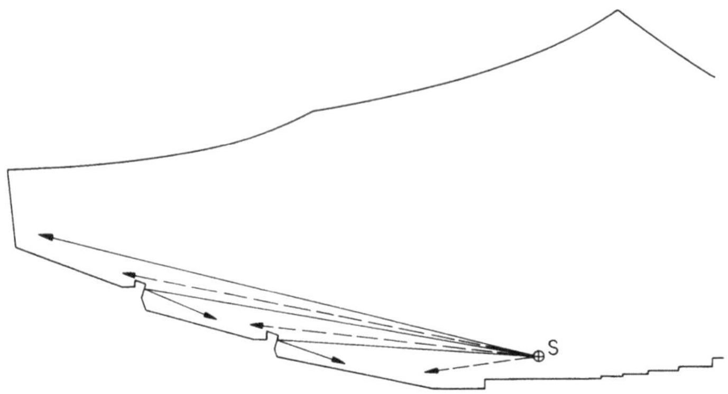

Fig. 6.11. Ray constructions for the vineyard steps of the Berlin Philharmonie.

solid rays show that the back seats in each block profit by the sound reflections from the step lying behind, especially if the step is inclined forward slightly.

(No doubt a ray construction such as this, dealing with rather small reflecting surfaces, does not strictly comply with the requirements of geometrical acoustics, as outlined in the earlier chapters of this book; but it serves, nevertheless, to point the way toward reasonable approaches to concert hall design.)

Figure 6.12 makes it clear that the vineyard steps in the Berlin Philharmonie interrupt the seating area in the lateral direction as well as in the longitudinal direction. The unusually great width of the hall and the central position of the stage make these lateral interruptions especially important in providing early reflections both for the musicians and the audience seated in the middle of the hall.

Although, strictly speaking, it may not be a proper concern of this book, two non-acoustical aspects of the vineyard steps and the seating blocks may be mentioned. If the listener, because of the steps, does not see all the seat rows between himself and the stage, he gets the impression that he is closer to the performance. Furthermore, the members of an audience distributed in seating blocks, as in Berlin, neither feel themselves submerged in the multitude, as in the antique amphitheaters, nor 'socially classified' as in the Baroque theaters, where a view of the Royal Box was more important than the view of the stage. They feel a sense of belonging to a group that is neither too large nor too small, a relationship that may be especially appropriate for the society of today.

The division of the audience by vineyard steps and blocks of seating

Fig. 6.12. Photograph of the stage and most of the audience in front of the stage, in the Berlin Philharmonie (architects, Scharoun and Weber).

also has advantages for congresses, where provision must be made for discussion from the floor. This seating layout offers opportunities for the placement of microphones throughout the seating, and a speaker can be seen and aurally localized (see Section I. 5.2) from nearly all other seats much better than when the seating is spread over a large single surface.

Chapter I.7

Model Tests

I.7.1 Light-ray Models[1]

In Section I. 1.3 it was pointed out that the propagation laws for sound waves and for light waves are similar. In fact, in order to represent the behavior of sound at high frequencies, we took over the concept of the 'sound ray' from familiar light rays, which can readily be made visible. This analogy further suggests the possibility of studying the paths and reflections of sound rays in rooms by the use of light sources and optical mirrors, to replace the sound sources and acoustical reflectors.

For such studies, small-scale models of the room are sufficient; in fact, it is seldom necessary to model the entire room, since we are interested only in first-order (or, at most, second-order) reflections. Consequently we are usually interested in a *particular* reflecting surface and the particular corresponding part of the room where the reflection falls.

In many cases a small-scale model of the hall will be built for architectural (or fund-raising) reasons anyway, particularly for large projects. With proper planning and coordination, the same model may serve for acoustical/optical tests, to help guide the design of the hall.

In the light model, the interesting sound reflecting surfaces (and this

[1] The reader will find a number of useful papers on model studies of room acoustics in the following references: 'Use of acoustical scale-models for the study of studios and concert halls', *Tech. Report 3089–E*, October 1969, Technical Centre of the European Broadcasting Union, 32 Albert Lancaster Avenue, Brussels, 18, Belgium; *J. Acoust. Soc. Am.*, **47** (1970) 399; Mackenzie, R. (ed.), *Auditorium Acoustics*, Applied Science Publishers Ltd, London, 1975, pp. 55–118; Lamoral, R., *Acoustique et Architecture*, Masson, Paris, 1975, p. 60 ff.

usually means those on and near the ceiling) must be 'mirrorized', so that they reflect light rays specularly, with ideal geometrical reflection. This may be done either by gluing mirrored foil on the surface, by making the surface itself out of polished sheet metal (tubing in the case of pillars and columns), or by spraying the surface with glossy paint or varnish.

As mentioned above, the sound source is replaced by a light source. If the exact location of a particular reflection, or the path of a special ray, is of interest, it can be studied with a single incident ray only; recently laser beams have been used for this purpose, because their extreme brightness and the tight collimation of the beam makes possible a very sharp image.

Figure 7.1 shows a multiple light-ray path between a flat floor and a spherical ceiling, similar to the example mentioned in Section I. 4.3, and shown in Fig. 4.8. The floor and ceiling surfaces were mirrored, while the wall surfaces (of no interest here) were painted flat black; smoke was blown into the model to make the light rays visible.

One can go a step further and study the distribution of reflections from

Fig. 7.1. Light-ray model (1:25) of a room with a plane floor and spherical ceiling ($r=2h/3$).

an entire ceiling, moving the source around to all the locations of interest. Figure 7.2 shows a photograph of a model for the recently restored National Theater in Munich. Its former spherical ceiling was represented in the model by a concave mirror made of plaster with an evaporated metal film. The light source for this photograph was placed on the stage. It is evident that the light rays reflected from the ceiling illuminate preferentially the Royal Box, its surroundings, and the rear-most rows on the main floor.

Fig. 7.2. Optical investigation of the reflection from a spherical ceiling. The direct light from the source (in the middle of the stage) is prevented from reaching either the observer or the audience area by a screen.

For architects, such demonstrations are very convincing, and in the restoration the new ceiling was given a nearly flat shape, in order to distribute the sound reflections more uniformly to the audience. The visual appearance of the original dome was retained by skilful painting.

Vermeulen and De Boer[2] not only used the light-ray model to trace

[2] Vermeulen, R. and De Boer, I., *Philips Technische Rundschau*, 1 Jahrgang, 1936, p. 46.

the paths of geometrical reflections, but they also tried to evaluate the distribution of the sound intensity on the basis of the variation of brightness of the light reflections. For this purpose, they replaced the audience on the main floor with a sheet of ground glass.

Unfortunately for this experiment, the brightness of the light rays on the ground glass does not match the incident sound intensity. Instead, it diminishes with the cosine of the angle of incidence, and thus it vanishes for grazing incidence. Since in their experiment the ground glass represents the audience area, Vermeulen and De Boer compared this cosine-law behavior of the light reflections to the reduction of sound pressure striking the audience area at grazing incidence. But we know, from Section I. 6.3, that the behavior of sound in the latter case, and the reasons for it, are too different for such a comparison. On the other hand, ceiling reflections, as studied in Fig. 7.3, are not affected by this problem, because of the oblique incidence. Figure 7.3 shows such an arrangement for the investigation of the ceiling of the Meistersinger Hall in Nürnberg.

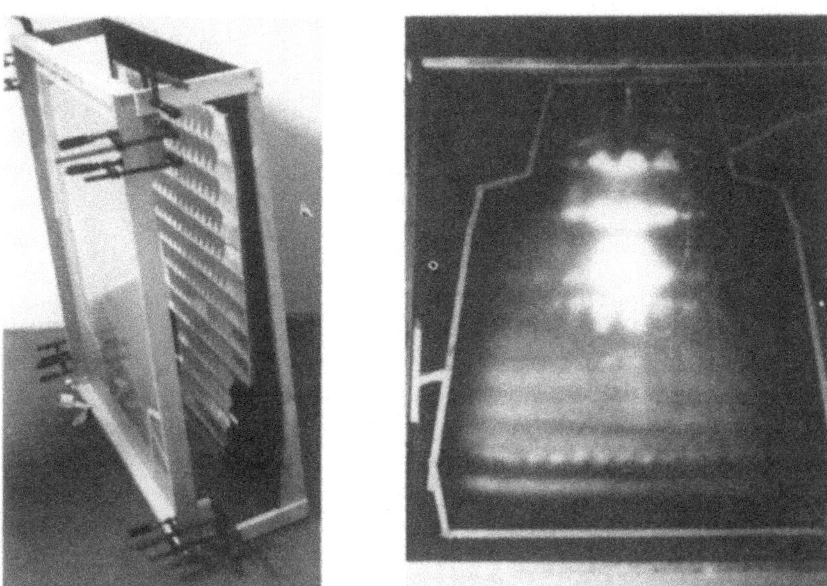

Fig. 7.3. Arrangement for the study of a diffuse-reflecting ceiling. The light source is placed on the stage; it is shaded so that its direct rays do not strike the screen. The purpose of the experiment is to study the distribution of brightness on the screen.

An essential disadvantage of all light models is that they cannot provide information concerning the time delay associated with a reflection; and time delays can be of critical importance in the subjective evaluation of sound reflections, particularly echoes. One might suppose that a brighter reflection corresponds to less ray divergence, and therefore a shorter path and a shorter time delay; but it could also mean the superposition of a number of different rays arriving at very different times. Thus, the brightness of an optical reflection by itself is of no help in identifying disturbing echoes. In order to get valid information on the time-of-arrival of reflections, the light-ray studies must be supplemented by measurements of the path lengths corresponding to the reflections under study.

Finally, the wavelengths of light are so short that they give a reasonable account of the behavior of sound waves only for very high frequencies; they do not exhibit the diffuse reflection and diffraction that are to be expected with low-frequency sound waves. For the evaluation of light-ray test results, one must always check to determine the frequency range to which they may be applied, as described in Section I. 1.5.

Optical model studies cannot yield any results that cannot, in principle, be found by ray constructions on the drawing board. But they can frequently produce the desired information for a variety of reflector shapes and source positions in a much shorter time, and can also provide an informative overall view that is not practical in ray diagramming. Certainly, they offer the least expensive method for ray studies.

I.7.2 Ray Diagramming with the Aid of a Computer

We have already mentioned, at the end of Section I. 5.3, that ray construction at the drawing board and light modelling of sound waves, can in principle be replaced by calculations made with the aid of a computer, even for reflections up to high orders. In computer calculations, all the sound rays and their reflections, as well as all the reflecting boundaries, are expressed by analytic representations of straight lines and (mostly) plane surfaces. The problem with computer solutions is how to present and to evaluate the results.

Krokstad *et al.*[1] have developed a computer program whereby many

---

[1] Krokstad, A., Strøm, S. and Sørsdal, S., *J. Sound Vib.*, **8** (1968) 118.

rays, radiated from a source in all directions at equal increments of solid angle, are traced to their ultimate intersection with the audience plane in discrete time intervals. In this program, the audience is treated as totally absorptive, even for grazing incidence; once a ray meets the audience it is forgotten. Figure 7.4 (top left) shows the geometrical data for the example. For simplicity, a rectangular room is chosen, with the source in the middle of a small wall, near the floor. The diagrams below show the intersection points of rays with the audience plane that occur in different time intervals after the arrival of the direct sound, as labelled. (Note that

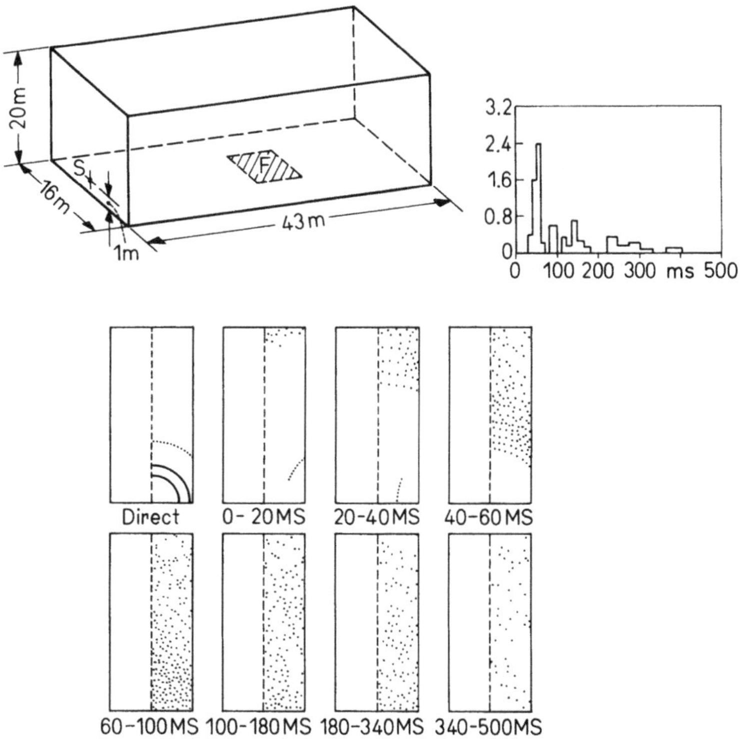

Fig. 7.4. Computer-calculated intersections of sound rays, from an omnidirectional source, with the audience plane. Top left: room dimensions for the example. Bottom: audience intersection points in different time intervals following the arrival of the direct sound. Top right: number of ray intersections with the surface element *F* in each time interval, as a function of time. (After Krokstad *et al.*[1])

the time intervals are not equal. The left-hand half of the room, shown blank in each case, would exhibit a mirror image of the right-hand half.) The circular wavefront of the direct sound is evidently constructed as in Section I. 1.3.

The fact that between and behind the 'direct sound' circles there appear large 'white areas', so that the direct sound seems a rather unimportant contribution to the sound field, is connected with a shortcoming of this kind of representation. It ignores the fact that the distances between the intersection points for two neighboring rays depends on the mean angle of incidence ϑ and increases with $1/\cos\vartheta$. (Here again, this cannot be regarded as the equivalent of the reduction of sound pressure amplitude for a wave travelling at grazing incidence over the audience.) Since the angle of incidence is always known, it may be possible—and better—to replace the intersection 'point' with an intersection 'line', having the length $1/\cos\vartheta$.

Those who are familiar with this peculiarity of the representation, however, will, nevertheless, be able to evaluate it adequately, at least with respect to changes in hall shape.

For instance, it is impressive that, during the time interval between 20 and 60 ms, the second-order reflections from the corner between the ceiling and the rear wall are beginning to reach the audience in the front half of the hall.

There is some question of the validity of plotting versus time the number of ray intersections with a small floor area F, as is done in the upper right of the figure. The distances between such intersections in the early time intervals is only a consequence of ignoring the influence of the angle of incidence.

Such a time function is not equivalent to an echogram, even though the later decay shown here depends on the sound absorption (introduced here in the assumption of a totally absorptive audience for all angles of incidence).

It may be mentioned here that Krokstad *et al.* have sometimes carried out the calculation represented in Fig. 7.4, top right, with the surface area F replaced by a hemisphere or a small cube above the floor. This appears more reasonable, although it is still far from being an adequate representation of the sound pressure at a listener's position.

On the other hand, it is easy to extend computer calculations to include the dissipation of sound energy during propagation between boundaries, to introduce values of sound absorption coefficients between the limits 0 and 1, and also to allow for diffuse reflections; the dependence of all of these effects on frequency can be included.

It is impossible to say at the present time how near to perfection computer models may come in the future. But, based on experience to date, one can expect that each new advance in calculation flexibility will be won at the cost of a certain arbitrariness; thus, for room-acoustics planning (see Section I. 7.5), the pure acoustical models may be more easily understood and also may be closer to reality. For systematic investigations, to study the effects of numerous small variations in a design, the computer simulations may be more useful.

I.7.3 Water-wave Models[1]

We have already seen in Fig. 1.2 (Section I. 1.3) that when we study wave motions at reduced scale, we can account for both diffraction and diffuse reflection. Furthermore, water-wave models avoid the shortcoming of light-ray models (namely, that 'everything happens at once'), since the propagation speed can be reduced to 20–30 cm s^{-1} (less than 1/1000 that of sound). Then the water-wave propagation is easily followed by the eye. These characteristics make water-wave models particularly attractive for demonstrations.

The apparatus is inexpensive. We need only a plane-parallel glass plate that can be illuminated with parallel rays from below. It must be framed to a certain depth so that the whole plate can be covered with water. According to Michel, a water depth of 7–8 mm is recommended. In this shallow water basin, sheet-metal models (like over-sized cookie-cutters) of the longitudinal sections or the ground plans of the rooms to be tested can be installed. To simulate sound-absorptive surfaces, or to avoid unwanted reflections at the frame boundaries, one can place a wedge-shaped strip of felt in front of the surface; the water waves will be absorbed in this strip as on a boggy shallow bank.

To excite a single water-wave impulse, one can either quickly withdraw an already submerged small object, or let fall a drop of water upon the surface. The illumination from below with plane-parallel light rays makes the water waves visible as on a screen, and the waves may be readily photographed. The sloping surfaces of the water waves cause

[1] As early as 1843, Scott Russel used water waves for room-acoustical investigations. One can find detailed descriptions in Michel, E., *Hörsamkeit grosser Räume*, Viewig, Braunschweig, 1921, and in Davis, A. H. and Kaye, G.W.C., *The Acoustics of Buildings*, Bell, London, 1927; in the latter work the water-wave model apparatus was called a 'ripple tank'.

refraction and produce stripes that are brighter or darker than the surrounding areas. The low propagation speed of the water waves makes it easy to distinguish reflections following one upon another. This also simplifies the photographing of successive instants with different time delays.

Figure 7.5 (bottom) shows a photograph of a water-wave model of a hall at the instant when a ceiling reflection approaches the floor,[2] in comparison with a Schlieren photograph (see Fig. 7.5, top, and Section I. 7.4).

This figure demonstrates a disadvantage of the water-wave model technique. Although only a single excitation impulse was used, we see several reflected wavefronts with different wavelengths, which become shorter in the direction of propagation. The short wavelengths are

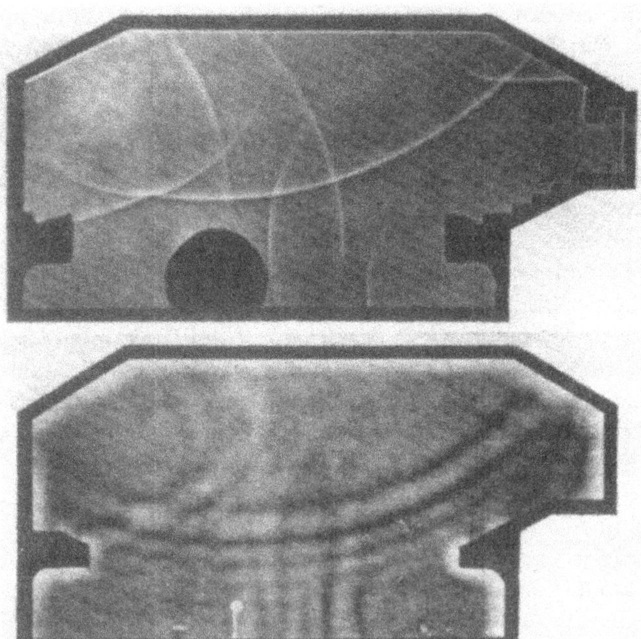

Fig. 7.5. Comparison of wave photographs of the same hall at the same instant. Top: Schlieren photograph. Bottom: photograph of a water-wave model. (After Davis and Kaye.[1])

[2] Taken from Davis and Kaye.[1]

travelling faster than the long ones. In Section I. 1.1 we mentioned that the propagation speed for water waves depends on the wavelength. But, for simplicity, we described there only the influence of gravity upon the waves: in that case, the *long* wavelengths travel faster. For very short wavelengths, by contrast, we must also take into account both the effects of surface tension, which increase with increasing wave curvature, and the corresponding effect on the acceleration forces. The result is that, as in Fig. 7.5, the waves with short wavelengths travel faster. Between those two regimes there is a zone where the two effects tend to cancel one another and the propagation speed is at a minimum. Fortunately, the wavelengths encountered in water-wave model testing mostly tend to fall into the middle zone, and the propagation speed is relatively constant. Nevertheless, the dependence of propagation speed on frequency (that is, dispersion) cannot be entirely neglected.

To reproduce diffraction and diffuse reflection phenomena exactly in the water-wave model, it is necessary that the ratio of the wavelength λ_m in the model to a characteristic length l_m in the model be the same as the ratio of these two lengths, λ_w and l_w in the full-scale situation:

$$\frac{\lambda_m}{l_m}=\frac{\lambda_w}{l_w} \tag{7.1}$$

or

$$\frac{\lambda_m}{\lambda_w}=\frac{l_m}{l_w}=\mu \tag{7.2}$$

The subscript w stands for *Wirklichkeit* = reality. The constant μ is the scale factor.

Since the wavelength of water waves is around 1–2 cm and we may be interested in sound wavelengths of around 1 m (so as to include the possibility of diffraction phenomena but to stay near the middle of the frequency range, say 300 Hz), a scale reduction of 1:100 to 1:50 is recommended.

It is unfortunate that the advantages of the water-wave model can be applied only to the study of two-dimensional problems: the propagation of sound waves in a single plane.

I.7.4 Sound Pulse Photography ('Schlieren' Photographs)

When we turn from water waves to sound waves, dispersion does not occur; the propagation speed is the same for all frequencies. We can profit from the fact that the wavefronts in airborne sound can also produce refraction in light rays, just as did the water-wavefronts in the ripple tank. Specifically, Toepler showed in 1864 that, when parallel light rays cross a sound field in a direction perpendicular to that of the sound rays, the part of the sound-wavefront that is met tangentially by the light rays produces two visible lines, close together, one light, the other dark, on a projection screen behind the sound field. There, the light image of the sound-wavefront may be studied directly or photographed.[1] This method, like the water-wave method, is restricted to the study of waves in two dimensions only; if the light rays pass a spherical wavefront twice, for example, the opposite refractions cancel one another and no visual image is produced.

Figure 7.5, top, shows a Schlieren photograph of wavefronts in a room with the same configuration, and at the same comparative instant, as the water-wave representation in the lower part of the figure. The superior clarity of the sound pulse photograph is evident.[2]

For such a sharp representation of the wavefront, a very short sound impulse is necessary; in this case, a spark was used with a length of 2·5 cm. The wavefront of the spark wave is cylindrical along the length of the spark, but spherical at the ends. The light rays that are tangential to the cylindrical portion of the spark-wavefront (in the direction along the spark axis) generate the aforementioned dark and light lines, each of about 1 mm thickness.

Since these images travel on the projection screen with the speed of sound, the exposure time used for photographing the image must be very short, of the order of 10^{-6} s, corresponding to a resolution of approximately 0·34 mm. In that way, the 1–2 mm wavefront images under study are sharply rendered, without blurring. For this purpose, Toepler and a number of his successors have used a second spark for the source of light for the Schlieren studies. The photographic screen (or plate) must be shielded from the direct light of the spark; otherwise, this intense illumination would wash out the faint refracted image of the sound pulse.

By changing the elapsed time between the spark that generates the sound pulse and the spark that illuminates the wavefront, one can observe different instants in the progress of the sound wave. Also, by

---

[1] In Germany, these images are called *Schlieren* (literally, 'streaks'), and the method is called Toepler's Schlieren Method.

[2] Again, taken from Davis A. H. and Kaye, G. W. C., *The Acoustics of Buildings*, Bell, London, 1927, Fig. 32.

generating several repetitions of the same sequence of sparks, one can prolong the exposure and produce a semi-permanent image on a fluorescent screen. For this purpose, it is important that the difference in time between the two sparks remains exactly the same; otherwise, the location of the wavefront image will not be stable and the picture will blur.

On the other hand, if the time between sparks is gradually increased in the course of a sequence of exposures, we get a stroboscopic effect, and the image of the sound-wavefront appears to propagate on the screen. Since we can, by this means, make the apparent propagation speed as slow as we like, we can regain the advantage of the water-wave model and readily follow the sound propagation progress by eye.

For room-acoustical model tests, Schlieren photography was first used by Wallace Clement Sabine, in 1912.[3] The basic principle has remained unchanged since that time. The spark that generates the sound impulse is surrounded by a reduced-scale wooden model in the shape of the room profile to be studied: it may be imagined as a small vertical wall following the outline of the ground plan (or the longitudinal or transverse section) in the architectural drawings of the hall.

In this case, the model may be smaller by a factor of ten than the corresponding water-wave model, a considerable additional advantage. This small scale is dictated, on the one hand, for photographic reasons (a small screen is easier to illuminate and produces sharper images), and on the other hand, by the fact that the sound waves are attenuated both by three-dimensional divergence and by losses in the medium during propagation, an effect that can no longer be neglected at the very high frequencies of the spark impulses (see Chapter IV.14, Volume 2). On the other hand, this is no real disadvantage, since in geometrical room acoustics we are interested only in the earliest reflections.

The extraordinary sharpness of the spark allows this reduction in size of the model without violating the law of similarity of wavelengths. The compression phase of the spark impulse has a duration of about 3×10^{-6} s (the subsequent rarefaction phase is longer, but it is of no interest in the present context). This duration corresponds to an impulse length l_m in the model of about 1 mm. The corresponding impulse length l_w in the original room is:

$$l_w = l_m / \mu \tag{7.3}$$

[3] Sabine, W. C., *Collected Papers on Acoustics*, No. 6, Harvard University Press, Cambridge, 1927.

Since the speed of sound is the same in both the model and the original room, the impulse duration in the original room is:

$$t_w = t_m/\mu \tag{7.4}$$

With $\mu = 1/500$, we get $t_w = 1{\cdot}5 \times 10^{-3}$ s, an impulse duration that is comparable to that of the pistol shots frequently used as the sound source for echograms in buildings. This implies that Schlieren photography, with the parameters assumed above, represents conditions in the original room for frequencies around 1000 Hz, corresponding to the maximum of the speech spectrum.

The clarity with which the Schlieren method depicts sound waves reflected from even rather small obstacles (columns, coffers, etc.), and its accurate representation of sound diffraction and of diffuse reflections, makes the method especially suitable for studying these details in room-acoustical investigations.

Figure 7.6, right, shows two Schlieren photographs made by Osswald.[4]

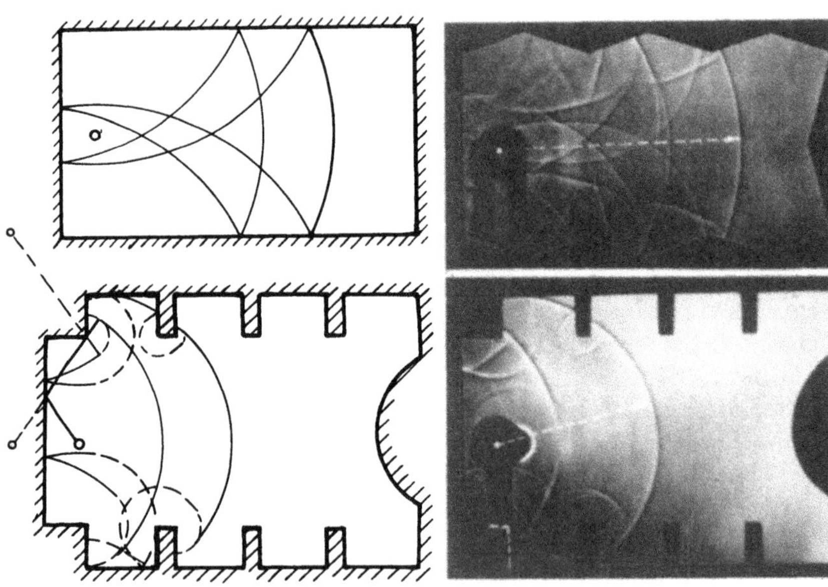

Fig. 7.6. Comparison of wavefront constructions (left) and Schlieren photographs (right) of two models with strongly modulated side-walls. (After Osswald.[4])

[4] Osswald, F. M., *Z. Tech. Phys.*, **17** (1936) 561.

In Fig. 7.6, top right, the walls of the room are zigzag in shape. The sketch at the left shows the wavefront construction for a similar configuration with plane walls. The Schlieren photograph exhibits clearly the effect of the diffuse reflecting walls in increasing the uniformity of filling the room with wavefronts.

Figure 7.6, bottom, shows the diffuse reflecting effects of coffered side-walls. One sees clearly the sources of the individual reflected waves; they fall into two groups. On the one hand, there is a group of waves corresponding to images in the plane (uncoffered) walls; they radiate from the image sources with the same radius of curvature as from the original source. On the other hand, there is a group of smaller circular wavefronts, centered on the forward edges of the coffers and the projecting edges at the front of the hall (which form strong shadows for the primary image waves). They remind us of the secondary sources of regenerated wavefronts required by Huygen's principle, constructed, in this sense, as dashed circles in the sketch at the left. The agreement with the Schlieren photograph at the right is good. It shows that it is sometimes possible to represent diffraction and diffuse reflections correctly in wavefront construction drawings, but the construction is in general quite tedious.

Finally, Schlieren photographs can be analysed with respect to the light intensity, in order to draw quantitative conclusions about the strength of the sound wavefronts.

I.7.5 Three-dimensional Models with Frequency-transformed Sound Field

Of the modelling methods mentioned so far, only the light-ray model is widely used nowadays, largely because of its simplicity. The restriction of the other two methods to two-dimensional wave propagation is a serious limitation for room-acoustics design; they have fallen from favor since the development of three-dimensional small-scale acoustical models.

Although the three-dimensional acoustical model does not permit us to visualize the sound propagation, we can measure and record the sound pressure at locations in the model corresponding to the same locations in the proposed room.

In this procedure, sound waves are radiated into the model room with wavelength λ_m which must be related to the wavelength λ_w in the 'original' room in the same ratio as the dimensions of the model and the original rooms, as we have mentioned in Section I. 7.3:

$$\frac{\lambda_m}{\lambda_w} = \frac{l_m}{l_w} = \mu \tag{7.5}$$

The relationship must hold throughout the audio-frequency range for the original room. This will be possible if the model-frequency $f_m = c_m/\lambda_m$ (c_m = speed of sound in the medium of the model) and the frequency in the original room $f_w = c_w/\lambda_w$ (c_w = speed of sound in the original room) are transformed according to:

$$\frac{f_m}{f_w} = \frac{c_m}{c_w} \cdot \frac{1}{\mu} \tag{7.6}$$

If the model room is filled with air, which is often the case, then $c_m = c_w$ and the frequency transformation is simplified to

$$\frac{f_m}{f_w} = \frac{1}{\mu} \tag{7.7}$$

With a scale factor of $\mu = 1/10$, all the frequencies used in the model must therefore be a factor of ten higher than the frequencies in the original room.

The frequency range of greatest interest in the original room, namely, 63–4000 Hz, is shifted to 630–40 000 Hz in the model testing.

Spandöck,[1] and later Jordan,[2] were among the first to prove that it is possible to carry out such model tests with electroacoustical sources and receivers. Not only can we record echograms for *objective* testing, using sparks for the sound impulse sources, but it is also possible to make a subjective evaluation of the sound field (reverberation, etc.) by the use of speed-scaling in the tape-recording/playback procedures; naturally, the same scale factor μ is involved.

The principle of scaled room-acoustical models was used by Meyer and Bohn[3] for studying the sound-reflection processes at special wall structures; and Reichardt[4] applied this model technique for testing the room shapes of both the Staatsoper and the 'Haus des Lehrers' in East Berlin, which we have already encountered in Fig. 3.8.

---

[1] Spandöck, F., *Ann. Phys. (Leipzig) V*, **20** (1934) 345.

[2] Jordan, V. L., *Elektroakustiske Undersögelser af Materialer og Modeller*, Reitzels Forlag/Axel Sandel, Kopenhagen, 1941.

[3] Meyer, E. and Bohn, L., *Acustica (Akust. Beiheft)*, **2** (1952) 194.

[4] Reichardt, W., *Wissenschaftl. Zeitschr. der Techn. Hochschule Dresden*, **5** (1955/56) 1.

Today it is seldom that a major concert hall is built without a room-acoustical model test of some kind.[5,6,7]

If we are interested only in the time sequence of early reflections from the room boundary surfaces in a frequency range of one to two octaves, it is sufficient to make the walls and ceiling of the model out of wood-chip panels, covered with impermeable plastic sheets, in order to minimize the sound absorption. The audience areas must be covered with porous material to simulate the sound absorption of the audience. For this purpose, it has been found that cardboard egg cartons (egg-side down!) are suitable; they even simulate the partially diffuse reflections from the listeners. Figure 7.7 shows an example of their use in a model at a scale of 1:16.

Fig. 7.7. Model of a multi-purpose hall at a scale of 1:16. Cardboard egg cartons are used to simulate the audience absorption.

[5] Reichardt, W., Budach, P. and Winkler, H., *Acustica*, **20** (1968) 149.
[6] Cremer, L. Keidel, L. and Müller, H. A., *Acustica*, **6** (1956) 466.
[7] Cremer, L., *Deutsche Bauzeitung*, (1965) 3.

Figure 7.8, top, shows an echogram recorded in the model of the Meistersinger Hall in Nürnberg; here the sound pressure is plotted against time. We see first (at the left) the direct sound (with some immediate reflections), then the typical gap before the arrival of reflections from the side walls and ceiling, and finally reflections of higher order with increasing density. As we have already discussed in Section I. 4.3, the occurrence, in this type of presentation, of pronounced peaks with large time delays following the direct sound would warn of the possibility of critical reflections that may be perceived as echoes.

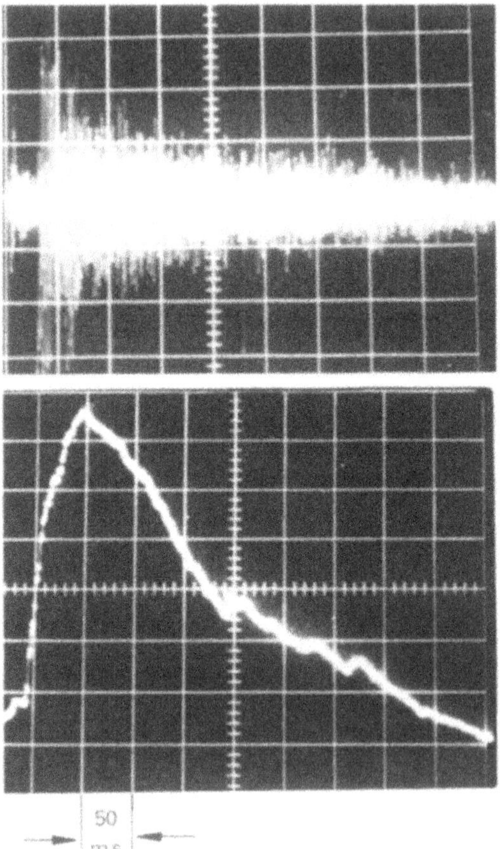

Fig. 7.8. Impulse responses in a model hall with a scale of 1:10 (the Meistersinger Hall, Nürnberg). Top: echogram showing sound pressure versus time. Bottom: squared sound pressure versus time (integrated by a circuit with 5 ms time constant).

These patterns of sound pressure versus time, if they are turned through 90°, resemble fir trees; accordingly, they are called fir-tree patterns. They can be recorded using normal condenser microphones in the (model) frequency range of interest, see Fig. 7.9, center. A spark-source is shown in Fig. 7.9, left; between the electrodes a small rod of graphite, for stabilizing the spark, is to be seen. Instead of using sparks for the sound source in the model, noise impulses radiated from an ultrasonic loudspeaker may also be used, see Fig. 7.9, far right.

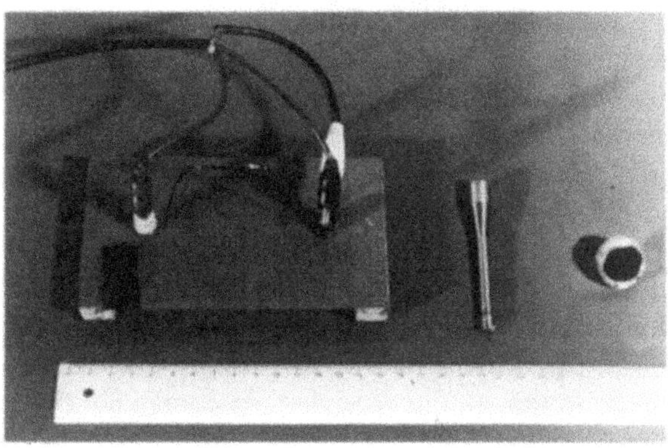

Fig. 7.9. Spark source (left half of photo), $\frac{1}{4}$in. condenser microphone, and ultrasonic loudspeaker (15–120 kHz).

In order to get a representation that corresponds to our subjective impression of the room impulse response, it is reasonable to rectify the sound pressure and to integrate the impulse over a number of closely spaced individual reflections (see Sections II. 7.2 and III. 1.4). In Fig. 7.8, bottom, the echogram that was shown at the top of the figure is squared (a simple way of rectifying the signal) and the spikes are smoothed out by an integrating circuit with a time constant of 5 ms.

The models for this kind of investigation, at a scale of 1:16, or even 1:10, are still rather large; thus, it is possible for the investigator actually to creep inside the model and to judge by eye which surfaces are likely to reflect useful sound and which may be detrimental because of excessive time delay (echoes).

A useful tool for this purpose is the 'time-delay tape-measure', on which (taking the model scale factor into proper account) the centimeter scale is replaced by a corresponding scale of transit time for sound propagation (marked in milliseconds).

In use, one first spans this tape measure from the sound source location to the receiver point of interest in the audience, and then increases the length of the tape by an amount corresponding to the limit of perceptibility in the model (17 m in the original room, for speech, or $\Delta t = 50$ ms). Now, with one end of this lengthened tape fixed at the source position and the other end at the receiver position, one investigates which room surfaces can be touched with the tape. Any surface that can be touched is a candidate for reflecting useful sound, provided that it is angled in the right direction. Surfaces that lie beyond the reach of the tape will send reflections to the receiving point with time delays that exceed the limit of perceptibility. This is a quick method of evaluating in three dimensions the important sound-reflecting surfaces in a proposed hall. A comparable analysis on the drawing board would be much more arduous and not so convincing.

The same technique can be used to identify the offending reflecting surface when one finds a pronounced echo in the fir-tree pattern. In this case, the length of the tape measure is chosen to be the sum of the time delay for sound direct from source to receiver and the time delay between the direct sound at the receiving position and the arrival of the problem echo.

In general, we are interested only in reflections arriving within about 100 ms of the direct sound (full scale), and thus it is not necessary that the absorption coefficients of the reflective surfaces be accurately simulated in the model.

On the other hand, it is possible to make a room acoustical model in such a way that its reverberation time (see Section II. 1.1) corresponds to that of the original, at least in the most interesting mid-frequency region. In this case, it is essential that the absorption coefficients of the materials used in the model be properly scaled to the coefficients of the corresponding surfaces in the original room. This is also necessary if the effect of coupling the main room to an adjacent room is to be studied.

Figure 7.10 shows the decay of the logarithm of sound pressure (the so-called 'sound pressure level', see Section III. 1.2) versus time, recorded in a model of a multi-purpose room which was coupled to an adjacent large lobby. If this decay curve obeyed the rules for simple, uncoupled rooms, it would be a straight line. Instead, we can see first a rather rapid decay corresponding to rapid absorption of sound energy in the main room, followed by a slower decay corresponding to the longer re-

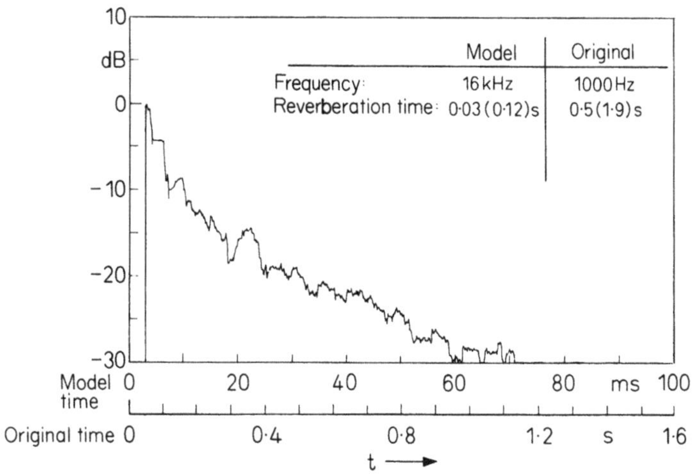

Fig. 7.10. Decay of sound pressure level in the model room of Fig. 7.7, when coupled to an adjacent lobby; scale = 1:16.

verberation in the adjacent lobby. The model scale was 1:16; the record was made at 8000 Hz, corresponding to 500 Hz in the original rooms.

Spandöck[8] and his collaborators[9] even attempted to model sound absorption coefficients over the entire audible frequency range. For this purpose, they developed a comprehensive table of absorption coefficients of model materials and surface treatments, such that one may choose materials for the model that exhibit the same absorption coefficients at the model frequencies f_m as are to be expected later in the full-scale building at frequencies f_w.

A serious difficulty arises for model experiments in which the influence of higher order reflections is of interest. This concerns the rapid attenuation of high-frequency sound as it propagates through the air, between reflections at the boundaries. (See Section II. 1.6, and Chapter IV.14, Volume 2.)

So long as we are content to record impulse responses only, this anomalously high attenuation with distance can be compensated for by

[8] Spandöck, F., *Proc. 5th. ICA, Liège, 7–14 Sept., 1965*, Vol. 2.

[9] Brebeck, D., *The theory and practice of sound and ultra-sound absorption of materials with reference to the building of acoustically similar models at the scale of 1:10*, Dissertation TH München, 1966.

gradually increasing the gain of the microphone circuit after the initial sound pulse. This is especially easy if the frequency regime of interest is limited.

But Spandöck was also interested in testing the model room with continuously running speech and music. For this purpose, the frequency-scaling law of eqn. (7.7) can be fulfilled if one tape-records the desired signals at normal recording speed and plays them back into loudspeakers in the model at a speed that is faster by a factor of $1/\mu$. Microphones at various audience locations in the model pick up these frequency-transformed signals, along with the acoustical response of the model room, and this 'hall-processed' signal is tape recorded at the same scaled-up recording speed. In order to evaluate the effect of the room acoustics of the model on the original signal, one listens to the recording of the model room sound played back at the normal tape speed, which restores all frequencies and time delays to their normal values.

In order to achieve a sound attenuation for high frequencies in the model that simulates the sound attenuation in the original room, we may make use of the fact that the anomalously high attenuation at high frequencies nearly disappears in conditions of low relative humidity (see Section IV. 14.6). Spandöck showed that, for a model scale of 1:10, filling the model with extremely dry air (an expensive process!) yields model sound attenuation that matches the sound attenuation in the original room over such a broad frequency range that it is possible to evaluate the acoustical conditions for running speech and music at various locations in the (model) room. He even tried to simulate an orchestra on stage by using several loudspeakers for the different instrumental groups.

When Spandöck first proposed his model room tests with frequency transformation, recorded disks were the only recording means available; with the development of magnetic tape recording, the available range of speed transformation was greatly increased. But today, we can record signals and room responses with the use of computers, transforming the analog sound pressure signals into digital signals, and vice versa. This not only allows us to meet easily the requirements for the time- and frequency-scaling transformations, but also we do not have to transmit these scaled signals into a three-dimensional model of the hall at all. We need only a recording of the impulse sound pressure response of such a model, whereby a microphone in an audience location in the model picks up the succession of room reflections following a voltage impulse to the loudspeaker. The computer convolves this room response with the speech or music, so that we may hear these signals, with the room acoustics of the model added, but in the time scale of the original hall.[10]

These idealized (but expensive!) model testing procedures may be of special value for fundamental research into room-acoustics problems. But, even if we could satisfactorily solve the many problems of electro-acoustical representation of room acoustics (which we will deal with in Section III. 2.5), it is doubtful that such procedures could be helpful in planning a new concert hall. For such purposes, the model would have to match with great accuracy all the acoustically significant details of the proposed hall. Typically, not only the details, but even larger elements in the hall, change several times during the course of the design. Once the architectural design of the building is settled, however, there is no time for finishing the acoustical model to the required accuracy, running the tests and evaluating the results. By that time, furthermore, there is no opportunity to make significant changes in the hall, even if the model tests warn of acoustical problems.

Therefore, the practical application of acoustical model tests is mostly restricted to a study of echograms in the mid-frequency range.

[10] Schroeder, M. R., *Proc. 6th. ICA, Tokyo, 1968*, Paper GP-6-1; see also *J. Acoust. Soc. Am.*, **47** (1970) 424 and *J. Acoust. Soc. Japan*, **1** (1980) 71.

Part II:
Statistical Room Acoustics

Chapter II.1
Reverberation

II.1.1 The Duration of Reverberation and the Reverberation Time

The physical laws presented in Part I, describing the geometry of sound propagation and the rules of sound reflection and absorption, make it possible, in principle at least, to predict, by ray diagramming for each location of a sound source and receiver, the paths by which sound may travel from one location to the other, as well as the corresponding time delays, and even, to some degree, the strength of the arriving sound impulses.

But it is evident that the procedure would be very tedious, much too tedious, in fact, considering the lack of precision associated with each step in the operation. Even the ray construction for a single pair of locations requires a tremendous effort. For the analysis of an entire room, one would have to study quite a number of listener locations, since in every room there are differences in the sound at different seats—a fact that is well known to every concert and theater goer, and which he takes into account, together with the price, whenever he buys a ticket.

Despite this place-to-place variation in the sound of a hall, it is, nevertheless, possible to form an average total aural impression, and from this viewpoint one may justifiably speak of the 'acoustics of the hall'.

For the designing architect, also, it is desirable to be able to characterize the sound of a proposed (or already existing) hall with a single number, which is relevant to at least some aspect of the acoustical quality, and which can be used as a simple overall rating for the hall.

However, nothing would be gained if this single figure were merely an average of the acoustical ratings for different seat locations. What we are interested in is a quantity that is found by experience to be the same at all listening positions (excluding places with obvious abnormalities) and that takes into account (i.e. depends on) all portions of the room and its boundaries.

Clearly, the theoretical basis for such a quantity can only be statistical in nature. And, as always in statistics, an individual evaluation may depart somewhat from the statistical average, since the general assumptions, without which no statistical description is possible, may not be valid for the individual case.

We must ask ourselves, then, what physical process could be a suitable basis for the characteristic property of the room that is generally meant by the physically and linguistically dubious term, its 'acoustics'.

It is the great merit of the American physicist, Wallace Clement Sabine,[1] to have recognized and proved that 'reverberation' (that is, the more or less long persistence of sound that is heard in a room after the sound source is suddenly stopped) is a quantity well suited to characterizing an important aspect of the acoustics of the room. (We have already mentioned reverberation in Section I.4.3, Fig. 4.7, where we described the flutter echo that can appear between parallel walls, noting that the flutter changes to reverberation when all the boundaries of the space are reflective.)

Sabine measured with a stopwatch the duration of audibility of the reverberation and determined that this quantity has the same value at all locations in the room. Thus, it is an acoustical quantity that is characteristic of the room as a whole.

But the duration of reverberation is not a property of the room alone; it also depends on the strength of the sound source. A pistol shot will reverberate longer than a snap of the fingers.

Sabine studied the influence of the source strength using organ pipes as sources, first one, then two, then four, always blown with the same air pressure. These changes did *not* double or quadruple the duration of reverberation; instead, the influence of the number of pipes was much smaller. Moreover, a change from one to two pipes gave the same absolute increase in duration as a change from two to four pipes. From this he concluded that the sound energy in a room decreases from four to two in the same time as it decreases from two to one. Generally, then, in equal time intervals the energy decreases by the same fraction of its initial value. But this means that the instantaneous *loss* of energy is

[1] Sabine, W. C., *The American Architect*, 1900; see also *Collected Papers on Acoustics*, No. 1, Harvard University Press, Cambridge, 1923.

always a constant percentage of the instantaneous *amount* of energy.

This mathematical relationship always occurs in nature when any store of energy (here, a room filled with sound) is emptied into any sink of energy (here, the sound absorbing boundaries and contents of the room). The process is described by a decreasing exponential function. If in this case we denote the energy density (i.e. the energy per unit volume in the room) by E, and its decrement during the time differential $\mathrm{d}t$ by $(-\mathrm{d}E)$, then Sabine's observation can be described mathematically by the differential equation:

$$-\mathrm{d}E/E = \mathrm{d}t/\tau \qquad (1.1)$$

where τ is the (still undetermined) time constant for the energy in this decay process. (Here, τ is related to the energy and not, as is customary, to a field quantity like the sound pressure.)

Integration of eqn. (1.1) leads immediately to:

$$\ln E = \ln E_0 - t/\tau \qquad (1.2)$$[2]

where we have chosen the constant of integration ($\ln E_0$) so that E_0 represents the initial energy density at time $t=0$. The solution (1.2) may also be written in exponential form:

$$E = E_0 e^{-t/\tau} \qquad (1.3)$$

From this we see that the energy theoretically never disappears altogether; thus, the objective physical process has no end (and therefore no 'duration'). The finite audible duration of reverberation t_s observed by Sabine is the consequence of a threshold E_s of energy density below which our ears are not sensitive enough to hear the sound. (The subscript 's' comes from the German word *Schwelle*, meaning threshold.)

The duration of reverberation is found from eqn. (1.2):

$$t_s = \tau \ln(E_0/E_s) \qquad (1.4)$$

Both the initial energy density and the threshold energy density affect the duration. Furthermore, it may be that the threshold depends not on the sensitivity of our ears but on other noise in the background, which may mask the decaying reverberant signal (see Section III. 1.2).

It requires considerable experience to assess the background noise accurately. In general we become aware of it only when it begins to

[2] 'ln' means the logarithm to the base *e*.

interfere with conversation; but by then it is already rather high. Less often, we notice how great the difference is between daytime and night-time noise in a large city, even when the windows are closed. But the constant noise we are forced to live with in cities becomes most apparent by contrast, when the opportunity arises for a stroll in the nerve-soothing, windless quiet of a forest.

While the duration of reverberation depends on the initial and final levels of the audible sound, the time constant τ is a quantity that characterizes the room; and this is the quantity that we are looking for. It can be expressed in terms of the time $(t_2 - t_1)$ that it takes for the energy density to diminish from E_1 to E_2:

$$\tau = \frac{t_2 - t_1}{\ln (E_1/E_2)} \tag{1.5}$$

If we accept for E_1/E_2 the ratio 4:1, which was the largest ratio used by Sabine in his organ pipe experiments, the time difference $(t_2 - t_1)$ would be rather short compared with the audible duration of reverberation. In concert halls, the maximum energy ratios may be as high as 1 000 000:1, where the lower value is near the masking noise level. In measuring the duration of reverberation, Sabine must have observed similarly large ratios of energy density,[3] and this may be the reason that he adopted, for the quantity to characterize the reverberation of a room, the time required for the energy to drop to one-millionth of its initial value. He called this quantity the 'reverberation time' T; it actually defines the *rate of decay* of sound, and must be carefully distinguished from the duration of reverberation.

By introducing $E_1/E_2 = 10^6$ into eqn. (1.5), we get the relation between τ and T:

$$\tau = T/6 \ln 10 = T/13{\cdot}8 \tag{1.6}$$

If we use the reverberation time T instead of τ, eqns. (1.3) and (1.2) take the forms:

$$E = E_0 \times 10^{-6t/T} \tag{1.7}$$

and:

$$\log E = \log E_0 - 6t/T \qquad (1.8)^4$$

[3] P. E. Sabine, a relative of Wallace Clement, confirms that this was, indeed, the case: see his *Acoustics and Architecture*, McGraw-Hill, New York, 1932, p. 77.

[4] 'log' means the logarithm to the base 10.

II.1.2 Linear Decay of Reverberant Energy Level

In the early days of room acoustics, it was believed that the logarithm of the sound pressure (or its square, proportional to sound energy density) is a valid measure of the subjective sensation of loudness (see Section III. 1.2):

$$L = 10 \log (E/E_s) \tag{1.9}$$

When we introduce the concept of loudness level (defined, so far, only for a frequency of 1000 Hz), the audible reverberation process appears with a linear slope, given by:

$$L = L_0 - (60t/T) \tag{1.10}$$

where L_0 ($= 10 \log E_0/E_s$) is the amount, supposedly, by which the initial sensation exceeds the threshold.

We shall learn in Section III. 1.2 that this definition is free from self-contradiction only in the frequency range from about 600 to 4000 Hz; furthermore, the notion of loudness level, as defined in eqn. (1.9), does not, in fact, correspond to people's subjective sensation of loudness.

Nevertheless, the linear decay of loudness is a much better approximation to our subjective loudness sensation than the exponential decay function, so long as we can still hear the decaying sound near the end of the duration of reverberation and the threshold of audibility is clearly exceeded.

Quite apart from the psychological aspects, however, expressing the reverberant decay in logarithmic form simplifies the overall description of the process, since we can characterize it by the slope of a straight line fitted to the fluctuating, but generally decreasing, curve of loudness level as a function of time. From this standpoint, the quantity L formally defined by eqn. (1.9), means only a logarithmic evaluation of a given energy density with respect to a reference quantity, in this case the threshold energy density E_s; nothing at all need be said about 'loudness'.

Again and again we shall find this kind of evaluation useful in acoustics. We always call such a logarithmic quantity a 'level'; and when we use logarithms to the base 10 for the energy ratios and multiply the result by 10, the unit of level is the 'decibel' (abbreviation dB).[1] Originally, logarithmic energy ratios were expressed without the factor of

[1] Thus, it would be appropriate to add the unit symbol 'dB' after eqns. (1.9) and (1.10).

10 as simply log (E_1/E_2); the unit was the bel, in honor of Alexander Graham Bell, the inventor of the electromagnetic telephone. But when it was found that 1/10th of a bel corresponds approximately to our threshold of discrimination for different sound levels, this smaller 'one-tenth unit' of level was adopted and given the name 'decibel'.

In Chapter II.4 we shall discuss a number of methods for recording the decaying sound levels directly; such recordings demonstrate that the reverberation decay appears as a nearly straight line.

As an example, Fig. 1.1 shows sound level recordings made during a concert in the old Berlin Philharmonic Hall.[2] It is clear that the reverberation may actually be heard in musical performances. To be sure, there are only a few musical compositions, such as Beethoven's Coriolanus Overture illustrated here, where suddenly stopped fortissimo chords are followed by pauses of sufficient length to permit us to hear the reverberant decay undisturbed by other sounds. But it would be wrong to conclude that the reverberant properties of a room are important *only* then; they determine the decays of *all* sounds even though, because of

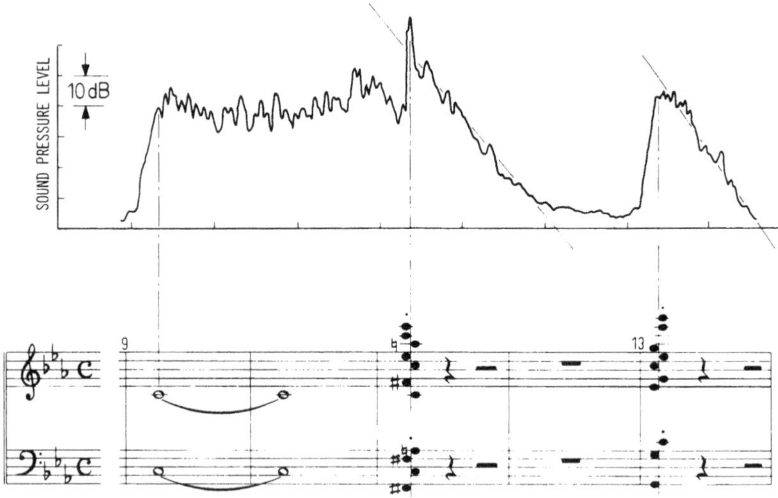

Fig. 1.1. Sound level recording of reverberation processes during a concert (Beethoven's Coriolanus Overture, Op. 62, measures 9–13).

[2] Meyer, E. and Jordan, W., *Elektr. Nachr.-Techn.*, **12** (1935) 217.

masking by the continuing music, we do not consciously hear them. (To attend to *all* of the sound decays, we would have to re-educate our ears in an anti-musical way, so as to ignore the main contours of the musical performance, thereby defeating aesthetic skills that were probably acquired in an earlier time of our lives which we can no longer remember.)

But even if we are not consciously aware of the reverberation of a musical signal, we nevertheless perceive its effect as either a support or a distortion of the music. Both are possible, because of what we have said in Part I about 'useful' and 'detrimental' reflections. Too long a reverberation will degrade the intelligibility of speech and the clarity and precision of music. On the other hand, not all reverberation is harmful, for there are also *useful* reflections—and reverberation is, after all, made up of a large number of single reflections. We shall discuss in Section III. 3.3 the optimal reverberation time for various occasions.

II.1.3 Reverberation Following Steady-state and Impulsive Excitation

The graphic record of Fig. 1.1 is interesting because it shows that the same kind of sound level decay occurs after short chords (the second peak) as after long-held chords. It happens even when there is no accent at the end of the chord (as there is in bar 11 of the Beethoven); this was the case with Sabine's organ tones. We refer to the latter as 'steady-state' excitation; the short chord is called an impulsive excitation, similar to those described in Section I. 4.2.

In contrast to the general impression given by Fig. 1.1, we now consider the possibility that the two different kinds of excitation may give rise to different kinds of reverberation. We shall see that this can, in fact, happen if the energy decays are not pure exponential functions.

Let us designate the energy density decay following steady-state excitation by $E_{-(t)}$ and the decay following a short power impulse $P\Delta t$ by $E'_{(t)}$.

In order to derive the relation between E_- and E', we first consider a third process, the initial increase of energy density following the start of a steady-state sound source; we designate this energy density growth as E_+.

To deal with all these processes, we extend eqn. (1.1) to include a term accounting for a supply of sound power, $P_{(t)}$.

We assume that, beginning at time $t=0$, a constant sound power P is radiated into a room whose volume is V. Then the total energy contained in the room $VE_{(t)}$ will increase by $V\mathrm{d}E$ during the time interval $\mathrm{d}t$ provided that the added energy $P\mathrm{d}t$ exceeds the energy lost in the same time interval by absorption $E\mathrm{d}t/\tau$ (see eqn. (1.1)). Thus, we can write an energy balance:

$$V\mathrm{d}E = P\ \mathrm{d}t - VE\ \mathrm{d}t/\tau$$

and a power balance:

$$V\ \mathrm{d}E/\mathrm{d}t = P - VE/\tau \qquad (1.11)$$

The last equation shows that, as the energy density approaches the asymptotic value

$$E_\infty = P\,\tau/V = PT/13{\cdot}8V \qquad (1.12)$$

the left-hand side of eqn. (1.11) vanishes, so the increment of E disappears and a steady state is attained; energy is being absorbed exactly as fast as it is supplied.

With respect to the reverberation decay that is to follow this state, E_∞ is the same as the initial energy density that we previously called E_0, corresponding to time t_0. (Here we call it E_∞ because, strictly speaking, it is reached only after an infinitely long time.)

If we express $E_{+(t)}$ during the build-up process in terms of the difference ΔE between the instantaneous energy and the final value:

$$E_+ = E_\infty - \Delta E \qquad (1.13)$$

then in eqn. (1.11) the steady-state terms vanish and we get for ΔE the same differential equation, (1.1), as we found for E during the reverberation decay:

$$-\mathrm{d}\Delta E/\mathrm{d}t = \Delta E/\tau \qquad (1.14)$$

For ΔE, we have the initial condition, at the beginning of the build-up, that the difference $\Delta E(0)$ between E_∞ and the instantaneous energy density, $E(0)=0$, is E_∞. Therefore, the growth of E_+ and the decay of E_- are complementary functions; combining eqn. (1.13) with the solution for ΔE, as in eqn. (1.3), we get:

$$E_+ = E_\infty(1 - e^{-t/\tau}) \qquad (1.15)$$

But if we plot the logarithm of E_+ (the growth curve), it is by no means a straight line. The E_+ function spends more of the time in the range of high values and less time at low values; and this is even more true for the logarithmic plot, since the nature of the logarithmic function is to compress high values and expand low values.

Figure 1.2 compares the time dependence for the different growth and decay processes. At the top, we see that the power supplied to the room is constant, from the instant it suddenly begins to the instant it suddenly stops. In the center of the figure is the energy density E, which increases according to eqn. (1.15) and decreases according to eqn. (1.3). The complementarity of the transients is evident: if the curve were cut in two where the power stops, the ending could be inverted and fitted perfectly to the beginning.[1] At the bottom of the figure is the logarithmic function

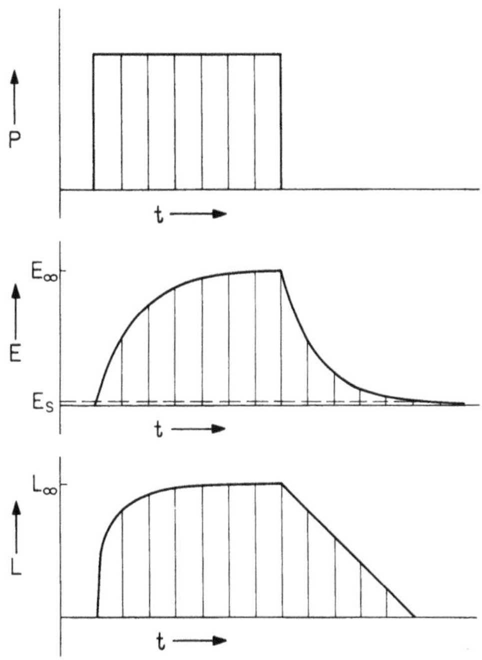

Fig. 1.2. Time dependence of the growth and decay functions for steady-state excitation. Top: the sound power supplied to the room. Middle: the energy density. Bottom: the energy density level.

[1] According to P. E. Sabine (*Acoustics and Architecture*, McGraw-Hill, New York, 1932, p. 84), M. J. O. Strutt has shown that this complementarity holds even for the general build-up and decay of sound.

of E, the energy-intensity level L (see eqn. (1.9)); the initial transient is short, the final transient is long. We can see the same contrast in Fig. 1.1.

In a certain sense, the quite different rates of change of the logarithmic transients during build-up and decay correspond better to our subjective impressions than the complementary (comparable duration) behavior of the energy transients $E_{+(t)}$ and $E_{-(t)}$.

However, when we sometimes hear a step-like growth of sound following the fortissimo onset of a musical chord, we should not overlook the fact that this effect is real; it is primarily due to the actual stepwise arrival of the direct sound and the first reflections, and only secondarily to our non-linear (but by no means logarithmic) sensation scale of loudness.

We can, however, regard the 'steady-state' excitation as a continuous, connected series of short impulses $P\mathrm{d}t_0$ just as a board fence can be seen as being made up of its individual planks; the top of Fig. 1.2 illustrates this. Each such impulse successively initiates its own differential reverberation process, starting at $t=t_0$. Since E' was defined as the reverberation following a power impulse $P\Delta t$, we must reduce the quantity E' in the ratio $\mathrm{d}t_0/\Delta t$ and must further take into account the different times t_0 of the respective differential impulses; so we get for $\mathrm{d}E'$:

$$\mathrm{d}E' = E'_{(t-t_0)}\,\mathrm{d}t_0/\Delta t$$

The sum of these individual reverberations (that is, their integral) gives the build-up of sound:

$$E_{+(t)} = \frac{1}{\Delta t}\int_0^t E'_{(t-t_0)}\mathrm{d}t_0 = \frac{1}{\Delta t}\int_0^t E'_{(\Theta)}\,\mathrm{d}\Theta \qquad (1.16)$$

where Θ is a dummy variable.

The steady-state condition is reached in the limit as $t\to\infty$:

$$E_\infty = \frac{1}{\Delta t}\int_0^\infty E'_{(\Theta)}\,\mathrm{d}\Theta \qquad (1.16a)$$

The subsequent reverberant decay is given by the difference:

$$E_{-(t)} = E_\infty - E_{+(t)} = \frac{1}{\Delta t}\int_t^\infty E'_{(\Theta)}\,\mathrm{d}\Theta \qquad (1.17)$$

In contrast to eqns. (1.11)–(1.15), the last three equations are not restricted to exponentially decaying reverberation. They assume only that the build-up and decay processes for steady-state excitation are always complementary, because the same reflections that must be added together during the build-up must be subtracted during the decay.

If we differentiate eqn. (1.17) with respect to time, we get the general relation:

$$-\mathrm{d}E_{-}/\mathrm{d}t = E'/\Delta t \tag{1.18}$$

If E_{-} is a decaying exponential function, then so is E', and with the same time constant. In all other cases, E_{-} and E' are different, as we will demonstrate below with several examples.

In the development of eqns. (1.11)–(1.16), we have relied on the fundamental assumption of statistical room acoustics, that energies (or powers) are to be added. This would not be valid, for example, in adding together pure tones or combinations of pure tones with fixed phase relations, for in that case the phase relations are very important, as we will see in Part IV (Volume 2).

But for measurements in room acoustics, the steady-state excitation typically makes use of random noise signals, and these are made up of a multitude of impulses with random time intervals; fixed phase relations never occur and all phase angles have the same probability. Under such conditions, it is possible to make statements only about statistical mean values, and, according to the general validity of the energy law, those correspond to simple addition of the component energies.

Unfortunately, we cannot observe the sound energy directly; we must measure the sound pressure and infer the energy. In studying room reverberation, for example, we record the decay of sound pressure level and not (as ideally we should) the mean energy. In this case, the sound decay is seriously affected by the instantaneous fluctuations that are implicit in random processes and are unavoidable; each resulting decay curve is different and depends both on the initial state of energy in the room and also on the instant at which the random signal is stopped.

Thus, repeated decays of $E_{-(t)}$ differ in detail; this is not the case for the decays $E'_{(t)}$, however, because they are generated by a precisely defined and accurately repeatable impulse.

Schroeder[2] has taken into account this essential difference between E_{-} and E' by proving that E_{-}, on the left-hand side of eqn. (1.17), is the

[2] Schroeder, M. R., *J. Acoust. Soc. Am.*, **37** (1965) 409.

equivalent of the ensemble average $\langle E_{-(t)}\rangle$ of an infinite number of reverberant decays with different 'stop-times', following steady-state excitation. Consequently, he has proposed that it is better to measure the steady-state function $E_{-(t)}$ by electronic integration of an impulse reverberation $E'_{(t)}$. We shall return to this method in Part IV.

At this point, we note only that most of the time-dependent signals that we encounter in room acoustics (speech and music) may be treated approximately as random noises, that is, as signals without defined phase relations.

The expression for E_{-}, implied in eqn. (1.15), becomes, when we take into consideration eqn. (1.12):

$$E_{-}=\frac{P\tau}{V}e^{-t/\tau} \tag{1.19}$$

Introducing this into eqn. (1.18), we get:

$$E'=\frac{P\Delta t}{V}e^{-t/\tau} \tag{1.20}$$

This means that we assume that the radiated energy $P\Delta t$ is equally distributed over the total room volume during a very short time compared with τ. In reality, the initial value of eqn. (1.20) is never reached, but nevertheless the subsequent reverberation corresponds essentially to an extrapolation back to $t=0$.

Equation (1.12) makes clear the proportionality between the steady-state energy density E_{∞} and the reverberation time T. If a room is damped (sound-absorptive material added) in order to decrease T, it is unavoidable that we also decrease the loudness of a steady-state sound, such as that shown in Fig. 1.1. Furthermore, eqn. (1.12) also shows that E_{∞} is inversely proportional to the volume V. As a consequence, the desired reverberation time also depends on the volume; if we are interested in a sufficiently high value of E_{∞} (adequate loudness), then T must increase with the volume.

The addition of the reverberant sound to the direct sound, like the use of the sustaining pedal on a piano, causes not only a smoothing of the individual musical syllables (impulses), but also an increase in the total strength of the sound energy density. Traditionally, this has been regarded as an advantage of long reverberation, which eqn. (1.12) makes especially clear for the case of steady sounds.

In large rooms, it is, in principle, possible to maintain the energy density without increasing the reverberation time, for example, by using more musicians or by having them play louder. Such strategies, of course, can never keep up with changes in room volume from that of a living room (perhaps $50\,m^3$) to a cathedral ($50\,000\,m^3$). But, according to experience, it is reasonable to expect an increase in sound power of the order:

$$P \sim V^{1/3} \tag{1.21}$$

This would mean a decrease in energy density inversely proportional to only $V^{2/3}$, if T remains constant. But it would be helpful, too, if T increased as:

$$T \sim V^{1/3} \tag{1.22}$$

Actually, as we have seen in Chapter I. 7, this generally follows from the rules of similarity; if we compare two rooms of the same shape and same materials but different dimensions, then all time-dependent processes take place on a larger time scale in the larger room, according to the increased typical room dimension which is proportional to $V^{1/3}$. This holds for both the reverberation and the reverberation time.

With this rule, we find a workable compromise—easy to keep in mind—between sound power, reverberation time and energy density, where all three make the same concession, since now E_∞ is inversely proportional to the cube root of the volume:

$$E_\infty \sim V^{-1/3} \tag{1.23}$$

II.1.4 The Equivalent Absorption Area

Although the volume of the room directly governs the reverberation time, this does not mean that we cannot change the reverberation time in a room with fixed volume, even over a rather wide range. We must now see how this is possible.

It is immediately evident that this is primarily a question of the amount of sound energy absorbed by the boundaries of the room and their finish materials. The materials and the details of construction determine uniquely the value of the sound absorption coefficient α of each surface, namely, the percentage of the incident energy that, at each encounter of a sound ray with a boundary, does *not* return to the room (see Chapter I. 6).

But the sound-absorptive capability of the wall surface increases as its surface area increases, so we characterize the total room absorption by the product:

$$A=\alpha S \tag{1.24}$$

Here A characterizes the amount of sound energy that is removed from the room in each second when one unit of energy falls with perpendicular incidence in each second on a wall with surface area S and absorption coefficient α.

Since α is dimensionless, A has the dimension of area. It is the actual area of the absorbing boundary surface in the case of total absorption, or (as W. C. Sabine expressed it) the equivalent absorption area of 'open window'. According to International Standards, A is called the 'equivalent absorption area'; sometimes, in the following text, we will shorten this to 'absorption area'.[1]

Sabine measured the absorption coefficients of numerous materials by direct comparison with the equivalent absorption of various areas of open window. Accordingly, it has sometimes been proposed to express the sound absorption capability of a room surface in *Sabines*, abbreviated to *Sab* when A is measured in m^2, and abbreviated to *sab* when A is measured in ft^2. However, apart from the understandable wish to honor the discoverer of the laws of reverberation, there is no advantage in introducing the Sabine units to replace the well-known area units.[2]

We must now account for what happens when the sound ray strikes a wall surface at *other* than perpendicular incidence, much the most frequent occurrence. The power absorbed by a boundary surface S changes with the angle of incidence ϑ. On the one hand, the absorbed power decreases because the surface intercepts only the projected area $S\cos\vartheta$ on the incident wavefront (see Fig. 1.3); thus, A changes to

$$A_{(\vartheta)}=\alpha S\cos\vartheta \tag{1.25}$$

In addition, we recall that the absorption coefficient α also depends on the angle of incidence, as was explained with examples in Section I. 6.2. In many cases, α increases with increasing ϑ, so that A decreases less than would be expected on the basis of the reduction in projected area. The two tendencies would exactly cancel one another if the dependence of α

[1] North American usage shortens the term even further and speaks of the 'absorption' of a room, A.

[2] In fact, North American practice *does* use the units of metric sabines and sabins for measures of room absorption, A.

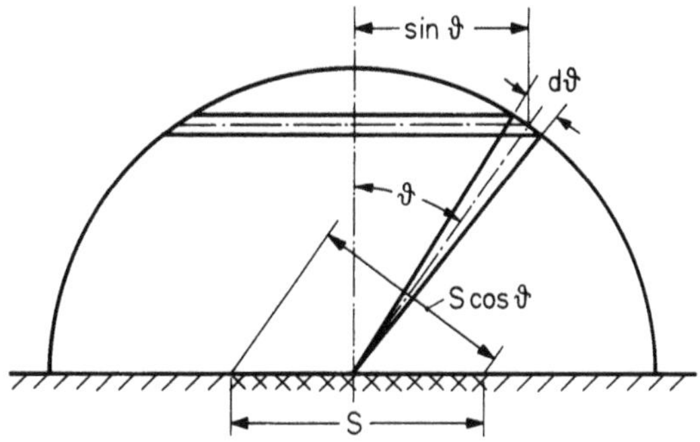

Fig. 1.3. Sketch for evaluation of the mean absorption coefficient for an omnidirectional distribution of sound incidence.

were according to:

$$\alpha_{(\vartheta)} = \frac{\alpha_{(0)}}{\cos \vartheta} \tag{1.26}$$

where $\alpha_{(0)}$ is the absorption coefficient for perpendicular incidence. Since, by definition α cannot exceed unity, but the right-hand side of eqn. (1.26) can approach infinity for angles near grazing, complete compensation is not possible. But we shall meet in the next chapter another situation with similar conditions that can be fulfilled.

Since in general the power absorbed at a surface depends on the angle of incidence, the mean power absorbed from the room depends on the distribution of the sound energy with respect to direction of incidence.

According to statistical theory, the most probable distribution of sound energy in a room is that for which the energy density is everywhere the same and for which every direction of energy flow is equally likely. Under these circumstances, grazing incidence occurs more frequently than perpendicular, or near-perpendicular, incidence.

As shown in Fig. 1.3, the increment dE that is incident on the surface S from the incremental angle between ϑ and $\vartheta + d\vartheta$, divided by the total energy density E, is equal to the area of the sphere-zone corresponding to this angle-increment, divided by the entire spherical surface area:

$$\frac{dE}{E} = \frac{2\pi r \sin\vartheta \, r d\vartheta}{4\pi r^2} = \tfrac{1}{2} \sin\vartheta \, d\vartheta \tag{1.27}$$

For plane waves, which figure more and more at greater distances from the source, the energy is propagated with sound velocity c. Thus, the intensity (i.e. the power per unit area) is given by:

$$J = Ec \tag{1.28}$$

and the differential increment corresponding to the angle between ϑ and $\vartheta + d\vartheta$ is given by:

$$dJ = c\ dE \tag{1.28a}$$

Thus, the differential power incident upon the surface S from the direction between ϑ and $\vartheta + d\vartheta$ is:

$$(cS \cos \vartheta)\ dE$$

and the absorbed power is

$$(\alpha\, cS \cos\vartheta)\ dE = \tfrac{1}{2}\ EcS\ \ (\alpha_{(\vartheta)} \cos \vartheta \sin \vartheta\ d\vartheta)$$

In order to get the total power $P_{S\alpha}$ absorbed by the surface S from all directions, we must integrate the above expression over the angular range (0 to $\pi/2$) within which the sound waves approach the surface:

$$P_{S\alpha} = \tfrac{1}{2}\ EcS \int_0^{\pi/2} \alpha_{(\vartheta)} \cos \vartheta \sin \vartheta\ d\vartheta \tag{1.29}$$

For the special case of an open window, that is, total absorption, we can set $\alpha = 1$ for all angles of incidence; the absorbed power equals the incident power:

$$P_S = \tfrac{1}{2}\ EcS \int_0^{\pi/2} \cos \vartheta \sin \vartheta\ d\vartheta = \tfrac{1}{4}\ EcS \tag{1.29a}$$

This relation was derived in 1903 by Franklin,[3] soon after Sabine's paper. For the extension to incomplete sound absorption, however, most authors[4] at first, following Sabine, did not attempt to account for the ϑ-dependence of α, but simply multiplied the result of eqn. (1.29a) by a 'mean' absorption coefficient α_m:

$$P_{S\alpha} = \tfrac{1}{4}\ EcS\ \alpha_m \tag{1.29b}$$

As Paris first pointed out[5] (it is evident here from a comparison of

[3] Franklin, W. S., *Phys. Rev.*, **XVI** (1903) 372.
[4] Eckhardt, E. A., *Journal of the Franklin Institute*, **195** (1923) 729.
[5] Paris, E. T., *Philos. Mag.*, **V** (1928) 489.

eqns. (1.29) and (1.29a)), this mean absorption coefficient is the weighted average given by the integration:

$$\alpha_m = \frac{\int_0^{\pi/2} \alpha \cos \vartheta \sin \vartheta \, d\vartheta}{\int_0^{\pi/2} \cos \vartheta \sin \vartheta \, d\vartheta} = 2\int_0^{\pi/2} \alpha \cos \vartheta \sin \vartheta \, d\vartheta \tag{1.30}$$

The weighting factors cos ϑ and sin ϑ account for the reduction of the projected surface area and the increased probability of incidence at higher angles of incidence.

If we write eqn. (1.30) in the form

$$\alpha_m = \int_0^{\pi/2} \alpha \sin 2\vartheta \, d\vartheta \tag{1.30a}$$

it becomes evident that the ϑ-region around 45° carries the greatest weight, because at that angle $\sin 2\vartheta = 1$. Therefore, it is a reasonable approximation to choose $\vartheta = 45°$ when calculating sound absorption coefficients, as we shall do in Part IV (Volume 2).

The weighting in eqn. (1.30) can be better appreciated by a comparison of two plots of x, as shown in Fig. 1.4. If the absorption coefficient as a function of incidence angle is given by the curve plotted on a linear angular scale from 0 to $\pi/2$, as at the top of Fig. 1.4, then we can wrap this curve into a circular cylinder with the diameter a and look at the plotted curve as projected on a vertical plane through the diameter (see Fig. 1.4, bottom). The abscissa then becomes

$$x = \tfrac{1}{2}a \,(1 - \cos 2\vartheta)$$

and, for this abscissa distribution, α_m becomes the arithmetic mean:

$$\alpha_m = \frac{1}{a}\int_0^a \alpha \, dx = \int_0^{\pi/2} \alpha \sin 2\vartheta \, d\vartheta \tag{1.30b}$$

which is identical with the form given in eqn. (1.30a).

Finally we shall also find the following equivalent form useful for

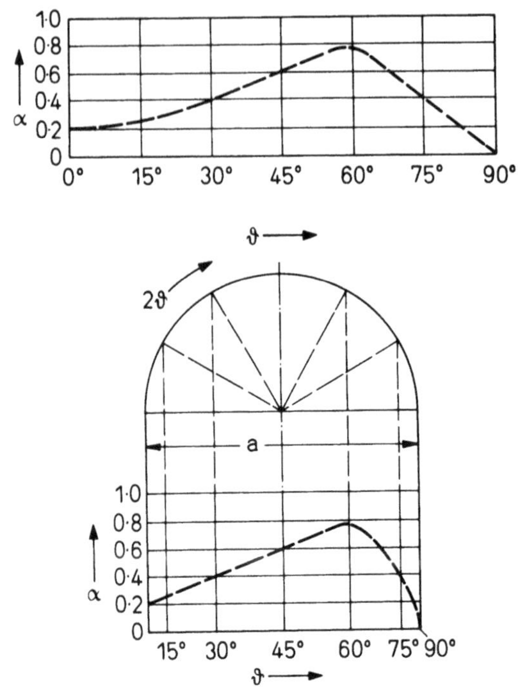

Fig. 1.4. Transformation of the dependence of the sound absorption coefficient on the angle of incidence (top) to the dependence according to the Paris formula (bottom). (The curve of $\alpha_{(\vartheta)}$ shown here is arbitrary but plausible.)

calculations, because $\alpha_{(\vartheta)}$ is often given in the form, $\alpha_{(\cos\ \vartheta)}$:

$$\alpha_{\mathrm{m}} = 2\int_0^1 \alpha_{(\cos\ \vartheta)} \cos\vartheta \; \mathrm{d}(\cos\vartheta) \qquad (1.30c)$$

In the following text, when we speak of the sound absorption coefficient we shall always understand this to be the mean coefficient given in eqn. (1.30), but we will drop the subscript 'm' for convenience.

In the general case, the boundaries of the room will be made of different materials, with different areas S_k and different absorption coefficients α_k. We get the total (*gesamte*) equivalent absorption area for the room from:

$$A_{\mathrm{ges}} = \sum_{k=1}^{n} \alpha_k S_k \qquad (1.31)$$

and the total absorbed sound power by:

$$P_\alpha = \frac{1}{4}Ec \sum \alpha_k S_k \tag{1.32}$$

In these formulae we can introduce a mean value of sound absorption coefficient, averaged arithmetically over all the room boundary surfaces:[6]

$$\bar{\alpha} = \frac{\sum \alpha_k S_k}{\sum S_k} \tag{1.33}$$

This mean coefficient is often used in general discussion to characterize with a single number the acoustical effect of the contents of a room.

II.1.5 Sabine's Formula and Its Inverse

When most of the sound energy losses occur at the boundaries of the room, which is very often the case, then P_α in eqn. (1.32) is the same as the power loss given by the last term in eqn. (1.11):

$$P_\alpha = \frac{VE}{\tau} \tag{1.34}$$

Thus, we get, via the time constant τ and eqn. (1.6), an expression for the reverberation time T:

$$T = (6 \ln 10)\tau = \frac{(6 \ln 10)\, 4V}{c \sum \alpha_k S_k} \tag{1.35}$$

By introducing $c = 340\ \mathrm{m\,s^{-1}}$ and expressing the other dimensions in meters, we find:[1]

$$T = 0{\cdot}163\, V / \sum \alpha_k S_k \tag{1.35a}$$

[6] Remember, each α_k is already a weighted average over angle-of-incidence; also α_k and $\bar{\alpha}$ are functions of frequency. In North America, the values of α_k for the frequencies 250, 500, 1000 and 2000 Hz are averaged and rounded to the nearest 0·05 to yield the noise reduction coefficient (NRC) of the material.

[1] The rounded value of $340\ \mathrm{m\,s^{-1}}$ is adopted here for convenience, see Section I.1.1 and Section IV. 1.4 (Volume 2) eqn. (1.12a). It leads to the constant 0·163 in the reverberation time equation, but it corresponds to a Celsius temperature of only 14·2°C. In American texts, it is common to find the constant 0·161, corresponding to a sound speed of 343·2 $\mathrm{m\,s^{-1}}$ and a temperature of 20°C. The reader is advised to choose in practice the constants appropriate to the temperature that applies to his problem.

This principal formula of statistical room acoustics was discovered by Sabine as early as 1898 (although his experiments led him to a slightly higher value for the constant). The formula is therefore known as the 'Sabine Equation'. It makes the fundamental statement that the reverberation time of a room is determined by the room volume and the total equivalent absorption area, and is independent of how the absorptive material is distributed around the room.

Such a statement, whose validity and limitations were experimentally studied by Sabine, can be based only on statistical assumptions. We recall the flutter echo as an obvious exception, and we shall deal with further restrictions on the validity of eqn. (1.35) in the next chapter.

But here we accept the Sabine Equation as a useful first approximation, noting its astonishing simplicity, compared with the complexity of the reverberation process that it describes.

The most important application of this formula may be its 'inverse', which makes it possible to evaluate particular absorption areas, or even absorption coefficients, by measuring reverberation times:

$$A_{\mathbf{ges}} = 0{\cdot}163\ V/T \tag{1.35b}$$

If we introduce into a reverberating room some additional absorptive material, then the original reverberation time T_0 of the empty room is reduced to T_Δ. We can calculate the absorption A_0 of the empty room from eqn. (1.35b) as:

$$A_0 = 0{\cdot}163\ V/T_0$$

and for the room with the added absorption:

$$A_0 + \Delta A = 0{\cdot}163\ V/T_\Delta$$

Eliminating A_0, we find that the amount of added absorption is directly proportional to the difference in the reciprocal reverberation times:

$$\Delta A = 0{\cdot}163 V\left(\frac{1}{T_\Delta} - \frac{1}{T_0}\right) \tag{1.36}$$

We have only to divide ΔA by the corresponding surface area S_x to get the mean value (in Sabine's sense) of the sound absorption coefficient of the added material:

$$\alpha_x = \Delta A / S_x \tag{1.37}$$

(We must distinguish between the absorb*ing* surface area, which in this case is S_x, and the equivalent absorp*tion* area, which is $S_x \alpha_x$.)

We have assumed here that the added absorptive material did not cover, and thus nullify the effect of, an area of the room that already had significant sound-absorptive properties. If this is not the case, then, instead of (1.37) we have:

$$\Delta A/S_x = \alpha_x - \alpha_0 \tag{1.37a}$$

where α_0 is the absorption coefficient of the bare boundary.

It is obvious that the evaluation of ΔA becomes more accurate the more ΔA exceeds A_0; this means that there is an advantage in using large samples of the material under test. In addition, the use of a large sample tends to even out the manufacturing differences in different parts of the sample.

On the other hand, with highly absorptive materials we must be careful not to use too large a sample, for then the assumptions of a statistical distribution of sound energy may be violated and the reverberation time could become so short that it is difficult to measure accurately. Since long reverberation times are easier to measure, the test room should not be too small (see Section IV. 11.2 (Volume 2) for further discussion). In any case, the mean absorption coefficient for the empty room should be as small as possible; therefore, rigid and impermeable walls are essential for the test room. Such rooms, which must also be insulated against the intrusion of exterior noise, are called 'reverberation rooms'. We shall come back to further recommendations for these rooms and the results of reverberation room tests in Chapters II.5 and II.6.

The reverberation room method of measuring sound absorption has a special advantage in comparison with other, possibly more exact, methods. Namely, it can be applied even in cases where the equivalent absorption area cannot be expressed, as in eqn. (1.31), by the product of an area and an absorption coefficient. This could occur, for example, for wall coverings where the material and shape change over distances comparable to the wavelength, and is even more pronounced for absorptive materials that are not extended surfaces but are individual objects, like furniture or people.

As every acoustically sensitive person knows, when one moves into a new dwelling the reverberation of the empty rooms becomes much shorter as the furniture is brought in, particularly if it is upholstered; the influence of the furniture cannot be neglected. But also important is the sound-absorptive clothing of the inhabitants of a dwelling, and particularly the audience in a concert hall. Musicians are keenly aware of the difference in sound between the nearly empty hall during rehearsal and the occupied hall during the concert.

Just as, in eqn. (1.31), we added together the contributions of the different parts of the room boundary to get the total equivalent absorption area, we can combine the contributions of n sound-absorptive objects (or people), each with its specific absorption area δA, as follows:

$$\Delta A = n\ \delta A \tag{1.38}$$

Again, the value of ΔA can be defined by two measurements of reverberation time, as in eqn. (1.36). This simple relationship is more accurate if the objects in question are located far apart, so that they do not influence one another.

Equation (1.38) may also be used if a large number of similar objects are located near one another, distributed in a two-dimensional pattern over a large area. In such cases, the specific absorption area for each object depends not only on the object itself but on the spacing from its neighbors. Then it is possible to divide the equivalent absorption area for the entire surface by the geometric area actually covered by the objects, in order to define an 'apparent absorption coefficient'.

If we combine eqns. (1.31) and (1.38), we get for the total absorption area:

$$A_{\text{ges}} = \sum \alpha_k S_k + \sum \delta A_l n_l \tag{1.39}$$

Similarly, we may split the reciprocal reverberation time into two parts:

$$\frac{1}{T} = \frac{1}{T_1} + \frac{1}{T_2} \tag{1.40}$$

The first term involves the reverberation time T_1 given by eqn. (1.35a), which is determined by the absorption of the room boundaries only; the second term involves the reverberation time T_2, which concerns only the absorption of the audience and the individual objects in the room, according to the formula:

$$T_2 = 0{\cdot}163 V / \sum \delta A_l n_l \tag{1.40a}$$

Assuming that the clothing of the audience leads to a more or less constant value for δA for different people, the latter reverberation time will be long or short depending on the ratio V/n, the room volume per seat, a quantity which is also important in building design for hygienic reasons. This ratio was first used in the calculation of reverberation time by Weisse.[2]

It is evident that the seats and the audience have greater influence on the reverberation in a room when the room volume per seat is small. In low cinemas, where the ceiling is scarcely higher than the screen, the seats and audience are of decisive importance; in cathedrals, hardly at all.

II.1.6 Influence of Dissipation during Sound Propagation

Although the incomplete reflection that occurs when sound strikes a room boundary (or an object) accounts for most of the energy loss during reverberant decays, the losses caused by dissipation during sound propagation through the air in the room also affect the reverberation time, particularly for high frequencies and for large room volumes. Indeed, the existence of that kind of energy loss was first discovered from measurements of reverberation time. We shall be concerned with the discovery, the investigation and the explanation of those losses in Section IV. 14 (Volume 2).

If the sound pressure in a plane wave is attenuated exponentially as it travels a distance x, according to:

$$p_{(x)} = p_0 e^{-\mu x} \tag{1.41a}$$

and the corresponding intensity is attenuated according to:

$$J_{(x)} = J_0 e^{-2\mu x} \tag{1.41b}$$

this means that the decay in sound intensity with respect to time is given by:

$$J_{(t)} = J_0 e^{-2\mu ct} = J_0 10^{-(\log e)\, 2\mu ct} \tag{1.41c}$$

Comparing this equation with the form we have chosen for eqn. (1.7), we can write:

$$J = J_0\ 10^{-6t/T} \tag{1.41d}$$

and it follows that the limiting case of reverberation time, T_3, where all the losses occur during propagation and none during reflection, is given by:

[2] Weisse, K., *Neugestaltung von Sälen für Tonfilm-Wiedergabe*, Berlin, 1939.

$$T_3 = 3/[(\log e)\ \mu c] = \frac{6{\cdot}9}{\mu c} = 2 \times 10^{-2}/\mu \qquad (1.42)$$

where μ is expressed in Np m^{-1} (1 Np = 8·7 dB).

If we take into account the room boundary losses and the audience losses, then all the corresponding reciprocal reverberation times must be added together, as we did in eqn. (1.40):

$$\frac{1}{T} = \frac{1}{T_1} + \frac{1}{T_2} + \frac{1}{T_3} \qquad (1.43)$$

This, together with eqns. (1.35) and (1.40a), yields:

$$\frac{1}{T} = \frac{c}{13{\cdot}8}\left[\frac{\sum \alpha_k S_k}{4V} + \frac{\sum \delta A_l n_l}{4V} + 2\mu\right]$$

or

$$T = 0{\cdot}163V \Bigg/ \left[\sum S_k \alpha_k + \sum n_l \delta A_l + 8\mu V\right] \qquad (1.44)$$

Thus, Sabine's Equation must be extended by adding in the denominator a term $8\mu V$ to account for energy dissipation during propagation of the sound through the medium.

The choice of the correct value for μ presents a slight problem, since it depends on the temperature and the humidity. It was the latter dependence that first attracted the notice of acousticians. P. E. Sabine[1] discovered the effect; Meyer[2] confirmed the effect quantitatively and showed that it becomes stronger with increasing frequency. The question that he raised, whether the humidity-dependent losses occur at the room boundaries or during propagation, was answered experimentally by Knudsen:[3] they are propagation losses. The physical explanation is due to Kneser,[4] which we shall discuss in Section IV. 14.6 (Volume 2). There we will also introduce Knudsen's measured values of μ for different frequencies and different values of relative humidity.[5]

For room-acoustics applications, that is, for a temperature of about 20°C, a range of relative humidity between 30 and 80%, and frequencies below 10 000 Hz, we may summarize the more recent measurements of

[1] Sabine, P. E., *Journal of the Franklin Institute*, **207** (1929) 341.
[2] Meyer, E., *Z. Techn. Phyz.*, **11** (1930) 258.
[3] Knudsen, V. O., *J. Acoust. Soc. Am.*, **3** (1931) 126.
[4] Kneser, H. O., *J. Acoust. Soc. Am.*, **5** (1933) 122.
[5] Knudsen, V. O., *J. Acoust. Soc. Am.*, **6** (1935) 201.

Evans and Bazley[6] by the following approximate formula for the dissipation coefficient, in $\mathrm{Np\,m^{-1}}$.

$$\mu = \frac{85}{\phi} f^2 \times 10^{-4} \tag{1.45a}$$

where ϕ is the percent relative humidity and f is the frequency in kHz. Transforming this equation to the corresponding reverberation time yields:

$$T_3 = 2{\cdot}4\ \phi/f^2 \tag{1.45b}$$

It must be emphasized that this equation holds only for a Celsius temperature of 20°C. But since in auditoriums the temperature is typically held between about 18 and 22°C for reasons of audience comfort, and since the dissipation term in eqn. (1.44) often amounts to only a small correction, this formula is an adequate first approximation. If we require a better approximation, the value of μ may be decreased by 4% for each Celsius-degree increase above 20°, or increased by 4% for each Celsius-degree decrease below 20°. Or, the value of T may be increased (or decreased) by the same percentage.

We can determine from eqn. (1.45) that the reverberation times corresponding to dissipation during propagation may, at high frequencies, be comparable to those resulting from reflection losses at the room boundaries; sometimes they may even be lower (i.e. the dissipation dominates the sound decay process). Therefore the influence of dissipation must be taken into account both in the planning of auditoriums and in reverberation time measurements in large rooms.

Although in the frequency range of audible sound the influence of dissipation during propagation will never become so great that the sound field in a room resembles outdoor propagation (permitting us to neglect the reflections from the boundaries), it may approach a condition such that only early reflections need to be accounted for and the late reflections lose their importance. In such cases, geometrical room-acoustics methods become easier and the energy distribution assumptions that underlie the statistical methods become doubtful. With increasing room volume and frequency, therefore, we must abandon the preferred methods of statistical room acoustics for those of deterministic geometrical room acoustics, in the sense of Part I.

---

[6] Evans, E. J. E. and Bazley E. N., *Acustica*, **6** (1956) 238.

Chapter II.2

Refinements of the Reverberation Formulae Based on Geometric Considerations

II.2.1 The Mean Free Path Length of a Sound Ray

The derivation of the statistical reverberation process which we presented in the last chapter, especially in Section II.1.4, not only is historically the oldest, but it also has the advantage of a simple and clear mathematical concept. Its chief problem is that it concentrates all the difficulties in the assumption that the total sound energy is, at every instant, distributed in a homogeneous and isotropic manner throughout the room; also, it does not give any insight into how some departure from this assumption (which can never be completely realized) would influence the reverberation.

Therefore, we shall supplement that derivation with another one that pays attention to the fate of the individual 'energy particles' on their to-and-fro paths along the sound rays; incidentally, this derivation was also used by W. C. Sabine.

For this purpose, we can leave out of account two processes that are common to all energy particles and rays. First is the divergence of the sound wave, which causes a diminution of the intensity inversely proportional to the square of the distance travelled in the time (ct). The second is a compensation of this diminution, caused by the fact that, as time goes on, more and more of the diverging rays overlap at each location.

L. Cremer would like to express his thanks here for valuable discussions concerning the contents of this chapter, both in person and in many letters, with W. B. Joyce at the Bell Laboratories, Murray Hill, New Jersey, and with H. Kuttruff, Aachen.

The first process results in a dependence of the energy density E in each diverging ray according to:

$$E = \frac{P/c}{4\pi(ct)^2} \tag{2.1}$$

The second results in a growth of the number of reflections ΔN during the time Δt which is proportional to the square of time, as we derived in Section I. 2.4, eqn. (2.3), for a rectangular room:

$$\Delta N = \frac{4\pi(ct)^2}{V} c\Delta t \tag{2.2}$$

Only at the beginning of an echogram (see Fig. 2.1) are the impulses of the direct sound and the first reflections separated from one another; the

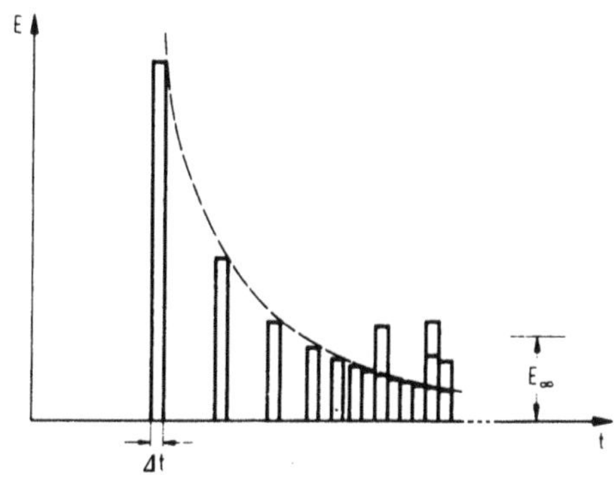

Fig. 2.1. Sketch of an echogram, without dissipation.

later reflections (regarded here as lossless) pile up and overlap, and (combining eqns. (2.1) and (2.2)) lead to the following value for the energy density:

$$E\Delta N = \frac{P\Delta t}{V} \tag{2.3}$$

We recognize this as the initial value of the impulse response, which we have already encountered in eqn. (1.20).

Furthermore, we may disregard the energy dissipation during propagation; the corresponding decay depends only on t and it may be superimposed on the formulae presented in Section II. 1.6.

Here we are interested only in the sound diminution due to absorption at the walls and by individual objects; in fact, even the latter are accounted for only qualitatively, while the quantitative relations for reflection and absorption apply to large surfaces only. Strictly speaking, only with specular reflections does a ray remain a ray; with diffusing elements, and in particular for reflection from individual objects, we must always expect a new splitting up of energy in different directions.

It would, however, severely restrict our later conclusions if we were to exclude this possibility here; indeed, we shall even go so far as to state (in Section II. 2.4) that diffuse reflection is the most effective condition for a statistical distribution of sound energy in a room.

Fortunately, it will be possible to include diffuse reflections in our considerations, although the concept of sound propagation in rays becomes ever less clear when we have to split the original ray energy again and again into parts.

For ray studies with computers (which have become more and more important even for statistical room acoustics) this energy-splitting of the individual rays is replaced by studying many rays, one following after the other with the same starting direction; they are assumed to be reflected in different directions at the room boundaries according to random rules, as in games of chance. For this reason, the procedure is called the 'Monte Carlo' method.[1]

When we plot the logarithm of the 'intensity' of the ray (disregarding its divergence) against the time or the equivalent path-length (ct), it becomes evident that we get a 'stair' instead of a straight line (see Fig. 2.2). In general, the steps of that stair will have different heights (risers)

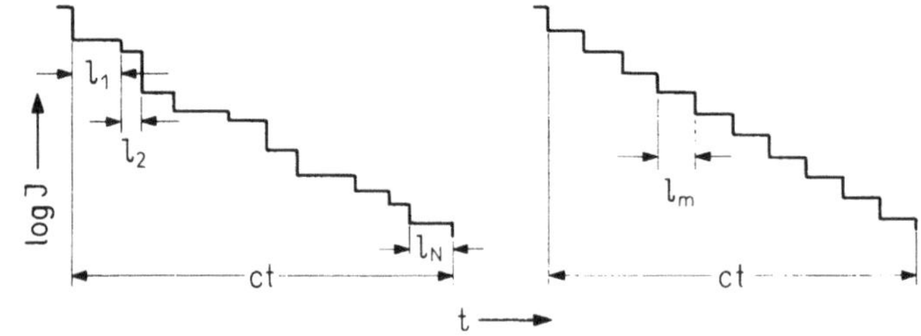

Fig. 2.2. The step-wise decay of the 'intensity' of a sound ray (left) and its replacement by a stair with steps equal in height and width (right).

[1] See, for instance, Kuttruff, H., *Room Acoustics*, 2nd. Edn. Applied Science Publishers Ltd, London, 1979, Section V.4.

and lengths (tread widths). The difference in the heights originates from the fact that the walls have different absorption coefficients; but these differences would occur even if all the walls were the same, because of the dependence of the absorption coefficient on the angle of incidence. The difference in the lengths of the steps arises because the transit times between encounters with the walls vary, as the sound ray travels different free paths between reflections. These paths depend on the shape of the room and on the distribution of the directions of the reflected sound, but not on the absorption coefficients of the reflecting surfaces. It is evident that the lengths of these paths increase with an increase in the typical room dimensions.

For a given ray, it seems reasonable that, for a large number N of reflections during a time t, all of the possible free path lengths appear so often, according to their corresponding probabilities, that we may divide the cumulative path-length (ct) by the number of reflections N so as to define a mean free path length, according to the equation:

$$\frac{1}{l_{\mathrm{m}}}=\frac{N}{ct} \tag{2.4}$$

For another ray, the distribution of the steps over time will be different. Since we measure the sound decay at a particular observation point, where numerous sound rays are always superimposed, the steps belonging to a single sound ray will not be observed. But if all these rays experience (on the average) the same number of reflections during the time of observation, they will have travelled the same 'mean free path'.

Sabine tried to evaluate this mean free path length from the reverberation that he measured in rooms for which he could calculate the equivalent absorption area on the basis of other experiments. For this purpose, he defined a mean sound absorption coefficient $\bar{\alpha}$ as the arithmetic mean of the different absorbing surfaces (see eqn. (1.33)):

$$\bar{\alpha}=\frac{\sum \alpha_k S_k}{S} \tag{2.5}$$

Thus, he replaced the variable steps in Fig. 2.2 (left) with the uniform steps of Fig. 2.2 (right) having the same mean slope. Furthermore, he approximated the step-wise decay with a continuous decay, according to the differential equation, (1.1). With $\bar{\alpha}$ as the relative energy loss at each step, and with ct/l_{m} steps during the period t, he derived the following differential equation for the decrease of the 'intensity' of the ray at an arbitrary observation point:

$$-\frac{\mathrm{d}J}{J}=\frac{\bar{\alpha}c\,\mathrm{d}t}{l_\mathrm{m}} \tag{2.6}$$

But this means that the relaxation time τ, and, therefore, according to eqn. (1.6), the reverberation time T, are proportional to the quotient of the mean free path length and the mean absorption coefficient:

$$T=\frac{6\ln 10}{c}\frac{l_\mathrm{m}}{\bar{\alpha}} \tag{2.7}$$

On the basis of measured reverberation times and calculated values of $\bar{\alpha}$, Sabine evaluated the mean free path length for two rectangular rooms of different size but similar shape:

$$l_\mathrm{m}=0{\cdot}62\sqrt[3]{V} \tag{2.8}$$

He expressed the length dimension by $\sqrt[3]{V}$, since, considering the very different shapes of auditoriums, theaters, churches, and so on, there was no reason to choose the dimension in any particular direction. But he did not rule out the possibility that the factor 0·62 in eqn. (2.8), which he determined for the ratios 6:3:2 of length:width:height, might depend on the particular shape.

If we compare eqn. (2.7) with Franklin's earlier derivation of eqn. (1.35):

$$T=\frac{6\ln 10}{c}\frac{4V}{S\bar{\alpha}} \tag{2.9}$$

we find that the dependence on room shape is given by the quotient of the volume V and the total surface area S as follows:

$$l_\mathrm{m}=\frac{4V}{S} \tag{2.10}$$

This is the same value that Clausius[2] found for the 'mean free path length' in his mechanical treatment of the theory of heat, by considering the motion of molecules of gas in an enclosure. But neither Sabine nor Franklin came to this conclusion from a comparison of eqns. (2.7) and (2.9). Equation (2.10) was introduced into room acoustics by Jaeger,[3] whose derivation followed the same course as that of Clausius.

[2] Clausius, R., *Die kinetische Theorie der Gase*, 2nd. edn., Braunschweig, 1889/91, p. 51.

[3] Jaeger, G., *Wiener Akad. Ber. Math. Nat. Kl.*, **120** (1911) 613. At the decisive point where he averages cos ϑ over the directions of interest, Jaeger refers to 'corresponding textbooks', and not to Clausius' original paper. His statement that the mean value of cos ϑ is 1/4 glosses over the facts that the true mean value is 1/2 and that only half the energy is directed towards the wall.

The relationship between Franklin's and Clausius' treatments is evident from the fact that the same integral for averaging over all directions appears (see eqn. (1.29a)). But Clausius started out pursuing an apparently quite different question: he was seeking the probability for a particle, propagating with the velocity c, to encounter a wall element $\mathrm{d}S$ during the time t. For this purpose, we may alternatively regard the particle to be at rest and the wall element to be moving with the same velocity in the opposite direction. It traces a course through the prismatic volume

$$c \cos \vartheta \, \mathrm{d}S \mathrm{d}t$$

where ϑ is the angle between the direction of motion and the inward normal to the wall surface. If the sound field is homogeneous everywhere in the room, then the probability that a particle meets $\mathrm{d}S$ at the incidence angle ϑ is equal to the ratio of that prismatic volume to the entire volume of the room:

$$c \cos \vartheta \, \mathrm{d}S \mathrm{d}t / V$$

But this is true only if the particle is progressing *toward* the wall, i.e. only if the cosine is positive; otherwise, the probability is zero.

Furthermore, the probability for an angle of incidence lying between ϑ and $\vartheta + \mathrm{d}\vartheta$ is, as in eqn. (1.27),

$$\frac{2\pi \sin \vartheta \, \mathrm{d}\vartheta}{4\pi} = \frac{1}{2} \sin \vartheta \, \mathrm{d}\vartheta$$

and thus, the probability for an element to meet $\mathrm{d}S$ within this range of angles of incidence is:

$$\frac{c \cos \vartheta \sin \vartheta \, \mathrm{d}\vartheta \, \mathrm{d}S \, \mathrm{d}t}{2V}$$

If we now ask for the probability that $\mathrm{d}S$ is met under any angle of incidence, we must integrate ϑ from 0 to $\pi/2$ (that is, over half the sphere):

$$\frac{c \mathrm{d}S \mathrm{d}t}{2V} \int_0^{\pi/2} \cos \vartheta \sin \vartheta \, \mathrm{d}\vartheta = \frac{c \mathrm{d}S \mathrm{d}t}{4V}$$

And if we proceed to the probability that any part of the entire surface is encountered, an integration over $\mathrm{d}S$ leads immediately to:

$$cS\mathrm{d}t/(4V)$$

When dt becomes the average time interval between the encounters of a sound ray with the room boundary, i.e. the mean free path transit time, l_m/c, then, naturally, this probability is 1, i.e. a certainty:

$$\left(\frac{cS}{4V}\right)\frac{l_m}{c}=1 \tag{2.10a}$$

We must note, now, that we have changed our definition of the mean of the free path. We started by studying the fate of a ray, or, equivalently, the path of an energy particle. This means that we defined the mean free path as a time average. Clausius' procedure averages all of the particles in the room during the same time interval. In statistics this is called an ensemble-average. Franklin's power-balance also involved a consideration of the ensemble of all energy particles. It is an important statement in statistics that, in many ideal cases, both kinds of averaging give the same mean values. Such systems are called 'ergodic'. But this agreement does not always occur. In Section II. 2.3, we shall treat sound fields in geometrically simple (rectangular and spherical) rooms with specularly reflecting walls where the mean free path length of a specific particle is not the same as the ensemble average. And in Section II. 2.4 we shall see that the most effective way to get an 'ergodic room' is with ideal diffuse reflection.

Kosten[4] has given another derivation of the mean free path length, in the sense of an ensemble average, that is geometrically lucid. If we subdivide the volume V of the room by tubes having a very small cross-section Δq and running in an arbitrary direction (see Fig. 2.3(a)), then we get the mean free path length for this direction of subdivision as the sum of the lengths of the tubes divided by their number. The first is equal to $V/\Delta q$, the second to $P/\Delta q$, where P is the projected area of the room normal to the chosen direction:

$$\bar{l}_{(\alpha)}=\frac{V}{P_{(\alpha)}} \tag{2.11}$$

This angle α (which describes the angle of projection in Kosten's derivation) must be distinguished from the angle ϑ which a ray (or tube) forms with the perpendicular to a surface.

[4] Kosten, C. W., *Acustica*, **10** (1960) 245. W. B. Joyce (*J. Acoust. Soc. Am.*, **58** (1975) 643) mentions a paper by E. Czuber (*Wiener Akad. Wiss. Math. Nat. Kl.*, *2A*, **90** (1884) 719), where he gave a purely geometrical proof of eqn. (2.10) for the ensemble average, some five years before Clausius' derivation of the mean free path length.

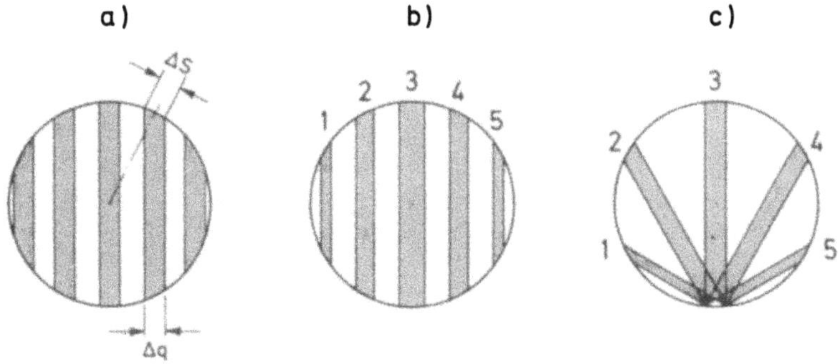

Fig. 2.3. A sphere divided by: (a) tubes of equal cross-section; (b) tubes of equal end-surface area; (c) same as (b) but with all tubes radiating from the same element ΔS.

In the case of a sphere of diameter D, for which the projected area is the same in all directions, this $\bar{l}_{(\alpha)}$ is the overall mean free path length and is equal to the value given by Clausius' formula:

$$l_{\mathrm{m}} = \bar{l}_{(\alpha)} = \frac{(\pi/6)D^3}{(\pi/4)D^2} = \frac{2}{3}D = \frac{4V}{S} \tag{2.11a}$$

It seems surprising that such an extreme focusing geometrical shape as the sphere should obey statistical laws. We know from Chapter I.3, that this by no means happens because of the regular reflections, which will more or less always occur in practice, and we will restate this in Section II. 2.3. But we will learn in Section II. 2.4 that it is Clausius' weighting factor, $\cos\vartheta$, that characterizes an isotropic, diffuse sound field and which makes the sphere especially suitable for statistical treatment.

At first glance, it appears that Kosten's derivation does not include this weighting factor. In fact, however, it shows up when we come to relate the equal cross-sections Δq of the tubes to the equal surface elements ΔS. Since Δq is equal to $\Delta S \cos\vartheta$, eqn. (2.11) may be written as

$$\bar{l}_{(\alpha)} = \frac{\sum l_k \cos\vartheta_k \,\Delta S}{\sum\limits_{S/2} \cos\vartheta_k \,\Delta S} = \frac{V}{\sum\limits_{s/2} \cos\vartheta_k \,\Delta S} \tag{2.11b}$$

Here the sums are to be calculated for only half the surface, in order to get the volume in the numerator and the projected area in the denominator.

Figure 2.3(a) shows tubes with equal Δq; Fig. 2.3(b) shows tubes with equal ΔS. For the special example of the sphere, and for this case only, we may rotate the tubes around the center so that they all start from the same surface element ΔS, as shown in Fig. 2.3(c). This corresponds to the other possibility, namely, that we can calculate the mean free path length according to the procedure of Kuttruff[5] and Joyce,[6] by averaging first over the angles of the rays leaving ΔS

$$l_{\mathrm{m}(\Delta S_k)} = 2\int_0^{\pi/2} l_{(\vartheta,\, \Delta S_k)} \cos\vartheta \sin\vartheta \, \mathrm{d}\vartheta \tag{2.12}$$

and then over the different wall elements. For the sphere, there is no difference between the wall elements, so that $l_{\mathrm{m}(\Delta S_k)}$ is the same as l_{m}. (Because of the presence of a weighting factor from the beginning, we may write l_{m} here instead of $\bar{l}$.) As may be seen in Fig. 2.6(b), below, for a sphere of diameter D, $l_{(\vartheta)}$ is $D\cos\vartheta$, so that we have again:

$$l_{\mathrm{m}} = 2D\int_0^{\pi/2} \cos^2\vartheta \sin\vartheta \, \mathrm{d}\vartheta = \frac{2}{3}D \tag{2.12a}$$

The most important and most frequently used example of a room for which the projected areas are different in different directions is the rectangular room. For that case:

$$P = l_y l_z \cos\alpha_x + l_z l_x \cos\alpha_y + l_x l_y \cos\alpha_z$$

and, according to eqn. (2.11),

$$\bar{l}_{(\alpha_x,\, \alpha_y,\, \alpha_z)} = \frac{1}{\dfrac{\cos\alpha_x}{l_x} + \dfrac{\cos\alpha_y}{l_y} + \dfrac{\cos\alpha_z}{l_z}} \tag{2.13}$$

In contrast to the case of the sphere, the cosines appear here in the denominator.

For the rectangular room, and for that case only, we get the same dependence on ϑ as here on α by following a ray that starts with the direction given by $\cos\vartheta_x$, $\cos\vartheta_y$ and the dependent $\cos\vartheta_z$. From the

[5] Kuttruff, H., *Room Acoustics*, 2nd. Edn., Applied Science Publishers Ltd, London, 1979, equation (V. 10). When Kuttruff also changes the sequence of integration in the following equation (V. 11), his procedure becomes more similar to that of Kosten.

[6] Joyce, W. B., *J. Acoust. Soc. Am.*, **58** (1975) 645, eqn. (3).

lower part of Fig. 2.6(a), we can easily see that the walls $x=0, l_x$, $y=0, l_y$, and $z=0, l_z$ are met again and again after free paths

$$l_x/\cos\vartheta_x,\ l_y/\cos\vartheta_y,\ l_z/\cos\vartheta_z$$

The combination of these reflections comprises a rather irregular sequence of free paths depending on the starting point. But we may add together the corresponding numbers of reflections (n_x, n_y, n_z) during time t to determine the resultant number n that defines a mean free path length for a given direction of the ray in the room:

$$\frac{ct}{l_{(\vartheta_x, \vartheta_y, \vartheta_z)}}=n=n_x+n_y+n_z$$

$$=\frac{ct\cos\vartheta_x}{l_x}+\frac{ct\cos\vartheta_y}{l_y}+\frac{ct\cos\vartheta_z}{l_z} \tag{2.13a}$$

(Note that we do not average the n's but add them together.)

The formal agreement between the expression for $\bar{l}_{(\alpha)}$ in (2.13) and $l_{(\vartheta)}$ in (2.13a) is not a consequence of the derivation, but is a special property of the rectangular room. Furthermore, here again $l_{(\vartheta_x, \vartheta_y, \vartheta_z)}$ is independent of the starting point and thus is also independent of the chosen surface element ΔS_k. Therefore, we can regard $l_{\mathrm{m}(\Delta S_k)}$ as equivalent to the mean free path length l_{m}, and use eqn. (2.12). By setting (2.13a) into (2.12), we find again Clausius' value, provided that we take into account that the weighting factor $\cos\vartheta$ must always correspond to the direction in which the ray starts out from a wall:

$$l_{\mathrm{m}}=2\int_0^{\pi/2}\frac{\sin\vartheta\,\mathrm{d}\vartheta}{\left(\frac{\cos\vartheta_x}{l_x}\right)\frac{1}{\cos\vartheta_x}+\left(\frac{\cos\vartheta_y}{l_y}\right)\frac{1}{\cos\vartheta_y}+\left(\frac{\cos\vartheta_z}{l_z}\right)\frac{1}{\cos\vartheta_z}}=\frac{4V}{S} \tag{2.12b}$$

We now proceed to rooms of arbitrary shape, and again we follow Kosten. As he explains (based on the corresponding two-dimensional problem), we must take into account the projected area (line) as a weighting factor in averaging $\bar{l}_{(\alpha)}$ (see eqn. (2.11)). This corresponds to subdividing the volume into tubes of equal cross-section Δq; it leads, if we carry it out for discrete angles α_k, to:

$$l_{\mathrm{m}}=\frac{\sum_n \bar{l}_{(\alpha_k)}P_{(\alpha_k)}}{\sum_n P_{(\alpha_k)}}=\frac{V}{\frac{1}{n}\sum_n P_{(\alpha_k)}} \tag{2.14}$$

or, expressed for a continuous change of α, to:

$$l_m = \frac{V}{\dfrac{1}{2\pi}\displaystyle\int_{2\pi} P_{(\alpha)}\,\mathrm{d}\Omega} \tag{2.14a}$$

Now we may always replace $P_{(\alpha)}$ by the integral $\int_{S/2}\cos\vartheta\,\mathrm{d}S$, where ϑ defines (as in (2.11b)) the orientation of $\mathrm{d}S$ to the direction of projection. When we substitute this integral into the denominator of (2.14a), we may change the sequence of integration and again replace $\cos\vartheta$ with its mean value 1/2; so we get once more Clausius' formula.

Finally, we can transform eqn. (2.14a) into another form. We have already introduced $P_{(\alpha)}=\sum\Delta q_{(\alpha)}$ in connection with eqn. (2.11); but this sum is identical with

$$(\sum\Delta q_{(\alpha)}\,l_{(\alpha)})/l_{(\alpha)} = (\sum\Delta V_{(\alpha)})/l_{(\alpha)}$$

and we may write this, instead of (2.14a):

$$\frac{1}{l_m} = \frac{1}{2\pi V}\sum\Delta V_{(\alpha)}\int_{2\pi}\mathrm{d}\Omega/l_{(\alpha)} \tag{2.14b}$$

or, in the integral form used by Joyce:[6]

$$\frac{1}{l_m} = \frac{1}{2\pi V}\int_V \mathrm{d}V\int_{2\pi}\mathrm{d}\Omega/l_{(\alpha)} \tag{2.14c}$$

As Bate and Pillow[7] have already demonstrated, this formula follows from physical considerations. We measure the decaying energy-density with a microphone at a particular location and later repeat this measurement at other locations in order to derive a space-average. In the theoretical limit we should average the decay rates over all points of the volume. At least the average over $\mathrm{d}V$ corresponds to our measuring procedure. Moreover, at every point sound rays from all directions are simultaneously recorded, and their next encounters with the room boundaries appear as steps. Their transit times, which are given here by l/c, govern the decay; therefore we must also integrate $\mathrm{d}\Omega/l_{(\alpha)}$.

As we have already proved in connection with Kosten's identical formula, this definition of l_m, where the inverses of the free path lengths are averaged, again yields Clausius' formula.

[7] Bate, A. E. and Pillow, M. E., *Proc. Phys. Soc. London*, **59** (1947) 535.

In the case of the rectangular room, this procedure is especially simple, because $l_{(\alpha)}$ is independent of the location, as it was above because of the choice of surface element. Therefore the average over volume vanishes and we get simply:

$$\frac{1}{l_{\mathrm{m}}}=\frac{1}{2\pi l_x l_y l_z}\int_{2\pi}(l_y l_z \cos\alpha_x + l_z l_x \cos\alpha_y + l_x l_y \cos\alpha_z)\,\mathrm{d}\Omega$$

$$=\frac{S}{4V} \tag{2.15}$$

The rectangular room and the sphere are examples where the mean free paths in different directions are different for specular reflection, and thus correspond to different decay exponents. But a sum of exponential decays with different decay coefficients never results in a unique exponential decay, as we shall discuss at the end of the next chapter. This arises only if all the decays, and thus the time averages of the mean free paths in all directions, are the same, i.e. the Clausius value. So the ergodicity of the problem amounts to a presupposition that we measure the same reverberation time at all points, and that purely geometric considerations, such as those of Kosten, result in the correct mean free path.

Kosten also explains that his procedure for evaluating $\bar{l}$ does not depend on the assumption that each surface element can send a ray directly to every other surface element, that is, that we are dealing with a so-called 'convex room'. (In the nomenclature of geometric room acoustics, it would be better to call such a room 'concave'.) Even for a room section like that shown at the top of Fig. 2.4, which might correspond to a concert hall with balcony, Kosten's derivation of l_{m} still holds. It is only necessary to regard the free paths above and below (at the right) as separate, and to give their projected area double weight, as suggested in Fig. 2.4 by the double thickness of the heavy line at the bottom of the figure.

Sometimes, however, such 'coupled rooms' are so low and so deep that it becomes questionable to treat them as a single room at all for a reverberation calculation (see Chapter II.3).

II.2.2 The Absorption Exponent

So far, we have demonstrated that the study of the average fate of the sound rays (or energy particles) under statistical conditions leads to the

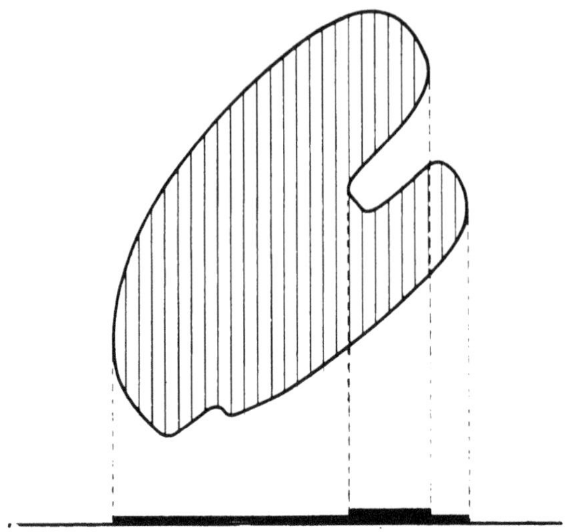

Fig. 2.4. Room with 'balcony', where the projected area (heavy line, below) must be partially doubled. (After Kosten.[4])

same formulae that we found by means of power balances in Section II.1.4. We shall now consider a point of view according to which the course of the sound rays corresponds better to the physical nature of reverberation, and which puts into evidence the idealized assumption that the 'intensity' of each ray is always diminished in the same ratio after tracing the same distance along a free path; this corresponds to step heights and widths of uniform size, as shown in Fig. 2.2 (right).

If we call the initial 'intensity' J_0, it will be reduced after the first reflection to $J_0(1-\alpha)$, after the second reflection to $J_0(1-\alpha)^2$, and after the Nth reflection to $J_0(1-\alpha)^N$. If we express N as in eqn. (2.4), we get for the 'intensity' of a ray at time t:

$$J_{(t)} = J_0(1-\alpha)^{ct/l_m} = J_0 e^{[\ln(1-\alpha)]ct/l_m} \quad (2.16)$$

in which we describe the step-wise decrease in intensity by a continuous function. Evidently, this corresponds to replacing the absorption coefficient α in eqn. (2.6) by the quantity

$$-\ln(1-\alpha) = \alpha' \quad (2.17)$$

which we may call the 'absorption exponent'; it describes the decay that occurs during the time required for a transit of the mean free path.

We can see, from an expansion of the logarithm in a power series:

$$\alpha' \doteq \alpha + \frac{\alpha^2}{2} + \frac{\alpha^3}{3} + \ldots$$

that α' approaches α for small values. But even for $\alpha = 0{\cdot}3$, the value of α' is 0·357, already a significant difference; for $\alpha = 0{\cdot}90$, the value of α' is 2·3 (more than double), and for α approaching 1, α' increases to infinity.

But this limiting case makes it clear how misleading the absorption coefficient α in the exponent can be. The exponential decay according to eqn. (2.6):

$$J_t = J_0 e^{-\alpha ct/l_m}$$

exhibits an impossible reverberation time of finite duration even if all the walls are totally absorbent ($\alpha = 1$); but if we replace the absorption coefficient α with the absorption exponent α' in eqn. (2.7):

$$T = \frac{6 \ln 10}{c} \frac{l_m}{[-\ln(1-\alpha)]} \tag{2.18}$$

we get the physically reasonable reverberation time of zero.[1]

The customary power balance in Sections II. 1.4 and II. 1.5 and the differential equation, (2.6), lead to reverberation times that are too high, because the power lost at each instant is treated as being always proportional to the total amount of energy remaining in the room. But this assumption is inconsistent with the concept of a discontinuous energy loss upon reflection. If, for instance, a plane wave train of finite length and constant power meets an absorbing wall, it gives up the same amount of power throughout the entire duration of the encounter, not a decreasing amount corresponding to the decreasing amount of energy left in the room. The energy particles, as they meet the wall, 'cannot know' how much energy the room has already lost.

On the other hand, we should note that, for sound radiation of sufficiently long duration, the expression for steady-state energy density, given by the build-up of the sound field:

$$VE_\infty = \frac{Pl_m}{c}[1 + (1-\alpha) + (1-\alpha)^2 + \ldots] = \frac{Pl_m}{c\alpha} = \frac{4}{c}\frac{PV}{S\alpha} \tag{2.19}$$

[1] This substitution was first introduced by Fokker, A. D., *Physica*, **4** (1924) 262; later by Waetzmann, E, and Schuster, K., in Müller-Pouillet, *Lehrb. d. Phys., 2. Auflage*, Band I, Teil 3, 1929, p. 457; and finally, by Eyring, C. F., *J. Acoust. Soc. Am.*, **1** (1930) 217.

does not require the replacement of α by α', because, in the stationary state, the step-wise process of sound absorption does not occur.

The assumptions for the statistical theory of reverberation are certainly better fulfilled the smaller the mean sound absorption coefficient. That is, the greater the number of reflections involved, the more we can expect adequate averaging over walls and directions. Therefore, it appears doubtful whether, in the context of statistical room acoustics, the replacement of α by α' has any practical significance at all.

We may also wonder whether the reverberation formulae lose their validity as soon as α and α' are significantly different. But if so, we must conclude from eqn. (2.17) that those formulae fail in normal concert halls: the mean absorption coefficient in such halls is typically about 0·3, and at some surfaces (for example, the audience) α approaches 1.

As we shall learn in Chapter IV. 11 (Volume 2), the wave-theoretical analysis of reverberation in a rectangular room, all of whose boundaries have uniform acoustical treatment, always leads to the absorption exponent α' in the damping constant, even though α' may be much greater than α. But the ray concept that we are using here also justifies the use of the absorption exponent α' in the statistical theory. It is, in fact, a matter of experience that exponential reverberation decays occur not only in cases where some of the absorption coefficients, but even where the mean absorption coefficient $\bar{\alpha}$, exceeds 0·3.

Now, however, we are confronted with the problem of how room boundaries with different absorption coefficients are to be considered, in order to arrive at the mean absorption coefficient. For this purpose, it might seem at first that it would be consistent practice, according to Sabine's formula (eqn. (1.35)), simply to replace α by α' at each boundary surface. This would change the reverberation time formula to:

$$T = \frac{0{\cdot}163\, V}{\sum \alpha'_k S_k} \tag{2.20}$$

a form that was recommended by Millington and Sette.[2,3]

A comparison between two reverberation tests, as discussed in Section II. 1.5, would thus exhibit the following difference in absorption

[2] Millington, G., *J. Acoust. Soc. Am.*, **4** (1932) 69.
[3] Sette, W. J., *J. Acoust. Soc. Am.*, **4** (1933) 160.

exponents:

$$(\alpha_x' - \alpha_0') = \frac{0{\cdot}163\ V}{S_x}\left(\frac{1}{T_\Delta} - \frac{1}{T_0}\right) \tag{2.21}$$

So long as we evaluate sound absorption data from measured reverberation times and use these data only for the calculation of reverberation times, this change would allow us to take over all the results that we have heretofore developed, using α instead of α'. The new equations, (2.20) and (2.21), would mean only that the quantities that we had earlier regarded as absorption coefficients (see eqns. (1.35a) and (1.36)) were actually absorption exponents. We would run into trouble only if the published tables of absorption data presented sometimes absorption coefficients and sometimes absorption exponents. For then we would not know whether to use these values in Sabine's or in Millington–Sette's formula.

These authors have also derived their new theoretical formulae from a consideration of the mean fate of sound rays, on the assumption that each ray encounters the different walls one after the other. Let us take as the simplest case the one-dimensional situation shown in Fig. 2.5(a), i.e. a

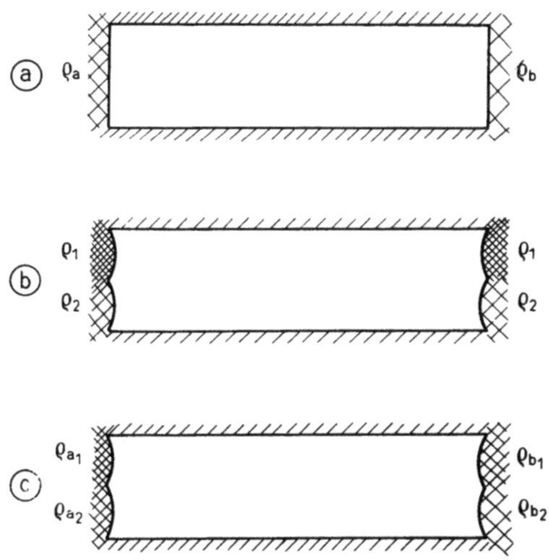

Fig. 2.5. Schematic sketch of the difference between: (a) surfaces encountered one after the other; (b) surfaces encountered side by side; and (c) surfaces encountered both one after another and side by side.

narrow tunnel with two different absorbing walls at the ends (absorption coefficients, α_a and α_b; reflection coefficients, $\rho_a=(1-\alpha_a)$ and $\rho_b=(1-\alpha_b)$, respectively), and no absorption at all on the side walls, ceiling and floor. Then we get for the decay of the intensity:

$$J_{(t)}=J_0(\rho_a\rho_b)^{ct/2l}=J_0e^{-\frac{1}{2}(\alpha'_a+\alpha'_b)ct/l} \tag{2.22}$$

Evidently, the mean absorption exponent here is the arithmetic mean of the two absorption exponents at the ends. We may equally say that the mean reflection coefficient ρ_m is the geometric mean of the reflection coefficients at the two end walls:

$$\rho_m=(\rho_a\rho_b)^{1/2} \tag{2.23}$$

In a room with n different surfaces of different sizes, it seems reasonable to assume that the probability for one of them to be met by a sound ray increases according to the ratio of its area to the total area of the room, S_k/S, as Sabine assumed for his averaging formula, eqn. (2.5). Millington and Sette further assume that all of these surfaces are encountered one after the other by each ray, as illustrated in Fig. 2.5(a). They thus developed, for a large number of reflections, the following formula for the decay of the ray intensity:

$$\begin{aligned} J_N &= J_0\rho_a^{NS_a/S}\rho_b^{NS_b/S}\ldots \\ &= J_0\exp[-N(\alpha_a'S_a+\alpha_b'S_b+\ldots)/S] \end{aligned} \tag{2.24}$$

This entails the same averaging procedure for absorption *exponents* as Sabine assumed for absorption *coefficients*:

$$\alpha'=\sum\alpha_k'S_k/S \tag{2.25}$$

Millington and Sette emphasize, as an advantage of their formula, that eqn. (2.20) can never lead to absorption coefficients greater than 1, no matter what differences in $(\alpha_x'-\alpha_0')$ may occur. But the formula also presents a corresponding difficulty: if any of the room surfaces, no matter how small, has an absorption coefficient of unity, then the calculated reverberation time is zero, certainly a contradiction of reality! If we open only a small window in a large reverberant room, such as a church, the corresponding influence on the reverberation time is rather small. Thus, the arguments of Millington and Sette cannot suddenly suspend the application of statistical methods and so are not a serious threat to the statistical reverberation theory of room acoustics.

This shortcoming of the Millington–Sette analysis for the case of an open window (of which Sabine even took a special advantage, for the purpose of defining a standard unit of sound absorption) is a consequence of the restrictive assumption that all rays share the same fate given by eqn. (2.24). This assumption demands that each ray not only meets each wall surface with a probability proportional to the size of that surface, but also (despite the fact that no account is taken of the unavoidable dependence of the absorption coefficients on the angle of incidence) at all possible angles of incidence, one surface after the other, in turn. This assumption is surely much more difficult to realize in practice than the homogeneous and isotropic distribution of sound energy required for the statistical analysis.

Furthermore, the sequence of the reflections in a real reverberant process should have no effect; but for the Millington–Sette theory this is evidently violated in the case of an open window: its ray-killing effect, according to eqn. (2.20), results in a zero reverberation time whether the encounter with the open window occurs as an early or a late 'reflection'.

For these reasons, Eyring[4] expressed doubt that there exist any real room situations at all for which the assumptions of the Millington–Sette analysis are fulfilled. (The one-dimensional case of Fig. 2.5(a) is, indeed, an exception.) In his earlier investigation, Eyring[5] took over Sabine's arithmetic mean of absorption coefficients according to eqn. (2.5), but substituted for Sabine's $\bar{\alpha}$ the corresponding absorption exponent in the reverberation time formula:[6]

$$T = \frac{0{\cdot}163V}{S[-\ln(1-\sum \alpha_k S_k/S)]} \tag{2.26}$$

This combination is not at all inconsistent; it corresponds, as Andrée[7] has explained, to the assumption that all the room boundary surfaces, after each mean free path transit, are encountered 'side by side', according to the arrangement shown in Fig. 2.5(b). Here the walls at the opposite ends of the tunnel are equal, with areas $S_a = S_b = S/2$; but they are split up into two equal parts: $s_1/2$ with the reflection coefficient ρ_1 and $s_2/2$ with the reflection coefficient ρ_2. After the first reflection, the

---

[4] Eyring, C. F., *J. Acoust. Soc. Am.*, **4** (1933) 178.

[5] Eyring, C. F., *J. Acoust. Soc. Am.*, **1** (1930) 217.

[6] Eyring emphasized in his later paper that he regarded this particular averaging as only one special possibility. But since he discussed only this case in his earlier paper, it has become usual to call eqn. (2.26) (and this formula only) the Eyring reverberation equation.

[7] Andrée, C. A., *J. Acoust. Soc. Am.*, **3** (1932) 535.

mean intensity is reduced to:

$$J_1 = J_0(\rho_1 s_1/S + \rho_2 s_2/S) \tag{2.27a}$$

It is clear from this arrangement that energy reflected from the single partial surface $s_1/2$ will subsequently meet both of the opposite partial areas $s_1/2$ and $s_2/2$. We must expect such mixing processes even more in three-dimensional rooms, where more and more surfaces are encountered 'side by side'.

After the second reflection, the intensity is decreased to

$$J_2 = J_0[(\rho_1 s_1/S)(\rho_1 s_1/S + \rho_2 s_2/S) + (\rho_2 s_2/S)(\rho_1 s_1/S + \rho_2 s_2/S)]$$
$$= J_0(\rho_1 s_1/S + \rho_2 s_2/S)^2 \tag{2.27b}$$

and, after the Nth reflection, to:

$$J_N = J_0(\rho_1 s_1/S + \rho_2 s_2/S)^N \tag{2.27c}$$

In this case, the fates of the various particles are not always characterized by an exponential decay of intensity; such decays happen only where the same reflection coefficients are repeated, for example, in the first and last terms of the binomial expansion:

$$(\rho_1 s_1/S),\ (\rho_1 s_1/S)^2, \dots\ (\rho_1 s_1/S)^N$$

or:

$$(\rho_2 s_2/S),\ (\rho_2 s_2/S)^2, \dots\ (\rho_2 s_2/S)^N$$

The very fact that the uniform exponential decay of the total energy may contain exponential decays with different damping constants proves, following up the implications of eqn. (2.46), below, that non-exponential decays are also possible.

It is also possible, as Kraak[8] and Kuttruff[9] have demonstrated, to start out with the terms of the binomial expansion of eqn. (2.27c), which may be derived from the probabilities (*W̲ahrscheinlichkeiten*) W for N reflections, that N_1 are reflected at S_1 and $(N - N_1)$ are reflected at S_2. For this, we find (from a formula of Newton):

$$W_{(N_1)} = \binom{N}{N_1}\left(\frac{S_1}{S}\right)^{N_1}\left(\frac{S_2}{S}\right)^{N-N_1} \tag{2.28a}$$

[8] Kraak, W., *Hochfrequ. u. Elektroak.*, **64** (1955) 90.
[9] Kuttruff, H., *Acustica*, **8** (1958) 273.

Here, the reverberation is composed of many decay processes that are, in general, not exponential:

$$J_N = J_0 \sum_{N_1=0}^{N} \left[\binom{N}{N_1} \left(\rho_1 \frac{S_1}{S} \right)^{N_1} \left(\rho_2 \frac{S_2}{S} \right)^{N-N_1} \right] \tag{2.28b}$$

but which, all together, according to the binomial theorem, lead to the exponential decay of eqn. (2.27c).

From the mathematical viewpoint, this representation is better adapted to statistical considerations than our simple splitting of eqn. (2.27c) into parts. But since it leads to the same result (the Eyring formula), it follows that it too is based on the assumption of sound rays meeting the absorptive surfaces side by side.

A comparison of eqn. (2.27c) with

$$J_N = J_0 \rho_m^N = J_0 e^{-\alpha_m' N}$$

shows that we make use of the mean absorption exponent

$$\alpha_m' = -\ln(\sum \rho_k S_k / S) = -\ln\left(1 - \frac{\sum \alpha_k S_k}{S}\right) \tag{2.29}$$

Extended to rooms with many different surfaces, the Eyring formula assumes that the energy reflected at one surface is again and again distributed to all the other surfaces in proportion to their sizes. This assumption is increasingly well fulfilled the more the different absorbing surfaces are distributed over the walls, ceiling and floor of the room, and the greater the degree of diffusion.

This ideal diffuse state is not always well realized in large rooms. Covering the floor and the rear wall of an auditorium with the sound-absorptive audience, while providing uniformly treated side-walls and ceiling, favors the condition that these surfaces are encountered by the sound rays 'one after the other' rather than 'side by side'.

If we wish to treat situations with only a single, very highly absorptive surface (as in courtyards or cathedrals with people crowding the floor), then Eyring's formula would yield a reverberation time that is much too long; the Millington–Sette formula, on the other hand, would predict zero reverberation time.

This latter statement should not be understood to mean that the direct sound will not be followed by a more or less long sequence of reflections; but it does preclude a three-dimensional reverberation. In these cases, we may have at most a two-dimensional reverberation, with divergence of sound rays into the third dimension unhindered; here, none of the formulae are valid. The Millington–Sette formula yields in this case a 'reasonable result' (namely, zero RT), indicating that the assumptions necessary for the three-dimensional statistical sound field are not fulfilled.

In room-acoustical practice, the so-called Eyring formula, where the mean absorption coefficient is replaced by the mean absorption exponent, is the only formula that is seriously recommended as an alternative to Sabine's simpler formula. It is not the change from α to α' that makes it more toilsome to work with the Eyring formula, but rather the additional calculation of the total surfaces. In Sabine's formula, only the expression $\Sigma\alpha_k S_k$ appears, and its value is not much changed even if all the 'hard surfaces' are ignored. In Eyring's formula, *all* the surfaces enter without weighting on account of their absorption.

On the other hand, Eyring's formula should not be blamed for the fact that, when it is used to evaluate reverberation time measurements, it can (like Sabine's formula, but not that of Millington–Sette) lead to values of the sound absorption coefficient greater than 1. From the reciprocal of eqn. (2.26), for the case in which the absorption area comprises the absorption of the empty room plus that of the material under test, $\Delta A = S_x\alpha_x$, it follows that:

$$\alpha_x = \frac{S}{S_x}\left(1 - e^{-0{\cdot}163\, V/(ST)}\right) - \frac{A_0}{S_x} \tag{2.30}$$

The probability that α_x exceeds unity is greater, the smaller S_x is compared with S. On the other hand, small sample sizes will produce only small changes in the reverberation time, leading to poor accuracy in the determination of the absorption coefficient. Furthermore, with small sample size, we must contend with the diffraction of sound waves into the material.

A reverberation formula that pretends to achieve accurate results, even under such unfavorable conditions, should be regarded with as much skepticism as a meter that cannot indicate an overload because the needle is already pinned!

The justifiable introduction of the absorption exponent has thus confronted us with the difficulty that we now have two different ways of averaging over the different surfaces in the room: one which is suitable if the sound rays encounter the surfaces 'one after the other', the second, if these surfaces are met 'side by side'. In practice, we may never encounter *only* one or the other of these two extremes.

It is, therefore, tempting to combine both possibilities, as suggested in Fig. 2.5(c), where the end surfaces S_a and S_b are subdivided, and all four partial surfaces are different. This results in an attenuation factor, corresponding to a path-length $2l$, of:

$$(\rho_{a_1}s_{a_1}/S_a+\rho_{a_2}s_{a_2}/S_a)(\rho_{b_1}s_{b_1}/S_b+\rho_{b_2}s_{b_2}/S_b)$$

Extending this concept to three-dimensional rooms, it implies a recommendation to divide the total surface area first into several large 'principal surfaces', which can be regarded as encountered by the sound rays one after the other ('in series'); and to subdivide these large surfaces into smaller areas with different absorption, which can be regarded as encountered side by side ('in parallel').

The smaller subdivisions are first averaged arithmetically, to determine the mean absorption coefficients for each of the principal surfaces. Those values are then transformed to the corresponding absorption exponents, and are averaged arithmetically:

$$\alpha_m'=\sum_q S_q\,[-\ln\,(1-\sum_k \alpha_{qk}\;s_{qk}/S_q)]/S \tag{2.31}$$

Even this combination formula will seldom be fully adequate, because the distinction between principal surfaces and subdivisions will not always be unequivocal. For large rectangular rooms, we may certainly regard the ceiling, the floor, the front and rear walls, and the side walls as principal surfaces. But even here it cannot always be expected that these surfaces are encountered one after the other, especially if we expect, on the other hand, that (because of divergence and diffusion) the sound rays are distributed equally over these areas.

But the combination formula (2.31) has the advantage that it yields reasonable results in the cases of small open windows and principal areas with high absorption. It embraces the formulae both of Eyring and of Millington–Sette where they are adequate, but it avoids their shortcomings. Moreover, it can never give results that are physically impossible. But the combination of these two limiting cases in eqn. (2.31) results in calculated reverberation times that depend upon the distribution of the absorptive materials in the room (see also Section II. 2.4).

II.2.3 The Influence of Direction in Regular Reflections

In agreement with the authors whom we have cited in the preceding section, we have not explicitly taken into account the dependence of the absorption coefficients on the angle of incidence in our discussion of the mean absorption exponent. But in Part I, Fig. 6.1, we learned not only that such a dependence exists, but also that the values of $\alpha_{(10^\circ)}$ and $\alpha_{(70^\circ)}$ for the same material may differ as widely as the absorption coefficients for different absorbing surfaces in a room.

Instead, in order to accommodate these differences, we made use of coefficients averaged over all angles of incidence, as given in Section II. 1.4 by Paris' Equation (1.30a):

$$\alpha_m = \int_0^{\pi/2} \alpha_{(\vartheta)} \sin 2\vartheta \, d\vartheta \tag{2.32}$$

The use of this formula implies the assumption (in the sense of the presuppositions of Eyring's formula) that each sound ray not only encounters all the room surfaces side by side, but also under all possible angles of incidence. It seems doubtful that this assumption can be fulfilled in reality.

With respect to the angles of incidence, it would seem to be physically more appropriate to assume that the different angles of incidence occur 'one after the other'. Thus, in Paris' formula we should average the absorption exponents, rather than the absorption coefficients:

$$\alpha_m' = \int_0^{\pi/2} \alpha'_{(\vartheta)} \sin 2\vartheta \, d\vartheta \tag{2.33}$$

But this averaging, if it is to be meaningful, requires a very large number of reflections during the observation time, a condition that is not likely to be met except for very small absorption exponents.

Furthermore, eqn. (2.33) can be used only in combination with Millington–Sette's method of averaging over different areas, because the sequence of averaging must be interchangeable if the procedure is to be applicable to all rooms.

In the last section, we distinguished between sequences of encounters that occur either 'one after the other' or 'side by side', but we were interested only in how this affects the absorption coefficients, that is, how it affects the heights of the steps in Fig. 2.2.

We shall now learn that averaging over the directions of sound ray incidence with respect to the mean absorption coefficients, and averaging over the various free paths may be inextricably connected for regular reflections. This is evident in the case of rooms with large plane surfaces, of which the rectangular room presents the simplest example.

We have already studied the ray-geometric properties of the rectangular room in Section I. 2.2, where we found that a specific sound ray encounters the respective walls always under the same angle of incidence. On the basis of the mirror principle, presented again in Fig. 2.6, top, in

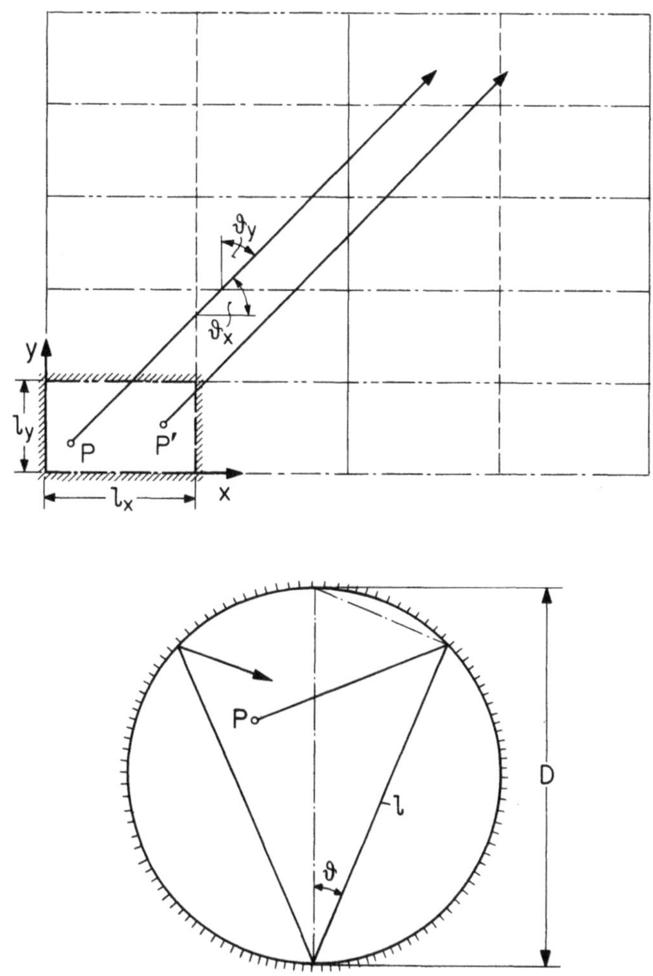

Fig. 2.6. Sketch for discussion of the relation between the length of the free path and the angle of incidence: top, in a rectangular room; bottom, in a sphere.

which all the image walls appear in the ground plan as parallel lines, we see that all the walls defined by x=const. are always met at the same angle ϑ_x and all the walls defined by y=const. are always met at the same angle ϑ_y. A section drawing of the room would show that all the walls defined by z=const. are always met at the same angle ϑ_z. (Naturally, only the projections of these angles appear in the plan and section.) But such a regular set of reflections cannot lead to an averaging over incidence angle if the reflections are specular.

From our studies of free paths in rectangular rooms, we know that opposite pairs of walls are always encountered after free paths with the lengths $l_x/\cos\vartheta_x$, $l_y/\cos\vartheta_y$ and $l_z/\cos\vartheta_z$. Each wall is encountered $N_x/2$, $N_y/2$ or $N_z/2$ times during t.

Now both relations appear combined when we ask for the resultant exponent with which the intensity of a ray in a given direction decreases. Here we regard the six different walls ($x=0$, l_x; $y=0$, l_y; and $z=0$, l_z) as homogeneous but with different absorption coefficients: α_{x0}, α_{xl_x}, α_{y0}, α_{yl_y}, α_{z0} and α_{zl_z}. If we compare:

$$J_N = J_0(1-\alpha_{x0})^{N_x/2}(1-\alpha_{xl})^{N_x/2}(1-\alpha_{y0})^{N_y/2}(1-\alpha_{yl})^{N_y/2}\ldots,\ \text{etc.}$$

with the general form:

$$J_N = J_0\, e^{-ct\alpha'_m/l_m}$$

we get:

$$\frac{\alpha'_m}{l_m} = \frac{(\alpha_{x0}'+\alpha_{xl}')\cos\vartheta_x}{2l_x} + \frac{(\alpha_{y0}'+\alpha_{yl}')\cos\vartheta_y}{2l_y} + \frac{(\alpha_{z0}'+\alpha_{zl}')\cos\vartheta_z}{2l_z} \qquad (2.34)$$

We may introduce $2l_xl_yl_z = 2V$ and thus transform eqn. (2.34) into:

$$\frac{\alpha'_m}{l_m} = \frac{\sum\alpha_k' S_k \cos\vartheta_k}{2V} \qquad (2.34a)$$

Furthermore, if we replace the mean free path length l_m with Clausius' value $4V/S$, we get:

$$\alpha'_m = 2\sum\alpha_k' \cos\vartheta_k S_k/S \qquad (2.34b)$$

If we now replace the quantity $\cos\vartheta$ by its mean value 1/2, we get the mean value of the absorption coefficient according to eqn. (2.25):

$$\alpha' = \sum\alpha_k'\, S_k/S$$

But our consideration of specular reflections shows that we must expect different damping constants for rays travelling in different directions; thus, no uniform exponential decay is possible if the sound rays are distributed over different directions in the initial state.

A uniform exponential decay is possible only if the dependence of α' on ϑ follows the rule:

$$\alpha'_{(\vartheta)} = \alpha'_{(0)}/\cos\vartheta \qquad (2.35)^1$$

that is, only if the dependence of absorption on ϑ exactly compensates that of the free paths.

Now, most absorptive wall coverings possess the property that the absorption coefficient increases with the angle of incidence (see Sections IV. 8.8 and IV. 11.5, Volume 2). We may, for instance, refer here to the experimental data of Section I. 6.2, Fig. 6.1, for the absorption of a wood fiber plate. If, as in Fig. 2.6, top, we compare the highest and lowest curves of that figure with those calculated from eqn. (2.35) (assuming the same values for $\alpha_{(0)}$), we find an agreement that, for the upper curve, is within the error of measurement, and, even for the lower curve, is within 10%. Thus, it seems conceivable that the essentially exponential decay in rectangular rooms is related to this frequently observed angular dependence of the absorption coefficients.

If we evaluate a reverberation process according to eqn. (2.34b), this means that, in the sense of Millington–Sette, we would get, under the condition of eqn. (2.35), a mean absorption exponent that is twice that for perpendicular incidence:

$$\alpha_m' = 2\alpha'_{(0)} \qquad (2.36)$$

At grazing incidence ($\vartheta \to 90°$), the validity of eqn. (2.35) is obviously questionable. As we have already mentioned in Section I. 6.3 (and will prove in Sections IV.7.4, IV.11.5 and IV.11.6, Volume 2), each absorption coefficient (as a function of the angle of incidence) decreases with increasing angle after having passed through a maximum, and vanishes altogether as ϑ approaches 90°. It is physically necessary that no absorbing surface can influence sound rays that are propagated for any considerable distance parallel to that surface. If, in a rectangular room, only the floor or the ceiling (or both) are absorptive, we can expect a uniform exponential decay only if eqn. (2.35) holds for the three-

[1] Cremer, H. and Cremer L., *Akust. Z.*, **2** (1937) 225.

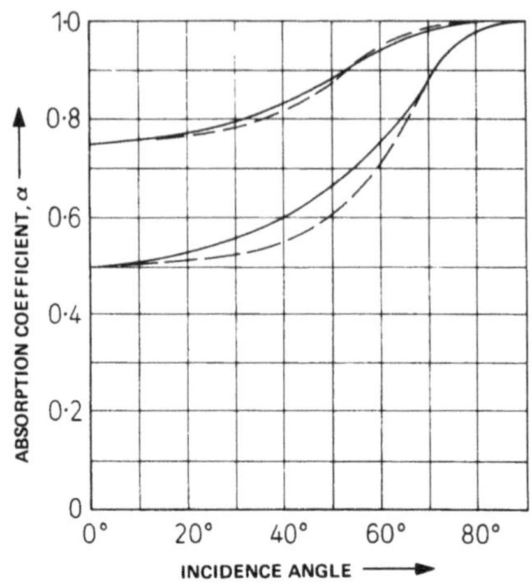

Fig. 2.7. Dependence of the absorption coefficient on the angle of incidence. ---- Measurements from Part I, Fig. 6.1, ——— calculated from eqn. (2.35). Lower curves, 1000 Hz; upper curves, 10000 Hz.

dimensional reverberation. For the overwhelming majority of the sound rays, this will characterize the early part of the reverberation; but finally, the two-dimensional reverberation, with rays traveling parallel to the absorbing surfaces, will prevail.

This effect appears even more pronounced if, in addition, one (or even both) lateral walls are absorbent. Then only a one-dimensional reverberation remains at the end. Here, the step-wise character of the superposition of sound waves may become noticeable, and, for impulse excitation, we may hear the equally spaced reflections that we described in Section I. 4.3 as flutter echoes.

In order to distinguish the rule of eqn. (2.35) (which for brevity we may call the 'cos ϑ-law'[2]) from these exceptions, we may now consider a room with a triangular floor, where flutter echoes cannot appear. We choose a right equilateral triangle (see Part I, Fig. 2.6), for which the image rooms fully cover the ground plane without overlapping and the image walls appear as straight lines; in this case the image walls form three square networks: two are shifted with respect to one another by half their width and the third is rotated by 45°. Each wall is met under two different

[2] It might be more logical to call it the '$(1/\cos\vartheta)$-law', but this would be longer; and the name 'secant' for $1/\cos\vartheta$ is seldom used.

angles of incidence; there is, therefore, some averaging with respect to angle in any case, and the requirement for compliance with the 'cos ϑ-law' is less strong. But even here it is true that the decay given by:

$$J_{(t)}=J_0 \exp \{-ct[(\alpha'_{a(\vartheta_{a1})}\cos\vartheta_{a_1}+\alpha'_{a(\vartheta_{a2})}\cos\vartheta_{a_2})/(2l) + (\alpha'_{b(\vartheta_{b1})}\cos\vartheta_{b_1}+\alpha'_{b(\vartheta_{b2})}\cos\vartheta_{b_2})/(2l) + (\alpha'_{c(\vartheta_{c1})}\cos\vartheta_{c_1}+\alpha'_{c(\vartheta_{c2})}\cos\vartheta_{c_2})/(\sqrt{2}l)]\} \tag{2.37}$$

is independent of the angle of incidence if the absorption coefficients obey the 'cos ϑ-law'.

For the expression in brackets in eqn. (2.37), which corresponds to α'_{m}/l_{m} and which may be supplemented in the three-dimensional case by the term $(\alpha'_{z0}+\alpha'_{zl})\cos\vartheta_z/2h$ (where h is the height of the room), we may again adopt the form of eqn. (2.34b), since again the shorter free paths correspond to the larger surface (the hypotenuse). But this time, two different values of $(\alpha'\cos\vartheta)$ appear at each surface, and these are to be averaged with the same weight.

According to Hubert Cremer,[1] it is even possible to show that the 'cos ϑ-law' results in an exponential decay independent of the direction, in rooms where the boundaries are composed of plane surfaces S_k. He assumes only a homogeneous sound distribution and applies Clausius' statement that the probability for a reflection at S_k during dt is given by $S_k c\, \mathrm{d}t \cos\vartheta_k/V$. (See Section II. 2.1.) He does not, however, assume that the sound field is isotropic. He introduces the time $\tau_{k(\vartheta)}\,\mathrm{d}\vartheta$ as that time interval during t in which the sound arrives from the directions between ϑ and $\vartheta+\mathrm{d}\vartheta$. For an isotropic sound field, as assumed in Section II. 2.1, we had:

$$\tau_k=t\frac{2\pi\sin\vartheta}{4\pi}=t\frac{\sin\vartheta}{2} \tag{2.38}$$

But whatever the directional distribution may be (given by $\tau_{k(\vartheta)}$), it is always true that the sound propagates toward S_k for only half the time, t, so we may always expect:

$$\int_0^{\pi/2} \tau_{k(\vartheta)}\mathrm{d}\vartheta=t/2 \tag{2.39}$$

If we introduce this distribution in the exponent of the decay and integrate over all directions, we must also include the absorption exponent $\alpha'_{(\vartheta_k)}$ in the integral. Thus, we get for the exponent:

$$\frac{(6 \ln 10)t}{T} = \frac{c}{V} \sum S_k \int_0^{\pi/2} \alpha_k' \cos\vartheta \, \tau_{k(\vartheta)} \mathrm{d}\vartheta \qquad (2.40)$$

If we now assume the validity of the 'cos ϑ-law' we find, independent of the shape:

$$\frac{(6 \ln 10)t}{T} = \frac{ct}{2V} \sum \alpha_k'{}_{(0)} S_k$$

or, converted to the form of Millington–Sette:

$$T = 0{\cdot}163 \frac{V}{\sum 2\alpha_k'{}_{(0)} S_k} \qquad (2.41)$$

The essential idea of this generalization is that we can drop the very restrictive assumption of the isotropy of the sound field, provided that, instead, we assume the 'cos ϑ-law' for absorption. Generally, this means that if we follow, instead of a single ray, a bundle of (nearly) parallel rays (corresponding to a broad plane wave) which strike a surface S_k with the angle ϑ, then the energy loss upon reflection is independent of the direction of incidence because the decrease of the projected area in this direction by cos ϑ is compensated for by an increase of the absorption exponent according to $1/\cos\vartheta$.

Equation (2.41) again assumes that all of the different walls are encountered one after the other, but it does not require that this occurs for all possible angles of incidence.

If we were to seek the conditions required for the same independence of angle of incidence when the surfaces are encountered side by side, we would have to replace the absorption exponents with the original absorption coefficients. In this case the 'cos ϑ-law' must be changed to:

$$\alpha_{(\vartheta)} = \alpha_{(0)}/\cos\vartheta \qquad (2.42)$$

If this relation were actually realized, it would mean that (in the sense of our original power balance in Section II.1.4, eqn. (1.29)), the power absorbed by the surface S_k would be independent of the angle of incidence. In that case, we could again drop the requirement for an isotropic sound field.

But it is evident that such a rule is physically impossible to fulfil. The definition of α permits only $\alpha \lesseqgtr 1$, and this limit will always be reached at a finite angle of incidence. Nevertheless, any tendency for α to increase with the angle of incidence will also decrease the influence of a particular distribution of angles of incidence.

Finally, we consider an exception which demonstrates that, for curved wall surfaces, the 'cos ϑ-law' does not guarantee independence of the incidence angle. We can see this best for the sphere, which is extremely different in its behavior from rooms bounded by plane surfaces, as we saw in the discussion of focusing in Section I. 3.1.

In our calculation of the mean free path in Section II. 2.1, where (surprisingly) the sphere was no exception to the general rule, we stated that its free path is given by (see Fig. 2.6, bottom):

$$l_{\mathrm{m}} = D \cos \vartheta \tag{2.43}$$

where D is the diameter.

During the decay, a ray travelling in any direction will meet the surface of the sphere with the same angle, again and again. So the decay of a ray may be described by:

$$J_t = J_0 \exp[-\alpha'_{(\vartheta)}\, ct/(D \cos \vartheta)] \tag{2.44}$$

If we wish all the rays to have the same rate of decay, we would have to require, contrary to eqn. (2.35), the inverse rule:

$$\alpha'_{(\vartheta)} = \alpha'_{(0)} \cos \vartheta \tag{2.45}$$

But since, in general, absorption coefficients increase with increasing angle of incidence, we must conclude, with respect to reverberation in a spherical room (or in a half-sphere over a plane, which is more likely to occur in practice), that the 'intensities' of the rays diminish more rapidly the more nearly grazing the angle of incidence. This implies that the rays traveling radially will prevail in the final phase of the reverberation. We see, therefore, that covering a dome with sound-absorptive material (as discussed in Chapter I. 3) is most effective for eccentric points of excitation and observation. The formation of a focal zone at or near the center may be slightly weakened, but the focus may, nevertheless, become more noticeable on account of the rapid dying away of the other rays.

Even in non-spherical rooms, the rays may not all decay with the same rate; we always have to allow that the reverberation will be composed of a sum of dissimilarly decaying exponential functions.

$$\begin{gathered} E = E_1 e^{-2\delta_1 t} + E_2 e^{-2\delta_2 t} + \ldots E_n e^{-2\delta_n t} \\ \delta_1 > \delta_2 > \ldots \delta_n \end{gathered} \tag{2.46}$$

or, in Kuttruff's[3] more general continuous integral form:

$$E_{(t)} = \int_0^\infty H_{(\delta)} e^{-2\delta t} \mathrm{d}\delta$$

This function starts with the initial damping constant:

$$\bar{\delta} = \frac{\sum \delta_k E_k}{\sum E_k} \tag{2.47}$$

or:

$$\bar{\delta} = \frac{\int_0^\infty \delta H_{(\delta)}\, \mathrm{d}\delta}{\int_0^\infty H_{(\delta)}\, \mathrm{d}\delta} \tag{2.47a}$$

which represents the steepest slope, and ends with the smallest (last) δ_n in the summation (or integral).

A plot of the level (log E) over time no longer leads to a straight line, but instead (expressed paradoxically but clearly), to a 'sagging reverberation curve'. If one were to try to evaluate the reverberation time from such a decay curve by fitting a straight line to it, the result would always depend on the portion of the decay that one chooses to fit.

We shall learn in Chapter II. 7 (and especially in Part III) that the first part of the reverberation decay is the most important. But it would be a mistake to concentrate our attention on only a small initial part of the slope, because the smaller the observed level difference (or time interval) the less accurate the evaluation.

It is certain that if $E_{(t)}$ is to be described by eqn. (2.46), then the reverberation following a steady-state excitation $E_{-(t)}$ will differ significantly from the reverberation following an impulse excitation $E'_{(t)}$. If we regard eqn. (2.46) as an adequate description of the first case, then eqn. (1.18) gives for the second case:

$$E'_{(t)} = 2\Delta t[\delta_1 E_1 e^{-2\delta_1 t} + \ldots \delta_n E_n e^{-2\delta_n t}] \tag{2.48}$$

or:

$$E'_{(t)} = 2\Delta t \int_0^\infty \delta\; H_{(\delta)} e^{-2\delta t} \mathrm{d}\delta \tag{2.48a}$$

[3] Kuttruff, H. *Acustica*, **8** (1958) 273.

But this means that the curve of log E' versus time 'sags' even more for impulse excitation.

II.2.4 Ideal Diffuse Reflection and Its Consequences

Although specular reflections, as treated in the last section, will always crop up in practice more or less, the simplest and most important statistical theory of reverberation is based on the assumption of ideal diffuse reflection.

A reflection is already at least partly diffuse whenever it is not completely specular: namely, if the sound wave (ray) incident on a surface with angle ϑ is reflected not only in the direction $\vartheta' = -\vartheta$ but at least a little in other directions as well. This happens at a 'rough' surface, as sketched in Fig. 2.8(a).

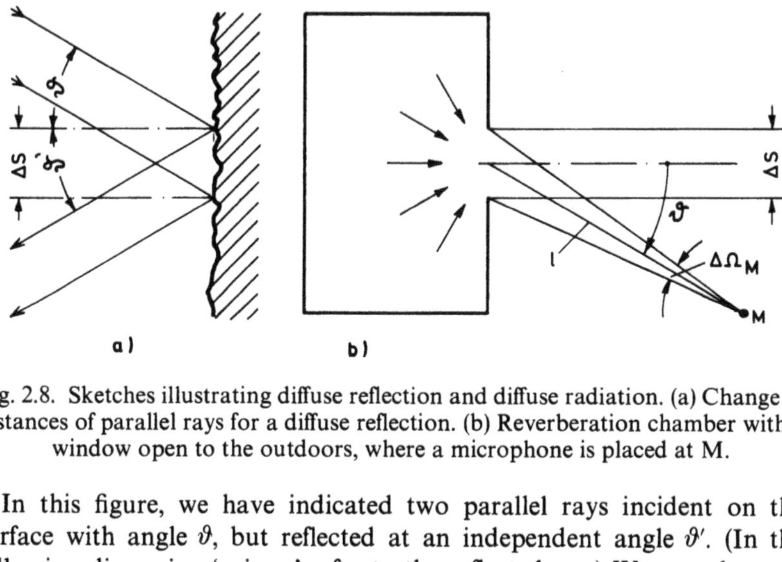

Fig. 2.8. Sketches illustrating diffuse reflection and diffuse radiation. (a) Change of distances of parallel rays for a diffuse reflection. (b) Reverberation chamber with a window open to the outdoors, where a microphone is placed at M.

In this figure, we have indicated two parallel rays incident on the surface with angle ϑ, but reflected at an independent angle ϑ'. (In the following discussion 'primes' refer to the reflected ray.) We note here an essential difference between specular and diffuse reflections. From the beginning, we have been accustomed to think that parallel sound rays 'preserve their distance' upon reflection: this means that the power and the intensity are both diminished upon reflection by the same ratio, given by the reflection coefficient ρ. (In the following discussion, we take it for granted that ρ may differ from surface to surface, and we consider either a particular surface or a reflection coefficient averaged over a number of surfaces.)

But if the angles ϑ and ϑ' are different, then the ratios of incident and reflected power and intensity will be different, even if there is no sound absorption associated with the reflection. For the increments of power and intensity, ΔP and ΔJ, incident upon and reflected from the surface element ΔS, the following relation holds:

$$\frac{\Delta P'}{\Delta P}=\frac{\Delta J' \cos \vartheta'}{\Delta J \cos \vartheta} \tag{2.49}$$

If we regard the sound rays as paths of energy-particles, the left side of this equation implies a splitting of the energy contained in the incident ray. (When we replace this energy-splitting concept with a computer analysis in which we follow individual rays, this ratio characterizes the *probability* for a change from ϑ to ϑ'.)

In principle, we have already become acquainted with this difference when we calculated (in Section II. 1.4) the power incident upon a surface. We assumed an equal angular distribution of 'intensity' over all directions, but we had to take into account that the area S of the surface could intercept only the projected area $S \cos \vartheta$ from the incident waves. So the directional distribution of the incident power decreased according to $\cos \vartheta$.

It will be convenient, in developing our present treatment of an 'ideal' diffuse reflection, to return to the assumption of an isotropic sound distribution in a room.

But we shall introduce notation that distinguishes between the power P and the power per unit of solid angle $\mathrm{d}P/\mathrm{d}\Omega$ and between the intensity J and the intensity per unit of solid angle $\mathrm{d}J/\mathrm{d}\Omega$. These distinctions, which are usual in radiology and optics, will be helpful in our statistical acoustics problems, as well.

Thus, we characterize the ideal, isotropic field of sound in a reverberant room by the statement that the 'angular current density' or 'radiance' $\mathrm{d}J/\mathrm{d}\Omega$ is equally distributed over all directions:

$$j=\frac{\mathrm{d}J}{\mathrm{d}\Omega}=\frac{Ec}{4\pi} \tag{2.50}$$ [1]

If we now assume that the reverberation room has a window open to the outdoors (or to a very large 'dead' room), we have an analogy to the

[1] We adopt the lower-case j for specific intensity from W. B. Joyce (*J. Acoust. Soc. Am.*, **58** (1975) 643), although in his paper j must be multiplied by the elementary energy of his energy particles, in order to equal $\mathrm{d}J/\mathrm{d}\Omega$.

heat radiation standard, namely, a small hole in the wall of a 'black cavity', for which eqn. (2.50) holds, also.

Since all of the following conclusions are based upon that analogy, we note here that for optical problems, a hole 1 mm in diameter represents more than 1000 wavelengths. If we consider a sound wave of frequency 3400 Hz in our reverberation chamber (i.e. near the upper limit of interest for room acoustics), a square window 2 m on a side would span only 20 wavelengths in each direction. Thus, we may be warned in advance that our extrapolation from optical studies to room-acoustical studies is no less doubtful for statistical theory than it was for the specular reflections considered in Part I. It is hardly surprising, therefore, that the rules derived from this analogy do not always agree with our acoustical measurements; what is astonishing is how well they agree most of the time!

For the problem of sound radiating from the reverberant room into the open air, we are interested only in the waves propagating toward the window; eqn. (2.50) holds for those waves, as well as for the waves just outside the window.

But if we ask what sound pressure a microphone will pick up (see Fig. 2.8(b)), we must remember that all the rays within the solid angle ($S \cos\vartheta/l^2$), under which the window appears from the observation point, have to be summed. We will call the corresponding quantity (which has the units, $\mathrm{W\,m^{-2}}$, of intensity) the 'ray-bundle intensity' and designate it by capital J^*; the asterisk is to remind us of the bundle. Thus, we get:

$$p^2 \sim j\Delta\Omega_{\mathrm{m}} = \frac{Ec}{4\pi}\,\frac{S\cos\vartheta}{l^2} = J^*_{(\vartheta)} \tag{2.51}$$

This ray bundle intensity decreases according to cos ϑ:

$$J^*_{(\vartheta)} = J^*_{(0)} \cos\vartheta \tag{2.51a}$$

a dependence that is known in optics as 'Lambert's Law'.[2] It makes no difference that the receiver (the microphone) measures the power incident on a surface extending over many wavelengths. But it would be more appropriate, here, to consider the 'ray strength', i.e. the power per unit of solid angle $\mathrm{d}P/\mathrm{d}\Omega$, radiated in different directions from the hole (window).

This quantity, corresponding to the incident power per unit of solid

[2] Pohl, R. W., *Einfuhrung in die Physik*, Springer Verlag, Berlin, Vol. 3, 2nd Edn., 1963, Section 36.

angle, also decreases with cos ϑ:

$$\frac{\mathrm{d}P}{\mathrm{d}\Omega}=I_{(\vartheta)}=\frac{2\,Ec}{4\pi}\,S\cos\vartheta=I_{(0)}\cos\vartheta \tag{2.52}$$ [3]

(The letter I is taken from the German National Standard, DIN 1304; we must distinguish carefully between I, in watts per steradian, and J, in watts per square meter. We have deliberately refrained from eliminating the factor 2 in the numerator to show that it is needed in order to get $Ec/4$ when integrating over the half-sphere.)

From this ray strength, the optical receiver intercepts a solid angle, $\Delta\Omega_{\mathrm{R}}=S_{\mathrm{R}}\cos\vartheta_{\mathrm{R}}/l^2$, and so measures the power:

$$P_{\mathrm{R}}=\frac{\mathrm{d}P}{\mathrm{d}\Omega}\Delta\Omega_{\mathrm{R}}=\frac{Ec}{2\pi}S\cos\vartheta\;S_{\mathrm{R}}\cos\vartheta_{\mathrm{R}}/l^2 \tag{2.53}$$

We shall make use of this relation again in discussions of statistical room acoustics, below.

Up until now, we have studied the ideal diffuse distribution (i.e. the 'Lambert distribution'), both for the incident sound field inside the reverberant chamber and for the sound field radiated out the window. But it is plausible that we can realize this ideal diffuse state best when a reflection from the wall exhibits the same ideal distribution over all directions.

In Section IV. 13.3 (Volume 2), we shall discuss how this can be achieved; here we discuss only some of the consequences. If we wish to include all of the possibilities between regular and diffuse reflection, we cannot restrict ourselves to rotational symmetry. The direction of the incident ray with its angular current density j must be characterized by an elevation angle ϑ and an azimuth angle ϕ. The same holds for characterizing the reflected ray, with its angular current density j' by the corresponding angles, ϑ' and ϕ'. Following Joyce,[4] we establish the relation between j and j' by means of a 'reflectivity matrix' R (a quasi-splitting of the continuous directions into discrete directions):

$$j'_{(\vartheta',\phi')}=\int_0^{2\pi}\int_0^{\pi/2}R_{(\vartheta,\phi,\vartheta',\phi')}\,j_{(\vartheta,\phi)}\cos\vartheta\sin\vartheta\,\mathrm{d}\vartheta\,\mathrm{d}\phi \tag{2.54}$$

[3] In Section 10 of the German edition, instead of eqns. (2.51) and (2.52), above, the equation $j_{r(\vartheta)}=J_{r(0)}\cos\vartheta$ was written by mistake, where J_r was regarded as proportional to the angular current density, as elsewhere in the literature.

[4] Joyce, W. B.,[1] especially Appendix A.

If we restrict the incident sound field to that of a single point source in an anechoic room, the sound meets a small area S at the distance r ($S \ll r^2$) with the intensity $J = P/4\pi r^2$, and we may remove $R \cos\vartheta$ from inside the integral in Joyce's defining equation, (2.54), (that integral then represents just this intensity) and get:

$$j'_{(\vartheta,\,\phi,\,\vartheta',\,\phi')} = R_{(\vartheta,\,\phi,\,\vartheta',\,\phi')} \cos\vartheta \, J_{(\vartheta,\phi)} = R \cos\vartheta \; P/4\pi r^2 \tag{2.54a}$$

If we further consider a receiver-point at a distance r' in the direction, ϑ', ϕ', we know from eqn. (2.51) that the square of the sound pressure is proportional to the bundle intensity

$$j' S \cos\vartheta'/r'^2$$

Thus, we find the following relationship between p^2 and P:

$$p^2_{(r',\,\vartheta',\,\phi')} \sim P_{(r,\,\vartheta,\,\phi)} \, R_{(\vartheta,\,\phi,\,\vartheta',\,\phi')} \frac{\cos\vartheta \cos\vartheta'}{(rr')^2} \tag{2.55}$$

According to the law of reciprocity, this relation must remain unchanged when the locations of the point-source and receiver-point are interchanged, that is, when (r, ϑ, ϕ) and (r', ϑ', ϕ') are interchanged. So this leads to the general condition for R, which Joyce calls the condition of 'detailed balance':

$$R_{(\vartheta,\,\phi)\,(\vartheta',\,\phi')} = R_{(\vartheta',\,\phi')\,(\vartheta,\,\phi)} \qquad (2.55\text{a})^5$$

A further important restriction is given by the power balance; it includes the calculation of the mean reflection coefficient, which here results from averaging over all directions of incidence (ϑ, ϕ) and all directions of reflection (ϑ', ϕ')

$$\rho_{\mathrm{mm}} = \frac{1}{\pi} \int_0^{2\pi} \int_0^{\pi/2} \cos\vartheta \sin\vartheta \, \mathrm{d}\vartheta \, \mathrm{d}\phi \left\{ \int_0^{2\pi} \int_0^{\pi/2} R \cos\vartheta' \sin\vartheta' \, \mathrm{d}\vartheta' \, \mathrm{d}\phi' \right\} \tag{2.56}$$

Here, we may regard the second integration (in brackets) as the definition of a partial mean reflection coefficient that is averaged only over the directions of reflection and is a function of the direction of incidence:

$$\rho_{\mathrm{m}(\vartheta,\,\phi)} = \int_0^{2\pi} \int_0^{\pi/2} R \cos\vartheta' \sin\vartheta' \, \mathrm{d}\vartheta' \, \mathrm{d}\phi' \tag{2.56a}$$

---

[5] Joyce, W. B., *J. Acoust. Soc. Am.*, **64** (1978) 1430.

If we now consider the limiting case in which all the power is reflected, this requires not only $\rho_{\mathrm{mm}}=1$, but also $\rho_{\mathrm{m}(\vartheta,\ \phi)}=1$. Thus, we get, instead of eqn. (2.56a):

$$1=\int_0^{2\pi}\int_0^{\pi/2} R\cos\vartheta'\sin\vartheta'\,\mathrm{d}\vartheta'\,\mathrm{d}\phi' \qquad (2.56\mathrm{b})$$

But, as Joyce[4] emphasized, and as follows from eqn. (2.55a), this also has the consequence:

$$1=\int_0^{2\pi}\int_0^{\pi/2} R\cos\vartheta\sin\vartheta\,\mathrm{d}\vartheta\,\mathrm{d}\phi \qquad (2.57)$$

This integral corresponds to an isotropic incidence of sound and shows that the result is independent of the direction of reflection; that is, the reflected angular current density is the same in all directions. No matter what diffusing properties the wall may have, an ideal diffuse incidence leads to an ideal diffuse reflection.

It is easy to see that this is true even for the case of a totally non-diffuse, specularly reflecting wall. But eqn. (2.57) states that *no* totally reflecting wall can change this ideal directional distribution. If this were not the case, it would mean that some ordering of the distribution of directions is possible. But if we trust, as Jaeger did, in the analogy with Clausius' kinetic theory of heat, this would also mean that it is impossible to have omni-directionality of pressure in a vessel, a condition which, as Joyce[4] emphasizes, can be regarded as a necessary consequence of the second law of thermodynamics. This analogy encourages us to conclude that we will always approach the ideal diffuse state, so long as the reflections are not completely specular but exhibit at least a bit of scattering, because even a small amount of diffuseness at the beginning tends to 'snowball' from reflection to reflection.

Thus, in principle, it may be correct when Joyce[6] conjectures that the small irregularities due to construction tolerances in room walls will ultimately lead to an ideal diffuse state. However, the question arises as to how much time (and, for exciting the reverberation, how much sound level) would be required in order to reach this asymptotic condition!

We know from the preceding section that this tendency toward diffuse

[6] Joyce, W. B., *J. Acoust. Soc. Am.*, **58** (1975) 646, second paragraph.

conditions cannot always be expected if sound absorption must be taken into account, at least not with specular reflections in certain kinds of rooms (rectangle, sphere). And since specular reflections cannot always be avoided, it is difficult to guess which tendency will win out: a trend toward the diffuse state or toward a sound field with preferred directions.

It is essential, of course, that the statistical theory be based on the first tendency. And it must be looked upon as a consistent and reasonable hypothesis when Kuttruff[7] assumes, first, that diffuse reflections *with* losses may exhibit directional distribution properties similar to those without losses, and, second, that the reflection coefficients are independent of direction:

$$\rho_{(\vartheta,\phi)} = \text{constant} = \rho \tag{2.58}$$

He has also found, on the basis of computer model tests with the Monte Carlo method under the assumption of a Lambert distribution for the reflected sound, that even with such extreme angular dependence for the absorption coefficients as (a):

$$\begin{aligned} \alpha &= 1 \text{ for } 0 < \vartheta < 60^\circ \\ &= 0 \text{ for } 60^\circ < \vartheta < 90^\circ \end{aligned}$$

and (b):

$$\begin{aligned} \alpha &= 0 \text{ for } 0 < \vartheta < 30^\circ \\ &= 1 \text{ for } 30^\circ < \vartheta < 90^\circ \end{aligned}$$

there is scarcely any influence on the reverberation time.

In order to treat the ideal diffuse sound field under the assumption of eqn. (2.58), Kuttruff[8] introduced another concept from the field of radiology into statistical room acoustics: the 'irradiation strength' B (*Bestrahlungsstärke*), that is, the power per unit of irradiated area, again an intensity-like quantity. For a plane wave with intensity J which encounters a surface S under the incidence angle ϑ, we get $B = J \cos\vartheta$.

The optical situation that we considered in eqn. (2.53) may now be transformed into a room-acoustical question: what increment of irradiation dB_1 is to be expected at a surface element s as a result of radiation from another surface element dS which receives the irradiation strength B_0? The latter may not correspond to an isotropic sound field, as in eqn. (2.53), where B_0 is given by $Ec/4$. But no matter how B_0 is

[7] Kuttruff, H., *Acustica*, **42** (1979) 187.
[8] Kuttruff, H., *Acustica*, **25** (1971) 333.

composed, it causes the surface element dS to radiate the power increment $dP = 2\rho B_0 \cos\vartheta' \, dS'$ in the direction ϑ', where the surface element s is located at the distance l, thus forming the angle ϑ between the ray and its perpendicular (see Fig. 2.9). Thus, we get:

$$dB_1 = \frac{B_{0(\mathfrak{r})}\rho_{(\mathfrak{r})} \cos\vartheta' \cos\vartheta \, dS}{2\pi l^2} \tag{2.59}$$

where the vector $\mathfrak{r}$ characterizes the location of dS.

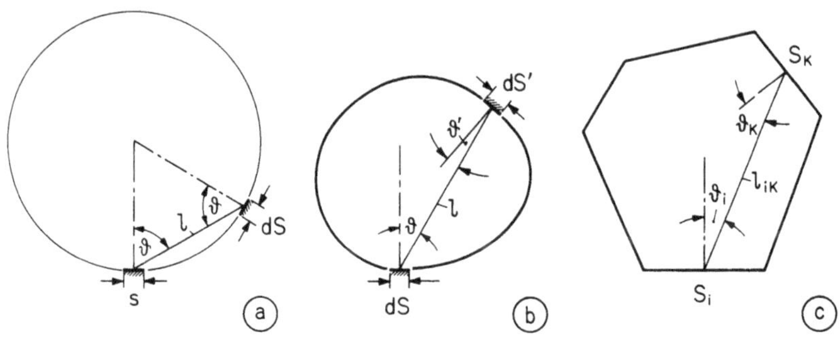

Fig. 2.9. Free path lengths l and their angles with respect to the perpendiculars to the sending and receiving surfaces: (a) in a sphere; (b) in an arbitrary 'convex' room; and (c) in a polyhedron. (Cases (b) and (c) are represented as two-dimensional problems.)

As a first example, we consider a sphere. Since in this case both angles are the same, we get:

$$dB_1 = \frac{B_{0(\mathfrak{r})}\rho_{(\mathfrak{r})} \cos^2\vartheta \, dS}{\pi l^2} \tag{2.59a}$$

and, if we integrate over all the surface elements:

$$B_1 = \frac{1}{\pi} \int_S \frac{B_{0(\mathfrak{r})} \rho_{(\mathfrak{r})} \cos^2\vartheta \, dS}{l^2} \tag{2.59b}$$

Since for every surface element on the inside of a sphere the free path is:

$$l = D \cos\vartheta \tag{2.60}$$

the $\cos^2\vartheta$ disappears and so, therefore, does the special choice of the observation location. The irradiation strength at the second reflection is everywhere the same:

$$B_1 = \frac{1}{\pi D^2} \int_S B_{0(\mathrm{r})}\ \rho_{(\mathrm{r})}\ \mathrm{d}S = \overline{B_{0(\mathrm{r})}\ \rho_{(\mathrm{r})}} \tag{2.61}$$

namely, the arithmetic mean of $B_0 \rho$. Even simpler are the formulae for the later reflections, where the preceding B_{N-1} are everywhere the same:

$$B_N = \bar{\rho}\, B_{N-1} \tag{2.61a}$$

The past history of the ray has no influence; the future ray becomes 'memoryless'.

Therefore, in optics, a hollow sphere, called Ulbricht's Sphere, is used to evaluate the radiation of a light source by measuring the irradiation strength at a small window which is shielded against direct rays from the source. For this measurement, care is taken that the value of ρ is constant.

Furthermore, we get the following expression for the irradiation strength B_N where we have to replace $B_{0(\mathrm{r})}$ in eqn. (2.59) by the constant value B_{N-1} and $\rho_{(\mathrm{r})}$ by ρ:

$$\mathrm{d}B_N = \frac{B_{N-1}\, \rho \cos^2\vartheta\ \mathrm{d}S}{\pi l^2}$$

If we divide $\mathrm{d}B_N$ by $\cos\vartheta$, we get a corresponding intensity $\mathrm{d}J_{N-1}$; and if we recognize (see Fig. 2.6 bottom) that

$$\frac{\cos\vartheta'\ \mathrm{d}S'}{l^2} = \mathrm{d}\Omega \tag{2.62}$$

corresponds to the element of solid angle with which $\mathrm{d}S$ appears at the observation surface s, we discover that the angular current density,

$$j_N = \frac{\mathrm{d}I_N}{\mathrm{d}\Omega} = \frac{B_{N-1}\, \rho}{\pi} \tag{2.63}$$

becomes independent of direction. Thus, we have the surprising result that even the sphere, which presents stronger focusing than any other room shape and could never contain a homogeneous and isotropic sound field with specular reflections, nevertheless achieves this condition with ideal diffuse reflections and a constant reflection coefficient.

Even when we change from a sphere to any other 'convex' room, represented as a two-dimensional problem as shown in Fig. 2.9(b), we again get rather simple relations if we assume ideal diffuse reflections. We have only to distinguish between the angles ϑ' for the direction of diffuse reflection, and ϑ for the direction of the new incidence (as we have already done in eqn. (2.59)), as well as the corresponding surface elements. Thus, instead of eqn. (2.59b), we get:

$$B_1 = \frac{1}{\pi} \int_S \frac{B_{0(\mathbf{r}')}\,\rho_{(\mathbf{r}')} \cos\vartheta \cos\vartheta' \,\mathrm{d}S'}{l^2} \tag{2.64}$$

If we integrate over all the irradiated surfaces:

$$\int_S B_1 \,\mathrm{d}S = \frac{1}{\pi} \int_S \int_S \frac{B_{0(\mathbf{r}')}\,\rho_{(\mathbf{r}')} \cos\vartheta \cos\vartheta' \,\mathrm{d}S' \,\mathrm{d}S}{l^2} \tag{2.64a}$$

and exchange the order of integration, introducing the solid angle:

$$\frac{\cos\vartheta \,\mathrm{d}S}{l^2} = \mathrm{d}\Omega'$$

under which the irradiated surface element $\mathrm{d}S$ appears at the diffusely reflecting surface element $\mathrm{d}S'$, we may substitute for the resulting integral:

$$\frac{1}{\pi} \int_S \frac{\cos\vartheta \cos\vartheta' \,\mathrm{d}S}{l^2} = \frac{1}{\pi} \int_S \cos\vartheta' \,\mathrm{d}\Omega' = \frac{1}{\pi} \int_0^{2\pi} \int_0^{\pi/2} \cos\vartheta' \sin\vartheta' \,\mathrm{d}\vartheta' \,\mathrm{d}\phi' = 1 \tag{2.65}$$

This reduces eqn. (2.64a) to:

$$\int_S B_1 \,\mathrm{d}S = \int_S B_{0(\mathbf{r}')}\,\rho_{(\mathbf{r}')} \,\mathrm{d}S' \tag{2.64b}$$

or (applied to the Nth reflection):

$$\bar{B}_N = \overline{B_{N-1}\rho}\,. \tag{2.64c}$$

Although this formula is still easy to comprehend, it no longer gives a constant irradiation strength; even for constant ρ we get only:

$$\bar{B}_N = \rho\,\bar{B}_{N-1} \tag{2.64d}$$

Accordingly, we no longer have an ideal diffuse sound field in an arbitrary convex room, as we had in the sphere, according to eqn. (2.63).

Up to this point, we have accounted for the reflections according to their order; but this no longer makes sense when we proceed to consider time-dependent reverberation. In this case we must account for the cotemporaneous irradiation strengths.

As in all theories of reverberation, we assume that all 'energy-like quantities' decrease exponentially with time. By introducing the mean free path length l_m and a quantity α'' that characterizes the sound absorption, which Kuttruff[9] introduced as the 'effective absorption exponent', we may assume:

$$B \sim e^{-\alpha'' ct/l_m} \tag{2.66}$$

At this point, Kuttruff considers the different paths l between the diffuse-reflecting elements and the irradiated locations by expressing the value of B at the first one as:

$$B_{(r',\ t-l/c)} = B_{(r',\ t)}\ e^{\alpha'' l/l_m} \tag{2.67}$$

With this device, he is able to drop the distinction between B_N and B_{N-1}.

Thus, he finds, as the condition for the time dependence assumed in eqn. (2.66), the integral equation:

$$B_{(r)} = \frac{1}{\pi} \int_S B_{(r')}\ e^{\alpha'' l/l_m}\ \rho_{(r')}\ \frac{\cos\vartheta \cos\vartheta'}{l^2} \mathrm{d}S' \tag{2.68}$$

Unfortunately, this general integral equation can be solved only under restrictive conditions. In order to adapt the solution to Eyring's procedure, Kuttruff replaced the free path length l in the exponent (and there only) by its mean value l_m. Then the factor $e^{\alpha'' l/l_m} = e^{\alpha''}$ can be taken outside the integral, so that we get the simplified integral equation:

$$e^{-\alpha''} B_{(r)} = \frac{1}{\pi} \int_S B_{(r')}\ \rho_{(r')}\ \frac{\cos\vartheta \cos\vartheta'}{l^2} \mathrm{d}S' \tag{2.69}$$

[9] Kuttruff, H., *Acustica*, **35** (1976) 141.

In the case of the spherical room, we get, analogous to eqn. (2.61):

$$e^{-\alpha''} B_{(\mathfrak{r})} = \frac{1}{S} \int_S B_{(\mathfrak{r}')} \rho_{(\mathfrak{r}')} \, dS' \tag{2.70}$$

which states that we have everywhere the same irradiation strength B. We can divide this into eqn. (2.70) to get Eyring's result for the absorption exponent:

$$e^{-\alpha''} = \bar{\rho}, \quad \alpha'' = -\ln \bar{\rho} = -\ln (1 - \bar{\alpha}) \tag{2.71}$$

Without doubt, this derivation is more appropriate to the three-dimensional reverberation problem than the earlier one-dimensional solution. (Eyring also discussed the reverberation for the case of the sphere, but he considered only propagation along radial rays; he could equally well have discussed the one-dimensional situation of Fig. 2.5(b).) Kuttruff's derivation yields Eyring's formula under the assumption of ideal diffuse reflections, but only in a very special three-dimensional room, a sphere.

For other convex rooms, we find, after integration over all the irradiated areas dS and taking eqn. (2.65) into account, the more general relation:

$$e^{-\alpha''} \int_S B_{(\mathfrak{r})} \, dS = \int_S B_{(\mathfrak{r}')} \rho_{(\mathfrak{r}')} \, dS' \tag{2.72}$$

or:

$$\alpha'' = -\ln \frac{\overline{B\rho}}{\bar{B}}$$

which includes eqn. (2.71) for constant irradiation strength; but the latter is realized only in the case of the sphere.

Despite the clarity of this weighting of the reflection coefficient by the irradiation strength, eqn. (2.72) still does not offer a direct method for calculation, since we do not know, in general, the distribution of the irradiation.

We find one possibility for calculating $B_{(\mathfrak{r})}$ if we replace the inner surface of a convex room by a finite number of partial surfaces with constant values of B and ρ; the latter may be arithmetic averages. For further simplification, we can treat these partial surfaces as planes (see Fig. 2.9(c)).

The integration in eqn. (2.69) then has to be carried out only for those partial surfaces S_k, for which $B_k \rho_k$ appears outside the integral as terms of a sum:

$$e^{-\alpha''} B_i = \sum B_k \rho_k \int_{S_k} \frac{\cos\vartheta_i \cos\vartheta_k}{\pi l_{ik}^2} \mathrm{d}S_k \tag{2.73}$$

and this equation must be averaged over S_i:

$$e^{-\alpha''} B_i = \sum B_k \rho_k \frac{1}{S_i} \int_{S_i} \int_{S_k} \frac{\cos\vartheta_i \cos\vartheta_k}{\pi l_{ik}^2} \mathrm{d}S_k \, \mathrm{d}S_i \tag{2.74}$$

If we introduce for the double integral, which depends on the shape and takes the Lambert distribution into account, the symbol p_{ik}:

$$p_{ik} = \frac{1}{S_i} \int_{S_i} \int_{S_k} \frac{\cos\vartheta_i \cos\vartheta_k}{\pi l_{ik}^2} \mathrm{d}S_k \, \mathrm{d}S_i \tag{2.75}$$

we may split eqn. (2.74) into n linear equations between the n irradiation strengths B_1 to B_n:

$$\begin{aligned} e^{-\alpha''} B_1 &= 0 + \rho_2 p_{12} B_2 + \ \dots \ + \rho_n p_{1n} B_n \\ e^{-\alpha''} B_2 &= \rho_1 p_{21} B_1 + 0 + \ \dots \ + \rho_n p_{2n} B_n \\ &\dots\dots\dots\dots\dots\dots\dots\dots\dots\dots \\ e^{-\alpha''} B_n &= \rho_1 p_{n1} B_1 + \rho_2 p_{n2} B_2 + \ \dots \ + 0 \end{aligned} \tag{2.76}$$

We make use, here, of the fact that all p_{ii} vanish

$$p_{ii} = 0 \tag{2.77}$$

since, for rays tangential to the same surface, the $\cos \vartheta_{i,k}$ are zero. (To what extent the implied assumption is physically justified, that waves propagating tangential to the absorbing surface do not lose energy to that surface, is not discussed here, since those limiting-case losses do not significantly affect the power balance.)

Such a linear dependence of one quantity on all the others appears in all vibration systems and communication networks. We shall meet it in a simpler form in the next chapter, in dealing with coupled rooms.

Such a system of linear equations is not inherently contradictory, but the contradiction is avoided only if the determinant of the coefficients vanishes. In order for this to happen however, the coefficients must contain a parameter that can be evaluated so as to fulfil this condition. As such a parameter we have at our disposal the factor:

$$\lambda = e^{-\alpha''} \tag{2.78}$$

which is called an 'eigen-value' of the system of equations. The values of λ are given by:

$$\begin{vmatrix} -\lambda & \rho_2\, p_{12} & \cdots & \rho_n\, p_{1n} \\ \rho_1\, p_{21} & -\lambda & \cdots & \rho_n\, p_{2n} \\ \cdots\cdots & \cdots\cdots & \cdots & \cdots\cdots \\ \rho_1\, p_{n1} & \rho_2\, p_{n2} & \cdots & -\lambda \end{vmatrix} = 0 \tag{2.79}$$

an equation of the nth order. From this we get n eigen-values and, correspondingly, n different damping constants, as in eqn. (2.46), and thus a 'sagging' reverberation curve.

The partial energy densities $E_1, \ldots, E_n$ depend on the initial distribution of the values of B_k. But this distribution will always asymptotically approach the distribution that corresponds to the smallest damping constant. If we regard that one as characterizing the exponential decay that we are looking for, it means we assume that this asymptotic state will be reached quickly. This may, indeed, be possible, but it would be very tedious to prove it.

Since, according to eqn. (2.76) the difference between the B_i depends not only on the room shape (p_{ik}) but also on the distribution of the absorption (ρ_k), all of the damping constants (and therefore also the smallest) depend on this distribution.

Gerlach and Mellert,[10] who have described the problem in another manner but with the same equation for δ_{min}, have presented their results with an example: a rectangular room with the ratios of length to width to height of 3:2:1. (This could be a concert hall 36 m long, 24 m wide and 12 m high.) Furthermore, they compared the reverberation times for three cases in which only one of the surfaces is absorptive, but in all cases the absorption coefficient averaged over all the surfaces is the same: $\bar{\alpha} = 0{\cdot}1$.

Thus, for all three cases they find the *same* reverberation time, but differing slightly according to Sabine's and Eyring's formulae:

$$T_S = 138\ (l_m/c)$$
$$T_E = 131\ (l_m/c)$$

[10] Gerlach, R. and Mellert, V., *Acustica*, **32** (1975) 211.

An evaluation according to the theory derived above, however, leads to the following results:

Covered surface, S	α	T
3 × 2	0·367	125 (l_m/c)
3 × 1	0·733	118 (l_m/c)
2 × 1	1·1 (!)	106 (l_m/c)

We would find the same tendency with the Millington–Sette formula, which here is identical to the combination equation, eqn. (2.31):

3 × 2	111 (l_m/c)
3 × 1	77 (l_m/c)

It cannot be applied to the third case with its assumed absorption coefficient of 1·1.

Since the calculation of p_{ik} is tedious, and since the highest eigen-mode cannot be explicitly represented, Kuttruff[9] developed an approximate formula, assuming that p_{ik} is proportional to S_k and that the B_i do not depart much from their common mean value:

$$B_i = C(S\bar{\rho} - S_i\rho_i)$$

(C is a constant that later disappears.)

In this way, he gets:

$$\alpha'' = \ln\frac{1}{\bar{\rho}}\left[1 + \frac{\sum \rho_i(\rho_i - \bar{\rho})\,S_i^2}{(\bar{\rho}S)^2}\right] \tag{2.80}$$

The first term corresponds to Eyring's formula.

Kuttruff also tried in several papers to take the differences of the free paths into account.[9,11,12] In these cases, the 'effective absorption coefficient' can be smaller than that of Eyring and could thus approach Sabine's coefficient.

For a special distribution of free paths treated by Schroeder,[13] we may even get Sabine's values exactly.

Since Sabine's formula represents by far the easiest treatment, it is

[11] Kuttruff, H., *Acustica*, **23** (1970) 238.

[12] Kuttruff, H., *Room Acoustics*, Applied Science Publishers Ltd, 2nd Edn, London, 1979.

[13] Schroeder, M. R., *Proc. 5th. ICA, Liège, Sept.* 7–14, 1965, Paper G-31.

encouraging that in some cases it yields not merely an approximation, but the true answer.

Joyce[14] has developed an approximate formula that combines the effects of different free paths and different distributions of absorption and irradiation; in this formula he presents correction terms to be added to the Sabine absorption coefficient. But since the angular current densities appear in this formula, the user is still confronted with the serious difficulty of evaluating their distribution.

Finally, it may be mentioned that both Kuttruff[12] and Joyce[14] have attacked the problem of partially diffuse and partially specular reflections (without taking into account the dependence of the absorption on the angle of incidence).

We must abstain from entering further into these interesting investigations and refer the reader to the recent literature.

In all of these extensions of the theory of reverberation, the question arises as to how long it takes until the assumed conditions are fulfilled. For reverberation measurements, we may have to wait until the decaying sound level is below the background noise. For subjective assessments, however, the important part of the decay appears to be the beginning; and here there may not be a statistical decay of energy at all, but rather a sequence of separate impulses, as shown in Fig. 2.1.

Thus, the 'refinements' of this chapter may contribute more to the physical understanding of the statistical theory of reverberation than they do to the design of good concert halls.[15]

---

[14] Joyce, W. B., *J. Acoust. Soc. Am.*, **67** (1980) 566, eqn. (10).

[15] Kuttruff, H. and Strassen Th., *Acustica*, **45** (1980) 246.

Chapter II.3

Coupled Rooms

II.3.1 Steady-state Conditions with an Open Coupling Area

The statistical analysis of a room, as developed in Chapter II. 1, is not applicable for rooms which, although they may comprise a single air volume, are nevertheless subdivided architecturally into a number of smaller subspaces. This happens, for instance, in churches where the high central nave is abutted by lower side aisles or by side chapels; also, in older court theaters there are often many rather deep boxes.

As a simple example, Fig. 3.1 shows the case where a large lecture hall with volume V_1 is coupled to a low, acoustically damped entry room of volume V_2 in such a way that the already small coupling area is further reduced by a deep beam, to S_{12}.

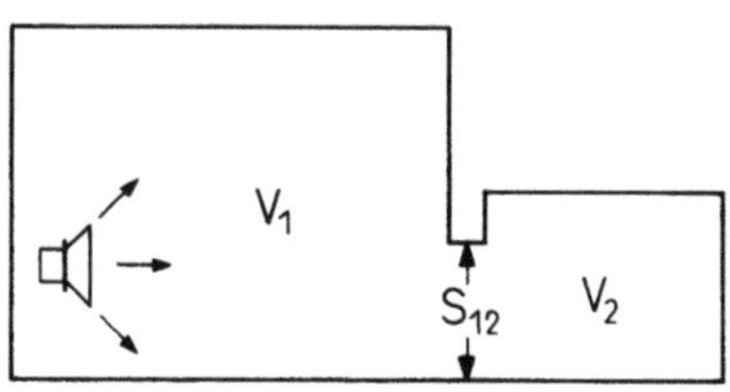

Fig. 3.1. Sketch of a section through two coupled rooms.

In this case, it cannot be expected that the sound power P_1 emitted in the lecture hall will be equally distributed throughout both rooms. In fact, experience shows that there is a remarkable decrease in loudness as one passes from the lecture hall into the entry room.

It is one of the advantages of statistical room acoustics that, even in such cases, the main acoustical features can be exhibited by distinguishing between the mean energy densities, E_1 and E_2, each equally distributed in the respective partial rooms. This analysis replaces the actual gradual transition between E_1 and E_2 with a step-wise change in energy density at the coupling surface S_{12}.

If we denote by A_{10} the equivalent absorption area of room 1 (including all surfaces and absorptive objects except S_{12}) and by A_{20} the corresponding quantity for room 2, we get (according to Section II.1.4) the respective amounts of sound power actually absorbed in the two rooms (assuming, as usual, diffuse sound fields):

$$(A_{10}E_1c/4) \quad \text{and} \quad (A_{20}E_2c/4)$$

In addition, the power transferred from room 1 to room 2 is:

$$(S_{12}E_1c/4)$$

and that from room 2 to room 1 is:

$$(S_{12}E_2c/4).$$

Thus, we can write power balances for the two rooms as follows:

$$P_1-(c/4)A_{10}E_1-(c/4)S_{12}E_1+(c/4)S_{12}E_2=0 \tag{3.1}$$

$$(c/4)S_{12}E_1-(c/4)A_{20}E_2-(c/4)S_{12}E_2=0 \tag{3.2}$$

By introducing

$$A_{11}=A_{10}+S_{12}$$

and

$$A_{22}=A_{20}+S_{12} \tag{3.3}$$

(which means that we include, as part of the total equivalent absorption areas of the rooms, the coupling area S_{12} with the absorption coefficient of unity) we may write eqns. (3.1) and (3.2) in the usual form for coupling problems:

$$\frac{4P_1}{c}= A_{11}E_1 - S_{12}E_2 \tag{3.4}$$

$$0= - S_{12}E_1+A_{22}E_2 \tag{3.5}$$

Solving these coupling equations leads to the energy density in room 1, containing the source:

$$E_1 = \frac{4P_1/c}{A_{11} - S_{12}^2/A_{22}} \tag{3.6}$$

If we had treated the two rooms as a single acoustical space we would have found the following expression for the common mean energy density:

$$E = \frac{4P_1/c}{A_{10} + A_{20}} \tag{3.7}$$

Taking into account eqn. (3.3), the denominator of eqn. (3.6) may be written:

$$\left(A_{10} + A_{20}\frac{S_{12}}{A_{22}}\right)$$

We see, comparing this with the denominator of eqn. (3.7), that the equivalent absorption area A_{20} of room 2 (not counting the coupling surface) does not enter into the energy balance with its full amount, but is diminished by the factor S_{12}/A_{22}, which we call the 'coupling factor' from room 2 to room 1:

$$k_2 = S_{12}/A_{22} \tag{3.8}$$

This factor, which characterizes the difference between the 'single-room' and 'coupled-room' analyses, depends not only on such geometric conditions as the ratio of the coupling area to the total area of room 2 but also on the absorption coefficients of all the surfaces in that room.

If $A_{20} \gg S_{12}$ (that is, if k_2 is very small compared with its maximum possible value of unity), the resultant absorption area for room 1 is $(A_{10} + S_{12})$. This means that the coupling area is to be regarded as an open window, which must be added to the rest of the absorption in room 1. Clearly it would be wrong, in this case, to add the much larger absorption area A_{20} to A_{10}, since it would be impossible to absorb more power from room 1 than enters the coupling area. This limiting case actually occurs for nearly all deep boxes in theaters and for the seating areas underneath deep balconies in auditoriums.

On the other hand, if $A_{20} \ll S_{12}$, the coupling factor S_{12}/A_{22} differs so little from unity that A_{20} can be added directly to A_{10}, in effect treating the two rooms as one.

A reliable decision as to how close we are to one or the other of these two limits requires an exact calculation of the coupling factor. But since the statistical theory of coupled rooms can give only an approximate answer anyway, we adopt the following rule-of-thumb: if the boundary area covered with absorptive materials in room 2 exceeds the coupling area, treat the coupling area as an open window; otherwise, treat room 2 as part of room 1. This rule-of-thumb also applies to the calculation of the reverberation time of room 1, as we shall see below.

The significance of the coupling factor becomes immediately evident in calculating the ratio of energy densities, based on eqn. (3.5):

$$\frac{E_2}{E_1} = \frac{S_{12}}{A_{22}} = \frac{S_{12}}{S_{12} + A_{20}} = k_2 \tag{3.9}$$

If $k_2 \approx 1$ (that is, if $S_{12} \gg A_{20}$), there will not be a significant decrease in loudness as we pass from room 1 to room 2. (The fact that balcony seats are sometimes preferred for their greater loudness may actually result from geometric room acoustics conditions and short delay times, rather than a simple energy balance; these matters lie beyond the reach of statistical room acoustics; see Chapter I.5.)

If, on the other hand, k_2 is small (that is, if $S_{12} \ll A_{20}$), the drop-off of energy density upon entering room 2 is quite evident, an effect that is frequently noticed upon stepping back under deep balconies in auditoriums. (In Section I. 5.6 we discussed how this effect can be minimized by geometrical means.)

The loss of loudness is usually not serious in such cases, because our ears can adapt to a lower loudness level. Indeed, the under-balcony seats may even have the acoustical advantage of great clarity of sound (again for geometrical acoustics reasons).

But the loss of loudness of the desired signal may become quite noticeable in comparison with the loudness of an intrusive noise that is produced within the coupled room, 2. Whereas the energy density of the signal in room 2 is decreased by a factor of S_{12}/A_{22}, compared with that in room 1, the energy density of the intrusive noise in room 2 is greater by a factor of A_{11}/A_{22} than the energy density that would be produced by the same noise source in room 1. This means that the ratio of signal energy density to noise energy density in room 2 is smaller by a factor of S_{12}/A_{11} than it would be in room 1 with the same noise source operating there.

This coupling factor concerns the coupling of sounds from room 1 to room 2, so we characterize it by k_1:

$$k_1 = S_{12}/A_{11} \tag{3.10}$$

Since A_{11} is usually very large compared to S_{12}, this coupling factor is usually small.

II.3.2 Steady-state Conditions for Coupling through a Partition Wall

The statistical treatment of coupled rooms may be extended to cases in which the coupling area is not an open window, but instead has a transmission coefficient $\tau < 1$, such as for example, doors, thin folding panels, curtains and even party walls and ceilings. In fact, these examples represent the chief field of application for the statistical theory of coupled rooms.

In this case, the power transmitted from room 1 to room 2 is given by $(cE_1/4)(\tau S_{12})$ and that transmitted from room 2 to room 1 by $(cE_2/4)(\tau S_{12})$. Thus, in eqns. (3.4) and (3.5) we have only to replace S_{12} with τS_{12} (assuming again that only room 1 contains a sound source):

$$4P_1/c = A_{11}E_1 - \tau S_{12}E_2 \tag{3.11}$$

$$0 = -\tau S_{12}E_1 + A_{22}E_2 \tag{3.12}$$

We can also take into account the possibility that the partition wall may, because of internal losses (dissipation), absorb more sound power at the surface facing the source than it transmits through to the other side. Then the absorption coefficient α and the transmission coefficient τ differ by an amount equal to the dissipation coefficient δ. A typical example would be a drapery or curtain; another would be a partition wall covered with an absorptive layer. We have then to add $\alpha_1 S_{12}$ to A_{10}, and $\alpha_2 S_{12}$ to A_{20} to get the total absorption areas A_{11} and A_{22} for the two rooms.

If τ differs only slightly from 1, we may adopt the same considerations as in Section II.3.1; in particular, the modified coupling factor of room 1 to room 2:

$$k_2 = \frac{\tau S_{12}}{A_{22}} = \frac{E_2}{E_1} \tag{3.13}$$

may be so large that we can treat both rooms approximately as one. Such large values for τ, except for large openings, are seldom encountered; they may, however, occur when the sound source is hidden from sight but is acoustically unhindered. In Section I. 3.4 we mentioned, as one of the possibilities for the avoidance of focusing, the replacement of a solid concave-curved wall surface with a finely perforated metal sheet. With 20% open area, such a panel is visually opaque but may be acoustically almost transparent, and thus provide practically no acoustical separation at all. (In Section IV. 9.1 (Volume 2), we will calculate the limit of this transmissibility.) The reason this possibility is not adopted more often is apparently the desire of the audience to see the musicians and of the musicians to be seen.

Richard Wagner, on the other hand, regarded the sight of the orchestra as a distraction from the (at least visually) more important performance on the stage. According to everything that he published about the orchestra pit in Bayreuth, this was the main reason for his covering the pit on the side toward the audience and painting the pit walls and ceiling black. This naturally changed the timbre of the orchestral instruments, especially those in the deep part of the pit below the stage, and gave to the performances at Bayreuth a very singular acoustical character; in addition, it allowed the singers to be heard to better advantage over the orchestra. No doubt Wagner perceived this difference in sound and took it into account in his compositions, but these acoustical effects were not the primary cause for his covering the orchestra; they must be regarded as secondary to the visual aspect.

This change of musical timbre at Bayreuth is unsuitable to the operas of other composers; moreover, the Bayreuth configuration inhibits musical communication, both among the orchestra musicians and between them and the singers, whose voices are sometimes completely covered up by the orchestra sound in the pit. These facts may explain why the Bayreuth solution was copied only once: at the Prinzregenten Theater in Munich; and even there it was later dropped.

Nevertheless, it would be possible to develop a pit covering, perhaps with perforated metal sheets, a lattice, or even with very lightweight, impermeable plastic membrane, so as to produce the visual effect that Wagner intended without suffering so much change of musical timbre. (In fact, Wieland and Wolfgang Wagner, with the acoustical advice of W. Gabler, introduced changes in parts of the Bayreuth pit covering, so as to increase the sound transmission (see Fig. 3.2). The resulting orchestral sound was noticeably richer in overtones and thus brighter. These changes encountered criticism, however, prompted primarily by piety for the genius of Richard Wagner. It is by no means certain that the creator of Bayreuth himself would have disapproved the changes, had he known about them!)

In general (that is, outside the theater), the separation of an adjoining room (room 2) from a main room (room 1) by a partition is done not only for visual reasons but also for acoustical privacy. In such cases, the transmission coefficient τ is so small, and thus the coupling factor k_2 is so small, that the energy feedback from room 2 into the source room 1 can be neglected. Even for a drapery, the value of τ may be only 0·1, while for a door or other thin wood panel τ may be at most 0·01. For real party walls or ceilings, τ is of the order of 10^{-4} to 10^{-5}, depending on the

Fig. 3.2. Orchestra pit and auditorium of Festival Theater at Bayreuth. The temporary experimental covering of a part of the pit with a perforated shield can be seen as a narrow strip at the lower right. Most of this perforated shield was later covered with solid panels (see extreme lower right), but some of the perforations are still visible.

weight per unit area, the thickness, and the number of plates. In order to cope with this wide range of values for τ, we use, instead of τ itself, the logarithm of its reciprocal, the so-called reduction index, R:

$$R = 10 \log \left(\frac{1}{\tau} \right) \text{dB} \qquad (3.14)^1$$

The larger the value of R, the better the sound isolation.

The unit 'dB' is added here, as in the case of loudness level (see Section II.1.2) to indicate the choice of the common logarithm of the ratio of powers and the factor of ten.

The relation between R and the sound pressure level difference between the source room 1 and the receiving room 2,

$$L_1 - L_2 = 10 \log \frac{E_1}{E_2} \text{ dB} \qquad (3.15)$$

if we take eqn. (3.13) into account, is given by:

$$R = L_1 - L_2 + 10 \log \frac{S_{12}}{A_{22}} \text{ dB} \qquad (3.16)$$

In order to evaluate the reduction index of a party wall or ceiling separating two reverberant rooms, it has long been customary in architectural acoustics, and has been internationally standardized, to radiate a sound broadband into the source room and to measure the resulting average sound pressure levels in both the source room and the receiving room.[2]

The measured level difference must then be corrected by substituting the appropriate numerical values in the last term in eqn. (3.16).

By this means it is easy to measure the coupling area S_{12} (in architectural acoustics designated S only); one can also determine the value of A_{22} (or A only) from the measured reverberation time in room 2. For this purpose, we use the simple Sabine formula (eqn. 1.35), since the last term in eqn. (3.16) is only a correction term and only rooms having small values of average absorption coefficient are to be used for such measurements. The values of reduction index so determined are plotted against a logarithmic frequency scale in the range between 100 and 3200 Hz.

Since this book deals primarily with room acoustics (that is, with the

---

[1] In North America, this quantity is called 'transmission loss', designated by *TL*

[2] ISO R 140, International Standards Organization, Geneva.

propagation of sound *in the source room*), we do not present here a collection of reduction index data for all the partition elements that may be of interest in auditoriums.[3]

It is, however, appropriate to discuss what values of reduction index might be suitable for use in concert halls. We have already mentioned in Section II. 1.1 that sound level differences of the order of 60 dB (the basis for the definition of reverberation time) occur frequently in orchestra concerts; when a chorus is also involved, even greater differences are encountered. Levels of 100 dB or more are not uncommon in the neighborhood of the sound sources.[4]

At the other extreme, we may occasionally find very low noise levels despite the presence of the audience, for example, when they become so rapt with the music that they literally hold their breath; such low noise levels are very seldom encountered in daily life. Therefore, it is reasonable to require that the noise of air-conditioning systems should not exceed 25 dB at middle and high frequencies (see Section III.3.4).

Now, if a concert hall is situated in an urban environment with heavy traffic, where the outdoor noise levels may easily exceed 85 dB, then we must provide a noise level difference of at least 60 dB between the outdoors and the inside of the auditorium; this isolation must be provided by the wall construction, taking into account all doors and other penetrations into the hall.

Let us suppose a hall volume of 10 000 m^3 and a reverberation time of 1·6 s; then, according to Sabine's formula, we have an equivalent absorption area of 1000 m^2. The coupling area to the outdoors may be of the order of 300 m^2, so the corresponding wall must therefore provide a reduction index of

$$R = 85 - 25 + 10\log\frac{300}{1000} = 55\,\text{dB}$$

---

[3] Interested readers may refer to the following publications where such data are tabulated:

Compendium of Materials for Noise Control, HEW Publication No. (NIOSH) 75–165, June 1975, US Department of Health, Education and Welfare.

Catalog of STC and IIC Ratings for Wall and Floor/Ceiling Assemblies, with TLC and ISPL Data Plots, no report number, September 1981, Office of Noise Control, California Department of Health Services, Berkeley, California, 94704.

Parkin, P. H., Purkis, H. J. and Scholes, W. E., *Field Measurement of Sound Insulation between Dwellings*, National Building Studies Research Paper No. 33, Her Majesty's Stationery Office, London, 1960.

[4] We do not take into account here the extremely high sound levels that can be generated with electroacoustical equipment, and that, with prolonged exposure, cause damage to the hearing of artists and listeners alike.

Such a high degree of sound isolation can seldom be achieved, even with a windowless brick wall 25 cm in thickness; instead, a double-wall construction is required. In fact, it is preferable for the auditorium to be surrounded by other quiet spaces, which serve as a buffer zone between the hall and the traffic noise. If these spaces are lobbies, circulation areas or stairways, however, it is inevitable that there may be speech, or perhaps even shouts, in those spaces during the performance. The level of speech in normally furnished rooms may reach 75 dB; thus, we will need a level difference (not reduction index) of about 50 dB between these surrounding spaces and the auditorium. This degree of isolation is readily possible with common wall constructions, particularly since the coupling area will usually be smaller than the equivalent absorption area of the auditorium.

The weakest links in the isolation structure are the doors. Fortunately, their coupling areas are small—or at least they should be kept small by design. But it may be wise to estimate for those coupling areas a total value of 1/20 of A, or about 50 m^2 in our example. In that case, the doors must provide a reduction index of 37 dB, which requires doors of special construction and careful gasketing, if single doors are used; it is readily achieved using double doors with a sound lock between them.

In our previous discussion, we have regarded the noise level in the source room (in this case the spaces surrounding the auditorium) as given, a procedure that corresponds to the usual treatment of noise isolation problems and measurement. In this case, the equivalent absorption area of the source room does not enter the picture. If, instead, we regard the sound power P_1 of the source as the given quantity, we get for the energy density in the adjacent room:

$$E_2 = \frac{4P_1 \tau S_{12}}{cA_{11}A_{22}} \tag{3.17}$$

Since A_{11} and A_{22} appear as a product, eqn. (3.17) is independent of the direction of sound transmission: a sound source of given power in room 1 produces the same level in room 2 as if the source were placed in room 2 and the level measured in room 1. This reciprocity, of course, demands that the transmission coefficient τ also be independent of the transmission direction: but we have already tacitly assumed this, since it is a consequence of the general law of reciprocity.

It is also possible to reduce the noise level in room 2 by increasing the absorption area in the source room. This could be accomplished in the spaces surrounding an auditorium by the ample provision of carpet, which has the additional advantage of reducing the noise of footsteps. The abundant use of draperies and sound-absorbing ceiling materials is also recommended.

The reciprocity of eqn. (3.17) concerns only the objective, physical sound energy in the neighboring room. The subjective impression may be rather different, since aural comparison with other sound sources plays an important role: in particular, a new signal may be totally masked by an existing noise level (see Section III. 1.3).

This may be demonstrated by the example shown in Fig. 3.3.[5] Here, two neighboring rooms with quite different equivalent absorption areas are considered: a bathroom and a bedroom. The first has a small area and hard surfaces, and thus a small absorption area A_1; the larger second room, because of the beds and other absorptive furniture, presents a

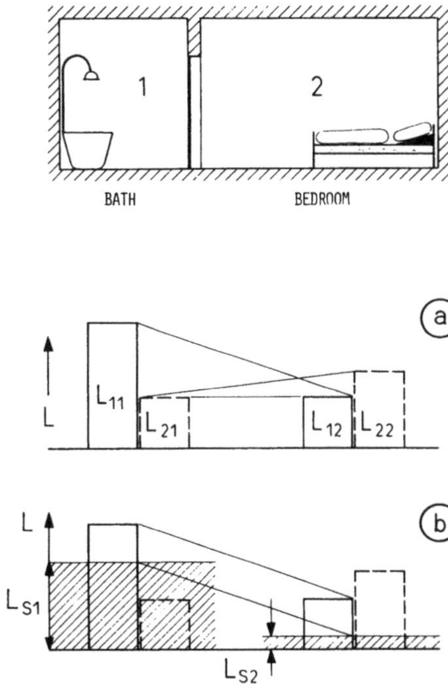

Fig. 3.3. The steady-state levels in two coupled rooms with quite different absorption areas (bathroom and bedroom). (a) Without background noise; (b) with background noise.

[5] Cremer, L., *Wärme- und Kältetechnik*, **40** (1938) Nos. 9 and 10.

large absorption area, so that, in fact, the homogeneous sound field assumed in the statistical theory may not actually be achieved. Nevertheless, we may state, according to eqn. (3.17), that if equal speech power is produced in room 1 or 2, then, although the source room speech levels L_{11} and L_{22} may be very different in the two rooms, the levels L_{12} and L_{21} produced in the neighboring rooms 2 and 1 may be nearly equal (see Fig. 3.3(a)). This means that with quiet background noise conditions in both rooms, the intruding speech from the adjacent room could be heard and understood equally well in both rooms.

However, if we now allow water to splash in the bathroom, we produce a noise level (*Störpegel*) L_{S1} in room 1; but this still may be less than the source-room (bathroom) speech level L_{11}, and the bathroom speech is not masked. The splashing noise is attenuated in passing from the bathroom to the bedroom, by the same amount as is the speech signal L_{11}, so the same signal-to-noise-level difference occurs in the bedroom, and the bathroom speech is clearly audible in the bedroom. By contrast, the level L_{21} produced in the bathroom by speech in the bedroom is completely masked by the splashing noise level L_{S1}.

The best protection against eavesdropping from adjacent rooms is the provision of a steady masking noise in those rooms. The noise of the typewriter in the reception room prevents the waiting callers from overhearing the telephone conversation in the inner office, even more effectively than the (often doubtful) isolation properties of the door or partition wall. Also, a doctor may protect his consulting room conversations from being overheard in the waiting room by providing in the waiting room some background music; it can be a very effective masking device even though it is rather quiet.

The reciprocity law may be violated, however, if there is an immediate influence of the source room on the radiated sound power. For example, a higher background noise level may cause the speaker to increase his voice effort and thus increase the intelligibility of his speech in a quiet neighboring room.

In quiet rooms, it would seem plausible that people adjust their speech power to achieve equal levels, that is, speak more quietly in a reverberant room than in a highly damped room. Strange to say, usually the opposite happens! A tiled bathroom or a bare staircase seems to invite the occupants to sing and shout, whereas a room with a sound-absorbing ceiling signals the desire for quiet and induces quieter behavior. Therefore, sound-absorbing ceilings in lobbies and staircases account for a significantly greater reduction in noise level than could be accounted for by their equivalent absorption area alone.

II.3.3 Reverberation in Coupled Rooms

We now proceed to non-stationary processes in coupled rooms; thus, we must drop the power balances given in eqns. (3.11) and (3.12). Instead, we assume that the introduction and removal of energy in both rooms leads to time-varying changes in the total energy contents, $(E_1 V_1)$ and $(E_2 V_2)$, of the two rooms. We represent the stopping of the second source by setting $P_1 = 0$ and attend to the reverberant sound decay only. Thus, instead of eqns. (3.11) and (3.12), we have the differential equations:

$$\frac{c}{4}(A_{11}E_1 - \tau S_{12}E_2) = -V_1 \frac{dE_1}{dt} \tag{3.18}$$

$$\frac{c}{4}(-\tau S_{12}E_1 + A_{22}E_2) = -V_2 \frac{dE_2}{dt} \tag{3.19}$$

Since these equations are linear, we set

$$E_{12} = E_{01,2} e^{-2\delta t} \tag{3.20}$$

which means that we assume that the reverberant process is made up of exponentially decaying functions. The quantity δ which characterizes the rate of decay of the sound pressure is called the damping constant. Since we must deal here with energies, which are proportional to the squares of the sound pressures, the quantity 2δ appears in the exponent. Between δ and Sabine's reverberation time, the following relations hold:

$$2\delta = \frac{6}{T} \ln 10; \qquad \delta = \frac{6{\cdot}9}{T} \tag{3.21}$$

By setting eqn. (3.20) into eqns. (3.18) and (3.19), and omitting the common factor $e^{-2\delta t}$, we get for E_1 and E_2, and so also for their initial values E_{01} and E_{02}, the linear equations:

$$\left(\frac{c}{4}A_{11} - 2\delta V_1\right)E_{01} - \frac{c}{4}\tau S_{12}E_{02} = 0 \tag{3.22}$$

$$-\frac{c}{4}\tau S_{12}E_{01} + \left(\frac{c}{4}A_{22} - 2\delta V_2\right)E_{02} = 0 \tag{3.23}$$

These two equations can be valid for a simple exponential decay with a single value of δ *only* if both equations result in the same ratio for E_{01}/E_{02}. But this requires that the determinant of the coefficients of E_{01} and E_{02} must vanish:

$$\begin{vmatrix} \left(\frac{c}{4}A_{11}-2\delta V_1\right) & -\frac{c}{4}\tau S_{12} \\ -\frac{c}{4}\tau S_{12} & \left(\frac{c}{4}A_{22}-2\delta V_2\right) \end{vmatrix} = 0 \tag{3.24}$$

The resulting so-called 'characteristic equation' for δ becomes simpler if we introduce the damping constants:

$$\delta_1 = \frac{cA_{11}}{8V_1}, \qquad \delta_2 = \frac{cA_{22}}{8V_2} \tag{3.25}$$

which correspond to the decays of the two rooms as they would be if the rooms were uncoupled and the quantity $\alpha_{1,2}S_{12}$ were included in their respective absorption areas.

Equation (3.24) may then be written:

$$\left(1-\frac{\delta}{\delta_1}\right)\left(1-\frac{\delta}{\delta_2}\right)-k_1k_2=0 \tag{3.26}$$

where k_1 and k_2 are the coupling factors according to eqns. (3.8) and (3.10), but multiplied by τ. Since only their product appears in eqn. (3.26), we may introduce their geometric mean

$$\kappa = \sqrt{k_1k_2} \tag{3.27}$$

and call it the mean coupling coefficient, as is usual in the theory of coupled oscillators.

Since eqn. (3.26) is quadratic in δ, we must expect two different damping constants; this is not surprising, since even uncoupled rooms would have two different values of δ in general. The corresponding eigenvalues δ_{I} and δ_{II} for the coupled rooms are:

$$\delta_{\mathrm{I,II}}=\tfrac{1}{2}(\delta_1+\delta_2)\mp\sqrt{\tfrac{1}{4}(\delta_1-\delta_2)^2+\kappa^2\delta_1\delta_2} \tag{3.28}$$

The difference between δ_{I} and δ_{II} is greater, the greater the coupling coefficient κ. If we assume $\delta_1<\delta_2$, we get:

$$\delta_{\mathrm{I}}<\delta_1<\delta_2<\delta_{\mathrm{II}}$$

With $\kappa \to 0$, δ_{I} approaches δ_1 from below, and δ_{II} approaches δ_2 from above.

If the coupling between rooms is provided by a partition wall or a door, the value of κ becomes so small that the differences $(\delta_1 - \delta_{\mathrm{I}})$ and $(\delta_{\mathrm{II}} - \delta_2)$ become negligible. In this case, we can neglect not only τS_{12} but even $\alpha_{1,2} S_{12}$ in the equations:

$$A_{11} = A_{10} + \alpha_1 S_{12}; \qquad A_{22} = A_{20} + \alpha_2 S_{12} \tag{3.29}$$

Thus, the damping constants δ_1 and δ_2 are hardly any different from:

$$\delta_{10} = \frac{cA_{10}}{8V_1}, \qquad \delta_{20} = \frac{cA_{20}}{8V_2} \tag{3.30}$$

which we found for impenetrable coupling surfaces.

If small rooms are coupled to a large room by open areas $(\tau = 1)$ (which was our original problem, see Fig. 3.1), κ is generally very small. Even if $k_2 = S_{12}/A_{22}$ is nearly unity, at least $k_1 = S_{12}/A_{11}$ will be very small because of the large value of A_{11}. For small κ, the differences between δ_{I} and δ_1, and between δ_{II} and δ_2 are always small. They are greatest for $\delta_1 = \delta_2$, but even then they are not larger than $\kappa\delta_1$. But if δ_1 and δ_2 themselves differ greatly—or, more precisely, if

$$\frac{(\delta_1 - \delta_2)^2}{4\delta_1\delta_2} = \frac{1}{4}\left(\sqrt{\frac{\delta_1}{\delta_2}} - \sqrt{\frac{\delta_2}{\delta_1}}\right)^2 \gg \kappa^2 \tag{3.31}$$

(a condition which is fulfilled for coupled theater boxes on account of the different volumes), then we may express the square root in eqn. (3.28) in a power series in (κ^2), neglect all but the first term, and get:

$$\delta_{\mathrm{I}} = \delta_1 - \kappa^2 \frac{\delta_1\delta_2}{\delta_2 - \delta_1}$$

$$\delta_{\mathrm{II}} = \delta_2 + \kappa^2 \frac{\delta_1\delta_2}{\delta_2 - \delta_1} \tag{3.32}$$

The presupposition (eqn. (3.31)) for this development shows that the correction terms are so small, compared with $\sqrt{\delta_1\delta_2}$, that in practice the damping constants of the given uncoupled rooms could be used as a good approximation. By 'given' we mean that S_{12} (or $\alpha_{1,2}S_{12}$) must be added to the other absorption areas, A_{10} and A_{20}, of the rooms. But this is just what we found to be expedient in our discussion of the steady-state condition.

A high degree of coupling is possible in room acoustics only if the partial absorption areas, A_{10} and A_{20}, are small compared to the coupling area: this condition would, therefore, begin to approach that of a single, rather reverberant room. In such a case, it is more suitable to express eqn. (3.28) in the form:

$$\delta_{\mathrm{I,II}}=\frac{1}{2}(\delta_1+\delta_2)\mp\sqrt{\frac{1}{4}(\delta_1+\delta_2)^2-(1-\kappa^2)\,\delta_1\delta_2} \tag{3.33}$$

and to develop the square root in a power series in $(1-\kappa^2)$. This results in:

$$\delta_{\mathrm{I}}\approx(1-\kappa^2)\frac{\delta_1\delta_2}{\delta_1+\delta_2} \tag{3.34}$$

and:

$$\delta_{\mathrm{II}}\approx\delta_1+\delta_2 \tag{3.35}$$

With $A_{10}\ll S_{12}$ and $A_{20}\ll S_{12}$, eqn. (3.34) approaches

$$\delta_{\mathrm{I}}\approx\frac{c}{8}\frac{A_{10}+A_{20}}{V_1+V_2} \tag{3.36}$$

This means that we get a damping constant corresponding to a single room having an equivalent absorption area of $(A_{10}+A_{20})$ and a volume of (V_1+V_2).

II.3.4 Examples of Reverberation in Coupled Rooms

For loosely coupled rooms (that is, for small κ), the damping constants are practically the same as for the uncoupled rooms (with $\alpha_{1,2}S_{12}$ included in their respective absorption areas); but this does *not* mean that the reverberation process is the same as when there is no coupling. Even for small τ, we may hear in the neighboring room (if we hear anything at all through the partition) the initial portion of the reverberation in the source room. Moreover, the decay corresponding to the neighboring room enters into the resulting reverberation in both rooms.

This resulting reverberation comprises in both rooms terms proportional to $e^{-2\delta_{\mathrm{I}}t}$ and $e^{-2\delta_{\mathrm{II}}t}$:

$$E_1=E_{\mathrm{I}1}e^{-2\delta_{\mathrm{I}}t}+E_{\mathrm{II}1}e^{-2\delta_{\mathrm{II}}t} \tag{3.37}$$

$$E_2=E_{\mathrm{I}2}e^{-2\delta_{\mathrm{I}}t}+E_{\mathrm{II}2}e^{-2\delta_{\mathrm{II}}t} \tag{3.38}$$

where the quantities E_{I1}, etc., refer to the initial values for the different exponential decays; they may even be negative. (For brevity here, we drop the additional subscript 0, used to signify the time $t=0$.) In both rooms, the reverberation ends with the exponential function having the smaller value of δ.

At the beginning, the decay processes are different. In the source room, the decay starts approximately as:

$$E_1 \approx E_{\text{I1}}(1-2\delta_{\text{I}}t)+E_{\text{II1}}(1-2\delta_{\text{II}}t)$$

That is, it begins as if with the damping constant:

$$\delta_{01}=\frac{E_{\text{I1}}\delta_{\text{I}}+E_{\text{II1}}\delta_{\text{II}}}{E_{\text{I1}}+E_{\text{II1}}} \tag{3.39}$$

For the second room, we would get the corresponding apparent damping constant:

$$\delta_{02}=\frac{E_{\text{I2}}\delta_{\text{I}}+E_{\text{II2}}\delta_{\text{II}}}{E_{\text{I2}}+E_{\text{II2}}} \tag{3.40}$$

Only if the initial values E_{01} and E_{02} in the two rooms fulfil the coupling equations, (3.22) and (3.23), with $\delta=\delta_{\text{I}}$ or $\delta=\delta_{\text{II}}$ (it would be possible *only* with those eigen-values), could we get the same purely exponential decay in both rooms. For *arbitrary* initial values of E_{01} and E_{02}, the pairs E_{I1}, E_{I2} and E_{II1}, E_{II2} must fulfil these conditions separately, for δ_{I} or δ_{II}, respectively. Therefore, we may eliminate two of these quantities; it appears reasonable to drop E_{I2} and E_{II1}, since they appear to be quantities introduced on account of the coupling. Furthermore, it is expedient in each case to use the condition that avoids the small differences in $|\delta_{\text{I}}-\delta_1|$ or $|\delta_{\text{II}}-\delta_2|$. Thus, we get, from eqn. (3.23):

$$E_{\text{I2}}=E_{\text{I1}}\frac{\frac{c}{4}\tau S_{12}}{\frac{c}{4}A_{22}-2\delta_{\text{I}}V_2}=E_{\text{I1}}\frac{k_2}{1-\delta_{\text{I}}/\delta_2} \tag{3.41}$$

and by (3.22):

$$E_{\text{II1}}=E_{\text{II2}}\frac{k_1}{1-\delta_{\text{II}}/\delta_1} \tag{3.42}$$

Since $\delta_1<\delta_2$ always implies $\delta_{\text{II}}>\delta_1$, and $\delta_1>\delta_2$ always implies $\delta_{\text{II}}<\delta_1$, these equations show that if E_{I2} and E_{I1} have the same sign, then E_{II1} and E_{II2} must have different signs. Therefore, we get the equations for the resulting reverberation processes by putting eqns. (3.41) and (3.42) into (3.37) and (3.38):

$$E_1 = E_{\mathrm{I}1} e^{-2\delta_{\mathrm{I}} t} + E_{\mathrm{II}2} \frac{k_1}{1-\delta_{\mathrm{II}}/\delta_1} e^{-2\delta_{\mathrm{II}} t} \tag{3.43}$$

$$E_2 = E_{\mathrm{I}1} \frac{k_2}{1-\delta_{\mathrm{I}}/\delta_2} e^{-2\delta_{\mathrm{I}} t} + E_{\mathrm{II}2} e^{-2\delta_{\mathrm{II}} t} \tag{3.44}$$

That is, whatever the signs of $E_{\mathrm{I}1}$ and $E_{\mathrm{II}2}$ may be, we get in one room the sum, and in the other room the difference, of the elementary exponential processes. Therefore, the final purely exponential process is asymptotically approached both from above and below.

We get $E_{\mathrm{I}1}$ and $E_{\mathrm{II}2}$ from the initial values E_{01} and E_{02} by the conditions:

$$E_{01} = E_{\mathrm{I}1} + E_{\mathrm{II}2} \frac{k_1}{1-\delta_{\mathrm{II}}/\delta_1}$$

$$E_{02} = E_{\mathrm{I}1} \frac{k_2}{1-\delta_{\mathrm{I}}/\delta_2} + E_{\mathrm{II}2}$$

with the following results:

$$E_{\mathrm{I}1} = \frac{E_{01} - E_{02} k_1/(1-\delta_{\mathrm{II}}/\delta_1)}{1-\kappa^2/(1-\delta_{\mathrm{I}}/\delta_2)(1-\delta_{\mathrm{II}}/\delta_1)} \tag{3.45}$$

$$E_{\mathrm{II}2} = \frac{E_{02} - E_{01} k_2/(1-\delta_{\mathrm{I}}/\delta_2)}{1-\kappa^2/(1-\delta_{\mathrm{I}}/\delta_2)(1-\delta_{\mathrm{II}}/\delta_1)} \tag{3.46}$$

The differences expressed in the numerators show that it is possible for the ratio of the initial energy densities, E_{01}/E_{02}, to be chosen so that one of the decay types vanishes.

If we introduce eqns. (3.45) and (3.46) into (3.43) and (3.44), then we can find the general solution for given values of E_{10} and E_{20}. But we are again interested only in the special solutions where a sound source radiates the power P_1 in the room 1.

Here we assume, first, that the reverberation follows a sufficiently long steady-state excitation that we can assume a stationary energy density, as we discussed in Sections II. 3.1 and II. 3.2. Thus, we can make use of all the approximations in those sections, for small κ and different damping constants, according to eqn. (3.31). Then we may assume:

$$E_{01}=\frac{4P_1}{cA_{11}} \qquad E_{02}=k_2E_{01}$$

$$E_{\text{I1}}=E_{01} \qquad E_{\text{II2}}=-k_2\frac{\delta_1}{\delta_2-\delta_1}E_{01} \quad \text{etc.}$$

Therefore, we can describe the reverberation processes by:

$$E_1 \approx \frac{4P_1}{cA_{11}}\left[e^{-2\delta_1 t}+\kappa^2\frac{\delta_1{}^2}{(\delta_2-\delta_1)^2}e^{-2\delta_2 t}\right] \tag{3.47}$$

$$E_2 \approx \frac{4P_1}{cA_{11}}k_2\left[\frac{\delta_2}{\delta_2-\delta_1}e^{-2\delta_1 t}-\frac{\delta_1}{\delta_2-\delta_1}e^{-2\delta_2 t}\right] \tag{3.48}$$

We gather from these equations that in the source room the second term (which is produced by the coupling) is rather small at the beginning, so that it has no influence on the initial level and the first portion of the decay. Only later does it become predominant on account of the slower decay. In the neighboring room, by contrast, we must take into account both terms from the beginning; and the second term exceeds the first very soon.

Furthermore, we see that, whatever the values for δ_1 and δ_2 may be, the first and second terms are always added for the source room; this means that in that room the asymptotic decay is always approached from above; therefore, in the neighboring room it is always approached from below. It is plausible that the initial stationary level in the source room is higher than that which would correspond to the second decay-type only.

From eqn. (3.40) we can deduce, making use of all the approximations, that the initial damping constant δ_{02} vanishes, so that the decay $E_{2(t)}$ in the neighboring room starts with a horizontal tangent (zero slope). This holds not only for small κ and different values for δ_1 and δ_2; it follows physically from the general initial conditions. When the source is stopped in room 1, only $\mathrm{d}E_1/\mathrm{d}t$ but not E_1, can suddenly change. In the neighboring room, the quantity (c/4 $\tau S_{12}E_1$) represents the power introduced there from the source room. But since that quantity cannot change suddenly, it follows that the decay $E_{(t)}$ can begin only with $\mathrm{d}E_2/\mathrm{d}t=0$.

We must now distinguish between the two extreme cases where $\delta_1 \gg \delta_2$ and $\delta_1 \ll \delta_2$; that is, where the source room is highly damped and the neighboring room is reverberant, or vice versa.

The first case may occur when we open the door from a living room, richly furnished with carpet and luxurious upholstery, into a bare entrance hall. If we shout in the living room, we hear first the reverberation with short decay of the living room; but when the sound level there has decreased sufficiently, we then notice only the longer reverberation of the entrance hall. The corresponding diagram of level versus time is plotted as a full line in Fig. 3.4, left. The dashed line corresponds to the decay of the level in the entrance hall; notice in this case the horizontal tangent at the beginning of the decay.

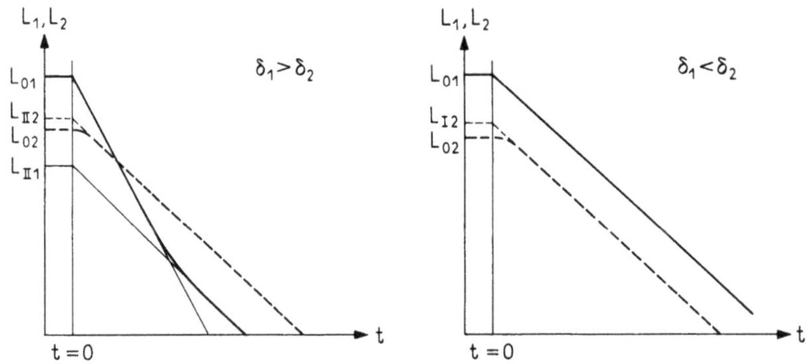

Fig. 3.4. Curves of sound level versus time for reverberation processes in coupled rooms, after stopping a steady-state source in room 1. Left: the source room is highly damped and the adjacent room is reverberant. Right: the adjacent room is highly damped and the source room is reverberant. Room 1———, room 2 – – –.

The other extreme case, $\delta_1 \ll \delta_2$, corresponds to our initial problem; that is, to theater boxes, seating areas under balconies, and side aisles connected to a high nave; all of these are relatively small, 'dead' spaces coupled to a larger 'live' space where the source of sound is located.

Again, Fig. 3.4 (right) shows the curves of level versus time; the full line corresponds to the source room, the dashed line to the neighboring room. In the source room, since even at the beginning of the decay its own longer reverberation process predominates, the shorter decay of the neighboring room is not heard at all. In the neighboring rooms, however, we find a process similar to that described above. The decay starts with a horizontal tangent and then diminishes according to the continuing but declining supply of energy from the source room. In practice, this first part of the decay is not apparent, since the necessary assumption of the statistical theory (i.e. a homogeneous, isotropic sound field) is not established until after many sound reflections. But the shift of the decay slope by a certain delay may be observed.

As a second example, we will discuss the reverberation process that follows an impulsive supply of energy $P_1\Delta t$ in the source room. Here we can make use of the relations with steady-state excitation already mentioned in Section II. 1.3. We have only to replace the quantity E_- in eqn. (1.18) with the expressions for E_1 and E_2 given in (3.47) and (3.48). We immediately find, for the reverberation processes following an impulsive excitation:

$$E_1' = \frac{4P_1\Delta t}{cA_{11}}\left(2\delta_1 e^{-2\delta_1 t} + \kappa^2\frac{2\delta_1^2\delta_2}{(\delta_2-\delta_1)^2}e^{-2\delta_2 t}\right)$$

$$= \frac{P_1\Delta t}{V_1}\left(e^{-2\delta_1 t} + \kappa^2\frac{\delta_1\delta_2}{(\delta_1-\delta_2)^2}e^{-2\delta_2 t}\right) \tag{3.49}$$

$$E_2' = \frac{P_1\Delta t}{V_1}k_2\frac{\delta_2}{\delta_2-\delta_1}\left(e^{-2\delta_1 t} - e^{-2\delta_2 t}\right) \tag{3.50}$$

Figure 3.5 shows the corresponding curves of level versus time, similar to those of Fig. 3.4. Especially striking is the difference with respect to E_2, which starts at zero with ($\log E_2 \rightarrow -\infty$). This corresponds to the system equations, (3.18) and (3.19), assuming that we add an impulsive source in room 1. During the impulse, only the derivatives of highest

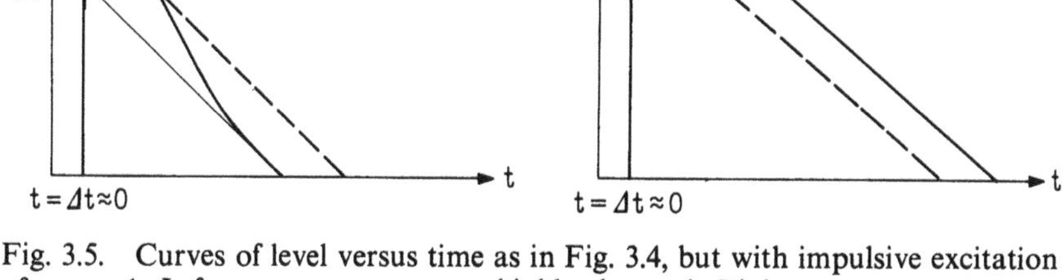

Fig. 3.5. Curves of level versus time as in Fig. 3.4, but with impulsive excitation of room 1. Left: source room more highly damped. Right: source room more reverberant than the neighboring room. Room 1———, room 2 – – –.

order (here dE/dt) play a role. But this means that E_2 does not change at all during the impulse, so it starts with the initial value:

$$E_{20}'=0, \tag{3.51}$$

whereas E_1 jumps from zero to:

$$E_{10}'=\frac{1}{V_1}\int_{\Delta t} P_1 \mathrm{d}t=\frac{\bar{P}_1 \Delta t}{V_1} \tag{3.52}$$

This corresponds to the same equal distribution of the supplied energy over the volume V_1 as we assumed in Section I. 1.3 for the volume V of a single room. In the present case, we cannot actually expect that this initial condition will be reached during the impulse duration; but it is reasonable to assume that this analysis describes fairly well the later part of the reverberation process.

We could have derived eqns. (3.49) and (3.50) in an alternative way by putting the initial values E_{20}' and E_{10}', given by eqns. (3.51) and (3.52), into (3.45) and (3.46) and considering the same simplifying approximations that we used earlier in (3.47) and (3.48).

Figure 3.5 (left) also shows that in the case $\delta_1 \gg \delta_2$ the reverberation in room 1 has changed; it is not a change in principle, but (at the right) in the level at which the bend in the curve occurs. Since $L_{\mathrm{II}1}$ is decreased by $10 \log (\delta_1/\delta_2)$, the level of the bend in the curve is correspondingly lowered. In practice, the bend may not appear in the part of the curve above the background noise that limits the recording. In the case of coupled rooms, it becomes very important to distinguish whether the reverberation follows steady-state or impulsive excitation, and, in the latter case, whether the reverberation was determined by direct recording or by 'backward-integration' (see Section II. 4.5) (which would give results corresponding to steady-state excitation).

A typical case of coupled rooms occurs in opera houses. The reverberation time of the auditorium is usually low (1·2 up to (at most) 1·6 s) because of the custom of fitting the greatest possible number of seats into the given room volume. On the performer's side of the proscenium, however, the stage house may have a reverberation time up to 3 s. It is paradoxical that open-air scenes, played on a bare stage, are given the most reverberant environment! In such cases, the sound events on the stage may produce in the auditorium slopes such as those in Fig. 3.4 and 3.5, right (the dashed lines). On the other hand, the pit orchestra, being located more in the auditorium, may produce curves like the solid lines in the left parts of the figures.

Such a difference, in fact, may even be attractive up to a certain point, so that experienced listeners just notice it, but others in the audience perceive it as a 'special' but undefined quality.

There have been occasional attempts to compensate for a too-short reverberation time in the auditorium by allowing the audience to hear the longer reverberation time of the stage house; but we have learned in this section that the decay process thus achieved is quite different from that which a longer reverberation time in the auditorium would give.

II.3.5 Electroacoustical Coupling between Rooms

Any consideration of coupled rooms would be incomplete today if we took into account only the 'natural' coupling, discussed up to this point, without also accounting for the possibility of 'artificial' coupling by means of microphones, amplifiers and loudspeakers.

The most frequently used applications of the latter occur with radio or television broadcasts and in cinema theaters. The sound is picked up by a microphone in one room and is transmitted, either 'live' or recorded, into another room (a living room or a cinema theater), where it is re-radiated by loudspeakers.

This kind of coupling exhibits two essential differences from natural coupling. First, the coupling is uni-directional; that is, with respect to our energy balance, room 1 transfers into room 2 an amount of power P_{21} that is proportional to E_1:

$$P_{21} = K_{21} E_1 \tag{3.53}$$

but room 2 transmits no power back to room 1 at all.

The second difference is that the power radiated into room 2 is not subtracted from the power P_1 radiated into room 1; that is, the electroacoustical equipment does not absorb sound energy in room 1 and thus has no influence on the value of A_1. (This is not as self-evident as it appears. At least we will discuss later an arrangement by which the electroacoustical equipment actually adds to P_1 and thereby reduces the value of A_1.)

The power P_{21} is taken from an independent power source, which is 'piloted' by the time-varying E_1; only in this manner is the uni-directional coupling possible. The constant K_{21}, which contains the gain of the amplifier chain, can be given any value we choose; therefore, the amount of power radiated into room 2 is independent of the amount of power produced in room 1.

The lack of any energy feedback from room 2 leads to a power balance equation for room 1 in which E_2 does not appear:

$$P_1 = \frac{c}{4} A_1 E_1 + V_1 \frac{\mathrm{d}E_1}{\mathrm{d}t} \tag{3.54}$$

After the sound source is stopped, we have only a single exponential decay with the damping constant:

$$\delta_1 = \frac{cA_1}{4V_1} \tag{3.55}$$

The initial energy density after long excitation is:

$$E_{01} = \frac{4P_1}{cA_1} \tag{3.56}$$

For impulsive excitation, we get (as above, in eqn. (3.52)):

$$E_{01}' = \frac{P_1 \Delta t}{V_1} \tag{3.57}$$

In the playback room, in both cases, the reverberation in room 1 appears as continuing but decreasing excitation. For steady-state excitation in room 1, we get the equation:

$$K_{21} E_{01} e^{-2\delta_1 t} = \frac{c}{4} A_2 E_2 + V_2 \frac{\mathrm{d}E_2}{\mathrm{d}t} \tag{3.58}$$

This reverberation forces a single decay process in room 2 with the same exponent:

$$E_{21} = E_{021} e^{-2\delta_1 t} \tag{3.59}$$

or, according to (3.58):

$$E_{21} = \frac{4K_{21}E_{01}}{cA_2} \frac{\delta_2}{\delta_2 - \delta_1} e^{-2\delta_1 t} \tag{3.60}$$

where we have again introduced δ_2 for the damping constant of room 2:

$$\delta_2 = \frac{cA_2}{8V_2} \tag{3.61}$$

But the initial energy density that appears in eqn. (3.60) is, in general, not that which exists at $t=0$ in room 2. If the reverberation follows steady-state excitation, we must expect there the energy density:

$$E_{02}=\frac{4K_{21}E_{01}}{cA_2} \tag{3.62}$$

If the decay follows an impulsive excitation, we have

$$E_{02}'=0 \tag{3.63}$$

In both cases, we must supplement eqn. (3.60) with a term that decays with the damping constant δ_2, in order to fulfil the initial conditions. Physically expressed, we must add to the 'forced-reverberation' term a 'free-reverberation' term.

Thus, we get a solution for steady-state excitation as follows:

$$E_2=\frac{4K_{21}E_{01}}{cA_2}\left[\frac{\delta_2}{\delta_2-\delta_1}e^{-2\delta_1 t}-\frac{\delta_1}{\delta_2-\delta_1}e^{-2\delta_2 t}\right] \tag{3.64}$$

In the case of impulsive excitation we get:

$$E_2'=\frac{4K_{21}E_{01}'}{cA_2}\frac{\delta_2}{\delta_2-\delta_1}\left[e^{-2\delta_1 t}-e^{-2\delta_2 t}\right] \tag{3.65}$$

(The expressions in brackets are the same as those in eqns. (3.48) and (3.50), showing that in those equations we have already neglected the feedback from the neighboring room.)

Note that eqns. (3.64) and (3.65), taking into account eqns. (3.56) and (3.57), obey the general condition of eqn. (1.18).

We distinguish again between the two possibilities, that room 1 or room 2 is the more reverberant. Figure 3.6 shows the curves of level versus time for these two cases (left and right); the situation with steady-state excitation is shown at the top, that with impulsive excitation at the bottom. The solid lines refer to the recording room, the dashed lines to the playback room, 2.

The case of $\delta_1<\delta_2$ (i.e. $T_1>T_2$, left) is the normal situation for the playback of concerts in living rooms. The reverberation heard there corresponds to the reverberation in the recording room (since the short time delay in room 2 is hardly noticeable). For the other case, $\delta_1>\delta_2$ (i.e. the reproduction of 'dry' music in a reverberant space, right), the short reverberation time of the recording room is not observable in room 2. For this reason, the changes in recorded reverberation that are some-

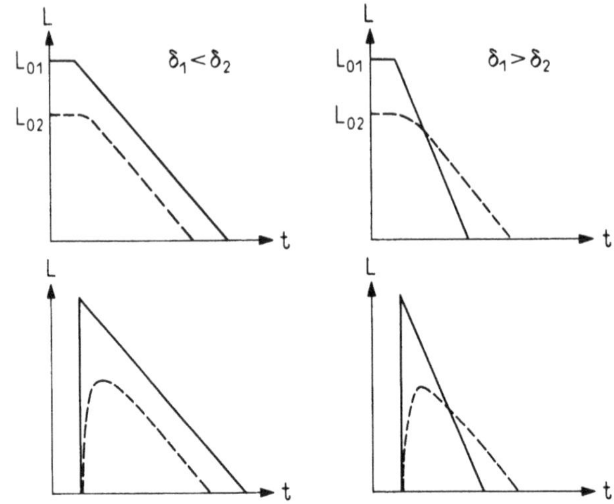

Fig. 3.6. Reverberation processes (level versus time) for unidirectional coupling between rooms. Left: source room more reverberant. Right: playback room more reverberant. Top: steady-state excitation. Bottom: impulsive excitation.

times introduced in an attempt to replace the missing visual scenery with 'acoustical scenery', are not heard, because they are swamped by the reverberation of the playback room.

Since such artificial manipulation of the recorded acoustics was regarded as a legitimate tool of the movie industry, it was assumed in the early days that it was necessary to make the acoustics of cinema theaters as 'dead' as possible. But this extreme practice was quickly given up, not only because cinema theaters are sometimes used for other purposes, but also because it was found that the loudspeaker sound is unnatural when the reflections from the walls and ceiling are missing. All the sound tends to be localized at the front, so that even when the reproduced sound contains reverberation, the listener never feels that he is in a reverberant space, nor even that he is looking into a reverberant space.

Certainly, a reverberant cinema theater would be unsuitable, because the speech would be unintelligible; so nowadays the acoustical requirements for a cinema theater are considered to be the same as for a lecture room or drama theater.

It is not always possible, on the stage, to adjust the acoustical impression to match the visual scene; sometimes, in fact, there is an obvious contradiction. The more-or-less absorptive painted canvas stage scenery, used for the backdrop and ceiling in the scene from 'Faust' in Auerbach's cellar, never succeeds in conveying the effect that the 'vault resounds' (Goethe: '*das Gewölbe widerschallt*'). On the other hand, when the shepherd is supposed to play his English horn in the open air in the third act of 'Tristan', he may be accompanied (in large stage houses) by a clear and most improbable reverberation.

As Fig. 3.6 shows, reverberation can never be removed at a later point in the transmission chain. Therefore, current recording practice (see Fig. 3.7, top) prefers a rather dead recording room, 1, and adds the desired reverberation later by playing the 'dry' performance into a reverberant room, 2, where it is recorded a second time.

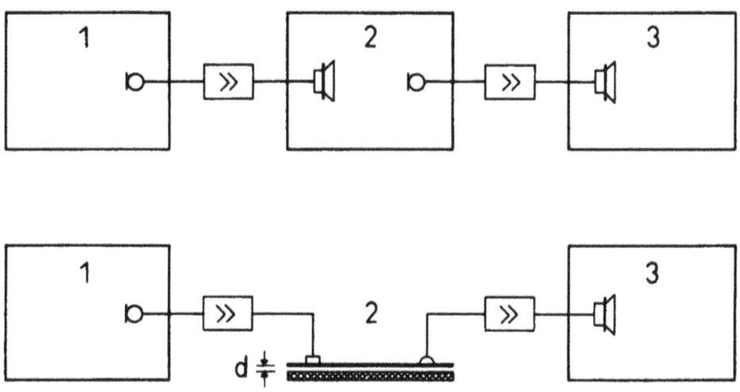

Fig. 3.7. Schematic representation of the principles of adding artificial reverberation to a 'dry' recording. Top: with a reverberation room added in the transmission chain. Bottom: with a reverberation plate added in the transmission chain.

This means a transition from the solid curves to the dashed curves in Fig. 3.6 (right). The playback room, 3, which is always more highly damped than room 2, does not cause a further change in the reverberation; here the process corresponds to the dashed curves in Fig. 3.6 (left).

The method of adding reverberation with the use of special reverberation rooms is expensive because it requires rather large rooms, with the provision of large amounts of variable absorptive material for the purpose of changing the reverberation time to the desired value. From the standpoint of statistical room acoustics, it appears possible, at first sight, to use small reverberation rooms with very hard walls for adding the desired reverberation to originally dry recordings; but such a conclusion gives insufficient consideration to the limits of statistical room acoustics as applied to small rooms. In small rooms, we must expect pronounced resonances at certain special low frequencies (so-called 'eigen-tones'), and this means a strong frequency dependence of both energy density and reverberation (to be discussed on a wave-theoretical basis in Section IV. 11.2, Volume 2). In larger rooms, the eigen-tones become so dense in the audible frequency range that these effects can be treated only by statistical methods, leading to the same results as we develop here with pure energy balance methods.

In order to avoid the expense of a large reverberation room, Kuhl[1] developed with surprising success the substitution of a flexible vibrating plate for the reverberation room. The plate is excited (i.e. driven into vibration) at one point by the original 'dry' signal; the 'reverberated' signal is picked up at another point on the plate (see Fig. 3.7, bottom). From the study of bells, it is well known that they ring with a slow decay; this is very desirable since their own reverberation replaces the missing reverberation out-of-doors. But, because of their great thickness, bells are tuned so high that their eigen-tones are widely separated on the frequency scale. For thin plates, however, the lowest eigen-tones (2–3 Hz) lie as low as those in a large hall. With flexural waves, the same order of magnitude separates the individual eigen-tones in the audible frequency range. Therefore, it is possible to simulate, using thin plates $2\,m^2$ or less in size, the reverberation of large rooms at low frequencies, even though the propagation laws for flexural waves are quite different, in principle, from those for airborne sound (see Section IV. 10.2, Volume 2). (This difference, in fact, makes it reasonable to supplement the reverberation of the plate at high frequencies with a reverberation room, which for this frequency range can be small.)

The vibrating plate offers another surprising advantage: its reverberation time can be changed without changing anything on the plate itself. It is sufficient, when a lower reverberation time is desired, to bring a porous, sound-absorbing layer more or less near to the plate (see Fig. 3.7, bottom); by changing the distance d between them, the reverberation can be changed from, say, 1 to 5 s (see Section IV. 10.8, Volume 2).

Combinations of spiral springs have also been used with success for adding artificial reverberation.[2,3]

Numerous efforts have been made to use artificial reverberation of

[1] Kuhl, W., *Rundfunktechnische Mitteilung*, **2** (1958) 111.

[2] Meinema, H. E., Johnson, H. A. and Laube, W. C. Jr., *J. Audio Eng. Soc.*, **9** (1961) 284.

[3] Kuttruff, H., *Room Acoustics*, 2nd. Edn., Applied Science Publishers Ltd, London, 1979; see Section X.5.

various kinds in order to adjust the acoustics of a drama theater to correspond with the different scenes, or to prolong the reverberation of a multi-purpose hall for special performances, or of an opera house for symphonic concerts. Here the sound is picked up in the auditorium and, after the addition of longer reverberation by means of an auxiliary reverberation room (or plate), it is re-radiated into the auditorium. In this case, the uni-directional coupling occurs in both directions, because now room 2 also transfers power back to the original room 1, proportional to E_2:

$$P_{12}=K_{12}E_2 \tag{3.66}$$

But the reciprocity that characterizes natural room coupling is completely missing, since K_{12} and K_{21} can be chosen arbitrarily according to the gains in the transmission chains, so that usually

$$K_{12}\neq K_{21} \tag{3.67}$$

However, so long as in the corresponding system equations

$$P_1=\frac{c}{4}A_1E_1+V_1\frac{\mathrm{d}E_1}{\mathrm{d}t}-K_{12}E_2$$

$$0=-K_{21}E_1+\frac{c}{4}A_2E_2+V_2\frac{\mathrm{d}E_2}{\mathrm{d}t} \tag{3.68}$$

the coupling efficiency

$$\kappa=\sqrt{\frac{K_{12}K_{21}}{A_1A_2}\frac{4}{c}} \tag{3.69}$$

is small in comparison with 1, we may expect the same behavior as in naturally coupled rooms; namely, in room 1 (the auditorium and stage) there will be a double-sloped decay curve, as shown by the solid curves in the upper parts of Figs. 3.4 and 3.5.

With such a system, it is possible that κ exceeds 1. But then the square root term in eqn. (3.28) exceeds the first term, and therefore one of the damping constants becomes negative; that is, instead of a decay of energy we find that the energy density increases.

This is evidently possible because each increase in E_1 leads to an increase in E_2; and this, in turn, again increases E_1, and so on, up to the limit of the independent power source.

We may, in fact, wonder why such a continuing increase in energy density does not always occur. However, a quantitative discussion of this instability is not possible with statistical methods alone, because the phases of the sound waves must be taken into account. Also, the directional characteristics of the microphones and loudspeakers in both rooms enter into the problem. All of these influences interact to cause the limit of instability to be reached first at a single frequency. The result, therefore, is not a general increase of the entire radiated signal, but only of a certain tone: the system starts to 'howl'.

This howling phenomenon is called 'electroacoustic feedback'; in fact, the feedback always exists, even below the limit of instability, but it is customary to use the expression only for the howling condition.

The feedback phenomenon does not require the presence of an artificial reverberation room (or plate) at all. It may also occur if the feedback path leads directly from the microphone to a loudspeaker in the auditorium itself, as in the case of a public address system (see Section I.5.2). In this case, eqns. (3.68) and (3.69) become (with $V_2 = 0$) only one equation:

$$P_1 = \frac{c}{4} A_1 E_1 + V_1 \frac{\mathrm{d}E_1}{\mathrm{d}t} - K_{11} E_1 \tag{3.71}$$

where $4K_{11}/c$ may partially compensate, or even exceed, the value of A_1.

Here, again, the statistical theory only allows this possibility; the actual calculation requires a treatment of the oscillatory character of the sound field parameters. But it should be emphasized that such electroacoustical feedback has been successfully used to prolong the reverberation in halls. The method developed by Parkin[4] (called 'assisted resonance') avoids the problem of exceeding the feedback instability limit at certain individual frequencies by using many microphone–amplifier–loudspeaker loops and inserting variously tuned Helmholtz resonators (see Section IV.9.2, Volume 2) into the different electroacoustic circuits, so that the gain of each frequency region can be adjusted (for non-howling and the desired reverberation time) independently of the others. In this manner it is possible to approach (but not exceed) the limit of instability at many frequencies simultaneously. By contrast, when we have broadband amplification, only one frequency governs this howling limit, while at other frequencies the gain is still far below the limit and there is no prolongation of the reverberation at all.

Parkin has used this method with great success to increase the

[4] Parkin, P. H. and Morgan, K., *J. Sound Vibr.*, **2** (1965) 74.

reverberation time of the Royal Festival Hall in London. In the initial installation, he used 90 channels of feedback, at low frequencies (70–340 Hz) only. Later, the number of channels was increased to 172 in order to extend the mid-frequency coverage up to 700 Hz and the low-frequency coverage down to 58 Hz.[5] This method, with its numerous channels of amplification, which must be constantly adjusted in order to assure stability, is naturally rather expensive.

The same holds for an alternative method developed by Franssen.[6] Again, many microphones and loudspeakers are used, distributed equally around the auditorium. But here the different channels are not assigned to different frequencies. Instead, the method relies on the fact that sound radiated from any one loudspeaker reaches all of the microphones and thus excites all of the other loudspeakers, and so on. Franssen's method is based on the assumption (which in practice proved correct) that all the possible feedback paths would have different howling frequencies. On the other hand, on account of the great number of parallel paths, he could set the gains far below the instability limit and still achieve a significant increase in reverberation time. A successful application of this principle has been installed in the Concert Hall in Stockholm.[7]

Both methods require careful attention and control by an experienced sound engineer during a performance. He is able to adjust the system so that it does not introduce any strange character to the sound, but the listeners sense an increased reverberation time and a greater spatial impression (see Section III. 2.5).

Since the phases, and therefore the lengths of the ray paths, establish the limit of instability in these feedback systems, one means of increasing the system gain without instability would be to use two microphones, alternately and smoothly switching back and forth between them, or by moving a single microphone smoothly back and forth. So far, neither of these methods has been very successful.

A more elegant and quite successful method was developed by Schroeder,[8] in which he makes use of a small frequency shift in the feedback circuit, so that the microphone picks up from its corresponding loudspeaker a sound with frequency slightly different than it transmitted

[5] Parkin, P. H., 'Assisted resonance', Chapt. 15 in *Auditorium Acoustics*, ed. R. Mackenzie, Applied Science Publishers Ltd., London, 1975. See also Kuttruff,[3] p. 298.

[6] Franssen, V. N., *Acustica*, **10** (1968) 315.

[7] Dahlstedt, S., *J. Audio Eng. Soc.*, **22** (1974) 627.

[8] Schroeder, M. R., *Proc. 3rd. ICA, Stuttgart*, 1959, Vol. II, Elsevier, Amsterdam, 1961, p. 771.

to that loudspeaker, thus interrupting the feedback loop for the potential howl frequency. This method can be analysed and explained only by wave-theoretical considerations.

Even the simple statistical theory, however, teaches us that every loudspeaker channel—operating below the instability limit—always results in an increase of the reverberation time, and thus complicates the problem if the room is already too reverberant. The notion that an electroacoustic reinforcement system invariably leads to an improvement is quite wrong. Rather, we should say that such a system will work better, the better the room is without it.

Chapter II.4

Measurement of Reverberation Time

II.4.1 The Subjective Method

Since the reverberation of a room can be perceived clearly by everybody, reverberation is one of the least controversial of all the room-acoustical criteria. Attention was focused very early on a simple and reliable method to measure the reverberation time *T*. Indeed, it can be said that modern room acoustics began in the 1890s with the attempt to describe reverberation quantitatively.[1] Moreover, a determination of the sound absorption of a material or an object in the reverberation chamber also requires a reliable and exact measurement of the reverberation time.

Today's definition of the reverberation time *T* as the time required after stopping a sound source for the energy to decrease to one-millionth of its original value (−60 dB) (see Section II.1.1) was already established by W. C. Sabine at the end of the last century. The reverberation time corresponds approximately to the duration of reverberation t_s that can be heard in a concert hall if the orchestra suddenly stops with a loud chord and the noise level in the concert hall is, as usual, around 25 dB(A).

As a matter of fact, Sabine's own measurement method was based on this definition of the reverberation time, since he measured with a stopwatch the duration, after interrupting a sound source, until the signal was no longer heard. Since the rooms that he studied had a relatively long reverberation, between 2 and 6 s, he achieved a satisfactory precision. His maximum error, which he carefully determined, was around

[1] Sabine, W. C., *Collected Papers on Acoustics*, Dover Publications, New York, 1964.

0·31 s when the reverberation time was approximately 4 s. For the average error, he found a value of 0·11 s.

His sound sources were organ pipes. By sounding first one, then two, then four pipes, he increased the sound pressure level each time by $\Delta L = 10 \log n = 10 \log 2 = 3$ dB. The audible duration of the reverberation was prolonged with a greater number of pipes, provided that the background noise remained the same. One determines the reverberation time T by increasing the number of pipes from 1 to 2 or to 4 and measuring the increase in the audible duration of reverberation Δt compared with that for one pipe:

$$T = \frac{60}{\Delta L} \Delta t \tag{4.1}$$

For $n = 2$, one gets $T = 20\,\Delta t$; for $n = 4$, $T = 10\,\Delta t$.

Measuring the reverberation time by this method, with a difference ΔL of only 6 dB (only 1/10 of the 'definition range' of 60 dB) requires a very risky extrapolation. Sabine used this method only to determine the sound power of the organ pipes in relation to the threshold defined by the background noise.

Today, all large concert halls, theaters, congress buildings and auditoria are equipped with loudspeaker systems, so that for reverberation time measurements it is easy to create a large level-difference simply by increasing the power that drives the loudspeakers. A ratio of 1:50 in the loudspeaker current, for example, leads to a level increase of $\Delta L = 34$ dB.[2]

In this way, a reverberation time above 2 s can be simply measured by using a stopwatch only and varying the loudness of the room excitation. It is best to vary the loudspeaker current in steps; the corresponding level differences, ΔL (i.e. 20 times the logarithm of the ratio of the current at each step to the minimum current), can be plotted against the measured reverberation duration t. In this way, one gets a straight line whose slope represents the reverberation time T. If the measurement results $\Delta L_1, t_1; \Delta L_2, t_2; \ldots \Delta L_n, t_n$ are scattered, so that it is doubtful how to approximate a straight line, one gets the exact reverberation time T by a least-squares fit, which results in:

$$T = 60 \frac{n \sum^{n} (t_i^2) - \left(\sum^{n} t_i \right)^2}{\sum^{n} \Delta L_i \sum^{n} t_i - n \sum^{n} (\Delta L_i t_i)} \tag{4.2}$$

[2] Strutt, M. I. O., *Elektr. Nachr. Techn.*, 7 (1930) 280.

Hunt[3] and others believe that it is sufficiently exact to draw the line best representing the decay curve simply 'by eye'. The variations incurred in this procedure are much smaller than those that result from changing the conditions of the experiment, e.g. the measurement position.

Loudspeakers have the advantage that they can be driven by a variety of signals, such as pure tones, chords, or mixed tones, simply by using special records or tapes, or by connecting the amplifiers to electronic signal generators. In any case, it is possible in this way to realize the desired filtering into individual frequency ranges more easily than by using organ pipes.

Despite this improvement, there is still a fundamental source of error in the subjective method of measuring reverberation time. It requires that not only the same initial level be strictly maintained within the frequency range of interest, but also the same threshold level, at least during one test series. This threshold level may actually be changing because of variations in the background noise that are hardly noticeable, even if one uses only the quiet night-time hours for measurement, as Sabine did.[4]

If one must measure not only during noisy hours but also under noisy-neighborhood conditions (traffic, industry, etc.), which exclude the possibility of a constant background noise, it would be advisable to radiate an artificial constant background noise during the measurements.

It would even be possible to determine the reverberation time by plotting the reverberation duration t against the background noise level, while the sound power of the measurement signal is held constant and only the background noise is changed. Finally, if signals with narrow frequency bands are used, it is possible to estimate the frequency dependence of the reverberation time, to a first approximation.[5]

---

[3] Hunt, F. V., *J. Acoust. Soc. Am.*, **5** (1933) 127.

[4] In Sabine's own words:

'The next experiment was on the determination of the absorption of sound by wood sheathing. It is not an easy matter to find conditions suitable for this experiment. Quite a little searching in the neighborhood of Boston failed to discover an entirely suitable room. The best one available adjoined a night lunch room. The night lunch was bought out for a couple of nights, and the experiment was tried. The work of both nights was much disturbed. The traffic past the building did not stop until nearly two o'clock, and began again about four. The interest of those passing by on foot throughout the night, and the necessity of repeated explanations to the police, greatly interfered with the work.'
Sabine, W. C., 'The Variation in Reverberation with Variation in Pitch'; in: *Collected Papers on Acoustics*, No. 2, Harvard University Press, Cambridge, 1927, (from *Proceedings of the American Academy of Arts and Sciences*, **XLII**, June (1906) No. 2).

[5] Gumlich, H., Studienarbeit, ITA, Berlin.

In fact, nowadays, no one relies on subjective methods for quantitative measurement of reverberation time. Nevertheless, it is common practice when an acoustical consultant visits an unfamiliar concert hall, for him to clap his hands for a rough subjective evaluation of the reverberation characteristics of the room, as well as to test for echoes and flutter echoes.

II.4.2 Objective Records of Reverberation Decays

Let us bear in mind the disadvantages of the subjective method: the observer must be trained; there must be complete quiet; short reverberation times cannot be measured; and one cannot determine by 'watching the reverberation' whether or not the decay is exponential over the entire range. This is important because the assumption of an exponential decay is the basis for the whole method. Thus, it is understandable that there were early attempts to replace the subjective method by an objective record.

Sabine, himself, attempted this, but he soon gave it up, realizing that in a room excited by organ pipes the decay was by no means exponential. Scientists of a generation later, with the instrumentation of modern acoustics, had the same experience using microphones, amplifiers and a moving-coil oscillograph for the registration of pure tone decays.[1] The fluctuations can be clearly seen in Fig. 4.1, which shows an oscillogram photographed by Meyer.

The reason for this fluctuation lies not in a gross inadequacy of the theory. It has more to do with the fact that the theory describes only the

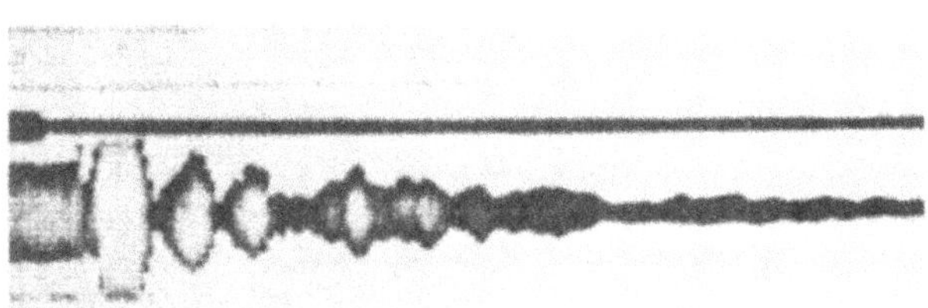

Fig. 4.1. Reverberation record of sound pressure versus time, for excitation with a pure tone.

[1] Meyer, E., *Elecktr. Nachr. Techn.*, **4** (1927) 135; Hollmann, H. E., *Elektr. Nachr. Techn.*, **4** (1927) 180; Trendelenburg, F., *Z. Techn. Phys.*, **8** (1927) 502; Chrisler, V. L., *J. Acoust. Soc. Am.*, **1** (1930) 418.

decay of the energy and not of the sound pressure or sound velocity, i.e. the field quantities that are measured by microphones. As we found in Part I, if we superimpose several wave trains, we can expect simple addition only for the energy, but not for pressure or velocity; the latter additions depend strongly on the phase angle. For example, only if all the waves have the same phase do the pressures add directly; in this case, the resulting sound pressure is much higher than would be expected by adding the energies. In such phenomena, which are referred to by the general term 'interference', it can happen that equal over- and under-pressures must be added together, thus resulting in complete cancellation of the pressure. Even if such extreme cases occur very seldom, the interference is often sufficient to create large fluctuations in a reverberation decay, thus making it practically impossible to derive quantitative results from the corresponding oscillogram.

There are various means of avoiding such fluctuations in a stationary sound field, and in such cases the decay curve exhibits smaller fluctuations.

The simplest method is not to use pure tones. On the other hand, single-frequency tones are desirable in order to determine the frequency dependence of the reverberation time, which is very important. But it requires only a small change of frequency to shift the interference pattern completely, while the absorption of a room changes only very slowly with frequency. Therefore, Meyer and Just[2] used, instead of pure tones, the so-called 'warble tones', whose frequency periodically fluctuates around a mean frequency.

The warble tone is characterized by its mean frequency (which determines its position on the frequency scale), its modulation swing (the difference between the highest and the lowest frequency within one modulation), and its modulation frequency (the reciprocal of the frequency-fluctuation period). The wider the swing, the more the interference pattern disappears. But the permissible swing is limited by the precision with which the frequency dependence of the reverberation time is to be measured.

Sine-wave generators used today for acoustical measurements are equipped with a warble-tone capability that allows the modulation swing and the modulation frequency to be freely changed.

Based on measurements between 250 and 2000 Hz, Hunt[3] recom-

---

[2] Meyer, E. and Just, P., *Elektr. Nachr. Techn.*, **5** (1928) 293; see also Just, P., *Schalltechnik*, **2** (1929) 5.

[3] Hunt, F. V., *J. Acoust. Soc. Am.*, **5** (1933) 127.

mended that a modulation swing $\Delta f/f$ of 20% is sufficient. Other experience, however, shows this value to be too low for low frequencies. This is regrettable, because at low frequencies vibrating panels at the room boundaries may lead to strong frequency dependence.

Later, following systematic investigation of the scatter of reverberation measurements, Meyer[4] used a constant modulation swing between 40 and 90 Hz; even then, he found that the low-frequency measurements required a larger swing than at high frequencies. Furthermore, the required modulation swing is smaller in large rooms than in small rooms. Both findings are in accord with the theoretical investigations discussed in Section IV. 11.2 (Volume 2).

Frequency modulation of a warble tone is not the only modulation that leads to a continued changing of the interferences. This can also be achieved by a periodic fluctuation of the pressure amplitude (amplitude modulation), though it is less effective. From the point of view of harmonic analysis, both types of modulation mean nothing less than the simultaneous radiation of various partial tones with a frequency separation that equals the modulation frequency.[5] If an organ is available, the same blurring of the interference pattern may be achieved by playing several neighboring keys at once.[6]

According to Barrow,[7] such a synthesis of simultaneous pure tones, which he called a 'multitone', is better than a warble tone. Therefore, he developed a multitone generator with eleven electrical resonant circuits excited by periodic electrical pulses.

If the number of single tones in a multitone is increased more and more, and the frequency spacing between them becomes smaller and smaller, we get a signal that approaches a continuous spectrum; we hear this signal subjectively as 'static' and call it 'Random Noise'.[8] If one uses electrical filters to select a certain frequency range out of this broadband signal, for example, a so-called third-octave band of noise (which, analogous to white light, creates no interference), this permits us to measure a rough frequency dependence. A disadvantage of statistical noise is its unavoidable fluctuation, so that—even though these signals

[4] Meyer, E., *Akust. Zh.*, **2** (1937) 179.

[5] Salinger, H., *Elektr. Nachr. Techn.*, **6** (1929) 293.

[6] Thienhaus, E. and Willms, W., *Musik und Kirche*, **5** (1933) 203.

[7] Barrow, W. L., *J. Acoust. Soc. Am.*, **10** (1939) 275.

[8] If random noise covers a wide frequency band, with equal power in equal frequency intervals, we call it 'white noise' because, like white light, it contains all frequency components in equal measure.

give linear reverberation decays—there is not much less data scatter than with warble tones.

Meyer and Just rectified the decaying signal and smoothed the rectified current by connecting in parallel a suitable condenser before the recording galvanometer. In this way, we can get a decay curve that is not only simpler to evaluate, but that also corresponds somewhat better to the functioning of the human ear.

A suitable choice for the 'inertia' of the recording equipment is one of the main secrets in measuring reverberation time. This inertia must be chosen in such a way as to allow—on the one hand—the measurement of very short reverberation times, but—on the other hand—sufficient smoothing of the warble tone periods or the random noise level fluctuations.

By simultaneous registration of time marks, it is possible to determine the reverberation time directly from the decay curve, preferably by plotting the signal values for successive instants in a logarithmic ordinate system. Then the measurement points scatter more or less around a falling straight line, which in most cases can be simply drawn or in doubtful cases can be calculated according to eqn. (4.2).

There is the disadvantage that registering the linear sound pressure allows an amplitude ratio of only around 10:1, or 20 dB, and, thus, only the initial part of the reverberation. Because of the typical 'sagging' of the reverberation curve, this leads to shorter reverberation times[9] than would be found with a full 60 dB decay. Meyer and Just achieved a fundamental improvement when they doubled this range by switching a relay during the registration, thereby increasing the sensitivity by ten times.

Not only are there technical difficulties due to interference, but there is also the possibility that the energy density in the room is not uniform. One can observe these differences by the fact that changes of the positions of the source and receiver lead to different levels under stationary conditions and often, also, to different reverberation times.

Due to the different phase angles of the source at the instant of switch-off in repeated reverberation measurements, all of the reverberation measurement methods discussed so far lead to varying results at the same measuring point. These so-called 'local' fluctuations are, in any case, smaller than those that result from changing the measurement

[9] Sabine, P. E., *J. Acoust. Soc. Am.*, **5** (1934) 220.

location in the room (or the source location) and which we call, following Eisenberg,[10] 'spatial fluctuations'.

Therefore, since the reverberation time is supposed to be a characteristic value of significance for the whole room, one must always make measurements at a number of locations, usually four to six.

One might consider avoiding such repeated measurements by the simultaneous use of several microphones. But so long as one simply adds the resulting alternating microphone currents, such a summation would also lead to interferences, and only the probability of the maximum deviation will be reduced. But a good smoothing of the interference fluctuations can be achieved by using, for example, only two microphones, provided that the signals are added after rectification and smoothing.[11] This, of course, has the disadvantage that it entails the additional cost of two or more complete amplifier–rectifier and filter systems.

Instead of this, it is possible to commutate two or more microphones in rapid alternation, with the help of a rotating switch, while using the same system of signal registration. The record thus achieved is equivalent to the more time- and work-consuming averaging of measurements with only one microphone at different locations.[12]

One can also introduce special measures at the sound source, in order to minimize the interference pattern and to increase the uniformity of the energy distribution. Analogous to the commutating of several microphones, one can use several sound sources at some distance apart, which can be switched on and off at will.[13]

At higher frequencies, a loudspeaker has a distinct directivity, so that a continuous rotation of the loudspeaker leads to a blurring of the standing waves and thus to an improvement in the energy distribution. Figure 4.2 shows a set of two loudspeakers used by the Bureau of Standards in Washington; they are installed in two large inclined plates, which are mounted like the vanes of a large fan on a vertical turning axle.[14] In relation to the dimensions of the room, these vanes are so large

---

[10] Eisenberg, A., Dissertation, TH Braunschweig, 1945.

[11] Kuntze, W., *Ann. Phys. (Leipzig)*, **5** (1930) 1058.

[12] Stanton, G. T., Schmid, F. C. and Brown, W. J., *J. Acoust. Soc. Am.*, **6** (1934) 95; Bedell, E. H. and Swartzel, K. D., *J. Acoust. Soc. Am.*, **6** (1935) 121. Knudsen at UCLA utilized one to four moving microphones to average the sound field (see Sabine, P. E., *Acoustics and Architecture*, McGraw-Hill, New York, 1932, p. 118), as did W. C. Sabine in his test room in the basement of Jefferson Lab. at Harvard University.

[13] Just, P., *Schalltechnik*, **2** (1929) 5.

[14] Chrisler, V. L., *J. Acoust. Soc. Am.*, **7** (1935) 84.

Fig. 4.2. Rotating mixing vane with built-in loudspeakers. (After Chrisler.[14])

that they are important as reflecting areas; thus, their rotation alone 'stirs' the sound field. They have this effect even without the loudspeakers, and thus also after switching off the loudspeakers for the reverberation measurement. These rotating vanes were introduced into building acoustics by W. C. Sabine;[15] they are especially suited to providing the prerequisite condition for the statistical reverberation theory: namely, the homogeneous and isotropic distribution of sound energy. Therefore, they have often been installed in reverberation chambers, particularly in North America.[16]

The rotating vanes are often combined with warble tone radiation. It has been proved that if one uses warble tones and rotating vanes, there is no need to change the location of the receiver. This, of course, holds only for small (e.g. 200 m^3) measuring rooms. Naturally, because of their required dimensions, rotating vanes can be used in large rooms only to a limited degree, so that in this case, it is necessary to carry out the reverberation measurements at various widely separated locations.

[15] Sabine, W. C., *The Brickbuilder*, **XXIV** (1915) No. 2; *Collected Papers on Acoustics*, No. 10, Dover Publications, New York, 1964.

[16] Schultz, T. J., *J. Acoust. Soc. Am.*, **45** (1969) No. 1; see particularly pages 27 and 28 and Fig. 9.

II. 4.3 Automatic Procedures with Objective Threshold Values

With the achievement of an exponential decay curve and the production of a photographic record, the problems of an objective reverberation time measurement were solved in principle; but this procedure is in most cases tedious and lengthy, especially because of the need for photographic development and the transformation to a logarithmic scale. Therefore, a number of laboratories attempted to replace the objective recording of the entire decay process by an objective measurement of the duration of the reverberation, using an automatic apparatus.[1]

The overall principle was the same at each laboratory. When the sound source is turned off, an electrically controlled clock or other time-measuring device is turned on; it is switched off again when the decaying signal in the receiving channel sinks below a certain predetermined level. But different means were developed for setting the objective threshold cut-off value. Meyer used at first a method that detected the instant when a glow-lamp was extinguished; later on, he adopted, as a means for switching the clock, the more precise start of a short-wave transmitter which was blocked at first because of a high negative grid-voltage.

In order to transform this measured reverberation *duration* into a reverberation *time*, the switching-off level must be preset as a precise fraction of the switching-on level.

Since in this procedure the time measurement is started by the same switch that stops the sound source, but is stopped by the decaying sound field, the unavoidable fluctuations of the decay curve can lead to a more or less premature stopping, compared with the instant at which the average energy level actually reaches the threshold. In order to eliminate these errors, the time measurement is begun, using the same apparatus, not at the instant of turning off the sound source, but after the level at the receiver has dropped by a certain amount.[2]

Scattered results can still occur, however, since the initial level at the microphone need not necessarily correspond to the average sound energy in the room. Especially in the neighborhood of the source, it tends to be

---

[1] Meyer, E., *Z. Techn. Phys.*, **11** (1930) 253; Strutt, M. I. O., *Elektr. Nachr. Techn.*, **7** (1930) 218; Wente, E. C. and Bedell, E. H., *J. Acoust. Soc. Am.*, **1** (1930) 422; Hopper, F. L., *J. Acoust. Soc. Am.*, **2** (1931) 499; Chrisler, V. L., Snyder, W. F. and Miller, C. E., *J. Acoust. Soc. Am.*, **3** (1931) 12.

[2] Strutt, M. I. O.,[1] and Hall, W. J. *J. Acoust. Soc. Am.*, **10** (1939) 302. This later paper also introduced the trick of blocking the start- and stop-relays, so as to prevent the fluctuations in the decay curve from stopping the count immediately after the first drop in level.

higher than would correspond to a uniform distribution of energy; in particular, steps can sometimes be observed in the decay curve that are attributable to the shutting off of the direct sound, as well as the sound from images of low order.

For this reason, the initial level at great distances must be somewhat lower than would correspond to the average energy.

In modern equipment for reverberation measurements, the switching is done by more precise integrated circuits. According to DIN 52216, a stabilized square-wave generator, which is used as the time-measuring device, is switched on after the energy has dropped 5 dB, and is switched off again after the energy has dropped another 30 dB. A selective switch permits this measurement to be carried out over different ranges of the decay, a feature which permits the evaluation of non-linear decay curves.

A great advantage of the direct measurement of reverberation time lies in the fact that, through an automatic, periodically repeated switching on and off of the sound source, or by automatically changing the microphone location or other room conditions, a great number of reverberation time measurements can be added together. Thus, one can automatically derive a mean value; this method was used by all of the authors mentioned in footnote 1.

An even further step in automation was achieved when the switching-off level was also used to turn on the source automatically for the next cycle; but before the rising level in the room reaches its steady-state value, the source is again turned off by a relay and the level decays once more. In this way, we produce a continuous alternation between the rise and decay transients, whose curve of level plotted against time has a sawtooth pattern. Hollmann and Schultes[3] were the first to make use of this process, which they appropriately called a room-acoustical 'relaxation oscillation'. For a trained ear, the relaxation *frequency* serves as a good indication of the reverberation time. But here, the transient range (between switching the source on and off) depends not only on the ratio of the chosen 'switching points' of energy E_1/E_2 but also on their relation to the stationary energy density E_∞ (which would depend on the strength of the sound source but is never reached); therefore, it is necessary to restrict the measurement to the portion of the cycle in which the sound is decaying. As we have pointed out previously, during the growth of sound in the room the necessary assumptions of statistical room acoustics are scarcely fulfilled at all. In particular, the influence of the direct sound is

[3] Hollmann, H. E. and Schultes, Th., *Elektr. Nachr. Techn.*, **8** (1931) 494 and 539.

so great that, in comparison with reverberation time, especially in highly damped rooms, the travel time between the source and receiver becomes significant.

As in all reverberation measurement procedures, the accuracy with which the reverberation time can be determined is greater, the greater the available range of decay. Hollmann and Schultes, who were using the ignition and extinction voltages of an electrical relaxation oscillator for their switching values, used only a 20 dB range at first. Holtsmark and Tandberg[4] used, instead, a potentiometer actuated by a switching relay to change the amplifier gain; this not only permitted a larger range between the limits, but also provided a simple means of establishing those limits. Hunt[5] improved the procedure even further by varying the lower switching level (as much as 80 dB, in steps of 5 dB), and thus was able to determine numerous points for each reverberation decay process. Figure 4.3 shows a series of data points determined by this method, in which each point was determined by 400 individual measurements. (The apparatus automatically added together groups of ten reverberation durations; for each of four different microphone positions, there were ten

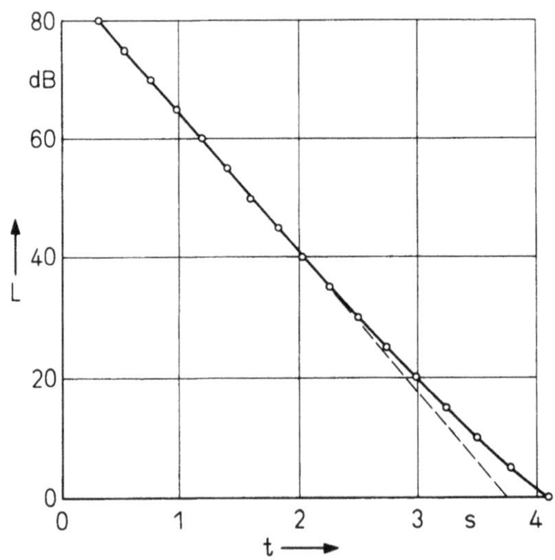

Fig. 4.3. Pointwise registration of a reverberation decay. (After Hunt.[5])

[4] Holtsmark, J. and Tandberg, V., *Elektr. Nachr. Techn.*, **10** (1938) 389.
[5] Hunt, F. V., *J. Acoust. Soc. Am.*, **8** (1936) 34.

such measurement series.) Hunt's data may be seen as a confirmation that the 'sagging' of the reverberation curves, mentioned previously, is a genuine phenomenon. It must be emphasized, in view of the comments in Section II.2.3, that these results were obtained in a rectangular reverberation room with only $57\,m^3$ volume, and were achieved without any diffusing vanes in the room.

Naturally, one would usually not make such a large number of measurements for a single point. For the process described above, Hunt has stated that with only 40 individual measurements (actually, four measurements with different microphone positions, where ten decays were automatically recorded) and the use of warble tones, he could reduce the average error of the results to only 10 ms.

For characterizing the acoustics of an auditorium in terms of the reverberation time, such precision is not needed. However, for the quantitative measurement of absorptive materials in reverberation rooms, this precision may be necessary.

II.4.4 Integrating Methods

The objective reverberation duration measurements that we have described in the last section suffer from uncertainty in the threshold values, the crossing of which is always somewhat random. Such random results must always be compensated for by taking a very large number of individual measurements. Therefore, at about the same time, procedures were developed which from the beginning took into account the unavoidable fluctuations in the reverberation curve. They sought to overcome the fluctuations by integrating the signal over a more or less long interval in the reverberation process.

We begin with the procedure that is most similar to the last described method. For determining the time function of a periodic voltage an old procedure from electrotechnology was adopted, analogous to the stroboscopic methods used nowadays to 'freeze' the motion of periodically vibrating or rotating objects by illuminating the object only once in each period at a precise phase-instant of its motion. If the period of illumination exactly matches that of the motion, the object appears to stand still. Similarly, the instantaneous voltage of a periodically repeated waveform can be measured by connecting a measuring device to the circuit briefly at the same phase instant in each period so as to read the voltage at that point in the waveform; by advancing the phase-instant forward step-wise, one can plot the waveform, point by point, throughout the entire period. Norris and Andrée[1] used this method for sampling a periodic reverberation process with the help of a circular contact switch. A thermal element at the amplifier output served to rectify and smooth the signal; direct current impulses from the thermal element were used to drive a galvanometer. Although, for the original purpose of this procedure, the sampling interval was made as short as possible in order to determine the fine structure of the alternating voltage, it was here an advantage to choose a somewhat longer sampling interval, at least as long as the warble period, so as to average out the fluctuations that are not of interest. Norris and Andrée clearly showed that sampled intervals of width Δt at different times in the reverberation process define an exponential decay of the pointer reading, from which the reverberation decay curve can be determined.

By plotting a galvanometer indication against time one can cover only a small range of levels. Hunt[2] was able to increase this range by adjusting a calibrated potentiometer in the amplifier chain in such a way that the galvanometer indication always stays the same; by this procedure, the observation range is limited by the range of the potentiometer, not the galvanometer. We will see very clearly the usefulness of this principle in the next section.

With this procedure, Hunt also avoided the danger of overloading the amplifier. Overloading can easily introduce serious errors in the reverberation curve, since for large input voltages the output voltage is no longer proportional to the input but instead becomes saturated. An indication of the presence of such an error may be a bending over (concave downward) of the upper part of the reverberation curve.

Knudsen[3] made use of a similar possibility of improvement by adopting as a level indicator the output not of a thermal element, which responds to the mean value of the current, but of a glow lamp, which responds to the peak value.

Hunt carried out at that time the first detailed investigation of the scatter of reverberation measurements. He showed that his series of sample points corresponded better to a decaying straight line if he used a larger warble swing and a longer sampling interval. Even so, there remained certain long-term fluctuations. These fluctuations do not create

[1] Norris, R. F. and Andrée, C. A., *J. Acoust. Soc. Am.*, **1** (1930) 306; also Norris, R. F., *J. Acoust. Soc. Am.*, **3** (1931) 361.

[2] Hunt, F. V., *J. Acoust. Soc. Am.*, **5** (1933) 127.

[3] Knudsen, V. O., *J. Acoust. Soc. Am.*, **5** (1933) 112.

any problem for the evaluation of a single case; even without the use of eqn. (4.2) it was always possible to draw a straight line through the data points. But the average deviation of the data points from this straight line gave, according to his experience, a good indication of the uncertainty of the measurement results.

If the requirements for measurement precision are not great, one can restrict the number of measuring points in this procedure and thereby make the average sampling interval rather large. The consequence of this simplification is that only one integration is made, by means of the thermal element, and the result is given by a galvanometer indicator whose displacement a is proportional to the reverberation time:

$$a \sim E_0 \int_0^\infty 10^{-6t/T} \mathrm{d}t = \frac{E_0 T}{6 \ln 10} \tag{4.3}$$

One needs only to set the steady-state energy density to the particular value E_0 (or, alternatively, to see that the sensitivity of the receiver is so adjusted that the indicating instrument in the steady-state condition shows a particular calibration indication) in order that a single reverberation decay leads to an approximate measurement of the reverberation time. Hartmann and Sommer[4] developed on this basis a simply operated, portable measurement apparatus. There remains, however, the need for an electrical sound source, since at the beginning of the reverberation measurement the sound source must be switched off and the receiver switched on. The authors calibrated their own apparatus with the help of the discharge of a capacitor, whose decay is essentially exponential and whose time constant can be accurately calculated from the capacitance and resistance.

One can use this comparison of the room reverberation with an equivalent condenser discharge not only in the sense of a substitution but also in the form of a simultaneous compensation. Olson and Kreuzer[5] have developed such a 'reverberation time measurement bridge'. The balance was adjusted with the help of a galvanometer in the bridge circuit.

Hollmann and Schultes[6] refined the balance criteria by introducing, instead of a bridge galvanometer, a circuit that compares the sum of the

---

[4] Hartmann, C. A. and Sommer, G., *Veröff. a.d.Geb. d.Nachr. Techn.*, **1** (1931) 173.
[5] Olson, H. F. and Kreuzer, B., *J. Acoust. Soc. Am.*, **2** (1930) 78.
[6] Hollmann, H. E. and Schultes, Th., *Elektr. Nachr. Techn.*, **8** (1931) 387.

positive deviations with the sum of the negative deviations of the room decay curve from the exponential condenser discharge, such that when these two sums are equal, the condenser decay most nearly equals the room decay.

Despite their fundamental advantage and simplicity of operation, the methods of integrating over the entire decay process have not become widely used, chiefly because only the first 20 dB of decay can be evaluated, and, furthermore, the beginning of the decay is given the greatest weight. (In accordance with the matters discussed in Section II. 5.2, however, this does not constitute a particular disadvantage nowadays.)

The principle of comparing a room decay with the condenser discharge was later introduced in combination with the so-called 'exponential vacuum tube', which is an amplifier tube whose transconductance increases exponentially with the grid voltage. Tuttle and Lamson[7] derived this negative voltage from a condenser discharge, with the result that the amplification increased exponentially while the reverberation decreased exponentially. They adjusted the time constant of the discharge so that the display on a cathode ray oscilloscope remained as constant as possible.

The same principle of compensating the reverberation decay by an exponential change of amplification was also used by Watson;[8] through the continuous variation of the setting of a motor-driven logarithmic potentiometer, he found that, by controlling the rotation speed, the change of gain can be adjusted to correspond to the reverberation decay. Watson's purpose in using this method was not to measure the reverberation time, since he had already measured this in advance by a method that will be described in the next section. Rather, he was interested in a precise study of the deviations of the actual reverberation decay from that of an ideal exponential decay. These deviations take three different forms:

(1) fluctuations caused by interference, which can be reduced by use of warble tones or filtered noise signals;
(2) steps in the decay curve, which can be explained by the disappearance of the direct sound and the sound from low order image sources; and

[7] Tuttle, W. N. and Lamson, H. W., *J. Acoust. Soc. Am.*, **10** (1938) 84.
[8] Watson, R. B., *J. Acoust. Soc. Am.*, **18** (1946) 119.

(3) sagging of the reverberation decay curve, which is always a sign of the coexistence of several reverberation processes with different reverberation times.

II.4.5 The Graphic Level Recorder

Since, for the evaluation of reverberation processes, we are interested not only in the mean value of the reverberation time but also in the deviations that give some insight into the validity of the measurement and into the room-acoustical characteristics, the development of reverberation time measurement apparatus went back to recording the entire decay process. This became practical with the development of graphic level recorders that incorporate logarithmic plotting of the decay; furthermore, one avoids the necessity for development in a photographic darkroom.

Not only reverberation time measurements, but also a number of other kinds of recording in the field of electroacoustics, require apparatus that can display the logarithm of the input voltage. This goal can be realized, in general, by feedback from the output to control the amplification of the instrument. From the standpoint of automatic control technology, two groups of devices should be distinguished.

In one, the rectified output voltage itself is the recorded quantity, and, therefore, it varies within certain limits.

A good example of this type, developed by Meyer and Keidel,[1] is shown in Fig. 4.4. In this case, a fluid potentiometer, in which the exponential variation of the cross-section can be clearly seen, is equipped

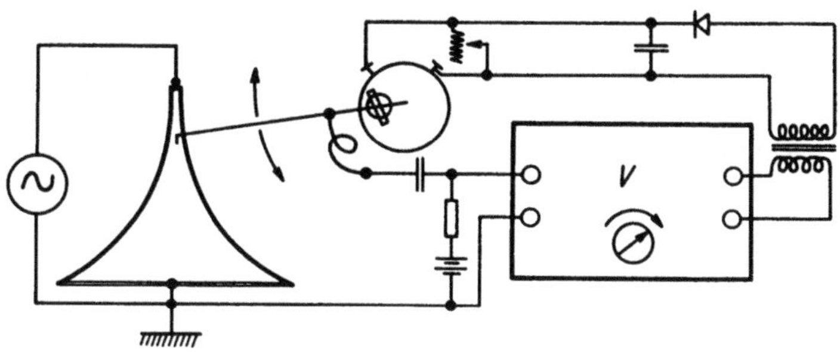

Fig. 4.4. Graphic level recorder with fluid potentiometer. (After Meyer and Keidel.[1])

[1] Meyer, E. and Keidel, L., *Elektr. Nachr. Techn.*, **12** (1935) 37.

with an easily movable electrical contact by means of which a partial voltage, $u_2 = \phi u_1$, is selected from the input voltage. The voltage fraction ϕ is itself a function of u_2, because the position of the contact is controlled by the rectified output voltage, which is proportional to u_2. If, now, $u_2 \sim u_1$, one gets for the dependence $\phi_{(u_2)}$:

$$\phi = u_2/u_1 \sim u_2 \times 10^{-u_2} \tag{4.4}$$

The fractional voltage, therefore, is not exactly logarithmic, but this can be corrected by modifying the shape of the fluid potentiometer.[2]

The output current passes through another moving coil system, to which is attached a mirror that controls a light ray in the vertical direction, which, in turn, meets a rotating fluorescent screen. The trace of the decay remains on the screen long enough for the operator to match it with a rotatable linear cursor, and to measure the steepness of the reverberation curve. For a more precise determination, of course, one can replace the fluorescent screen with light-sensitive paper. Fig. 1.1 was registered in this way.

The same regulation function, which in Fig. 4.4 was achieved by the fluid potentiometer with a movable pointer, can also be realized with the aforementioned 'exponential vacuum tube'. In this case, the rectified output voltage is added to the grid voltage in such a sense that with increasing output voltage, the steepness is decreased. This method was originally used by Ballantine[3] and was subsequently adapted and improved by Holle and Lübcke.[4]

With this apparatus, a range of levels spanning about 80 dB can be measured, and the writing speed is fast enough to give error-free measurements of reverberation times as short as 0·2 s.

As successful as these results may be, a second method of control has replaced the first method for practical use (at least until now). In this method, the output voltage is always kept at the same value; for this purpose, a more rugged control device is used. Stanton and Tweedale[5] arranged that the contact of a slidewire potentiometer be moved by a motor until the output of the amplifier reaches a certain reference value.

[2] In a later development, Keidel replaced the fluid potentiometer having an exponentially shaped casing by multiple electrodes connected by suitable fixed resistors. (Keidel, L., *Akust. Zh.*, **4** (1939) 169.)

[3] Ballantine, St., *J. Acoust. Soc. Am.*, **5** (1933) 10.

[4] Holle, W. and Lübcke, E., *Hochfrequ. u. Elektroak.*, **48** (1936) 41.

[5] Stanton, G. T. and Tweedale, J. E., *J. Acoust. Soc. Am.*, **3** (1932) 371.

If, for a short time, there are higher or lower output currents, a polarizing relay is switched on that causes the motor to run in one or the other direction. In this case, the position of the contact on a precise logarithmic potentiometer determines an output that is proportional to the logarithm of the input voltage.

It is clear that the apparatus in which this control principle was first embodied (which was used by the authors only for recording the loudness of railroad trains) had too great an inertia to record short reverberant decay curves accurately. But the principle was used in a number of other institutes, and was constructively adapted to the purpose at hand. Above all, it was necessary to overcome the inertia of the mass of the motor. This was done by running the motor continuously with a fixed rotational speed. By the use of a magnet, the contact of the potentiometer was pressed now on one side, now on the other side of the rotating disk, depending on the deviation of the input voltage from the reference voltage.

This principle can be seen in Fig. 4.5, which shows the circuit diagram of a graphic level recorder constructed by G. Neumann, Berlin, on the basis of a principle proposed by v. Braunmühl and Weber.[6] The poten-

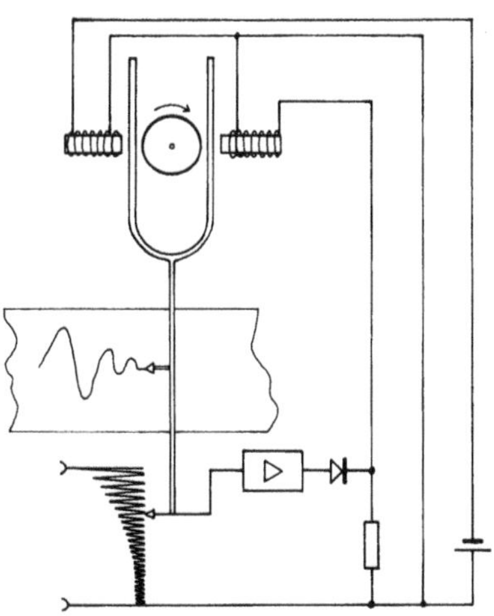

Fig. 4.5. Principle of the graphic level recorder from the firm of G. Neumann, Berlin.

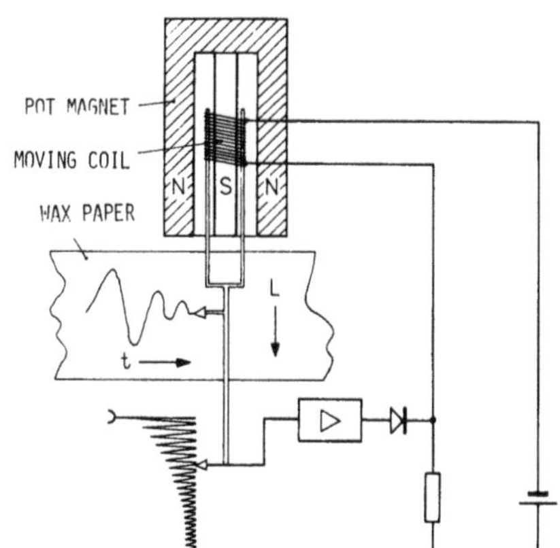

Fig. 4.6. Principle of the graphic level recorder from the firm of Brüel and Kjaer, Copenhagen.

[6] v. Braunmühl, H. E. and Weber, W., *Elektr. Nachr. Techn.*, **12** (1935) 223.

tiometer contact and a writing pen are connected together with a fork, which, in the resting condition, slips freely on both sides of a disk mounted on the axle of a motor. This 'rest condition' requires that both electrical coils act with the same force on the fork. If the voltage at the input is increased, one of the coils gets a higher current, the fork is pulled against the rotating disk and is moved in one direction or the other until, once again, both coils have the same current, at which point the fork is released from contact with the disk.

In spite of the sensitivity and the high writing speed of this instrument, which allows the measurement of reverberation times down to 0·2 s, the force reserve is large enough to overcome not only the friction of the potentiometer but also that of the writing device, here a stylus scratching a trace on the moving strip of waxed paper. Thus, one gets an immediately evaluable record of the slope of the decay, and also a permanent recording of the decay. This Neumann recorder was rugged and was readily portable, so that it became, for a long time in Germany, a part of the standard equipment for field measurements.

The level recorders developed at the same time in the United States were the same in principle; they differed only in certain details.[7,8] The 'High Speed Level Recorder' developed by Wente *et al.* had a variable rotation speed for the moving disk, and thus it allowed a better accommodation to the fluctuations in the input signal by selecting the writing speed.

Thus, for example, long reverberation times can be registered with a low rotation (writing) speed and with reduced control velocity, in order to smooth the fluctuations of the decay curve. With this apparatus, which was able to register in one slope a decay of over 90 dB, Bedell and Swartzel made their fundamental investigations (mentioned in Section II.4.2) of the accuracy of reverberation measurements. To this end, they compared the different aforementioned means for smoothing the graphic level record and improving the statistical energy distribution. In particular, they showed that the scatter in the measurement results decreases with an increase in the measurement range (in dB). Finally, they were able, by the repetition of the reverberation experiment over a 90 dB range with changed amplification, to record an extraordinary reverberation decay over 125 dB, which, moreover, was a completely straight line. If

[7] Stanton, G. T., Schmid, F. C. and Brown, W. J., Jr., *J. Acoust. Soc. Am.*, **6** (1934) 95.

[8] Wente, E. C., Bedell, E. H. and Swartzel, K. D., *J. Acoust. Soc. Am.*, **6** (1935) 121; also Bedell, E. H. and Swartzel, K. D., *J. Acoust. Soc. Am.*, **6** (1935) 130.

one compares this astonishing achievement with the early experiments of W. C. Sabine, who used for his evaluations a range of only 6 dB, one can appreciate the remarkable development of acoustical measurement techniques.

Instead of using a motor, which always requires changing a rotary motion into a linear motion, one can use a direct electromechanical transformation to move the potentiometer contact and the writing pen back and forth. The contact and the writing pen are attached to a coil that moves like the voice-coil of an electrodynamic loudspeaker in a very deep pot-magnet, as shown in Fig. 4.6. Surely, it required ingenious, constructive work to develop a reliable graphic level recorder based on these initially very simple-appearing principles. Such an instrument was developed by P. Brüel,[9] which allows a change of the paper speed between 0·003 and 100 $\mathrm{mm\,s^{-1}}$, and of the pen-writing speed between 50 and 1000 $\mathrm{mm\,s^{-1}}$ (corresponding to 50 and 1000 $\mathrm{dB\,s^{-1}}$).

Finally, even the small task of determining the average slope of the reverberation decay can be simplified by using a transparent plastic protractor.[10] It is recommended that the first 5 dB of the decay be neglected, because in this range the statistical conditions of room acoustics are far from being fulfilled. Therefore, in DIN 52216, the range between -5 and -35 dB below the steady-state level is recommended for the evaluation of the reverberation time. Furthermore, one must observe whether the lower level of the decay is influenced by the background noise, a condition that can be detected by a tendency for the reverberation curve to bend toward the horizontal. In the latest described apparatus, there are a series of parallel lines in the waxed paper that serve as an ordinate raster for the sound level. If one designates the angle between the average slope of the decay curve and these horizontal lines by α, the dB range on 50 mm-wide paper by R, and the paper speed in $\mathrm{mm\,s^{-1}}$ by v, then one gets the reverberation time from the following equation:

$$T = (3000/Rv)\ \mathrm{ctg}\ \alpha \tag{4.5}$$

[9] This graphic level recorder is produced by the firm of Brüel and Kjaer in Naerum, near Copenhagen. The excellent characteristics of this and other instruments developed by P. Brüel for room acoustics and building acoustics measurements may be a result of the fact that for several years he was director of the acoustical laboratories at the Chalmers Institute of Technology in Gothenberg, Sweden.

[10] Morrical, K. D., *J. Acoust. Soc. Am.*, **10** (1939) 300.

If one determines the relative error by differentiation of eqn. (4.5):

$$\Delta T/T = -\Delta\alpha/(\sin\alpha\cos\alpha) \tag{4.5a}$$

one notices that this error is least for an angle of 45°. Therefore, it is good practice to select the value of *Rv* in such a way that, if possible, this angle is around 45°, but in any case lies between 20° and 70°.

A more practical and more precise tool for the evaluation of decay slopes was developed by Lauber and Hesse;[11] in their invention, several parallel lines are engraved on a transparent and rotatable disk. The registered decay on the paper strip is centered between these parallel lines and the reverberation is read from a window on the periphery of the disk, where a reverberation time scale is marked.

Furthermore, this same tool can also transform the reverberation time data into electrical impulses, which can be used for automatic averaging in a computer.

Concerning the precision with which such lines can be fitted to the decay slope, it is very similar to that of Hunt's 'series of points'; that is, the errors that can be made in this evaluation are, according to experience, much smaller than the differences between individual decay records, measured under different conditions.

Although the registration of decay processes with the use of the level recorder allows in most cases a sufficiently accurate evaluation of the reverberation time, we must not forget that a strict reproducibility of the decay curves is not possible, as Fig. 4.7 shows. The variations, as we have stated several times, are caused by the fact that the amplitude and phase angles of the eigen-modes of the room, at the instant when the sound is switched off, are always different. If a room is excited by random noise, the amplitude and phase conditions are random. In order to even out these fluctuations, which do not depend on the room characteristics, the reverberation must be registered several times for the same positions of source and microphone, and from these decays an average slope is determined:

$$\langle p_{(t)}^2 \rangle = \lim_{N\to\infty} \frac{1}{N} \sum p_{N(t)}^2 \tag{4.6}$$

Under certain conditions with respect to the randomness of the switch-off point, a curve can be determined that depends only on the room characteristics.

---

[11] Lauber, A. and Hesse, K., *Techn. Mitt., PTT*, 1959, No. 6.

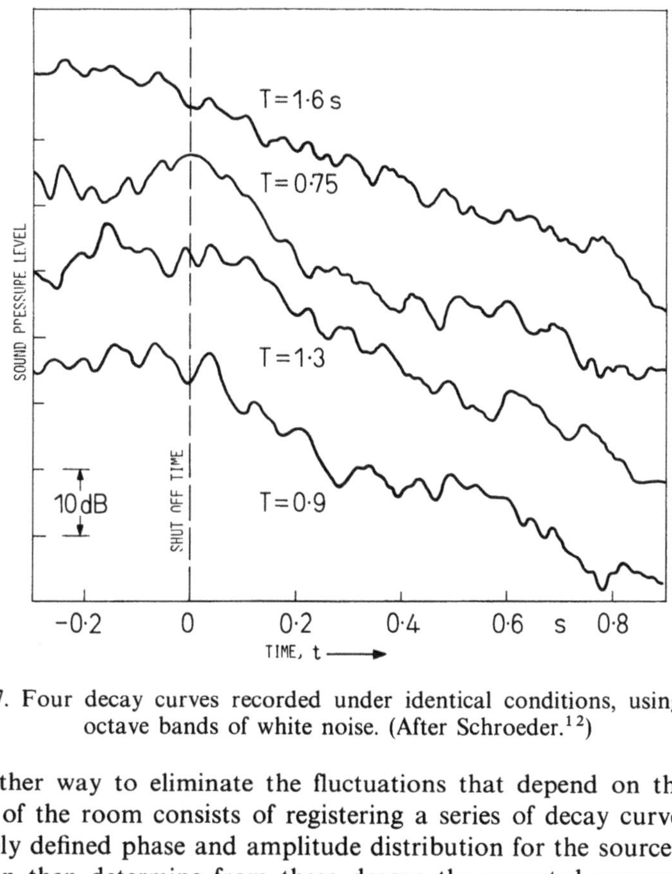

Fig. 4.7. Four decay curves recorded under identical conditions, using third-octave bands of white noise. (After Schroeder.[12])

Another way to eliminate the fluctuations that depend on the excitation of the room consists of registering a series of decay curves with precisely defined phase and amplitude distribution for the source signal; one can then determine from these decays the expected average slope ('ensemble average'). One source signal for which these conditions are fulfilled is the impulse (pistol shot, spark source, bursting balloon, etc.).

In Section II. 1.3, eqn. (1.17), it was mentioned that the decay can be described by

$$E_{-(t)} = \frac{1}{\Delta t} \int_t^{\infty} E'_{(\Theta)} \, \mathrm{d}\Theta \tag{4.7}$$

where $E'_{(\Theta)}$ means the energy density at time Θ after an impulse excitation with an impulse duration of Δt. Schroeder[12] was able to show

[12] Schroeder, M. R., *J. Acoust. Soc. Am.*, **37** (1965) 409.

theoretically that $E_{-(t)}$ is the desired ensemble average of all possible decays. But this means that an excitation-independent decay curve of a room can be determined if one integrates a single impulse response from the instant of interest to infinity.

Fig. 4.8 shows a picture of such an integrated impulse response, in comparison with the measured decay curve. Even from a very rough decay process, one can determine a clear and easily evaluated curve.

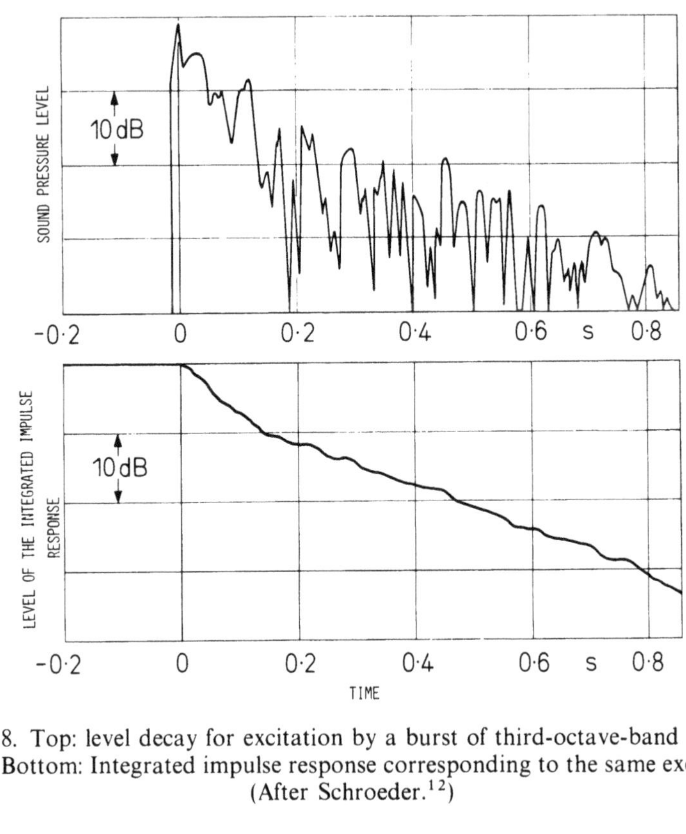

Fig. 4.8. Top: level decay for excitation by a burst of third-octave-band random noise. Bottom: Integrated impulse response corresponding to the same excitation. (After Schroeder.[12])

This backward-integration can be calculated 'by hand' from the decay curve registered in the customary manner.

Another method already described by Schroeder consists of playing backwards a tape recording of a decay process initiated by an impulse (noise burst or pistol shot), and integrating the squared signal with an RC circuit. The voltage on the capacitor constitutes the desired value

$\langle p^2_{(t)} \rangle$. The increasing voltage can be registered with the aid of a graphic level recorder.

Simplest is the measurement of the integrated impulse response by means of a computer, which allows us to store the digitalized signal and to integrate it backwards by digital means.

Another procedure for measuring the reverberation time by using the integrated impulse response was described, independently of one another and nearly at the same time, by Kürer and Kurze[13,14] on the one hand, and by Kuttruff and Jusofie[15] on the other.

To avoid the backwards integration, which is awkward to handle from the technical measurement viewpoint, it was recommended in both cases to write the integral of eqn. (4.7) in the following way:

$$E_{-(t)} = \frac{1}{\Delta t}\left(\int_0^\infty E'_{(\Theta)}\,\mathrm{d}\Theta - \int_0^t E'_{(\Theta)}\,\mathrm{d}\Theta \right) \tag{4.8}$$

This means that the backwards-integrated impulse response can be obtained by exciting the room and determining the first term of the equation. After that, the room is excited once more and the second term of the equation is subtracted from the first. This can be done by the electronic equipment shown in Fig. 4.9 in the following way. During the first decay process, the capacitor is charged; and with the second impulse, in response to the decaying received signal, it is gradually discharged.

The voltage on the capacitor during the discharge corresponds to the

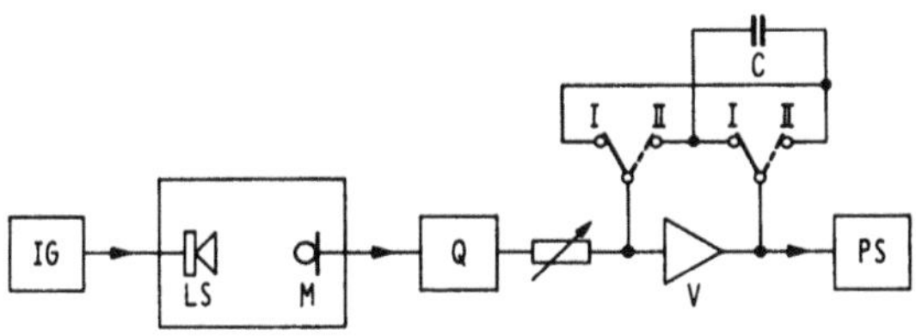

Fig. 4.9. Block diagram for recording an integrated impulse response. IG = impulse generator; LS = loudspeaker; M = microphone; Q = squaring device; V = amplifier; C = condenser for integration; PS = graphic level recorder. (After Kuttruff and Jusofie.[15])

[13] Kürer, R., paper presented in Warsaw, April, 1967; published in *Archiwum Akustyki*, **2**, **3** (1968) 153.
[14] Kürer, R. and Kurze, U., *Acustica*, **19** (1967/68) 313.
[15] Kuttruff, H. and Jusofie, M. J., *Acustica*, **19** (1967/68) 56.

desired reverberation function. This discharge function can be registered on a graphic level recorder; it represents the ensemble-average reverberation decay.

II.4.6 Frequency Filtering in the Receiver

Recording the reverberant decay curve not only gives an overview of the variations in the curve, but also reduces the requirements on the sound source. The power of the source needs neither to be constant nor to be switched off at a precise instant.

So far, as with the subjective reverberation measurements, the frequency dependence of the reverberation has been determined by the output of the sound source. One way was to filter a continuous noise signal by electrical means, so that only a certain frequency range is passed; outside this range, the signal is highly attenuated. This electrical filtering takes on a more general significance in room-acoustical measurement practice, because it can be done not only in the sound source but also in the sound receiver circuits.

Filtering in the receiving circuit decreases the effect of the background noise and thus makes it possible to extend the measurement range to lower sound pressures.

But such filters always have a certain rise-time, which becomes shorter the wider the bandwidth of the filter. Since for reverberation measurements it is sufficient, in general, to use third-octave steps, the bandwidth of the filter need not be less than one-third octave.

The corresponding rise and decay times, in this case, remain far shorter than the reverberation time to be measured.

Switching a third-octave (or octave) filter into the receiving circuit allows us to measure the frequency dependence of reverberation using any sound source whatever; the only requirement for the sound source is that, in the frequency range of interest, it must have enough power to provide a suitably long range of decay. By this means, we are able to use a very primitive sound source (mentioned in Section II.4.1), namely the handy starting pistol.[1] The economy of means that can be realized in this

[1] L. Keibs (*Nachrichtentechnik*, **6** (1956) 347) has made a careful study of the differences between the reverberation times measured with warble tones, with third-octave-band random noise, and with pistol shots (apart from the dependence given in eqn. (1.18)) and of the influence of the bandwidth of the filters.

manner is illustrated in Fig. 4.10, which compares the pistol source with the source equipment commonly used today.

Finally, it should be mentioned that the reverberation time of occupied rooms (i.e. concert halls with audience and orchestra) can be determined from recordings of music. The curve shown in Fig. 1.1 was determined by this means. In this case, one must take care that the music used for the decay does not include instruments whose reverberation is longer than that of the room. Kettledrums especially can lead to erroneous reverberation time measurements at low frequencies for this reason.[2,3]

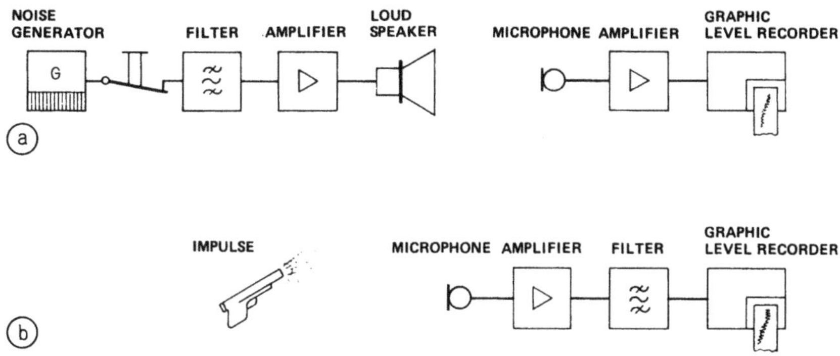

Fig. 4.10. Block diagrams for reverberation recording: random noise versus pistol shot.

It is best to make magnetic tape recordings during the concert and to analyse the reverberation time afterward in the laboratory, by using various octave or third-octave filters.[4] The room-acoustics observer does not even need to visit the hall under study; he can, with the help of readily available disk recordings or radio broadcasts, study the reverberation of the important churches and concert halls of the world by using suitable cut-off chords in the music. Even the frequency dependence can be determined with the use of octave filters and a graphic level recorder.[2] In this procedure, one can expect that the reverberation of the recording-room has the longest decay of all the elements in the transmission chain. Sometimes, however, one has to take into account

[2] Meyer, E. and Jordan, W., *Elektr. Nachr. Techn.*, **12** (1935) 213.

[3] Schultz, T. J., *J. Audio Eng. Soc.*, **11** (1963) 307.

[4] Eisenberg, A., Dissertation, Braunschweig, 1945.

the possibility that the dynamic range of the signal has been compressed, either automatically or by the recording engineer 'riding gain'. Another way in which the reverberation may not be true to that of the hall is a result of overloading at the transmitter or the receiver. Nowadays especially, one has to make certain that the recording has no artificially added reverberation (see Section II. 3.5), which often happens if the recording studio is rather 'dead'. Furthermore, one can expect in recordings from occupied halls a decay range of only about 25–30 dB.

Even though these 'long-distance' reverberation measurements do not represent the most accurate method, they are, on the other hand, the most convenient. They highlight how much the cost of reverberation analysis has been reduced since the time of Sabine.

Chapter II.5

Measurement of Sound Absorption in the Reverberation Room

II.5.1 Evaluation of the Measured Reverberation Time

For calculation of the reverberation time of a room it is necessary to know the absorption coefficients of the boundary surfaces. The absorption coefficient α depends on both the angle of incidence and the frequency of the sound. Only the frequency dependence must be explicitly known for the calculation. One method for determining the absorption coefficient, averaged over the angle of incidence but dependent on frequency, is provided by the reverberation room measurement procedure. Starting from the formulae derived in Section II. 1.5, this method consists of measuring the reverberation time of the room, both with (T_Δ) and without (T_0) the test sample, and calculating the absorption coefficient by using the following formula:

$$\alpha_x = \frac{0{\cdot}163\,V}{S_x}\left(\frac{1}{T_\Delta} - \frac{1}{T_0}\right) \tag{5.1}$$

where V is the volume of the reverberation room in cubic meters and S_x is the area of the test sample in square meters.

Despite the technical improvement in the instruments for reverberation measurement, which leads to very low scatter in repeated measurements, the precision with which absorption coefficients can be determined in a reverberation room is by no means satisfactory. The reason for this is only partly that the value of α_x is determined from the difference in the reciprocals of the two reverberation times; this causes the relative error $\delta\alpha_x/\alpha_x$ in determining the absorption coefficient to be larger than that in determining the reverberation time $\delta T/T$,[1,2] which we assume will be the same for both T_0 and T_Δ:

$$\frac{\delta\alpha_x}{\alpha_x} = \frac{\delta T}{T}\,\frac{T_0+T_\Delta}{T_0-T_\Delta} = \frac{\delta T}{T}\,\frac{2A_0+\Delta A}{\Delta A} \tag{5.2}$$

If the equivalent absorption area ΔA of the test sample is only as large as the absorption area of the room without the sample, one gets $\delta\alpha_x/\alpha_x = 3\,\delta T/T$. That is, with a relative error of 3% in the reverberation time measurement, which Steffen[1] says is typical of the normal measurement procedure, the relative error in α_x amounts to 9%. One should at least try to achieve the ratio $\Delta A/A_0 = 1$, in the most important frequency range. Naturally, it is better to make $\Delta A = 2\,A_0$, because then $\delta\alpha_x/\alpha_x$ is reduced to only 6%; but this is possible only for sufficiently large values of α_x. The absolute error in α_x, according to eqn. (5.2), does not decrease below a certain limit, even if α_x becomes very small:

$$\lim_{\alpha_x\to 0} \delta\alpha_x = \lim \frac{\delta T}{T}\,\frac{(2A_0+S_x\alpha_x)}{S_x} = \frac{\delta T}{T}\,\frac{2A_0}{S_x} \tag{5.3}$$

If one accepts the basic rule that S_x should be at least equal to $2A_0$, then the absolute error in α, for small values of α, is no greater than the relative error in T. Apart from the recommendation that the difference in the reverberation times, T_0 and T_Δ, should be sufficiently large, the statistical theory does not offer any further guidance for the testing laboratory. The shape of the room and the distribution of the test samples do not figure in the results.

But the better the measurement techniques become, the more one learns that things are not so simple as the statistical theory implies; in other words, the predictions of the theory can be only approximately fulfilled. The scatter of measurement results for a certain sound-absorptive configuration, measured in different laboratories, is unfortunately disappointingly large.[3]

Now, one may ask the reasons for this large amount of scatter. To

---

[1] Steffen, E., *Hochfrequ. u. Elektroak.*, **67** (1958) 73.

[2] Waterhouse, R. V., *Proc. 3rd. ICA Congress, Stuttgart, 1959*, Elsevier, Amsterdam, 1961.

[3] As demonstrated in comparison measurements between different laboratories over the years: Sabine, P. E., *J. Acoust. Soc. Am.*, **6** (1935) 239; Meyer, E. and Schoch, A., *Akust. Zh.*, **4** (1939) 61; Eisenberg, A., *Akust. Beih.*, **2** (1952) 108; Kosten, C. W., *Proc. 3rd. ICA Congress, Stuttgart, 1959*, Elsevier, Amsterdam, 1961, and *Acustica*, **11** (1961) 400.

investigate this question, Meyer and Kuttruff[4] measured the absorption coefficient of a sample of material in different model scale rooms and compared the results with the 'true' absorption coefficient defined by the Paris formula (eqn. (1.30)). Figure 5.1 shows an example of their results.

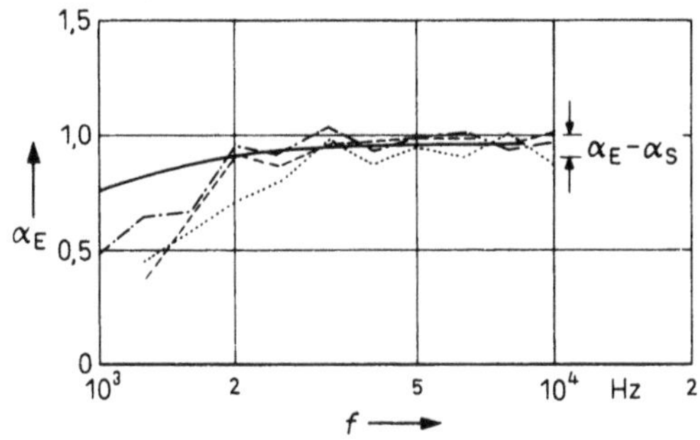

Fig. 5.1. Frequency dependence of the sound absorption coefficient of a porous sheet measured in a scale model reverberation room, where the test sample covered the entire floor. Solid line: the 'true value' calculated from the measured impedance. Other lines: measured values of α_E, calculated from the Eyring formula, in which the number of means for realizing a diffuse sound field were successively increased, from the dotted line, to the dashed line, to the dot-dashed line. (After Meyer and Kuttruff.[4])

The solid line gives the frequency dependence of the 'true' absorption coefficient, the other curves relate to the measurements in the model scale rooms. They were calculated by using the Eyring formula, or more correctly, its inverse (see eqn. (2.30)):

$$A = S(1 - e^{-0{\cdot}163\, V/ST}) \tag{5.4}$$

From the difference between the relation for the empty reverberation room:

$$A_0 = S(1 - e^{-0{\cdot}163\, V/ST_0})$$

and that for the room including the test sample:

$$A_0 + \Delta A = S(1 - e^{-0{\cdot}163\, V/ST_\Delta})$$

one gets the desired absorption coefficient α_E which in this case is

[4] Meyer, E. and Kuttruff, H., *Göttinger Nachr. Math. Phys. Kl*, (1958) No. 6, 97.

identified by the subscript E (Eyring):

$$\alpha_E = \frac{S}{S_x}(e^{-0{\cdot}163\,V/ST_0} - e^{-0{\cdot}163\,V/ST_\Delta}) \qquad (5.5)$$

If we develop the exponential function in a power series of the exponent and consider only the first three terms, which surely is allowed for all cases in which the assumptions of the statistical theory are fulfilled, one gets the simplified formula

$$\alpha_E = \frac{0{\cdot}163\,V}{S_x}\left(\frac{1}{T_\Delta} - \frac{1}{T_0}\right) - \frac{(0{\cdot}163\,V)^2}{2\,SS_x}\left(\frac{1}{T_\Delta^2} - \frac{1}{T_0^2}\right) \qquad (5.5a)$$

(The advantage of this simplification was considerably greater in the days before hand calculators became so widespread!) Here, the first term, corresponding to eqn. (5.1), represents the α of the simple Sabine formula, which will now be characterized by the subscript S (Sabine). With the introduction of this value, we can transform eqn. (5.5a) into the form of a correction to the value of α_S:

$$\alpha_E = \alpha_S\left[1 - \frac{0{\cdot}163\,V}{2\,S}\left(\frac{1}{T_\Delta} + \frac{1}{T_0}\right)\right] \qquad (5.5b)^1$$

This can be further transformed into:

$$\alpha_E = \alpha_S\left(1 - \frac{S_x}{2\,S}\,\alpha_S - \frac{A_0}{S}\right) \qquad (5.5c)$$

If, as in all reverberation room studies, the conditions $\alpha_S S_x \ll 2S$ and $A_0 \ll S$ are fulfilled, the correction terms amount to only a few percent; for this reason the standardization committees have decided, taking the other measurement uncertainties into account, not to heed these corrections at all, and to evaluate reverberation room tests fundamentally with the use of the Sabine formula.

It may be mentioned here that Kuttruff and Strassen[5] regard this as a fundamental shortcoming of the standardized rules for reverberation room measurements, which could be easily avoided.

In the experiments of Meyer and Kuttruff, the following relations were fulfilled:

$$S_x/S \leqq 0{\cdot}13; \quad A_0/S = 0{\cdot}03$$

[5] *Acustica*, **45** (1980) 246.

With $\alpha_S = 1$, this leads to a difference of $\alpha_S - \alpha_E \approx 0{\cdot}09$. This difference is shown here at the right in Fig. 5.1; it is large enough to justify the statement that the corrections given in eqn. (5.5c) correspond to the true physical situation, because if one were to shift all the curves higher by this amount, the asymptotic agreement with the 'true' absorption coefficient would be worse.

We must question, in those cases for which the Sabine and the Eyring formulae give significantly different results, whether the necessary assumptions of statistical room acoustics that underlie both formulae can be fulfilled, especially if we consider the kinds of room shape that Meyer and Kuttruff had to use to get this agreement. These measures, translated into real reverberant rooms, would be in several respects so expensive that their introduction into technical room-acoustical practice—which must surely take matters of economics into account—can hardly even be considered.

The investigations of Meyer and Kuttruff showed that, in general, one must give up the attempt to measure in a reverberation room the 'true' absorption coefficient corresponding to a statistical sound distribution. On the other hand, under the practical conditions in which the measured absorption coefficients will be applied, the necessary statistical assumptions are also not fulfilled.

This does not mean that we must give up the goal of getting comparable measurement results from one laboratory to another; rather, the freedom implied by the Sabine and the Eyring formulae must be restricted by standardizing the size and the shape of the reverberation rooms, the area of the test sample, the measurement apparatus, and the evaluation procedure.

II.5.2 The Influence of the Test Room

The first question concerns the necessary volume of the reverberation room. Because of cost considerations, one is interested, in the first place in having as small a room as possible. From the standpoint of statistical theory, a small room does not appear especially unfavorable, because the number of reflections per unit time that contribute to the average value for the absorption coefficient is very large. But, as we have already mentioned, in very small rooms the statistical considerations that are based on purely geometrical ray constructions and energy summation are no longer valid. For such problems, the wave nature of the sound, with its interference patterns and resonances, must be taken into consideration, as we will do in Chapters IV.11 and IV.12 (Volume 2).

From the wave-theoretical standpoint, reverberation consists of the superposition of a great number of independent vibration patterns, the so-called eigen-modes. These can be considered statistically only if the number of the eigen-modes and of the incidence angles for an individual eigen-mode is large enough. If these conditions are not fulfilled one can observe widely different reverberation times in one and the same reverberation room, where only the locations of the source and receiver are changed. Because the number of eigen-modes in a given frequency band becomes greater the larger the room, the scatter in the reverberation time data for a large room is always smaller than for a small room.

Knowing this relationship, Meyer,[1] as early as 1937, compared reverberation time measurements made in the 160 m^3 reverberation room of the Heinrich Hertz Institute and in the enormous 13 000 m^3 intake surge-chamber of the Walchensee power station. In both rooms, he determined the scatter of the reverberation times at different warble-tone center frequencies in terms of the cumulative distribution of the mean reverberation times, normalized to unity. Figure 5.2 presents some of his results. The upper two curves correspond to the Institute room, the lower two curves to the surge-chamber; low-frequency results are shown at the left, high frequencies at the right.

One can see in the upper curves that, even in the small reverberation room, as the frequency increases the steepness of the curve also increases, meaning that the scatter becomes smaller. The measured values for the surge chamber (lower curves) clearly fall much closer together. Therefore, at least in one respect, the suitability of larger rooms was demonstrated.

Since the wavelengths of the eigen-modes of a room are proportional to the linear dimensions, there is a relation between the volume of the room and the lowest frequency f_g (or the largest wavelength λ_g) for which a reverberation room is usable:

$$\lambda_g \sim \sqrt[3]{V}$$
$$f_g \sim \frac{1}{\sqrt[3]{V}} \tag{5.6}$$

[1]Meyer, E., *Akust. Zh.*, **2** (1937) 179.

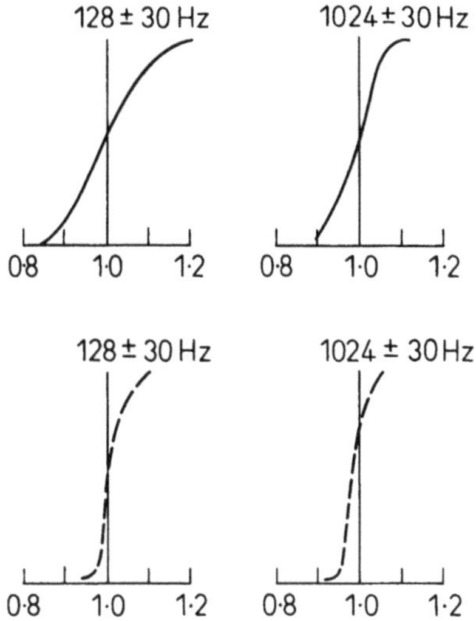

Fig. 5.2. Cumulative distributions of reverberation time measurements. Top: in a 160 m^3 reverberation room. Bottom: in the 13 000 m^3 surge-chamber of the Walchensee power station. (After Meyer.[1])

In Fig. 5.1, this frequency limit can be clearly seen, where the agreement between the measured values and the 'true' value is first reached at about 2000 Hz. Since the volume of the model reverberation room was only 0·5 m^3, Meyer and Kuttruff concluded from this and many similar experiments that the relation between the limiting frequency and the volume, which is given by eqn. (5.6), can be calculated by the easily remembered rule-of-thumb:

$$\sqrt[3]{V} f_g = 1000 \tag{5.6a}$$

with V in m^3 and f_g in Hz.

According to this relation, for measurements at frequencies as low as 100 Hz a room volume of 1000 m^3 is required.

But the use of larger rooms is also restricted by a limit that is determined by the maximum usable upper frequency. With increasing room volume and increasing frequency, according to the comments of Section II. 1.6, the dissipation of sound energy during transit between reflections increases, and more and more determines the reverberation time.

At first, it was believed that this influence did not have to be taken into account, since it falls out when the difference (eqn. (5.1)) is calculated, provided that the experimenter is careful to avoid changes in the temperature and relative humidity during the measurements.

But even if one accepts this assumption, the shortening of the reverberation time has an influence on the accuracy of the measurement, because, according to eqn. (1.44), an amount equal to $8\mu V$ must be added to the absorption area of the empty room, and in this case the relative error according to eqn. (5.2) is increased to:

$$\frac{\delta\alpha_x}{\alpha_x} = \frac{\delta T}{T}\,\frac{2A_0 + 8\,\mu V + \Delta A}{\Delta A} \tag{5.7}$$

Now, the sound-pressure-related damping constant μ, at a frequency of 6000 Hz (roughly the highest measurement frequency of interest) and an average relative humidity of 40%, amounts to approximately $7{\cdot}65 \times 10^{-3}$ neper m^{-1}. But this means that in a 1000 m^3 room the absorption area of the test sample must be increased by about 60 m^2, in order to avoid increasing the relative error in determining α.

This estimate shows the great influence of the magnitude of the damping constant μ. Even small, almost unavoidable, changes in temperature and relative humidity, if one does not take them into account, can cause significant errors in determining α. Furthermore, without an expensive air-conditioning system, an acceptable constancy of temperature and relative humidity cannot be maintained. Even opening the test-room door in order to install or remove the test sample can change the room conditions, and the installation of a moist or hygroscopic test specimen may change the relative humidity of the air in the reverberation room.

It has proved useful in eqn. (5.1) to take into account the possible difference $(\mu_0 - \mu_\Delta)$ in the damping constant between the empty room and the room with the test sample:

$$\alpha_x = \frac{0{\cdot}163\,V}{S_x}\left(\frac{1}{T_\Delta} - \frac{1}{T_0}\right) + \frac{8\,V}{S_x}(\mu_0 - \mu_\Delta) \tag{5.8}$$

In the Eyring formula, the expression $S\mu V$ does not appear as an addition to the absorption area because this formula assumes that the absorption, in contrast to the dissipation, occurs in discrete steps as the sound energy encounters the room boundaries. In practice, however, the same additive correction term appears here, corresponding to a series development, so long as α_S is measured without a correction:

$$\alpha_E = \alpha_S\left(1 - \frac{S_x}{2S}\alpha_S - \frac{A_0}{S}\right) + \frac{8V}{S_x}(\mu_0 - \mu_\Delta) \tag{5.9}$$

The essential difference between the step-wise absorption and the continuous dissipation makes it inappropriate to regard the term $8\mu V$ as simply a further contribution to the equivalent absorption area, although according to its dimensions this could be done.

Moreover, the term 'absorption' should be used in room acoustics only for energy losses occurring at reflections, while dissipation signifies transformation of acoustical energy into heat in general.

Unfortunately, the necessity to take the energy dissipation during propagation into account in the reverberation formula causes some uncertainty in its evaluation, because very different values for μ are to be found in the literature. Therefore, one should try to keep the correction term $8V(\mu_0 - \mu_\Delta)$ smaller than $1/5\,\Delta A$. If we assume that changes in μ do not exceed 20%, this leads to the conclusion that ΔA should be equal to at least $8\mu V$.

Even though, in a $1000\,m^3$ room, good measurements can be made at 100 Hz, there will be difficulties at 6000 Hz. But, naturally, it is desirable to be able to carry out the entire measurement series in the same room. Therefore, a room with a volume of $200\,m^3$ seems to be a reasonable compromise. The term $8\mu V$ in this case drops to $12\,m^2$ at the upper frequency limit, so that a test area of $12\,m^2$ can be regarded as sufficiently large. On the other hand the lower frequency limit, according to the Meyer–Kuttruff formula, increases only from 100 to 170 Hz. Furthermore, one can hope to overcome to some extent the expected measurement and evaluation errors by performing a large number of individual measurements.

If the large wavelengths at low frequencies and the energy dissipation at high frequencies were the only difficulties, one would at least expect good agreement in the mid-frequency range among the measurement results from different laboratories. Unfortunately, this is not the case; one cannot even say that the very low and very high frequency data are characterized by a significantly greater scatter than at mid-frequencies. Two principal causes seem to account for these discrepancies. The main condition for the validity of the statistical theory, namely, a homogeneous and isotropic distribution of sound energy in the room, is not fulfilled. Furthermore, the averaging over incidence angle does not correspond to the basic assumptions. Therefore, we must agree not only on the volume, but also on the shape of the room and on its furnishings.

At first, rectangular-shaped rooms were used because they were simple to build. Certainly, these rooms were preferable to cylindrical or spherical rooms, because at least the room could be uniformly filled with sound.

On the other hand, we have seen that a rectangular room has a particular distribution of incidence angles, particularly when lightly damped sound waves remain because the test sample is mounted only on the floor, as was customary in the beginning.

To avoid wave propagation parallel to the test sample, the reverberation room for the National Physical Laboratories at Teddington (London), as early as 1933, was provided with walls which, though plane, were angled so that no two room boundaries were parallel, not even the floor and ceiling.

The advantage of non-parallel boundaries for significantly improving the measurement conditions has been further demonstrated by the use of a pyramidal room.[2] Even when the test sample is mounted on only one boundary, this reverberation room gives the same results as a cubic room six times larger, where the test sample is distributed over three mutually perpendicular boundaries. Surprisingly, the measured values of α from the pyramidal chamber were systematically lower than those measured in normal reverberation rooms.

The scattering of sound in all directions is surely improved if each of the six boundary surfaces of a rectangular room is divided into smaller, angled surfaces. Pursuing this concept further, one may cover the reverberation room walls with convex cylindrical diffusing elements (so-called poly-cylinders), whose usefulness Meyer mentioned in 1937, in his paper about the model reverberation room,[1] see Fig. 5.3.

Even though Meyer stated at that time that the cylinders shown in Fig. 5.3 have hardly any effect on low-frequency reflections (with large wavelengths), this shaping was widely used to increase the diffusion in broadcast studios (see Part IV, Fig. 13.2 (Volume 2)) and, later on, also in reverberation rooms. In Germany, for example, the National Bureau of Standards (PTB) in Braunschweig has such a reverberation room which, according to Venzke,[3] yields linear reverberation decays that are easy to evaluate. There, the cylindrical axes of the poly-cylinders, which cover two adjacent walls and the ceiling, are all perpendicular to one another.

[2] Müller, H. A., *Proc. 3rd. ICA Congress, Stuttgart, 1959*, Vol. II, Elsevier, Amsterdam, 1961.

[3] Venzke, G., *Acustica*, **6** (1956) 1.

Fig. 5.3. Scale model reverberation room, with sound scattering ceiling and wall elements. (After Meyer.[1])

The agreement shown in Fig. 5.1 between the 'true' absorption coefficients and the model measurements was achieved by Meyer and Kuttruff with the use of cylindrical diffusers. Figure 5.1 shows, in particular, that by covering only two walls with such diffusers (dotted line), which means that two plane-parallel walls remain, the 'true' values are reached only at much higher frequencies than if three boundaries are covered (dashed line); the additional covering of two more boundaries, which are opposite to already diffuse walls does not further decrease the limiting frequency (dot-dashed line). Incidentally, in this case the width of the diffusers corresponds to about $1\frac{1}{2}$ wavelengths at the limiting frequency and the depth was about $\frac{1}{4}$ of the width. One should deliberately vary the dimensions from one cylinder to another in order to avoid correspondence with only certain wavelengths.

There is also a third way to influence the distribution of sound, namely, to hang oblique reflecting surfaces in the room, a measure that was already introduced by W. C. Sabine with his mixing vanes; in addition Sabine caused these vanes to rotate.[4]

---

[4] Sabine, P. E., *Acoustics and Architecture*, McGraw-Hill, New York, 1932; see especially pages 98–101.

Early model tests by Steffen[5] found no essential difference between stationary and rotating vanes. This observation is in conflict with recent studies in the United States.[6] Lubman, in fact, has devised a 'figure of merit' for rotating diffusers, based on the reduction in spatial variance of the measured sound pressure:

$$M = \left(V_R^2\right)_{\text{stat.}} \Big/ \left(V_R^2\right)_{\text{rot.}}$$

where V_R^2 is the mean-squared deviation of p^2 from its room-average value and the subscripts 'stat.' and 'rot.' refer to conditions with the diffusing vane stationary and rotating. It is very important to note, however, that a rotating diffuser is most helpful when dealing with pure tones or narrow bands of noise, for which the reduction in V_R^2 is dramatic; practical values of M from 2 to 4 are typical but values as high as 10 have been reported. In addition, a rotating diffuser provides a worthwhile reduction in the variance of sound power output with changes in source position. However, if the source spectrum is broad and flat the figure of merit will be low; a rotating diffuser is relatively ineffective for reducing the spatial variance for broadland signals.

Furthermore, Meyer and Kuttruff, in their model tests with non-rotating diffusing panels, found that two such vanes, which were so large as to practically divide the test room in two, were not so favorable as 24 randomly arranged scattering surfaces (slightly curved, if possible), each with an area of λ_g^2. The efficiency of diffusion decreased if they used a greater number of, but correspondingly smaller reflectors. With such diffusing elements one can achieve the necessary diffusion even in a rectangular room.

Diffusers hanging in a room, however, have a disadvantage, namely, that they shorten the mean free path length, and thus shorten the reverberation time, since the absorption coefficient of the diffusers may be at least as great as that of the bare walls.

In the interest of accuracy in determining α, it is desirable that the empty room have a reverberation time as long as possible. Therefore, the walls, floors and ceilings of reverberation rooms are preferably constructed of materials as heavy and non-porous as possible.

---

[5] Steffen, E., *Hochfrequ. u. Elektroak.*, **67** (1958) 73.

[6] Lubman, D., *J. Acoust. Soc. Am.*, **56** (1974) 523. See also, Lubman, D., *J. Sound Vibr.*, **16** (1971) 43; Tichy, J. and Baade, P. K., *J. Acoust. Soc. Am.*, **56** (1974) 137; Ebbing, C. E., *J. Sound Vibr.*, **16** (1971) 99.

Unfortunately, only the reverberation time at low frequencies can be significantly increased by this means, because at high frequencies the reverberation time is limited by the unavoidable absorption of the walls (see Section II. 6.2) and energy dissipation during propagation. These effects lead to the typical steeply falling curve of reverberation time versus frequency for a reverberation room.

This steep slope can lead to a sagging of the reverberation decay curve, even within the bandwidth of a warble tone or a band of noise, because the reverberation is made up of components having very different decay rates.

Still more with broadband sources, especially impulses, in spite of filtering in the receiver circuit the slower-decaying lower frequencies can determine the reverberation more and more toward the end of the decay, so that one gets strongly sagging decay curves and thus false absorption data. This was one of the reasons that led to the exclusion of impulses for reverberation room experiments.

For all these reasons, it has been found necessary, at least for the standardized $200\,m^3$ room, to set the following minimum values for reverberation time in the empty room, which, it will be noticed, do not increase below 500 Hz:[7]

f	125	250	500	1000	2000	4000	Hz
T_0	5·0	5·0	5·0	4·5	3·5	2·0	s

Reverberation times that are too large also have disadvantages, especially at low frequencies. The lower the damping of a system, the more noticeable the individual eigen-modes become. In a reverberant room, this shows up as strong beats in the pressure-time decay created by neighboring eigen-frequencies, a tendency, on the average, to sagging reverberation decays, and, finally, a strong dependence of the reverberation time on frequency as well as on the position of the sound source and receiver.

Therefore, it is no disadvantage if some of the walls and ceiling are relatively thin and, therefore, as we will show in the next chapter, are absorptive at low frequencies; the poly-cylinder construction is an example. However, this absorption should be distributed over a wide frequency range and must not be effective at only a single frequency, as with a resonator. Resonators are useful only if it is necessary to damp a

---

[7] DIN 52212: *Measurement of the absorption coefficient in a reverberation room.* See also ASTM C 423–77.

particular strong room resonance. Therefore, we have the further general recommendation that the absorptive effect of the walls must be so 'tuned' that one gets a curve of $T_{(f)}$ as flat as possible.

II. 5.3 The Influence of the Test Sample

The next problem concerns the size and the arrangement of the test sample. Here again, one tries to keep the cost of the test material as low as possible. But we have previously recommended that the additional absorption area ΔA that is introduced by the test sample should be at least equal to the absorption area A_0 of the empty reverberation room. According to this rule, the minimum area S_x for the test sample would depend on its absorption coefficient. But this is a function of frequency and it is difficult to say at which frequency the absorption is smallest; it is even more difficult to guess the value of the coefficient, since this is the very quantity to be measured.

Even aside from this difficulty, it would be unnecessarily expensive to base the measurement area S_x on the *lowest* value of α, because this value is generally of least interest. Moreover, in this case, the requirement for precision is not so great as for higher values of the coefficient. Also, even if it were feasible to recommend a different area for every test sample, such an individual adjustment would create a great complication for the laboratory and for the client. Therefore, in the earliest drafts of the measurement standards, a standard test area of 10–12 m^2 was prescribed.

Only if one wants to measure an arrangement whose absorption coefficient is obviously very low (for example, different floor surfaces, plasters or similar hard materials), may it be left to the discretion of the tester to use a larger area for the test sample. In this case, he must also take into account the absorption coefficient of the area of the test room that is covered by the test sample, a matter that is normally neglected:

$$\Delta A = S_x(\alpha_x - \alpha_0) \tag{5.10}$$

One is at first inclined to calculate the value of α_0 by finding A_0 from the reverberation time of the empty room and dividing it by the total surface area of the room boundaries. The reverberation time of the empty room, however, depends not only on the sound absorption of the boundaries, but also, at least at low frequencies, on the tendency of the walls, the ceiling, the installed loudspeakers, the doors and heating radiators to vibrate and on the losses through ventilation openings; in short, the empty-room absorption depends on things whose absorptive effect remains unchanged even after the test sample is installed. Therefore, one can more easily establish an inequality:

$$\alpha_0 < A_0/S \tag{5.11}$$

In such cases, the necessary determination of α_0 is a very tedious procedure, which is possible only by comparison measurements between rooms with different large but similar surfaces, in which the walls and ceilings hardly vibrate at all and the remaining objects are all kept the same.

The prescription of a standard area for the test sample is expedient, not only for the reasons described above, but also because it is found that the absorption coefficient α_x depends on the sample size S_x.

The systematic investigations that were carried out by Meyer as early as 1937 showed that the common procedure of spreading the test sample on the floor of the reverberation room leads to values of α_x that decrease with increasing S_x, up to the point where the test sample covers the entire floor, while the value of T asymptotically approaches a constant value. This result, which formally contradicts the statistical theory, can easily be explained by the earlier-mentioned non-uniform directional distribution in a rectangular room. The sound rays that encounter the test sample quickly disappear; eventually, only those rays persist that are parallel to the test sample and are hardly affected by it; they determine the asymptotic residual reverberation time.

To exclude rays that propagate parallel to the test sample, Meyer and Schoch[1] proposed, during a German 'round robin' test in 1939, that the test sample be distributed on three mutually perpendicular room surfaces. This procedure seems simpler than the use of reverberation rooms with oblique angles or with diffuse reflecting surfaces.

Later on, the width of the individual pieces of the test sample was standardized at 1·5 m. This was an attempt to reduce the data scatter caused by the so-called 'edge effect', which had been theoretically well-known for a long time,[2] and was confirmed by the investigations of

[1]Meyer, E. and Schoch, A., *Akust. Zh.*, **4** (1939) 61.

[2]Sabine, P. E., *J. Acoust. Soc. Am.*, **5** (1934) 220; Morse, P. M. and Bolt, R. H., *Rev. Mod. Phys.*, **17** (1944) 69; Pellam, J. R., *J. Acoust. Soc. Am.*, **11** (1940) 396. For circular test objects, the edge effect was investigated in detail, both theoretically and experimentally, by R. K. Cook (*Proc. 3rd. ICA, Stuttgart, 1959*, Elsevier, Amsterdam, 1961).

Kuhl.[3] Although a satisfactory explanation of this effect, which depends on diffraction of the sound waves at the boundary of the test specimen, can only be given on a wave-theoretical basis, it is apparent that the sound energy that enters the material close to the boundary of the test sample comes not only from above the sample but also from the side. Such a sidewise transport of energy is self-explanatory if, for example, the test material lies like a mattress on the floor and thus presents additional absorbing areas on the sides. This side absorption can be hindered in absorption measurements by framing the test sample with gypsum board or plywood. But even if the sides of the test sample are so covered, the remaining boundary between a fully reflecting and an absorbing area leads to an energy transport from the side that increases with the edge-length L.

The measured absorption is composed, therefore, of one part that increases with increasing area and another part that increases with increasing boundary length:

$$\Delta A = \alpha_{x\infty} S_x + K\,L_x \tag{5.12}$$

Here, we understand by $\alpha_{x\infty}$ the coefficient for an edge-free, infinitely extended test sample. The constant K, which is significant for the edge effect, has the dimension of length in this notation, and it thus seems natural to relate it to the wavelength. This leads to writing (5.12) as follows:

$$\Delta A = \alpha_{x\infty} S_x + \kappa\,\lambda\,L_x \tag{5.12a}$$

Also, it can be expected from the basic equations of the wave motion of sound that the aforementioned energy flow can be detected at greater distances from the boundary, the larger the wavelength. But because the 'edge effect' depends on the material, and in particular on $\alpha_{x\infty}$, which characterizes its absorption capability, and since this quantity is itself frequency-dependent, the constant κ already includes the frequency, and thus the wavelength. Therefore, the frequency must not appear explicitly in the factor λ that appears in eqn. (5.12a). But when the material has practically reached its highest value of $\alpha_{x\infty}$, κ no longer increases and then a decrease of the 'edge effect' with increasing frequency can be expected. This was proved by Kuhl, by a very careful evaluation of his own results and the results of others.

[3] Kuhl, W., *Acustica* (*Beihefte*), **10** (1969) 264. Kuhl also discusses in detail the contributions of other authors on this effect.

The splitting implied in eqn. (5.12) assumes a broad reflecting area next to the boundary. If, instead, the edge of the test sample is in direct contact with a vertical wall of the room, the 'edge effect' is completely suppressed for that edge, because of the image of the test sample in the wall. This was clearly shown by Meyer and Kuttruff in their model tests.[4]

Meyer and Kuttruff concluded from this result that the true absorption coefficient $\alpha_{x\infty}$ could be determined only if one excludes any influence of wave bending, for example by covering an entire wall or floor with the test sample. (Strictly speaking, this must be a rectangular area bounded by plane, vertical, totally reflecting walls. Walls meeting at either an acute or an obtuse angle lead to multiple imaging; and a wall covered with poly-cylinders is not a plane mirror surface at all.)

Now, it may be asked whether the 'true' absorption coefficient, $\alpha_{x\infty}$, has any meaning in practical applications. At least in most applications there *are* free edges, so it is not without interest to try to capture this 'edge effect' through measurements.

Therefore, it has been recommended by various people that measurements are made of test samples with different edge lengths, and that the two constants $\alpha_{x\infty}$ and κ from these measurements are evaluated. As logical as this proposal may be from the physical point of view, it misses the point of the goal and the possibilities of technological testing. The test procedures must be as simple as possible, so that not only typical constructions can be measured but also any construction that might occur in practice. Furthermore, if we must keep track of both $\alpha_{x\infty}$ and κ the application of the data would become much more complicated without significantly increasing the accuracy of the calculation, because the edge effect depends on the acoustical properties of the neighboring areas. Measurements in the reverberation room are relevant only to acoustically hard neighboring surfaces.

These difficulties do not mean that we have to give up the hope of comparable results from different laboratories; we have only to agree on a certain ratio of the edge length to the area of the test sample.

Since the ratio L_x/S_x has the dimension of L^{-1}, and since this ratio must be compared with the wavelength, which is independent of the room size, it is reasonable to prescribe the test sample area and edge length in absolute terms. Therefore, it was agreed to adopt standardized

---

[4]Meyer, E. and Kuttruff, H., *Göttinger Nachr. Math. Phys. Kl*, (1958) No. 6.

sizes, with tolerances that are dependent on the material and the room:[5] namely, a test area of about 4 m × 3 m (corresponding to $S_x = 12\,\mathrm{m}^2$ and $L_x = 14\,\mathrm{m}$), which must be set approximately in the middle of the floor or a wall. Practically, this was a return to the original arrangement, but now with the means, even for this one-sided mounting, of creating the effect of a diffuse ('all-sided') sound field.

II.5.4 Requirements for Source and Receiver

Only loudspeakers are used as the sound source for reverberation room investigations nowadays, because by this means one can generate in the easiest way any desired frequency combination, especially third-octave bands of random noise or warble tones. Standardized values have been chosen for the warble tones; namely, a warble frequency of about 6 Hz and a warble swing of about ± 10% of the mean frequency below 500 Hz, and ± 50 Hz above 500 Hz. The mean frequencies, for both warble tones and bands of noise, have been established by DIN 52212 as the following preferred series of third-octaves:[1]

100	200	400	800	1600	3200	6400	
125	250	500	1000	2000	4000		Hz
160	320	640	1250	2500	5000		

The underlined frequencies correspond almost exactly to those used by W. C. Sabine in his investigations. In cases where one can expect a monotonic dependence on frequency, the reverberation measurements can be carried out using only the octave series of frequencies.

In general, however, one should measure an unknown material in third-octave steps, so that any possible, and often useful, resonant peak of the absorbing device is measured exactly. But this is not possible if the resonance peak is so narrow that it prolongs the reverberation (see Section IV.9.5, Volume 2).

---

[5] DIN 52212, 2nd. Edn., Jan. 1961, *Bestimmung des Schallabsorptions-grades im Hallraum* (*Determination of sound absorption coefficients in a reverberation room*); ANSI/ASTM C 423–77, *Standard test method for sound absorption and sound absorption coefficients by the reverberation room method*; ISO/R 354–1963, *Measurement of absorption coefficient in a reverberation room.*

[1] DIN 52212, 2nd Edn., Jan. 1961, *Bestimmung des Schallabsorptions-grades im Hallraum* (*Determination of sound absorption coefficients in a reverberation room*).

On the other hand, one cannot determine the frequency dependence in a reverberation room in too much detail, because a statistical signal can be achieved only when the bandwidth is wide enough (i.e. not, for example, a pure tone). Only in this case are enough eigen-modes excited.

Even when using warble tones that do not cover a full third-octave band, one can get a 'gapless' picture with a procedure that was tested and especially recommended by Meyer and Schoch. It consists of using a warble tone whose mean frequency continuously glides upward; reverberation decay measurements are then made only at half-tone steps. In this manner, one gets a series of reverberation measurements that are closely spaced in frequency. The scatter at low frequencies is significantly greater than at high frequencies, though the transition is gradual. But it would be wrong to consider these fluctuations as a property of the material being measured. It is much more useful, as Meyer and Schoch did, to lay an average curve (of reverberation versus frequency) through the scattered data points and to evaluate the curve at third-octave frequencies.

This procedure does not require any more time than measurements only at third-octave frequencies, because in that case one must perform six measurements at each frequency, with different source or receiver positions or some other variation, in carrying out the reverberation time measurements.

Since it is desirable to produce as uniform a sound field as possible, and since all loudspeakers exhibit an increasing directivity at higher frequencies, the use of several loudspeakers pointed in different directions is necessary for high-frequency measurements. For example, a device was developed by Harz and Kösters[2] for monitoring the signal in broadcast studios, in which a number of individual loudspeakers were mounted on the faces of a polyhedron. Figure 5.4 shows such a system with 32 loudspeakers, developed by L. Keidel, in one corner of the former reverberation room in the Technical University of Berlin, where, incidentally, a triangular ground plan was tried.

At low frequencies, one can achieve uniform sound radiation with only one system because of the small dimensions of the loudspeaker with respect to the wavelength. Here, however, there exists another danger, namely, that the loudspeaker does not excite all of the eigen-modes in the excited frequency range, because there is a pressure minimum for certain

[2] Harz, H. and Kösters, H., *Hausmitteil. NWDR*, **3** (1951) 205.

Fig. 5.4. Reverberation room corner with loudspeaker for low and high frequencies and microphone with omnidirectional characteristics.

modes at the loudspeaker location. This danger cannot be overcome by using several loudspeaker positions. Such combinations can lead to the failure to excite certain eigen-modes or, in other words, to the predominance of a certain directional distribution. One can overcome this problem much better by using only one loudspeaker, but placing it where all the eigen-modes of the room exhibit a maximum sound pressure, a so-called pressure antinode, namely in the corner of the room. As an example, Fig. 5.4 also shows a low-frequency loudspeaker, with the back closed, in the corner of a reverberation room.

The same considerations apply for microphones; but in this case it is not necessary to have separate units for high and low frequencies, because sufficiently sensitive microphones are available with such small dimensions (see Fig. 5.4) that uniform sensitivity is achieved even at the higher measurement frequencies, at any rate, up to 4000 Hz.

With respect to the microphone position, different rules are to be followed for low and for high frequencies. It is an advantage to place the microphone in a room corner for low frequencies; the distance from the corner, however, must be smaller than about 1/10 of a wavelength if the microphone is to occupy a pressure antinode of all the excited eigenmodes.

This arrangement means, then, that one has to realize the six variations of the measurement conditions by other means, since the placement of both the microphone and loudspeaker in corners is not always feasible, or at least is awkward.

But if the microphone is not placed directly in a corner, then it is necessary to stay at least one-half wavelength from all the walls, because in the region between $\lambda/10$ and $\lambda/2$ from the wall, one must expect especially strong destructive interference effects for sound at certain incidence angles; such sounds are then not included in the measurement.

Of the methods of measuring reverberation time that were discussed in Chapter II.4, only those are used in reverberation room experiments that can provide sufficient accuracy. A prime example is the registering of reverberation decays with the help of a level recorder, as described in Section II.4.5, where a writing speed of at least $300\,\mathrm{dB\,s^{-1}}$ may be required. Also, as described in Section II.4.2, a procedure of automatically switching a stopwatch (or other counting device) on and off when certain levels are reached is widely used. In the latter case, one must determine whether the decay curves are 'sagging'; this can be done by switching to different level ranges and seeing whether the respective decay times are the same. The method of the integrated impulse response is also often used for determining reverberation time.

Because practically every reverberation decay curve 'sags' if the observation time is long enough, it is necessary to agree on a certain range of levels in the decay to be used for the evaluation. The automatic procedure requires that the 'switch-on' level for the measurements is considerably lower than the steady-state level. But, when using level recorders, it is recommended that the part of the decay immediately after switching off the sound source is not used for establishing a straight line approximation to the decay slope, because at the switch-off time, even with warble tones, there can be sudden drops or increases in level. Therefore, one starts the decay slope evaluation at a point 5 dB below the stationary level. With both procedures, one must then observe the decay over a range of 30 dB, that is, to a level of -35 dB, if one is to achieve a sufficient level difference to average out the unavoidable fluctuations. However, an increase in this observed range is not at all recommended, because in any reverberation process there may always remain slowly decaying components that correspond to a non-uniform directivity distribution.

If, within the standardized range from -5 to -35 dB, there is a significant sagging of the decay curve, this recording should not be used for determining the reverberation time. Instead, one should attempt to relocate the source and receiver or to change the arrangement of the test sample in order to get a better registration of the decay process. Kuttruff, on the other hand, specifically recommends the evaluation of sagging decay curves, because in the initial slope of the decay the rapidly decaying components can be clearly seen. This is surely correct. Yet strongly sagging decay curves in a reverberation room are often a sign of insufficient averaging over the angles of incidence. This would be good reason to check the entire experimental set-up.

Finally, a word about the presentation of the results. For the reader, a plotted curve is easier to read than a table. Therefore, every test report should present the test results by means of curves.

To facilitate comparing the curves with one another, the same scales should be used for the ordinates and the abscissae throughout. On the abscissae, a logarithmic frequency scale is always used. In Germany, one-third octave corresponds to 5 mm.

Because sound-absorptive treatments are used also to decrease the sound level in noisy environments, and since one can describe the expected reduction in the noise level in decibels by $10 \log (A/A_0)$, the absorption coefficient was also originally displayed on the ordinate with a logarithmic scale. Today, however, there is international agreement that α and δA should be displayed on a linear scale. In Germany, in accordance with international agreements, an increase of 0·1 in α corresponds to a distance of 5 mm; for values of δA, 10 mm corresponds to a change of 0·1 m^2.

The graphs must allow enough space to plot values of α_S larger than 1 (say, up to 1·2), because the use of Sabine's formula, as well as wave-bending effects, can lead to such values. For this reason, it is not appropriate to express α_S in percent. Nevertheless, the display in percent should be kept for explicitly expressing differences in the values of α measured under certain angles of incidence, for example, in an impedance tube. By contrast, the absorption coefficient measured in a reverberation room and evaluated according to Sabine's formula, should be designated by α_S.

The German standard DIN 52212 includes a standardized data sheet to be used for presenting the test measurement results and the most important room conditions, as well as space for a drawing of the test objects.

Chapter II.6

Absorption for the Control of Reverberation

II.6.1 General Remarks

The Sabine reverberation formula derived in Section II. 1.5

$$T=\frac{0{\cdot}163\,V}{A} \tag{6.1}$$

is surprisingly simple. The reverberation time depends only on the volume V (in m^3) and the equivalent sound-absorption area A (in m^2). The latter can be calculated from the individual areas of the room boundaries S_k and their absorption coefficients α_k:

$$A=\sum_k S_k\alpha_k \tag{6.2}$$

If one calculates the volume of a 'highly fissured' room, such as a theater with galleries, balconies, or *loges* and deep boxes, one must not include the volume of these boxes and *loges*, etc., in the room volume. If the absorptive area inside these sub-volumes is greater than the coupling area to the auditorium, the latter can be entered into the equation as fully absorbent, provided that the sub-volume is much smaller than the auditorium. If this condition is *not* fulfilled, one has to handle the coupled volumes as described in Section II. 3.1.

The areas of the individual room boundary surfaces S_k can be easily determined; one has only to take care that walls whose roughness is small compared with the wavelength are included according to their projected area. As a guideline, one can say that 'roughnesses' (for example, moldings, ledges or shoulder profiles) up to a depth of 1 m can normally be neglected.

The chief difficulty in calculating the reverberation time arises in deciding what values to use for the α_k. Tables of absorption coefficients have been published in the literature since W. C. Sabine's first measurements. But these values are based on such different measurement arrangements and such different principles of evaluation that their comparability is very much in question. Not only that, but as far as the test sample and the method of mounting are concerned, these are often unstated.

In Germany, therefore, the various testing institutes have carried out absorption measurements in accordance with standard procedures,[1] and the measured results are published in a Sound Absorption Coefficient Table by the German Standardization Organization.[2] For this reason, no tabulation of absorption coefficients is included in this book.

Besides, there are many test reports for commercial sound-absorbing panels and other products, which are, despite numerous brand names that can scarcely be remembered, physically very similar to one another. No matter how valuable their development has been for noise control purposes, these special absorptive wall and ceiling coverings (e.g. mineral wool or glass fiber tiles or boards) are seldom applied in the interiors of large, architecturally important spaces like auditoriums, concert halls or churches. Most of the surface areas in such buildings are covered with plaster, wooden panels, curtains, tapestries or other traditional room finishes, whose absorption coefficients have only very seldom been measured.

The architect will always try, even in planning for sound absorption with wall or ceiling coverings, to develop solutions in accordance with

[1] DIN 52212, 2nd. Edn., Jan. 1961, *Bestimmung des Schallabsorptions-grades im Hallraum* (*Determination of sound absorption coefficients in a reverberation room*). Corresponding documents in English are: ISO/R 354–1963, *Measurement of absorption coefficients in a reverberation room*; ANSI/ASTM C 423–77, *Standard test method for sound absorption and sound absorption coefficients by the reverberation room method.*

[2] *Deutscher Normenausschuss*, *Schluckgradtabelle*, Beuth-Vertrieb GmbH Berlin, Köln, Frankfurt, 1968. The absorption coefficients for commercially available North American sound absorption products were published annually for many years by an industry organization whose name changed from time to time. The last edition was published in 1974 under the auspices of the Acoustical and Insulating Materials Association, 205 West Touhy Avenue, Park Ridge, Illinois, 60068. Since the data are not subject to rapid change, this last edition contains data that are still useful. In addition, a collection of sound absorption data for various products and arrangements is tabulated in: *Compendium of Materials for Noise Control*, June 1975, US. Department of Health, Education and Welfare, NIOSH, Division of Laboratories and Criteria Development, Cincinnati, Ohio, 45202 (HEW Publication No. (NIOSH) 75–165).

his overall architectural concept. Moreover, it often happens that the architect wants the acoustical treatment to be scarcely visible, if at all. This point of view regards the acoustical treatment as a mere technical device of low priority, like the plumbing pipes and the electrical wiring. But from the aesthetic standpoint of 'honest construction', meaning that the basic construction principles of the building should show through, it is hard to understand why one should not also 'see' the acoustical significance of a wall or ceiling covering.

In the framework of this book, which is dedicated to fundamental principles, it is almost obligatory to present an overview of the physical laws that determine sound absorption, in order to help the reader assess the sound absorption of various kinds of room boundaries. This would seem to be more important than merely listing an assortment of measured results.

Accordingly, after discussing the unavoidable absorption at a sound reflection, we will describe sound-absorbing treatments that are primarily effective at high, medium, or low frequencies (so-called 'high-, medium- or low-frequency absorbers'). Finally, we will consider sound-absorptive objects, such as resonators, pieces of furniture, or persons, that cannot be characterized by a projected surface area. In this chapter, more emphasis is placed on phenomenological descriptions intended to further the understanding of the process, rather than on strict mathematical theory, which is saved for the wave-theoretical treatment in Part IV (Volume 2). Finally, an example of precalculation of reverberation time is given.

II.6.2 Unavoidable Sound Absorption

We begin with the losses that are unavoidable in the reflection of sound, and the related question: what are the lowest possible sound absorption coefficients? This question is of interest because the goal of room acoustics is by no means always to absorb sound. Some surfaces, for example, the baldachino over a pulpit, sound-reflecting panels above an orchestra, and, in low rooms, the entire ceiling, should reflect as much sound as possible, which means that their absorption coefficients should be as low as possible. Furthermore, in many cases the audience introduces so much absorption into the room that it is important for the other room surfaces to present a minimum of absorption.

An experimental determination of the lowest achievable sound absorption coefficients is given by the reverberation times of reverberant rooms that are built of heavy, flat, unporous materials with the aim of holding the absorption to a minimum. So far as we know the longest reverberation time is found in the largest reverberation room in Japan[1] ($V = 512\,m^3$); taking into account the dissipation of energy during propagation and assuming that the equivalent absorption area is uniformly distributed over all the room boundaries, we find the following reverberation times and absorption coefficients at the principal octave frequencies:

f:	125	250	500	1000	2000	4000 Hz	
T_{60}	48·3	36·8	22·4	13·6	12·1	9·4 s	
α_S:	0·0045	0·0059	0·0097	0·016	0·018	0·023	(a)

(In the following discussion, we will not list the frequencies again because the six absorption coefficients, listed side-by-side, are always related to these same frequencies. If the values for the coefficients are unknown at certain frequencies, this will be indicated by dashes.)

This means that at each reflection, on the average 0·45 to 2·3% of the incident energy is lost. Since sound penetrates even very heavy walls, implying an energy loss from the room, it is not astonishing that a certain part of the incident energy is not reflected. If, for example, the sound insulation of a wall (i.e. ten times the logarithm of the ratio of the incident to the transmitted energy) amounts to 40 dB, then one-ten-thousandth of the energy is transmitted through the wall to the outside; the transmission factor τ is 10^{-4}. Still, this energy loss is not sufficient to explain the values of α tabulated above. Because the sound insulation of walls increases with frequency, the absorption due to this loss should decrease with frequency; in fact, the opposite is true.

But there are other kinds of effective losses. The air layer immediately next to the wall is not free to move parallel to the wall; it 'sticks to' the wall. In an 'acoustical boundary layer'[2] that is only about 0·1 to 1 mm thick, depending on the frequency, frictional losses are caused by the viscosity of the air.

Another cause of energy loss that is not so familiar from our daily experience but is nevertheless just as effective, arises from the fact that not only does the tangential motion disappear, but also the temperature near the wall is nearly constant because of the heat conductivity and the heat capacity of the wall. The compressions caused by the sound pressure are, therefore, isothermal (which means that in every phase of over-

[1] The first-named author is indebted for this information to a communication from the Japanese Industrial Standards Committee.

[2] Cremer, L., *AEÜ*, **2** (1948) 136.

pressure heat is transferred to the wall, and in every phase of underpressure heat is taken from the wall). But at a short distance from the wall, just outside the boundary layer mentioned above, the compressions take place adiabatically, without any interchange of energy with the surroundings. Within the boundary layer, therefore, there must be a transition from one limiting case to the other; and thermodynamics teaches us that under such conditions there is more heat lost in the overpressure phase than is recovered in the underpressure phase. This implies energy losses, which, like frictional losses, result in a continuous heating of the air at the reflecting wall, even if it can hardly be detected.

Both losses can be described by a formula for an absorption coefficient averaged over all incidence angles, as derived in Section IV. 7.8 (Volume 2), eqn. (7.74):

$$\alpha = 1{\cdot}8 \times 10^{-4} \sqrt{f} \tag{6.3}$$

where f is the frequency in Hz. At the principle octave frequencies this leads to the following values for the absorption coefficient:

$$\alpha_S = 0{\cdot}0020 \quad 0{\cdot}0028 \quad 0{\cdot}0040 \quad 0{\cdot}0057 \quad 0{\cdot}0080 \quad 0{\cdot}011 \qquad \text{(b)}$$

These values are only about half those in row (a).

Because of the nature of the boundary layer losses, these losses will increase with increasing surface area, and from this point of view one would have to calculate the actual area very precisely when using the theoretical formula on which row (b) is based. Even boundary roughnesses with a height of 0·1 mm can significantly increase the losses within the boundary layer, at least at high frequencies. This may be the reason for the somewhat higher absorption coefficients measured for a rough plaster surface, compared to smooth plaster. But even if the coefficients of row (b) were three times higher, they would still be so small that the resulting absorption area, in comparison with the absorption of other typical surfaces in a reverberation time calculation for an auditorium, hardly plays any role at all.

In cases where these small numbers do attain some importance, for example, in large, bare churches with non-porous masonry, it is recommended that only the large profiles (of the order of one meter) be taken into account, but with roughly rounded coefficients that are about twice as great as in row (a):

$$\alpha_S = 0{\cdot}01 \quad 0{\cdot}01 \quad 0{\cdot}02 \quad 0{\cdot}02 \quad 0{\cdot}03 \quad 0{\cdot}04 \qquad \text{(c)}$$

The differences in absorption between the usual surface treatments lie, according to experience, so close to the limit that can be discriminated in reverberation room studies that it is hardly worthwhile to distinguish between these surfaces.

II.6.3 Porous High-frequency Absorbers

Porous materials into which air can enter are very effective sound absorbers. To this group belong textiles, felts and even masonry with open pores.[1] It is quite possible that the discrepancy between the calculated and measured absorption coefficients for a rigid wall is caused by pores in the structure of the bricks or the plaster coat, even if they cannot be seen with the naked eye. Since porous materials often have a rough surface, people tend to mix up these two characteristics.

The porousness of a material is described quantitatively by the so-called porosity factor σ which is the ratio between the volume of the interior pores to the total volume.[2] The porosity of a material with a rigid skeleton is acoustically effective only if the holes within the material are interconnected. This would be true, for example, when many granules are cemented together with a binding agent, or in the case of fiber layers. Typical examples are panels made of pumice-stone concrete, or of mineral or vegetable fibers. If the pores are closed, as in aerated concrete, the air cannot enter the material (see Section IV.8.3, Fig. 8.2, Volume 2).

A determination of the acoustically interesting porosity, therefore, cannot be made by simply weighing the material, but only with a procedure that depends upon the degree of interconnection of the pores, for example, by immersing the material in a suitable fluid, or, even better, by assessing the ability of the material to take in gas through the open

[1] With respect to sound-absorptive porous minerals, it is interesting to note that a commercially produced product, intended to simulate the appearance of masonry in monumental buildings while providing useful sound absorption, was developed in America in the early years of this century, with consulting advice from W. C. Sabine. (Beranek, L. L. and Kopec, J. W., *J. Acoust. Soc. Am.*, **69** (1981) 1.) The name of the product was Akoustolith, manufactured by Guastavino. It was rather widely used—so much so, in fact, that in subsequent years it has challenged the ingenuity of acousticians who wish to nullify its absorptive properties in buildings where a long reverberation is appropriate! (See Newman, R. B. and Ferguson, J. G., Jr., *J. Acoust. Soc. Am.*, (1982) to be published.)

[2] In both Germany and North America, it has been customary to call this ratio 'porosity'. But this seems improper, since porosity is a *quality* of the material. In order to quantify that quality, we propose here to use the term 'porosity factor'.

pores. If one compresses the gas in a closed volume in which the test sample is contained, one can evaluate, by the change in the quantity of gas required for a certain increase in pressure, the effective volume of the test sample into which the gas can penetrate. The procedure is described in Section IV. 8.3, Fig. 8.12.

If a sound wave encounters a porous wall, the air vibrating in the direction of propagation penetrates into the porous material at the boundary surface. Since the cross-sectional area by which the air can enter is restricted by the skeleton of the porous material, the particle velocity v_i of the air inside the material is increased. This increase in particle velocity with respect to the exterior velocity v_e is inversely proportional to the porosity factor if we have a pipe structure as shown in Part IV, Fig. 8.2, right:

$$v_i = \frac{1}{\sigma} v_e \tag{6.4}$$

If the structure of the material is as shown in Part IV, Fig. 8.2 (middle), where relatively large volumes are inter-connected by very narrow channels, one finds at certain points an even greater air velocity. This means that the porosity alone is not sufficient to determine the air velocity to be expected inside the structure. Zwicker *et al.*[3] have introduced the 'structure factor' χ for a quantitative description of the influence of the structure of a material. It describes the extent to which the particle velocity inside the structure is increased, compared with what can be expected on the basis of the porosity factor alone (see eqn. (6.4)). This structure factor, therefore, cannot be smaller than unity. Normally, it lies between 1 and 4, but values as high as 20 have been observed. For simple structures it can be calculated; an experimental evaluation is also possible (see Section IV.8.3).

The sound absorption coefficient for an infinitely thick layer of material with a rigid skeleton, a structure factor χ, and a porosity factor σ, is given for normal incidence in the limit of high frequencies by:

$$\alpha_{(0)} = 1 \Big/ \left[\frac{1}{2} + \frac{1}{4}\left(\frac{\sigma}{\sqrt{\chi}} + \frac{\sqrt{\chi}}{\sigma} \right) \right] \tag{6.4a}$$

The formula is derived from the fact that the resistance that must be overcome by the sound wave in the interior of the porous layer is larger

[3] Zwicker, C., van den Eijk, J. and Kosten, C. W., *Physica* (*Utrecht*), **VIII** (1941) 469.

than outside; the increased resistance is attributable to the narrowing of the cross-section of the pores and the correspondingly increased velocity. If the resistance of the absorber equals that in the free field, no reflection takes place; one says, in this case, that the absorbing layer is optimally matched. The more these resistances differ from one another, that is, the greater their 'mismatch', the stronger will be the sound reflection from the material (see Section IV. 2.2).

For small values of $\sigma/\sqrt{\chi}$, one gets for normal incidence:

$$\alpha_{(0)}=4\,\sigma/\sqrt{\chi} \tag{6.4b}$$

for diffuse incident sound we get about twice that value:

$$\alpha_{\mathrm{m}}=8\,\sigma/\sqrt{\chi} \tag{6.4c}$$

This formula states that, with a structure factor of $\chi=4$, it would be enough to assume a porosity factor of only $\sigma=0{\cdot}005$ in order to explain the absorption coefficient of 0·02 measured in a reverberation room, as shown in data set (a). Such a small porosity cannot even be seen with the naked eye.

For most masonry building materials (bricks or blocks), the porosity factors described in the literature are higher. For sandstone, values between 0·02 and 0·06 are reported, and for bricks, values between 0·25 and 0·30. Even if it is questionable whether these are the acoustically effective porosities, it is known that one can blow out a candle through a brick tile, and this is in accord with the absorption coefficients measured for an unplastered brick wall:

$$\alpha_{\mathrm{S}}=0{\cdot}15 \quad 0{\cdot}19 \quad 0{\cdot}21 \quad 0{\cdot}28 \quad 0{\cdot}38 \quad 0{\cdot}46 \tag{d}$$

According to eqn. (6.4), with $\sigma=0{\cdot}25$ and $\chi=4$, one would expect for normal incidence a sound absorption coefficient of $\alpha_{(0)}=0{\cdot}5$ for all frequencies, and for diffuse incidence an even higher coefficient.

Until now, we have not taken into account the most influential quantity in the acoustical behavior of a porous material, namely, the friction inside the pores. This can be approximated quantitatively with a static experiment, by measuring the pressure difference Δp necessary to force a flow with velocity v through the test sample. In Section IV. 8.4 and Fig. 8.9 an apparatus for this kind of measurement is described.

The quotient $\Delta p/v$ is called the 'flow-resistance'. However, this quantity is not a constant of the material, because it depends on the thickness d of the test sample. Only the specific flow resistance

$$\Xi = \frac{\Delta p}{v\,d} \tag{6.5}$$

is a material constant. The dimension of this quantity in the CGS system is $\mathrm{g\,cm^{-3}s^{-1}}$; it is often called Rayl $\mathrm{cm^{-1}}$, in honor of Lord Rayleigh. The Rayl is thus the unit for all p/v ratios, such as the characteristic impedance, the wall impedance, the flow resistance, etc. (Recent national and international standards have dropped the use of the Rayl; the current SI unit has no name.)

One can rightly assume that this quantity also depends on the absolute structure of the porous material; but it is independent of both the structure factor χ and the porosity factor σ.

One has only to imagine a structure (see Part IV, Fig. 8.1, right) scaled down in the ratio $m{:}1$. This process does not change χ or σ, but it undoubtedly increases the flow resistance by the factor m^2, as shown in Section IV. 8.4.

It is obvious that the friction offers a higher resistance to the sound entering the material than only the inner inertia determined by the increase of velocity.

If we have a pressure wave with sinusoidal time dependence, i.e. a pure tone of frequency f, then the inertial resistance inside the porous material, related to the exterior particle velocity, amounts to

$$\frac{\Delta p}{v} = \frac{\chi}{\sigma}\rho\, d\, 2\pi f \tag{6.6a}$$

(ρ = density of air).

The necessary alternating pressure differential Δp must be higher, the faster the masses of air are moved back and forth, i.e. the higher the frequency. Adding the effect of the flow resistance to the inertial resistance, and taking into account the temporal relation of these effects acting at different instants, that is, the phase difference of 90°, one gets (see Section IV. 8.5, Volume 2):

$$\frac{\Delta p}{v} = \sqrt{\left(2\pi f \rho \frac{\chi}{\sigma} d\right)^2 + (\Xi\, d)^2} \tag{6.6b}$$

At very high frequencies, the friction plays a minor role. Since eqns. (6.4) were derived without consideration of the friction, they represent limiting values for high frequencies. On the other hand, most brick and stone walls have such a high flow resistance that the inertia can be neglected compared to the flow resistance.

At lower frequencies, the inertial resistance becomes smaller and smaller. Therefore, the flow resistance and the compression of the gas inside the absorber become primarily responsible for its absorption properties. Theory (Section IV. 8.8) shows that then the absorption coefficient is determined by the dimensionless quantity

$$\nu = 2\pi f \rho \kappa \sigma / \Xi \tag{6.7a}$$

where $\kappa = 1{\cdot}4$ (the ratio between the specific heat capacities of air), and where the frequency and the specific flow resistance enter as a ratio. Thereby, one gets for normal incidence the simple equation:

$$\alpha_{(0)} = 1 \Big/\!\!\Big/ \left[\frac{1}{2} + \frac{1}{2\sqrt{2\nu}}\right] \tag{6.7b}$$

The absorption coefficient for diffuse incidence of sound is derived by using the Paris equation (see Section II. 1.4, eqn. (1.30)); it is approximately

$$\alpha_{\mathrm{m}} = 5{\cdot}6\sqrt{\nu}[1 - 0{\cdot}7\sqrt{\nu}\ln(1/\nu)] \tag{6.7c}$$

In Part IV, Fig. 8.21 shows the theoretically derived frequency dependence of the average absorption coefficient, where the frequency is expressed by the dimensionless parameter ν. Using $\rho = 0{\cdot}0012\,\mathrm{g\,cm^{-3}}$ for air, this parameter has the value $\nu = 0{\cdot}01\ \sigma f / \Xi$, with f in Hertz and Ξ in Rayl cm^{-1}.

One gets the values of absorption coefficient in data set (d) with good agreement if one sets $\Xi/\sigma = 2500$ Rayl cm^{-1}, that is, with $\sigma = 0{\cdot}3$, $\Xi = 750$ Rayl cm^{-1}:

$$\alpha_{\mathrm{S}} = 0{\cdot}11 \quad 0{\cdot}15 \quad 0{\cdot}20 \quad 0{\cdot}27 \quad 0{\cdot}35 \quad 0{\cdot}45 \tag{e}$$

Because the flow resistance of pure brick is much higher, according to Raisch (see Section IV. 8.4) even as much as 130 000 Rayl cm^{-1}, one must assume that in this case, it is not the bricks themselves but the mortar joints that are responsible for admitting the sound; the flow resistance is decreased by the ratio of the mortar joint area to the total area. Such low flow resistance is plausible because of the large pores in the mortar joints, expecially when one considers that the bricks will later be plastered.

This corresponds to the results of later experiments carried out in the

Institute for Technical Acoustics (ITA) of the Technical University, Berlin; these tests showed that the absorption coefficient of brick walls with well-filled mortar joints is scarcely higher than 0·05. The absorption of unplastered brick walls depends very much on the skill of the bricklayers; and here one faces the paradox that careless work gives better acoustical results, if one is interested in achieving high absorption. Reverberation time calculations cannot, of course, be based on the results of accidentally poor workmanship.

Building stones made of pumice or brick chips are highly porous due to the fabrication process. They are, however, so unattractive that they cannot often be used for an unplastered wall inside a building. However, one could consider improving the surface appearance by grinding, painting or glazing it without essentially changing the acoustical efficiency.

In contrast to bricks, building stones made of pumice or brick chips are coarse-grained; therefore, their flow resistance is low. So long as the porous layer is assumed to be 'infinitely' thick, this is of no importance; the equations cited so far are valid under this condition. But if the admitted sound passes through and is reflected back into the porous layer, which happens, for example, with a layer of absorbing material in front of a rigid wall, then the considerations so far are valid only if the initially admitted energy does not return to the surface. If one requires, for example, that the pressure amplitude of a sound wave, entering a porous layer of thickness d, be reduced to 4% of the initial value after total reflection at the back (i.e. after travelling a distance $2d$), calculation[4] shows that this requirement can be fulfilled for interestingly large values of Ξ/f if

$$d > 900/\sqrt{\Xi \sigma f} \tag{6.8}$$

(d in cm, Ξ in Rayl cm^{-1}, f in Hz).

To meet this requirement with a thin layer, it is necessary to have a large flow resistance. The thickness of the layer can be less for the absorption of high frequencies than for low frequencies. On the other hand, in order to improve the 'impedance match', which allows the incident sound to penetrate the surface, it is preferable if the specific flow resistance is as low as possible. It follows that neither very large nor very small Ξ-values are good; instead, there is an optimal range. This op-

[4] Cremer, L., *Elektr. Nachr. Techn.*, **12** (1953) 333. See also Section IV. 8.6.

timum obviously depends on the thickness of the layer. For thicker layers, smaller values of Ξ can be afforded, from the standpoint of impedance matching. Therefore, it is better to look for an optimal value for the product (Ξd) meaning the flow resistance R of the whole layer. But, as we will see in Section IV. 8.6, this optimal value is not unique. It depends on whether one wants a larger absorption at higher or at lower frequencies. In any case, the range

$$2\,\rho c < R < 4\,\rho c \tag{6.9}$$

or about

$$80 < R < 160\,\text{Rayl}$$

can be recommended as useful.

That the flow resistance of pumice and brick chip walls is very low is best shown by the low sound insulation of such walls when they are not plastered.[5]

There is no reason that high sound absorption on the front face of a simple homogeneous wall (in other words, large α) should necessarily lead to high transmissibility (in other words, high τ). As a matter of principle, it is quite possible to give a porous wall with the usual thickness of 24 cm such a high flow resistance that, on the one hand, there is sufficiently good impedance match at the front surface to get an absorption coefficient as high as 0·5, but, on the other hand, there is enough dissipation in the interior that the energy leaving the back surface amounts to less than $\tau = 10^{-5}$ of the incident energy; therefore, the sound insulation may exceed 50 dB.

In practice, nobody makes use of this possibility; a distinction is always made between the sound-isolating wall, which belongs to the basic building structure, and its absorptive property, which belongs to the interior decoration. For acoustical planning in general, it is surely an advantage that the different tasks of room acoustics (good hearing) and building acoustics (noise isolation) are assigned to different elements of the building.

Since porous layers are usually mounted either directly against, or with only a small separation from, a rigid, sound-reflecting wall, we will be concerned with that configuration here. Accordingly, we will assume that the wall behind the porous layer is always totally reflecting.

[5] See, for example, Kristen, Th. and Brandt, H., *Fortschritte und Forschung im Bauwesen*, **D2** (1952) 16.

It can easily be seen that a return of the incident wave to the room, after penetrating the porous layer, leads to a degradation of the absorption. Especially if the flow resistance is very low, the wave reflected back into the room consists mainly of sound reflected at the back of the layer which may be weakened only slightly.

Finite thickness of the porous layer does not necessarily lead to a reduction of the absorption efficiency. This initially surprising result comes from the wave nature of sound, which requires us to consider the phases of the superposed entering and reflected waves.

Especially in the immediate vicinity of a rigid wall, there is a standing wave field where the particle velocity goes to zero immediately at the wall and also at the so-called nodes, located at distances of 1, 2, 3, etc., half-wavelengths in front of the wall. Between these nodes, at the antinodes, the sound velocity takes on maximum values, called 'bellies' in German (see Section IV. 2.1).

Because the sound absorption of porous layers is based on the relative motion between the air inside the pores and the skeleton material, it can be understood that a thin porous layer in the immediate vicinity of the wall, where the particle velocity is very small, has only a minor absorptive effect. On the other hand, at a distance of a quarter wavelength in front of the wall, where the particle velocity is at its peak, there is significantly higher dissipation, and so also greater sound absorption.

Such interferences are also developed within the porous layer, but according to the sound velocity inside the material, which is lowered on account of the structural factor:

$$c' = c/\sqrt{\chi} \tag{6.10}$$

Here, also, calculation shows (see Section IV. 8.7) that the layer has maximum absorption if its thickness d equals one-quarter wavelength of the sound within the layer, or an odd multiple thereof. Thus, there are absorption maxima at the frequencies

$$f_n = (2n-1)c/(4d\sqrt{\chi}); \qquad n = 1, 2, 3 \ldots \tag{6.11a}$$

The first maximum is to be expected at the frequency

$$f_1 = c/(4d\sqrt{\chi}) \tag{6.11b}$$

The frequencies of the absorption maxima, therefore, give a clue to the magnitude of the structure factor. On the other hand, between these frequencies, at

$$f_n' = nc/(2d\sqrt{\chi}) \tag{6.11c}$$

there are minima of the absorption coefficient.

The appearance of the layer thickness d in this equation points out clearly that the absorption coefficient is not only a matter of the material, including its structure, but it also depends on the dimensions. These dimensions should be expressed in terms of their relation to the wavelength ($\lambda' = c'/f$), in accordance with the general modelling rule of acoustics.

As an example, Fig. 8.18 in Part IV shows the absorption coefficient of an 11 cm pumice block wall mounted against a rigid surface, as measured in the reverberation room.

Since, in general, porous layers with such great thicknesses are not available (they are mostly only 2–4 cm), and since structure factors are very rarely greater than unity, the first maximum of absorption of a porous layer usually lies in the high-frequency range, above 2000 Hz. The second absorption maximum then occurs above the frequency range of interest.

In discussing unplastered, porous brick walls, we have learned all the essential physical characteristics of a porous absorptive layer with a rigid skeleton. Since in these cases, the purely geometrical properties are of primary importance, our conclusions are also valid for quite different room finishes, of such different nature as porous plaster, carpets, draperies, soft boards or wood-wool acoustic boards, mineral and vegetable fiber layers, and, lately, plastic foams.

Nowadays, the porosity of a masonry wall is seldom used for sound absorption directly. Usually such walls are plastered, so that sound cannot penetrate; if a sound-absorbing wall treatment should be desirable afterwards, it is mounted on the plaster. From the acoustical point of view, a large part of the possible effect is thereby given away for nothing. In such cases, the porous layer, which is mostly only 2 cm thick, achieves its maximum effect at much higher frequencies than if the same layer were mounted directly on the porous, unplastered masonry wall.

In the last case, the surface absorptive layer could even be much thinner. It only needs to present a good appearance without essentially hampering the penetration of sound. This, however, entails the possibility of a direct air flow, and may thus lead to an increased deposit of dust.

This is especially undesirable if, because of different flow resistances or rates of heat conduction, the dust layer is uneven; after a while, the underlying construction becomes visible. Also, requirements for heat insulation may necessitate plastering, especially at exterior walls.

One should consider, however, before plastering a wall, whether this step can be omitted altogether because a sound-absorbing covering may be necessary later, and in that case the sound absorption property of the wall itself can be used to good advantage.

In many cases, for example, in industrial halls, spraying with paint is enough. Usually this leaves the pores open, while painting with a brush, especially with oil paint, will close the pores. Also, a thin porous surface layer can be made by broom application of a coat of coarse-grain plaster.

If one exploits the porosity of a masonry wall for sound absorption, one must then consider that its sound insulation will be relatively low, unless one closes the back of the wall with plaster.

Since in many cases a plastered but sound-absorbing masonry wall is desired, plasters have been developed by different manufacturers that are supposedly sufficiently porous to have significant sound-absorbing properties and are even transparent enough to take advantage of the porosity of the wall behind. Special mention is made here of three different products which were developed some time ago and which demonstrate the existing possibilities.

BASF succeeded in developing a porous cement plaster, named 'Iporit-Putz', by adding a foaming ingredient in the mixer. In the reverberation chamber of the Technical University of Braunschweig, A. Eisenberg measured the following absorption coefficients for a 1·5 cm thick layer applied to a rigid wall:

$$\alpha_S = 0{\cdot}08 \quad 0{\cdot}10 \quad 0{\cdot}20 \quad 0{\cdot}29 \quad 0{\cdot}39 \quad — \tag{f}$$

Obviously, the porosity of the underlying structure does not add to the absorption of this 1·5 cm layer, because the frequency dependence corresponds well with eqn. (6.7c) for large layer thicknesses, if one assumes a proper value of 2500 Rayl cm^{-1} for Ξ/σ. This shows that the ratio of flow resistance to porosity factor exceeds the optimum.

However, in order to achieve the absorption coefficients mentioned above, it takes an expert plasterer who mixes the ingredients in the correct proportion and takes the mixture from the mixer at just the right moment. Here, again, one can see how the desired acoustical effect depends on the expert's work, as was mentioned earlier in regard to the construction of a masonry wall.[6]

Porous plasters with cement as a binding medium have the great advantage that they offer a strong, rigid surface. They can even be used in places where they may be touched by people or struck by objects. Finally, they are scarcely more expensive than normal plasters.

By contrast, a rather fragile plaster should be mentioned which consists of asbestos fibers in a binding medium; it was developed as a sprayable plaster in the United Kingdom. However, it is substantially more expensive and, because of its soft skeleton material, it is easily damaged, even after it has dried. On the other hand, however, the manufacturing process guarantees a well-defined structure whose absorption coefficient is very high, even in thin applications of the material. Figure 6.1 shows the absorption coefficients measured for a 2 cm layer in

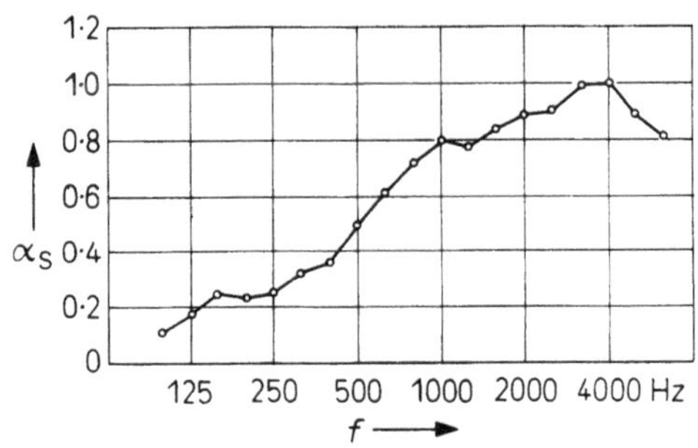

Fig. 6.1. Absorption coefficient for a 2 cm sprayed asbestos layer with a weight of 3·2 kgm^{-2}, measured in a reverberant room.

the reverberation room of the Institute for Technical Acoustics (ITA), Berlin. The steep increase with frequency, as well as the attainment of maximum absorption at 4000 Hz, speak for a favorable value of Ξ/σ and a structure factor that hardly differs from 1.

A third example will be mentioned later which offers a somewhat lower α-value, but at lower cost, namely, the Novolan plaster, consisting of rockwool in a binding medium, that was developed by Grünzweig &

[6] Because one can seldom count on such expert workmanship, the results of acoustic plasters applied in the field are often unreliable, and many acoustical consultants in North America routinely advise against the use of these plasters.

Hartmann. We will consider its application in an arrangement where its higher flow resistance is put to especially good use.

In this connection, we have found from our experience as consultants to architects that high absorption values are normally not as important as the workmanship (such as the quality of the plastering) and, last but not least, the surface appearance. In the last respect, a sound-absorbing plaster that can be used everywhere has not yet been developed. However, without underestimating the purely technological difficulties, it can be said physically that a sound-absorptive structure with a porosity factor corresponding to an optimal flow resistance but with a smooth surface does not entail requirements that intrinsically contradict each other. Certainly, architects would prefer a solution with such a sound-absorbing plaster for reducing the indoor noise in manufacturing halls, offices, corridors and staircases, because it can be adjusted to every kind of shape and can also be colored to one's desire.

Any particular shade can be achieved by color additives to all the kinds of plaster discussed so far. Furthermore, as mentioned before, light spraying with color is permissible, but not normal painting.

A thicker layer of plaster shows higher absorption coefficients at low frequencies if the flow resistance is properly chosen; for example, the curve in Fig. 6.1 would be shifted one octave toward lower frequencies if the thickness of the layer were doubled.

A still more uniform frequency characteristic can be achieved by applying the plaster not immediately on a rigid surface, but at as large a distance as possible; for instance, an expanded metal lath can be hung below a bare floor slab and plastered with the absorptive plaster. However, care must then be taken that the space between slab and plaster is not colder than the room below, because otherwise air convection will cause an uneven dust layer and corresponding ugly patterns in the ceiling. This leads also to acoustical problems, not because the dust closes the pores (dust nearly always is porous in itself!) but because the ceiling must then be painted, often without considering the loss of its acoustical effectiveness.

The fact that a thin porous layer applied directly to a rigid surface absorbs sound significantly only at very high frequencies is typical, especially for carpets. Even for a velour carpet, von Braunmühl got only the following absorption coefficients:

$$\alpha_S = 0{\cdot}05 \quad 0{\cdot}06 \quad 0{\cdot}10 \quad 0{\cdot}24 \quad 0{\cdot}42 \quad 0{\cdot}60 \qquad \text{(g)}$$

The fact that, in spite of this, the subjectively perceived acoustical change is remarkable when one carpets a reverberant foyer depends on the change of 'sound color', to which we are more sensitive than to an absolute change of reverberation and sound level. A room damped only by thick carpets sounds 'boomy'.

On the other hand, even thinner porous layers, like velvet 'wallpaper' or flocking, are almost without effect. But even this can be advantageous. For example, in the reconstruction of the Cuvilliés-Theater and the National Theater in Munich, a textile covering for the walls in both halls was desired for architectural reasons, but was undesirable from an acoustical standpoint. The choice of a thin damask fabric and a silk wall covering, respectively, cemented directly to the walls, avoided the loss of the desired clear sound color in these rooms. Similarly, in the Herkulessaal, Munich, some of the very valuable tapestries were covered on the back with a relatively heavy impermeable plastic sheet, in order to decrease their sound-absorbing properties. These tapestries have natural folds that always trap a certain air cushion behind them, even if mounted directly on the wall.

But in general, one wants to achieve with draperies (the second 'household remedy' of acoustical furnishing!) the greatest and most frequency-independent sound absorption possible. We have already shown that the maximum efficiency can be expected if the material is located at a point of maximum particle velocity, and that this will be found where the distance from the wall is an uneven multiple of a quarter-wavelength:

$$d=(2n-1)\frac{\lambda}{4}, \qquad n=1,\ 2,\ 3,\ \ldots \tag{6.12a}$$

This equation is valid only for normal incidence. If the sound is incident with an angle ϑ, the maximum particle velocities are found further apart:

$$d=(2n-1)\frac{\lambda}{4\cos\vartheta} \tag{6.12b}$$

For diffuse sound incidence these maxima are averaged, so that one can say, roughly, that above the frequency

$$f_1=\frac{c}{4d}=\frac{8500}{d}\ \text{(Hz)} \tag{6.12c}$$

a uniform absorption is reached; here d is measured in cm.

Furthermore, draperies are usually hung in pleats, so that another averaging of the maxima is thereby gained. Hanging the draperies in pleats increases the absorption efficiency, not so much by an increase in the absorbing surface area, but by an increase of the effective flow resistance.

The optimal flow resistance R for free-hanging curtains is lower than for homogeneous layers mounted in front of rigid walls, namely

$$\rho c < R < 2\,\rho c \tag{6.13a}$$

or

$$40 < R < 80 \text{ Rayl} \tag{6.13b}$$

This can be explained by the fact that there is no part of the material located in a velocity node, which would make it an ineffective absorber.

The skeleton of a porous body cannot always be so rigid that it is unmoved by the forces of friction and the direct-acting pressure fluctuations.[7] If, for example, a free-hanging textile is, on the one hand, very dense and thus has a relatively high flow resistance, and, on the other hand, is very light, then the pressure difference acting on the fabric can more easily move the mass of the material back and forth at low frequencies than it can force the air through the pores.

The inertia of the material does not act as an additional obstacle, like the air within the pores, but as a parallel possibility for giving way to the air motion. Because the relative motion inside the pores is thereby reduced, the dissipation of sound energy decreases at low frequencies in the case of light textiles.

This effect of 'short circuiting' the flow resistance of the textile material by a 'parallel inertia resistance' can be seen most easily in the case of free-hanging curtains, such as a stage drapery, the curtains for visual partitioning of rooms, or curtains in recording studios that are used only as a sound absorber.

In the latter applications, the absorption coefficient α is not of primary interest, but only that portion of it that represents sound energy transformation into heat, that is, the dissipation coefficient δ. The energy passing completely through the material, corresponding to the transmission coefficient τ, remains in the room and therefore has no effect as far as the sound absorption of room energy is concerned (see eqns. (I.6.1)–(I.6.4)).

The measurement of free-hanging textile materials in a reverberation

[7] Wintergerst, E., *Schalltechnik*, **6** (1933) 5.

room yields the dissipation coefficient δ_S; in eqn. (5.1), the test area S must be counted twice because it is struck by sound waves from both sides:

$$\delta_S = \frac{0{\cdot}163V}{2S}\left(\frac{1}{T_\Delta} - \frac{1}{T_0}\right) \tag{6.14}$$

The increment of equivalent absorption area added to a room by a piece of free-hanging fabric whose (one-side) area is S is thus $2S\delta_S$.

By calculation, one finds that the dissipation factor for a material with a rigid skeleton (i.e. a fabric that is heavy enough) and a flow resistance R, for an average angle of incidence of 45°, is (see Section IV. 7.6):

$$\delta_{(45^\circ)} = 1\bigg/\left[1 + \frac{1}{2}\left(\frac{R}{2\sqrt{2}\rho c} + \frac{2\sqrt{2}\rho c}{R}\right)\right] \tag{6.15}$$

The optimal matching is, therefore, at

$$R = 2\sqrt{2}\rho c \approx 120 \text{ Rayl} \tag{6.16}$$

and it leads to a maximum value for the dissipation factor of

$$\delta_{max} = 0{\cdot}5 \tag{6.17}$$

The equations show, furthermore, that in this case one-quarter of the incident energy is reflected and one-quarter passes through. If the flow resistance is smaller, the reflected energy is less but the transmitted energy is greater, and vice versa. It cannot be expected for a thin layer that all the energy enters but none is transmitted, that is, that δ could ever be equal to one.

Occasionally, one would like a textile to act only as a visual, but not as an acoustical, boundary, so that the transmitted sound energy, upon returning, reaches the room again practically undiminished. In this case, one must use curtains with pores as large as possible, such as marquisette, screen tulle and the like. With such very low flow resistance, eqn. (6.15) can be simplified:

$$\delta_{(45^\circ)} = 0{\cdot}7\frac{R}{\rho c} = 0{\cdot}017\,R \tag{6.18}$$

with R and ρc in cgs Rayls.

We must now account for the effect of the motion of the textile mass by an equation, more to clarify the tendency of this influence than for quantitative calculation. It involves the addition of a term that decreases both with increasing frequency and increasing mass per unit area M (Part IV, eqn. (7.50a)):

$$\delta_{(45^\circ)} = 1\Big/\left[1 + \frac{1}{2}\left(\frac{R}{2\sqrt{2}\,\rho c} + \frac{2\sqrt{2}\,\rho c}{R}\right) + \frac{\sqrt{2}\,R\rho c}{(2\pi f M)^2}\right] \tag{6.19}$$

The motion of the porous layer is not always a disadvantage; there are also exceptions here, as in the case of thin layers in front of a rigid wall. Here, the mass of the textile, together with the air cushion behind it, can constitute a resonant system with a low resonance frequency; it absorbs sound especially well at this resonance frequency.

We will discuss this possibility of sound absorption in detail in connection with perforated plates and non-porous plates.

Even for a homogeneous porous layer before a rigid wall, the motion of the skeleton can play a major role, especially if its surface is non-porous, as is the case with some plastic membranes and rubber textiles.[8,9] However, it would lead us too far astray to go into the details of all these possibilities here, since it is hardly feasible to give general rules of application; in these cases, one must calculate or measure each individual configuration.

On the other hand, we cannot conclude this survey of porous layers without describing at least briefly the most essential properties of mats and plates that have been specially developed as porous layers.

We mention first the oldest materials, namely, soft fiber boards pressed from wood shavings together with a binding medium. They have the advantage of being very light and are self-supporting, so that they can be easily mounted on ceilings and walls by dry installation. Their disadvantage is a relatively high specific flow resistance, and therefore poor impedance matching. Plates with a thickness of 1–2 cm can be calculated according to eqn. (6.7) under the assumption that the reflections from the back surface need not be taken into consideration.

If we find, in spite of this, a significant improvement in absorption at low frequencies when these boards are mounted not directly against the ceiling but on a lattice, so as to form an air cushion with the wall, this depends again on the motion of the board, not on its porosity.

Another common improvement can only be mentioned here; it consists

---

[8] Zwicker, C. and Kosten, C. W., *Sound Absorbing Materials*, Elsevier, Amsterdam, 1949, p. 116ff.

[9] Venzke, G., *Acustica*, **8** (1958) 295.

of spot-drilling or grooving the board. By this means the entrance of sound into the porous inner layer is improved. It is surprising that these holes and grooves, whose projected area is sometimes only 10% of the entire area, are sufficient to yield relatively high absorption coefficients for the board, even when the remaining area is covered with a completely impermeable and therefore easily cleaned layer. At certain frequencies, very significant improvements can be achieved by this means; for methodical reasons these will be discussed in the following section. However, as an example, Fig. 6.2 shows the frequency dependence of the

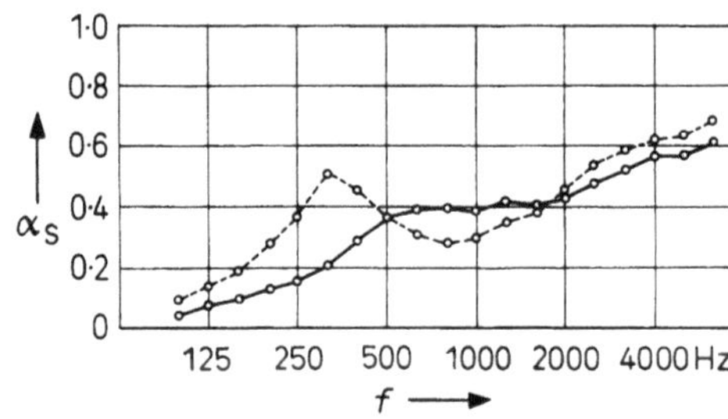

Fig. 6.2. Absorption coefficient for a 1·2 cm soft board, measured in a reverberant room. ——— Immediately in front of a rigid wall; - - - at 2·5 cm distance from the rigid wall.

absorption coefficient for a soft board with checkered grooves, in one case glued directly to the wall and in the second case supported at 2·5 cm distance from the wall. In a room where it is important not to change the tone color and which contains no other elements for absorbing the low frequencies, the additional costs of providing an air cushion behind the board are worthwhile.

Secondly, mention should be made of the soft boards pressed from excelsior or wood-wool. As a matter of fact, these materials generally belong to the basic building construction, for example, when they are used as permanent forms for concrete; in most applications they are plastered. In this case, everything that was said above about the covering or the porous plastering of masonry walls applies here as well. Since the 1930s the firm of Heraklith AG has manufactured, for its sound-absorbing qualities, a wood-wool soft board pressed with a magnesite binding medium, with a carefully finished surface. It is called 'Herakustik' plate and can be used for covering the walls directly without plastering. Although in the case of soft boards, we found the specific flow resistance to be in general well above the optimal value, it is somewhat too low in the case of these coarse-grain structures, especially for thin sheets. This can be clearly seen from the appearance of resonance peaks. Figure 6.3 gives the absorption coefficient for a 2·5 cm wood-wool slab in front of a non-porous wall at distances of 0, 3 and 5 cm.

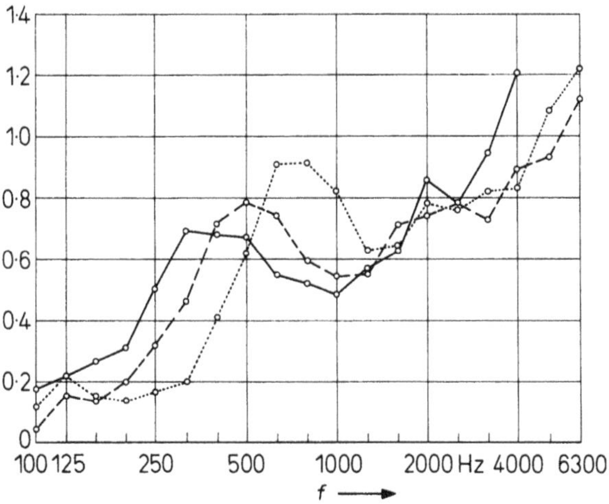

Fig. 6.3. Absorption coefficient (ordinate) for wood-wool plates of different thickness, measured in a reverberant room. ..... Plates of 2·5 cm thickness in front of a non-porous wall; – – – plates of 2·5 cm thickness at 3 cm distance in front of a rigid wall; ——— plates of 5 cm thickness in front of a non-porous wall.

The location of the first peak below 1000 Hz corresponds to a structure factor of around $\chi = 3$.

The mats and plates made from organic and mineral fibers are the best fitted for the task of sound absorption and are also the cheapest.

The latter, best-known in Germany under the names of basalt wool, glass wool or rock wool, have the advantage of non-flammability and a high melting temperature. Their disadvantage, namely that they may cause severe itching for the people who handle them, requires certain care in their installation.

Statements suggesting that these fibers, which are for the most part strongly interwoven, may eventually sift out and be suspended as very fine particles in the air, where they become detrimental to the health, seem to us to be unfounded; the same may be said of the occasional claim that glass fibers tend to disintegrate into dust with the passing of time.

In any event, the first possibility can be avoided by covering the material with a fabric having fine pores and not too high a flow resistance, or with a thin very light impermeable membrane, not to mention the coverings otherwise necessary for decorative reasons, which will be discussed later.

The detachment of fiber particles from mineral fiber plates pressed with a binding medium is almost impossible; naturally they are more expensive than the loose mats or mats stitched on paper, but they are much more suitable for routine construction and installation.

When fire-resistance plays only a minor role, the market offers for the same purpose sound-absorptive mats made from seaweed, coconut fiber or other organic fibers, whose flammability is reduced to a large degree by chemical impregnation.

Finally, there are mats made of metal or plastic fibers; however, they have larger pores and the flow resistance is therefore far below the optimal value.

The acoustical effectiveness of a mineral fiber mat only 5 cm thick with optimal flow resistance is shown in Fig. 6.4 for a bonded mineral-wool plate. The frequency dependence shown there represents the result of an international round robin test carried out by Technical Commission 43 of ISO, who paid strict attention to their measurement directives.[10] The

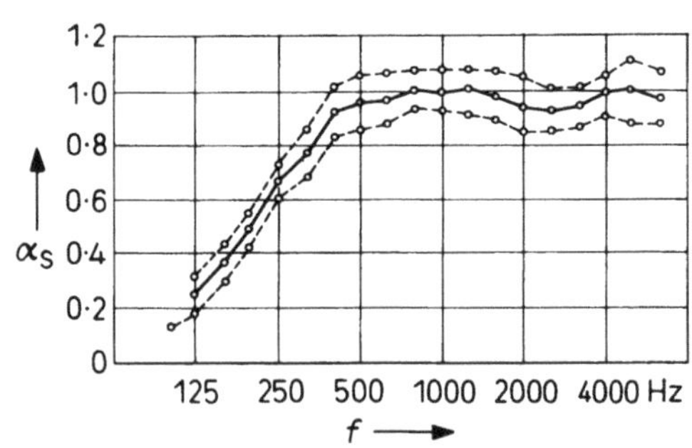

Fig. 6.4. Absorption coefficient for a mineral wool plate (Sillan SP 100), 5 cm thick, measured in different reverberation rooms. ——— Average value for the individual measurements; – – – standard deviations for the individual measurements.

[10] This comparative study was very carefully analysed by C. W. Kosten; see *Proc. 3rd. ICA, Stuttgart*, 1959, Vol. II, Elsevier, Amsterdam, 1961; also *Acustica*, **10** (1960) 400.

solid line gives the average of the α-values measured in the various reverberation rooms, the dashed lines above and below represent the standard deviations:

$$\sigma_\alpha = \pm\sqrt{\frac{\sum^{n}(\alpha-\alpha_n)^2}{n-1}} \tag{6.20}$$

A total of $n=8$ institutes participated in this comparison.

(We must mention again that values of α_S greater than 1 appear in the test results because of the assumptions and approximations of Sabine's formula. This is not an error in measurement and therefore such values may be used later in precalculating the reverberation time.)

Finally, we mention as the last group (because it was developed last) the rubber foams. Their porosity consists mainly of air bubbles separated by thin skins.[11] By mechanically breaking these skins after the foam is formed, more or less porous layers (in the acoustical sense) are created. Thereby it can happen that the flow resistance remains large, so that a layer of only 1 cm thickness exhibits a frequency dependence typical of 'infinite' layer thickness, according to eqn. (6.7c). Figure 6.5 shows such an example, in which the curve according to eqn. (6.7c) is drawn for a value of $\Xi/\sigma = 4000$ Rayl cm^{-1}. There are, however, frequency dependence curves, even for the same foam thickness and installation in front of a rigid wall, that indicate significantly lower values of flow

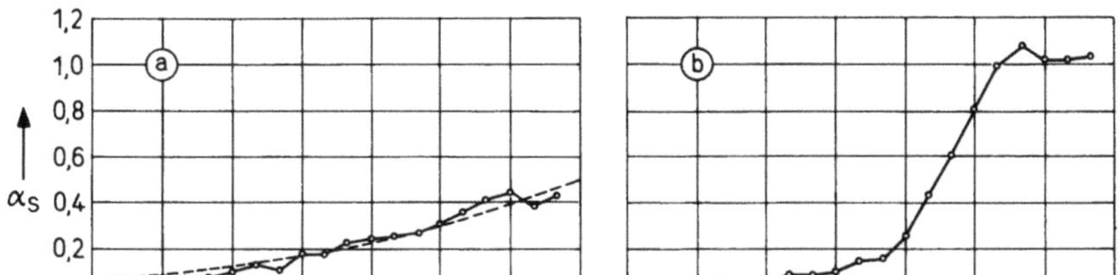

Fig. 6.5. Absorption coefficient for a plastic foam sheet with a thickness of 1 cm, measured in a reverberation room. (a)——Foam with high flow resistance (– – – corresponds to the calculation); (b) foam with low flow resistance.

[11] Venzke, G., *Acustica*, **8** (1958) 295.

resistance, and yet the material in both cases had the same specification by the manufacturer. This comparison suggests the wide range of development possibilities; but it also warns of the necessity of specifying porous layers not only by their chemical properties but by their acoustical characteristics as well. In this respect, especially for the specific flow resistance, the desired value *must* be guaranteed within certain tolerances.

Recently, attempts have been made to develop closed-cell porous foam materials that are sound absorptive,[12] since there are some applications where materials with open pores cannot be used. One thinks of operating theaters, kitchens, wet rooms and so on, in which until now we have had to make do with porous materials enclosed in light, and therefore sound transparent, but impermeable plastic membranes. Because there is no relative motion between the air and the structure in materials with closed pores, other mechanisms must be found for transforming the sound energy into heat. For example, losses caused by the deformation of the structure can be used. A material is produced by the firm of E. Freudenberg, consisting of a 12 mm thick closed-cell polyethylene foam with inclusions of small lead balls ('beebees') and polyurethane foam, that gives absorption coefficients as shown in Fig. 6.6. The lead balls create

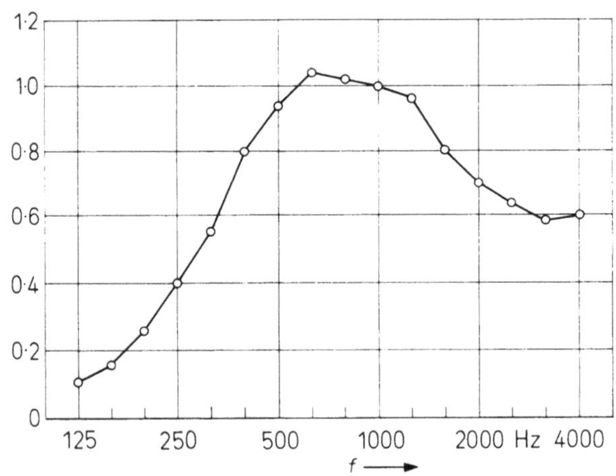

Fig. 6.6. Absorption coefficient (ordinate) for a layer of closed-cell plastic foam with small lead balls included ('*Frelen*').

[12] Kurtze, G., *Akustische Anwendungsmöglichkeiten von geschlossenzelligem Polyäthylenschaum*, Vortrag vor der Schweizerischen Akustischen Gesellschaft, Zürich, 1976.

a large relative motion within the foam; this motion leads to very significant sound absorption at mid-frequencies—and leads us directly to the following section on 'Medium-frequency Absorbers'.

II.6.4 Medium-frequency Absorbers

The porous absorbers discussed so far, however different their skeleton material, exhibit in common the property that their absorption coefficients decrease at low frequencies, and often at medium frequencies. If occasionally there were maxima in the absorption curves, these were followed at higher frequencies by only small dips and other maxima, so one can say that the systems considered so far were, in short, high-frequency absorbers.

Now, because of unavoidable absorption at the walls and the dissipation of energy during sound propagation, there is already a tendency for rooms to absorb sound more at high frequencies than at low frequencies. This is especially true in an auditorium, where the porous clothing of the audience absorbs high-frequency sound very effectively (see Section II. 6.6). Therefore, in auditoriums, theaters, and particularly in concert halls, there is little need to provide additional absorption at high frequencies. In fact, care must be taken that the reverberation time does not fall below the value determined by the unavoidable absorption at high frequencies.

In this section, we will discuss measures for decreasing the α-values at high frequencies, creating by this means systems that we may call 'medium-frequency absorbers'. In the next section, we go even further to make systems that absorb sound mainly at low frequencies.

The means for changing high-frequency absorbers into medium-frequency absorbers were originally adopted for quite different reasons. We have already called attention to the general disadvantage of all porous layers, namely, that their functionally necessary porous surfaces must not be covered with paint. But in fact the urgency to repaint these surfaces in subsequent renovations is great, since, because of their porous and generally rough surfaces, they invariably collect dust. In an attempt to make it easier to clean and repaint such materials without altogether destroying their porosity, the surface density was increased and holes and grooves were made in the surface to admit the sound. This was tried first for the soft-board plates. The measurements showed that this procedure decreased the sound absorption at high frequencies but not in the medium-frequency range that is of most interest; indeed, in this range the absorption often increased astonishingly.

The same effect can be achieved by covering fiber mats with perforated or slotted plates, thus making the treatment architecturally more acceptable. (The tactic of covering the intrinsically cheap absorptive material with expensive perforated plates, which moreover require additional substructure for support, illustrates very clearly the general rule that the price of a sound-absorbing wall or ceiling treatment is usually determined not by acoustical but by architectural requirements.)

We have already mentioned in Section I. 3.4 the acoustical transparency of perforated plates and explained this effect by the phenomenon of diffraction, which is best described by regarding each small hole as a new sound source. If the holes are spaced closely enough, the concerted action of these small sound sources will generate behind the plate a wave which, as far as directivity is concerned, corresponds to the original incident wave.

One can visualize this process by thinking that the sound wave encounters the perforated plate and moves the air back and forth through the holes. But then the particle velocity inside the holes must be essentially larger than that in front of the plate. The ratio of open area for the perforated sheet, which corresponds to the porosity factor introduced above, is also called σ. The air inside the holes must therefore be accelerated by a factor of $1/\sigma$. The vibrating mass of the air plugs in the holes corresponds to a mass per unit area (of hole) of $\rho\, l_0$ (where l_0 is the length of the hole and ρ is the density of the air) and thus has the same inertia as if the mass were $1/\sigma$-times greater. Therefore, as described in detail in Section IV. 9.1 (Volume 2), the perforated plate acts on the incident wave like a layer with a mass of

$$m=\frac{\rho\, l_0}{\sigma} \tag{6.21a}$$

Since not only the air inside the holes but also close in front and behind is included in the motion on both sides, an 'orifice correction' must be added, so that one gets for the mass layer

$$m=\frac{\rho\,(l_0+2\Delta l)}{\sigma} \tag{6.21b}$$

Although the density of air ($0{\cdot}0012\,\mathrm{g\,cm^{-3}}$ at room temperature) is three orders of magnitude lower than that of any light rigid material, this mass layer in front of the absorptive mat can be quite significant at high frequencies if σ is small enough (see Fig. 6.7).

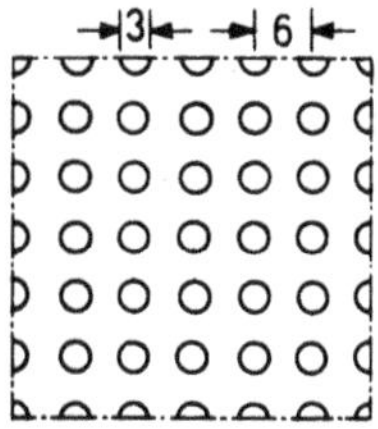

Hard fiber plate

Thickness (l_0):	3·5 mm
Hole radius (a):	1·5 mm
Perforation ratio (σ):	0·20
Mass layer (m):	0·035 kg m^{-2}

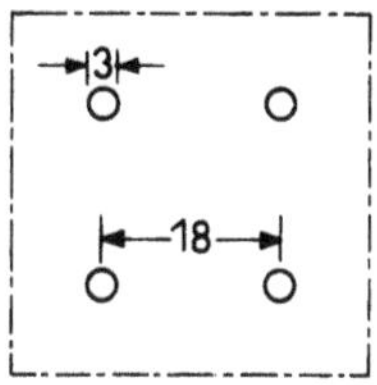

Hard fiber plate

Thickness (l_0):	3·5 mm
Hole radius (a):	1·5 mm
Perforation ratio (σ):	0·02
Mass layer (m):	0·325 kg m^{-2}

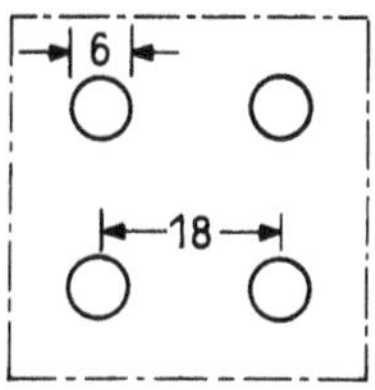

Gypsum board plate

Thickness (l_0):	9·5 mm
Hole radius (a):	3 mm
Perforation ratio (σ):	0·09
Mass layer (m):	0·196 kg m^{-2}

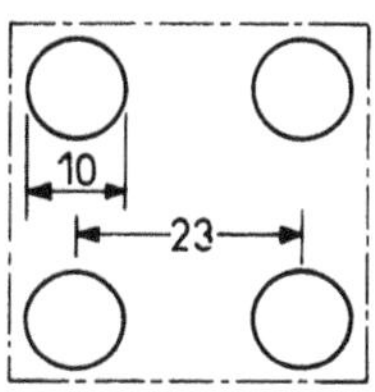

Gypsum board plate

Thickness (l_0):	9·5 mm
Hole radius (a):	5 mm
Perforation ratio (σ):	0·15
Mass layer (m):	0·140 kg m^{-2}

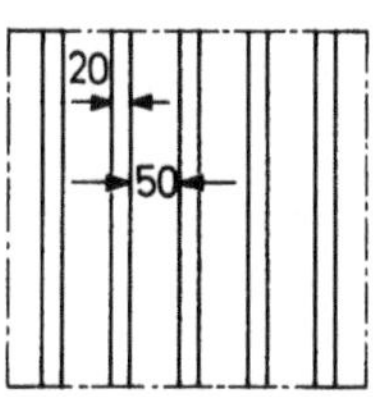

Lath grille

Thickness (l_0):	25 mm
Slit width (b):	20 mm
Distance of slits (B): (center-to-center)	70 mm
Open area ratio (σ):	0·29
Mass layer (m):	0·157 kg m^{-2}

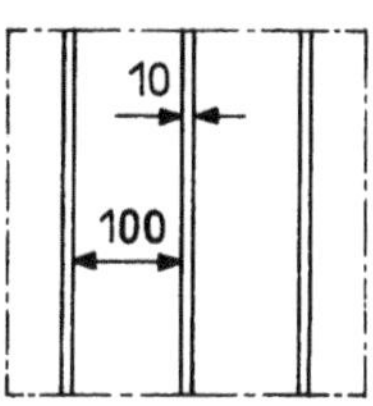

Wood board covering with open slit

Thickness (l_0):	10 mm
Slit width (b):	10 mm
Distance of slits (B):	110 mm
Open area ratio (σ):	0·09
Mass layer (m):	0·320 kg m^{-2}

Fig. 6.7. Examples of commercially available perforated plates.

In the case of circular holes the orifice correction (see Section IV. 9.1) amounts to

$$2\Delta l = 1{\cdot}6a \tag{6.22}$$

where a is the radius of the hole. Because the diameter $2a$ is often as great as the thickness of the plate l_0, this correction is significant. Furthermore, it is obvious that this correction, which depends on constriction of the flow in front of and behind the hole, must diminish if σ approaches its highest value 1, in which case there is no constriction at all. This correction, which is only seldom of importance for commercial perforated plates, may be found in Section IV. 9.1.

The dependence of the orifice correction on the open area ratio is even greater in the case of slits; in this case it is given by the ratio b/B (b = the width of the slit and B = the on-center spacing of the slits). This is because the flow velocity for circular holes expands in two dimensions and therefore decreases inversely with the square of the distance from the center of the hole. In the case of slits, however, only a one-dimensional expansion of the flow is possible, and therefore the flow velocity decreases only inversely with the first power of distance from the center of the slit. Thus the influence of the 'neighborhood' is greater in the latter case.

For the same reason, the orifice correction must change if there is a rigid wall relatively close behind the slits at a distance d, as when one covers a very thin porous layer with a perforated sheet. Figure 9.2 of Part IV shows the orifice correction (for both sides) calculated by Kosten and Smith.[1] This correction is normalized to the slit width b and is plotted against the ratio $1/\sigma = B/b$; the ratio d/b is introduced as a parameter. Clearly, if the bridge between the slits is very large, one must pay attention to the increase of the orifice correction with decreasing distance to the wall.

Using formulae (6.21)–(6.22) with Fig. 9.2 of Part IV, we can calculate the equivalent mass layers. Some results of this calculation for typical plates marketed in Germany are shown in Fig. 6.7.

In general, plates with circular holes are cheaper but, on the other hand, they are less pleasant to look at, especially if viewed from a short distance. Their appearance can cause a certain feeling of dizziness, due to the uncertainty in stereo-optical appearance caused when the two eyes look at different groups of periodic holes; this is even more annoying than in the case of a grating.

[1] Kosten, C. W. and Smith, I. M. A., *Acustica*, **1** (1951) 114.

The slit configuration can be a grille made of many individual bars or slats, which in turn can be veneered with various woods. Therefore such designs are preferred for elegant and formal banqueting rooms. One must not be misled by the relatively large σ (e.g. 0·2) into supposing that a grille, because of the greater slit length l_0, would lead to a much higher equivalent mass than a thin plate with the same (optical) perforation.

If the covered areas are kept dark and not illuminated, the eye sees only the surface of the covering as a whole, even with an optical transparency of 20 or 30%. In principle, the passage for sound remains open even if one cannot see through because the path is 'bent', as shown in Fig. 6.8 (center) for a slat profile developed in Copenhagen.[2] However, the invisibility of the layer in the rear can be achieved only by increasing the effective length of the slits and thus increasing the equivalent mass.

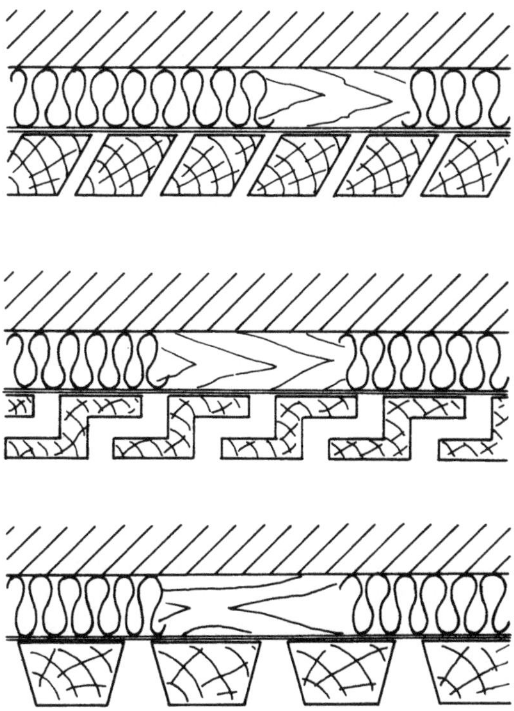

Fig. 6.8. Different kinds of wood lath systems with open gaps. (Center: after Ingerslev.[2])

[2] Ingerslev, F., *Akustik*, Teknisk Vorlag, Kopenhagen, 1949, p. 214.

Each doubling of this equivalent mass leads to a halving of the cutoff frequency, above which most of the sound is reflected rather than absorbed. This can be seen from the following equation, given in Section IV. 7.6.

$$\tau = 1 \Bigg/ \left[1 + \left(\frac{\pi f m}{\rho c} \right)^2 \cos^2 \vartheta \right] \tag{6.23}$$

Figure 9.3 of Part IV shows the frequency dependence of τ for different values of the equivalent mass layer m as parameter.

But it would be wrong simply to reduce the absorption coefficient of the bare porous layers by this factor τ; τ only approximately represents the decrease at high frequencies of the absorption coefficient with decreasing transparency of the perforated plate. Quantitatively, the decrease is always a little less.

Above all, this equivalent mass, working together with the elasticity of the air contained in the porous layer, can constitute an oscillating system that resonates at a certain frequency τ at which it exhibits very high absorption. This is the reason for the increase of absorption in the medium-frequency range, mentioned above (see Section IV. 9.3).

In general, the resonance frequency of such a system is proportional to the square root of the ratio of stiffness to mass. Because the stiffness of an air cushion is inversely proportional to its thickness d, one gets for this resonance frequency an equation in whose denominator there appears the square root of the product of the equivalent mass (per unit area) and the distance from the wall d. For normal incidence, with d in cm and m in kgm^{-2}, it is:

$$f_{0(0)} = \frac{600}{\sqrt{d\,m}} \text{ Hz} \tag{6.24a}$$

At oblique incidence the frequency of maximum absorption increases by the factor 1/cos (see Section IV. 9.2). This means that for a random distribution of incidence angles the absorption peak is broadened; the highest value occurs at a frequency corresponding to an average angle of incidence of about 45°:

$$f_{0(45\ )} = \frac{850}{\sqrt{d\,m}} \text{ Hz} \tag{6.24b}$$

If, however, the lateral motion within the air layer is restricted, for example, by using an 'egg-crate grille' to subdivide the volume behind the plate, then the resonance frequency given by eqn. (6.24a) is valid for *all* angles of incidence, and the resonance frequency remains stable even for random incidence.

This same 'lateral decoupling' can also be achieved by filling the air cushion with a porous material. In this case the resonance frequency lies below the value given by the well-known rule-of-thumb (6.24a); the decrease occurs even for small flow resistance R because of the isothermal compression. However, the decrease in back volume due to the skeleton of the porous material has the opposite tendency. For a porosity factor of $\sigma = 0{\cdot}7$ the two effects would balance. For a system with porous material completely filling the air cushion:

$$f_0 = \frac{500}{\sqrt{\sigma\, d\, m}}\ \text{Hz} \qquad (6.24c)$$

Since, in general, the air cushion behind the perforated plate is only partly filled with the porous layer, we usually find ourselves between the conditions of eqns. (6.24b) and (6.24c). It has therefore become customary to use, as a well-proven approximation for this calculation, the 'in-between formula' (6.24a), even if there is no decoupling of the lateral motion of the air inside the air cushion.

Even better, of course, in this case would be a reverberation room measurement. Figure 6.9 shows a typical example. A broad peak can be seen at medium frequencies where the value $\alpha = 1$ is reached. (The fact that at high frequencies α decreases to only about 0·5 has nothing to do with the resonance properties that are of interest here.)

It is, however, not always possible to have at one's disposal a thick air cushion, such as is necessary, in combination with a perforated plate with a high porosity factor, in order to achieve a certain resonance frequency. According to eqn. (6.24), the same resonance frequency can be attained, even with a wall distance of only 1·5 cm, if we multiply the equivalent mass by a factor of 4. This in turn can be achieved by doubling the separation e between the holes in both directions. Alternatively, one can achieve the same effect without changing the frequency of the resonance peak, according to eqn. (6.24), by again doubling e and dividing the distance d by 4.

We suspect that this saving in construction space is gained at the expense of a certain disadvantage: namely, a strong frequency dependence of the absorption coefficient. The coefficient is determined by taking into account that the exciting sound pressure must overcome the

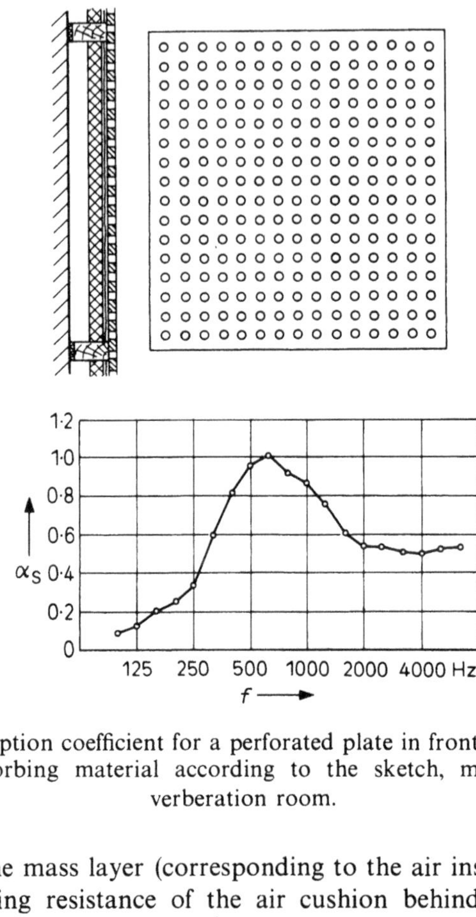

Fig. 6.9. Absorption coefficient for a perforated plate in front of an air cushion filled with absorbing material according to the sketch, measured in a reverberation room.

resistance of the mass layer (corresponding to the air inside the perforations), the spring resistance of the air cushion behind the perforated plate, and the flow resistance R. This leads, as shown in Section IV. 9.2 for normal incidence and adiabatic compression, to:

$$\alpha_{(0)}=\frac{4R}{\rho c}\Bigg/\left[\left(1+\frac{R}{\rho c}\right)^2+\left(\frac{c}{2\pi f_0 d}\right)^2\left(\frac{f}{f_0}-\frac{f_0}{f}\right)^2\right] \tag{6.25}$$

First it must be noted that the value of the maximum absorption coefficient at $f=f_0$ does not depend directly upon either f_0 or d, but, according to

$$\alpha_{\max}=\frac{4R/\rho c}{(1+R/\rho c)^2}=1\Bigg/\left[\frac{1}{2}+\frac{1}{4}\left(\frac{R}{\rho c}+\frac{\rho c}{R}\right)\right] \tag{6.26}$$

only on matching the flow resistance R with the impedance of the air ρc. On the other hand, the quantities f_0 and d (or alternatively, m and d) determine whether small or large changes of frequency ($\Delta f = f - f_0$) are necessary to decrease α from α_{max} to $\alpha_{max}/2$. If eqn. (6.25) is divided by (6.26), we get

$$\alpha/\alpha_{max} = 1 \bigg/ \left[1 + \left(\frac{f_0}{\Delta f_H}\right)^2 \left(\frac{f}{f_0} - \frac{f_0}{f}\right)^2\right] \tag{6.27}$$

From this equation we see that the interesting portion of the absorption coefficient curve is given by plotting the α-ratio against f/f_0 over the interval determined by $\Delta f_H/f_0$. Here we have introduced an additional parameter Δf_H. This parameter is called the half-power bandwidth, a name that is justified only so long as, by setting

$$\frac{1}{2}\left(\frac{f}{f_0} - \frac{f_0}{f}\right) \approx \frac{\Delta f}{f_0}$$

we can approximate eqn. (6.27) with a 'resonance function':

$$\alpha/\alpha_{max} = 1 \bigg/ \left[1 + \left(\frac{2\Delta f}{\Delta f_H}\right)^2\right] \tag{6.27a}$$

As a matter of fact, in this case a frequency deviation of $2\Delta f$, which is necessary on both sides of the absorption peak to reduce α to $\alpha_{max}/2$, is a half-power bandwidth. If, however, eqn. (6.27) deviates from eqn. (6.27a), the peak is always somewhat broader.

By comparing eqns. (6.25), (6.26) and (6.27), we get the value for this half-power bandwidth:

$$\Delta f_H = 2\pi\left(1 + \frac{R}{\rho c}\right)\frac{d}{c} f_0{}^2 \tag{6.28a}$$

or:

$$\Delta f_H = \frac{(R + \rho c)}{2\pi m} \tag{6.28b}$$

The latter equation is somewhat general because it is not tied to the assumption of an adiabatic compression. In any case, eqn. (6.28a) shows that the half-power bandwidth at a given f_0 increases with increasing distance d between the wall and the perforated plate.

Only in the exceptional case where it is desired to absorb a single frequency, which sometimes happens in noise control problems, is it immaterial whether the resonance frequency is attained by a large mass layer m and a small distance d or vice versa. In room acoustics, by contrast, our goal is usually to achieve equal absorption within a broad frequency range between a lower frequency f_1 and a higher frequency f_2. Even with a large wall distance d this requires a number of contiguous half-power bandwidths. Therefore one is forced to combine n differently tuned 'resonators' with half-power bandwidths Δf_H, in order to cover the whole frequency range in such a way that, for every frequency, α reaches α_{max} in at least one partial surface area:

$$n = \frac{f_2 - f_1}{\Delta f_H} \sim 1/d \tag{6.29}$$

If d is small, however, the corresponding required number of partial areas increases, and therefore so does the necessary total area S for equivalent absorption over a frequency range $(f_2 - f_1)$. One can see, therefore, that the necessary construction volume remains unchanged:

$$S\,d = \text{const.} \tag{6.30}$$

What we save in depth of air cushion must be made up in additional treated wall area. In each individual case, the choice depends on the available space. Certainly enlarging the covered area is more expensive than increasing the depth of the treatment, so it is recommended not to make the distance d for a sound-absorbing wall treatment too small; it should be at least 5 cm.

According to eqn. (6.26), α_{max} can reach the value of 1 if

$$R = \rho c = 41 \text{ Rayl} \tag{6.31}$$

For oblique incidence, this optimal flow resistance increases to

$$R = \rho c / \cos \vartheta \tag{6.31a}$$

which, for an average angle of incidence of 45°, amounts to 58 Rayls.

For determining the value of R, the same range is valid here as was suggested in eqn. (6.13) for mounting a fabric at a distance d in front of a rigid wall without covering it with a perforated plate.

As we have emphasized before, not only is the maximum absorption important but also the width of the peak. In this respect a large value of R would be favorable, as can be seen from eqn. (6.28). Therefore it is better to choose a flow resistance higher than suggested by eqn. (6.13), rather than lower.

In choosing the value of R from a construction point of view, one must differentiate between several cases (Part IV, Fig. 9.6). The simple equation

$$R = \Xi b \tag{6.32a}$$

is valid only if the porous layer of thickness b is installed at a certain distance from the rigid wall but is not in immediate contact with the perforated plate; the air must have the opportunity to move freely through the holes in order to fill the entire space behind the holes.

If the porous layer is mounted immediately on a rigid wall, a procedure which simplifies the installation, its flow resistance is effective with only one-third its full value:

$$R = \frac{1}{3} \Xi b \tag{6.32b}$$

because the particle velocity required for energy dissipation diminishes near the wall and becomes zero immediately at the wall. On the other hand, in this case a thicker layer has the advantage of lateral decoupling.

By contrast, if a thin porous layer is installed immediately at the perforated plate, the flow resistance will be increased by a factor of $1/\sigma$, which can be quite significant. In this case, only those parts of the layer that cover the holes and the close vicinity are acoustically effective; the remaining parts serve only for mounting.

A combination of the latter two cases is represented by spot-drilled or grooved soft-boards with few pores and sometimes with a painted surface.

Jordan[3] first proved, using a wood-fiber plate, that this painted surface acts like a perforated plate (i.e. like a mass) and thus leads to a resonance peak. Fig. 6.10 shows the frequency dependence of the absorption coefficient for a typical currently marketed, spot-drilled, soft-board plate, as measured in a reverberation room.

Obviously the manufacturer has succeeded quite well in matching the flow resistance of this plate; here the narrowing of the cross-section for air flow has to be taken into account. Because the cylindrical walls of the spot-drilled holes are acting as the entrance area into the porous material, only the thickness l of the non-porous cover should be used for calculation of the mass layer according to eqn. (6.21), for use in eqn. (6.24).

[3] Jordan, W., *Akustische Zeitschrift*, **5** (1940) 77.

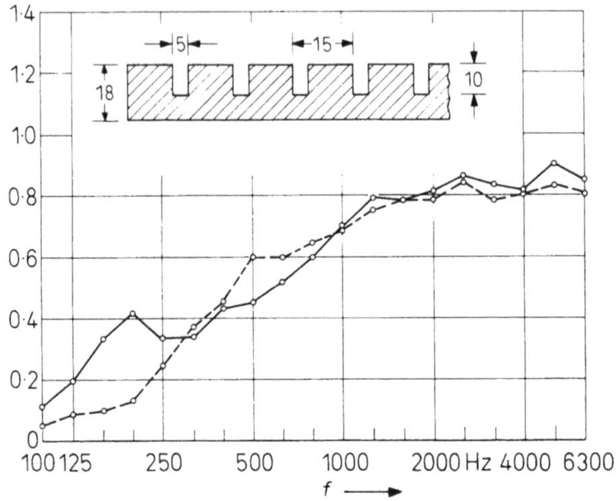

Fig. 6.10. Sound absorption coefficients (ordinate) for a grooved 18 mm soft-board plate (inset, dimensions in mm). --- Directly in front of a rigid wall; ——— at 5 cm distance in front of a rigid wall.

The same is true for the grooved soft-board plates with slitted surfaces often used today.

The predetermination, or even precalculation, of the flow resistance of a commercial sound-absorbing material is, at the present time, unfortunately quite uncertain. Fortunately, the choice of flow resistance does not very significantly influence the result.

The required flow resistances are relatively low. One can easily blow through an optimally matched material with the mouth. Often the resistance of the perforated plate is sufficient in itself to achieve high absorption. A back layer of fibrous material in this case need not be used.

As shown in Section IV. 9.3, the flow resistance of perforated plates can be calculated for certain cases. It is shown there that the flow resistance of a perforated plate with thickness l_0, holes of radius a, and a distance e between the holes amounts to

$$R = 0{\cdot}53 \frac{e^2 l_0}{a^3} \sqrt{f} \times 10^{-3} \text{ Rayl} \tag{6.33}$$

with the dimensions in cm and the frequency in Hz. This equation is valid if the radius of the holes is at least 1 mm.

It is quite possible to produce a matched flow resistance by using narrow holes. This method has been used by Rschevkin and Terossipjantz.[4]

The orifice corrections for slit plates have not yet been determined. One may, however, calculate the specific flow resistance for the most technically interesting 'broad' slits of width b:

$$R = 1{\cdot}65 \times 10^{-3} \frac{B(l_0 + 2\Delta l)}{b^2} \sqrt{f} \text{ Rayl} \tag{6.34}$$

where B is the center-to-center distance between the slits. The dimensions are in cm. The values for $2\Delta l$ may be taken from Fig. 9.2, Part IV.

If, for example, we consider a lattice made from ceiling laths with a cross-section of $2{\cdot}5 \times 5$ cm, placed edgewise at a distance of 2·5 cm in front of a rigid wall with a slit width of 0·5 cm, we get

$$R = 1{\cdot}65 \times 10^{-3} \frac{3(5 + 0{\cdot}5)}{0{\cdot}5^2} \sqrt{f} = 0{\cdot}11 \sqrt{f} \text{ Rayl} \tag{6.34a}$$

For 1000 Hz, $R = 3{\cdot}5$ Rayl.

This value is very low. One might be inclined to regard this lattice in front of a reflecting wall as only a visual covering, but this is surely not correct. According to eqns. (6.21b) and (6.24b), this system has a resonance frequency of 850 Hz; and we find from eqn. (6.34a) that R in this case is 3·5 Rayl. Using eqn. (6.26), we find an astonishing $\alpha_{max} = 0{\cdot}27$.

If, however, in the air cushion there is sound-absorbing material with approximately matched flow resistance, the small amount of absorption contributed by the lattice can be neglected in estimating the absorption coefficient.

It can be shown by experiment that the mass layers of perforated plates are equivalent to impermeable mass layers. But in that case one needs surface densities so low that they cannot be obtained in commercial plates but only in impermeable membranes, such as old-fashioned oilcloth and the currently available plastic membranes (e.g. PVC sheeting).

Fig. 6.11 shows the absorption coefficient measured in a reverberation room at different distances from a rigid wall. The space behind the membrane was completely filled with mineral wool, which prevented lateral coupling. The resonance frequencies are in accordance with eqn. (6.24c) with $\sigma = 0{\cdot}9$.

By using very thin lightweight membranes, the resonance frequency

[4] Rschevkin, S. N. and Terossipjantz, S. T., *J. Physics Acad. Sci. UdSSR*, **4** (1941) 45.

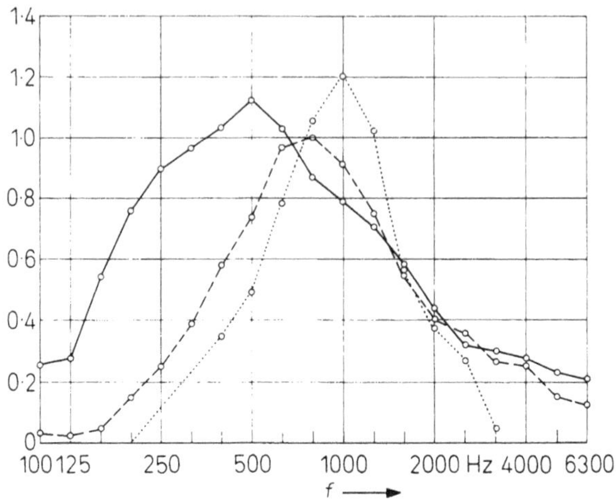

Fig. 6.11. Absorption coefficients (ordinate) for fiber mats covered with membranes (weight: $0{\cdot}15\,\mathrm{kg\,m^{-2}}$), measured in a reverberant room ···· 1·5 cm mineral wool; – – – 3 cm mineral wool; —— 7 cm mineral wool.

can be shifted above 2000 Hz so that even high frequencies are absorbed; by this means a 'hissing' of the room (caused by too long a reverberation at high frequencies) can be avoided.

Covering absorptive materials with membranes instead of perforated plates undoubtedly has the advantage that porosity is not necessary; therefore the sanitary objections to porous layers are largely overcome. Furthermore, the smooth surfaces are easily cleaned.

Because both light and heavy membranes may have the same decoration, and the observer cannot see the separation between the membrane and the rigid wall, one can, by using different membrane coverings and depths, combine the necessary acoustical diversity (e.g. various resonance frequencies, etc.) with an apparent architectural uniformity.

The same is true also for covering with perforated plates, if one neglects the slight shading of the plates that stems from the different numbers of holes; such differences hardly catch the eye if there is a uniform pattern covering the entire area.

Moreover, the perforated areas need not be laid out in a repetitive rectangular pattern. The very difference between perforated and unperforated areas may be used for an ornamental purpose.

The fact that unperforated plates may later be replaced by perforated ones, or heavy membranes by light ones, without essentially changing the architectural appearance, provides a welcome opportunity for subsequent corrections by the acoustician. Since there are always many absorptive influences that cannot be calculated, such opportunities for correction are not only desirable but they should even be planned from the beginning.

A layer of porous plaster also offers the possibility of such correction if it remains elastic after drying, since it can be partly covered with a thin layer of spackle if the need arises (Fig. 9.12, Part IV).

II.6.5 Low-Frequency Absorbers

The discussion in the last section, especially the simple resonance equation (6.24a), shows that we have only to choose a suitable value for the product

$$dm > 4 \tag{6.35}$$

(with d in cm and m in $\mathrm{kg\,m^{-2}}$) in order to get a resonant system that absorbs preferentially at frequencies below 300 Hz.

We will start now with the more rare case that the distance d is very large, in fact even with the limiting case in which the distance is so large that the stiffness of the air cushion behind the mass layer plays no role at all; only a very light inert mass remains. Equation (6.23) shows how much sound is transmitted through such a mass; this same equation also describes the absorption coefficient, if we neglect the losses inside the plate:

$$\alpha = 1 \Bigg/ \left[1 + \left(\frac{\pi f m}{\rho c}\right)^2 \cos^2\vartheta\right] \tag{6.36}$$

Every vibrating mass, therefore, is a pure low-frequency sound absorber because the 'absorption coefficient' becomes larger and larger, the lower the frequency.

Of course, mass values as low as that which we found for perforated coverings (Fig. 6.7) are unheard of for room surfaces. Rather, for reasons of sound insulation the walls must have such low τ-values that they are of no interest at all as sound-absorptive surfaces.

However, such high sound insulation does not always occur in buildings, especially in the case of windows and doors and occasionally with light wall and ceiling constructions adjacent to anterooms and attics. A simple window pane weighs only about $7{\cdot}5\,\mathrm{kg\,m^{-2}}$. The absorption coefficient for incident sound at an angle of 60° is given by eqn. (6.36): $\alpha_{(60°)}=0{\cdot}11$ at 100 Hz, and 0·33 at a frequency one octave lower.

It is no wonder, therefore, that churches with large window areas are bright not only for the eye but also for the ear: they are free of excessively long reverberation time at low frequencies, such as is found in rooms with very thick walls without windows. An example is given in Fig. 6.12, which shows the frequency dependence of the reverberation time measured in the crypt of the Völkerschlachtdenkmal in Leipzig.[1]

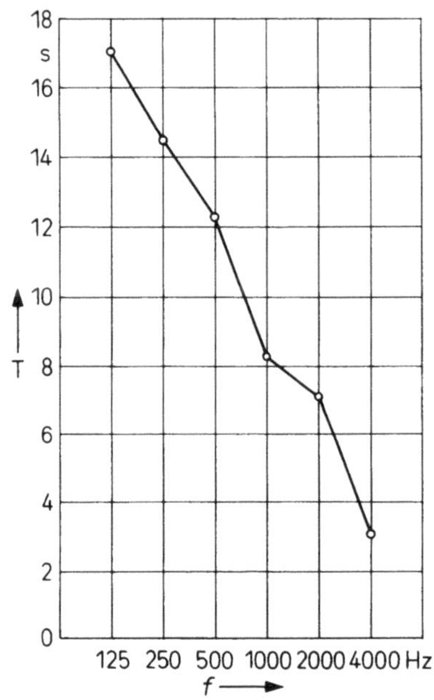

Fig. 6.12. Frequency dependence of the reverberation time of the crypt in the Völkerschlachtdenkmal in Leipzig.

Indeed, an increase in reverberation time at low frequencies is a sure clue to the construction of the walls and ceiling, whether they are heavy or light. Thus, for example, there is no doubt that the wooden ceiling covered with a thin layer of plaster in the Festspielhaus at Bayreuth, as well as the relatively thin construction of the side walls, prevents the reverberation time at low frequencies from increasing. By this means the tonal shift toward low frequencies, caused by the covered orchestra pit, is

[1] Meyer, E. and Cremer, L., *Z. f. techn. Phys.*, **14** (1933) 500.

advantageously compensated by the acoustics of the room. The temporary structure of this building was certainly better for the acoustics than the projected marble palace, which turned out to be too expensive to build.

In general it can be said that temporary buildings, especially wooden barracks, often have optimal absorption at low frequencies without any special additions.

The increase in absorption at low frequencies is especially surprising in circus tents. It leads to a very clear speech sound, to a blaring sound from the trumpets, and to a specially sharp sound from the whips. It might be possible in this case to calculate the absorption using eqn. (6.36), because the tent-fabric has no bending resistance.

Usually, however, eqn. (6.36) leads to values of α that are too low, probably because of the bending resistance of the construction. Like the stiffness of the air cushion discussed above, the bending resistance, acting together with the mass of the plate, constitutes a resonant system.

Since the construction data are quite different from case to case, however, it is hardly possible to give general rules for windows, doors and light walls.

On the other hand, the resonance frequencies for double windows can be readily determined, because in this case the elasticity depends mainly on the stiffness of the air cushion that couples the two masses of the window panes m_1 and m_2 (in $\mathrm{kg\,m^{-2}}$). Because this air cushion is not laterally decoupled, its stiffness depends on the angle of incidence, leading to a broad resonance curve at random incidence. For an average angle of incidence of 45°, eqn. (6.24) is valid if one introduces, instead of the surface densities m_1 or m_2, the following combination:

$$\frac{1}{m}=\frac{1}{m_1}+\frac{1}{m_2}$$

Thus, one gets:

$$f_{0\,(45°)}=850\sqrt{\frac{1}{d}\left(\frac{1}{m_1}+\frac{1}{m_2}\right)}\ \mathrm{Hz} \tag{6.37}$$

where the separation d is in cm.

It should be emphasized that the double window panes used today for heat insulation have a relatively high resonance frequency; for example, two 4 mm panes (surface density = 10·4 $\mathrm{kg\,m^{-2}}$) separated by 10 mm in the same frame have a resonance frequency $f_{0(45°)}$ of 370 Hz.

This resonance potential of double panes at 'medium frequencies' is especially serious as far as degrading the sound insulation is concerned.

How much it adds to the absorption of a room depends on what other energy losses are present, i.e. it depends on the difference between α and τ.

The value of α is based on energy losses at the boundaries, on frictional losses at the support points, and on the inner damping of the material—in short, on effects that, at least today but perhaps also as a matter of principle, cannot be described with simple equations.

There is, therefore, no possibility for evaluating the contributions of low-frequency absorbers in the construction other than to measure these elements in a reverberation room, or—even better—to measure the reverberation time in the building itself. This must be done at a very early stage of construction when, on the one hand, these elements are already present but, on the other hand, the reverberation time is not determined by objects that will be removed later (e.g. construction materials, scaffolds, etc.).

With such intermediate measurements, even the influence of air-conditioning openings, radiators and so on can be determined, which can hardly be calculated.

Our inability to precalculate the low-frequency absorption is technically acceptable only because the low-frequency absorbers in the construction are usually not nearly sufficient to decrease the reverberation time at low frequencies as much as is necessary; for the most part the intrinsic absorption of the structure is negligible compared to the low-frequency absorption that must be deliberately provided.

This is even more true if considerable design emphasis is put on achieving adequate sound insulation against noise from outdoors and from the foyers. In this case, even for the low-frequency absorbers, the previously mentioned separation of the functions of sound insulation and sound absorption is exploited: the sound insulation is achieved by the shell of the building, consisting of very heavy non-porous walls and ceilings, and the sound absorption is achieved by a significantly lighter inner skin with surface density m, mounted at a distance d in front of the basic wall and tuned according to eqn. (6.24).

Thereby, however, a certain mixing of the two goals is unavoidable: in a positive sense the additional inner skin increases the sound insulation at frequencies above its resonance; but, unfortunately, in a negative sense the resonance which is very important for the low-frequency sound absorption leads to a decrease of sound insulation in the frequency range of the resonance. But if the air cushion is deep and includes sound-absorbing elements, this decrease of the sound insulation is of minor importance.[2]

As an example of this kind of combination we may mention the floating floor, that is, a floor slab resting on a fiber mat or some other resilient material. If the mat is not too stiff itself, the stiffness is determined almost entirely by the trapped air. On the other hand, since the flow resistance of the mat leads to lateral decoupling, eqn. (6.24c) can be used. A 24 mm asphalt flooring ($m = 55{\cdot}2\,\text{kg}\,\text{m}^{-2}$) resting on a 0·8 cm mat with $\sigma = 0{\cdot}9$ leads to a resonance frequency of 80 Hz. It is known that the improvement in sound insulation provided by a floating floor starts only above this frequency, but experience shows that this does not cause any difficulties.

From a room acoustics point of view, we must consider that such an area, which is usually very large, absorbs very selectively at a certain frequency; according to eqn. (6.28), because of the very small mat thickness $d = 0{\cdot}8$ cm, we must expect a half-power bandwidth of only 3 Hz if the resistance matching is good, e.g. $R = 2\rho c$. Because of this small half-power bandwidth, the absorption efficiency measured within an octave band, or at least a third-octave band, is not very large. But since it is not yet proven whether, for musical performances, strong absorption in a narrow frequency range can cause tonal coloration, one would be well-advised to vary the tuning by variation of the thickness of the mats or the weight of the floor; similarly, it is advisable not to use ceilings of plaster on lath with the same thickness everywhere.

Since the acoustical requirements for variation of the thickness and the weight must not be visible inside the rooms, it leads to a more difficult and therefore more expensive construction. Therefore one must take care not to ask for more variation than is necessary to get the individual half-power bandwidths to overlap.

In any case, the floating floor shows us that separations less than 1 cm are not favorable. Resilient skins for walls, however, can easily have separations of 8 cm, and therefore 10 times the half-power bandwidth of a floating floor. To get a resonance frequency of 80 Hz in this case, according to eqn. (6.24c), one needs a surface density for the plate of only $6\,\text{kg}\,\text{m}^{-2}$ if $\sigma = 0{\cdot}9$; this can be provided by 1 cm plywood.

Of course, a combination of a thinner plate with a larger distance would be even more favorable. But with rigid plates it is hardly possible to get a weight less than about $3\,\text{kg}\,\text{m}^{-2}$. One might consider, however,

[2] See Cremer, L., *Schalltechnik*, **20** (1960) 10.

getting the necessary mass layer by using membranes or perforated plates. In that case two things must be taken into account: on the one hand, the air cushion acts only as a spring, so eqn. (6.24) is valid only if

$$d < \frac{\lambda_0}{12} < \frac{2800}{f_0} \tag{6.38}$$

(with d and λ_0 in cm, f_0 in Hz).

For lightweight perforated plates with widely spaced holes, the incident sound wave must overcome not only the inertia of the mass m_L of the air in the holes but also the inertia of the mass of the plate m_P, as in the case of freely hung textiles (see Section IV.2.5, Volume 2). Both inertias act as if they were 'in parallel', so that one has a 'resultant mass'

$$\frac{1}{m_{res}} = \frac{1}{m_P} + \frac{1}{m_L}$$

or

$$m_{res} = m_L \frac{m_P}{m_P + m_L}. \tag{6.39}$$

This means that the mass of the air in the holes m_L appears to be diminished.

In the case of heavy and rigid plates it must be remembered that they have their own stiffness, which leads to a (lowest) plate resonance frequency f_1 if the air cushion is so large that its spring effect can be neglected. One can approximate the resulting resonance frequency f_{res}, at which the maximum absorption efficiency can be expected, by combining the resonance frequency f_0 according to eqn. (6.24), where only the air cushion is taken into account, and the lowest resonance frequency of the plate f_1:

$$f_{res} = \sqrt{f_0^2 + f_1^2} \tag{6.40}$$

The plate frequency f_1 of a plate supported at its opposite ends at a distance L is given by

$$f_1 = 0{\cdot}45\ c_L\ h/L^2 \tag{6.41}$$

In this equation c_L is the longitudinal sound velocity in the plate material; for plywood this is 3000 ms^{-1}; h is the thickness of the plate. If one strikes the plate while changing the distance between it and a rigid wall, one first hears a high tone, which gradually becomes lower with increasing distance until finally the tone of the plate without the air cushion is reached.

One can perform this experiment qualitatively by using a piece of thin plywood or thick cardboard on a table top; leaving one end of the plate on the table, one strikes the plate while lifting the other end. A person doing this for the first time is astonished that the air between the table and the plate could influence the pitch of the tone at all. His knuckles feel only the stiffness of the plate; and with respect to his effort in lifting and lowering the plate the spring effect of the air does not play any role at all. In this slow movement the air has enough time to escape sideways, which is not possible with the rapidly vibrating sound waves. For these acoustical vibrations the air cushion is as stiff as that inside a cylinder closed by a piston, where adiabatic compression even replaces the isothermal compression.

In the case of porous plates (for example, a wood-wool slab plastered on the front), which can also be used as low-frequency absorbers, the distance between the wall and the back of the plate must not be used as distance from the wall d in the equations. The air enclosed in the open pores of the plate also acts as a spring, so that it is more correct to take the distance d between the wall and the plaster layer for use in eqn. (6.24). If the plaster layer itself already has a large enough stiffness, it can happen that increasing the thickness of the air cushion between the wood-wool slab and the wall behind does not cause a change at all, so long as this distance is large enough that the plate can vibrate.

If one uses a porous plaster here, one gets a system that combines the properties of a low-frequency absorber with those of a high-frequency absorber. Figure 6.13 shows the absorption coefficients, measured in a reverberation room, for a 2·5 cm wood-wool slab mounted at 3 cm distance from a rigid wall and covered with a 1 cm layer of porous plaster.

The porosity of this plaster may be recognized by the absorption at high frequencies. On the other hand, its flow resistance is great enough to cause a vibration of the whole plate at low frequencies and thus to create a resonance effect, at 100 Hz with rock wool behind and at 200 Hz without. This rock wool leads to lateral decoupling, so the absorption coefficient decreases only below the mentioned frequencies. In this way one gets a nearly constant absorption coefficient in the frequency range between 200 and 3200 Hz.

A similar behavior could also be achieved using softboard plates, either grooved or perforated, or even plain (Fig. 6.2).

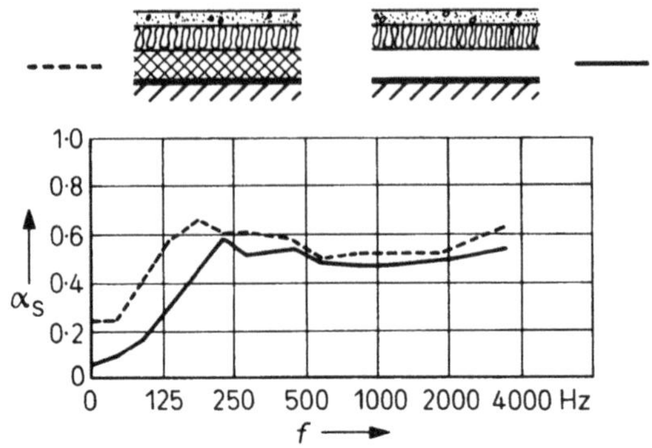

Fig. 6.13. Absorption coefficient for a wood-wool plate covered with a porous layer of plaster, at 3 cm distance in front of a rigid wall.—— Empty air cushion; - - - air cushion filled with rock wool.

Such frequency-independent absorption coefficients are especially desirable in foyers and staircases, where there are only few people and sparse furnishings. In auditoriums this kind of absorption is needed only where echoes caused by the interior shape must be avoided. In general the audience and the upholstered seats introduce so much absorption at high and medium frequencies that normally one is interested only in pure low-frequency absorbers.

Furthermore, a certain variation of the frequency dependence of the absorption coefficients of different wall and ceiling areas is desirable in order to make the reflection more diffuse and so to achieve a uniform sound distribution.

The necessity of taking into account the resonance frequencies of the plates themselves results in a significant increase in the number of data that are necessary to describe a construction. In addition to the material constants and the thicknesses of the individual layers, which describe the construction in a direction normal to the surface, one now also needs data that describe the dimensions of the construction in a direction tangential to the wall. Connected with this is the fact that the amplitude of vibration decreases from the centre of the plate toward its fixed boundaries; and this in turn means that the reflected airborne waves are not returned in a single direction, according to the rule of equal incidence and reflection angles, but are scattered in other directions as well. In short, plate resonators with a certain bending stiffness lead to an increase in the uniformity of sound distribution.

In discussing eqn. (6.40), we looked at the problem only from a two-dimensional point of view, assuming a plate which is supported at its opposite boundaries at a distance L. The lowest resonance frequency is then described by eqn. (6.41).

In general, however, a plate is fixed on all four sides and in the simplest case is only supported, not clamped. This means that in eqn. (6.41) the quantity $1/L^2$ can be replaced by:

$$\frac{1}{L^2}=\frac{1}{L_x^2}+\frac{1}{L_y^2} \qquad (6.42)$$

(where L_x and L_y are the dimensions of the plate in the two perpendicular directions).

Figure 10.8 in Section IV.10.5 shows a family of curves in the L_x–L_y plane, using this 'resultant length' as a parameter.

Finally, the manner of fixing the edges of the plate plays a role. It is known that the lowest resonance frequency of a bar is increased by a factor of 2·25 when it is clamped on both ends; this means a rise of more than an octave compared to a bar which is only supported. The kind of mounting achieved by screwing or nailing a plate lies between these two extreme cases and can hardly be calculated; the problem is particularly complicated because the vibration ability of the structure on which the plate is mounted is also influential. In short, when using plate absorbers one must have accurate measurements for the construction in question if one must know the exact resonance frequencies. This is especially important if a certain construction element is used repeatedly in a building, for example, if it covers the entire ceiling of a room.

Figure 6.14 shows the absorption coefficient for one module of a coffered ceiling, measured in a reverberant room. In spite of the different thicknesses of the plates and the different distances from the rigid backing, one finds a maximum absorption at 200 Hz. This maximum cannot be explained by the mass of the plates and the stiffness of the air cushions, but only by the resonance frequencies of the construction elements themselves, including the influence of the stiffening supporting bars.

This example shows how important it is in such cases for the acoustical consultant to avoid equal tuning; he must insist on a variation of construction, even if this is more expensive. The reason for choosing such a wooden structure as an example here is the well-known favorable influence on acoustics of wooden panelling, wooden stages and vibrating wooden floors. The corresponding absorption at low frequencies is a welcome complement to the medium- and high-frequency absorption of

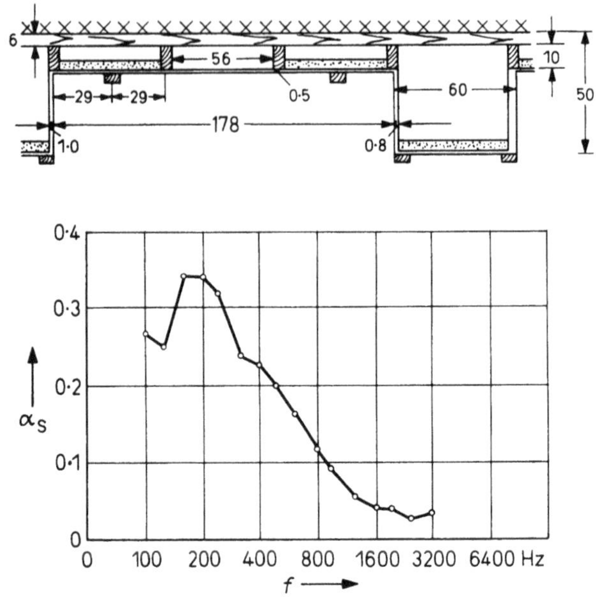

Fig. 6.14. Absorption coefficient for a coffered wooden ceiling, measured in a reverberation room.

the audience. This is especially true in the case of older concert halls which were planned and built without today's knowledge of room acoustics but nevertheless have good acoustics.

People have often believed that the favorable effect of wooden panels, whose vibration can even be felt with the finger tips when the orchestra plays fortissimo, may be compared to the much greater significance of the use of wood for musical instruments. In particular, it was supposed that the sound created on the stage is transmitted via wooden floors, wooden side-walls and eventually even the ceiling, so that the whole room is excited like the body of a violin, and thus radiates the sound created on the stage uniformly into the audience. Furthermore, because the sound velocity in rigid bodies is greater than in air, it was also supposed that this led to a shortening of the rise time. Specifically, it was thought that the structure-borne vibrations in the wood acted as 'forerunners', to bring the music to the listeners ahead of the airborne sound.

There could be some truth in this because wood by nature has a structure that combines great stiffness with small weight. On the one hand, this makes wood especially suitable as a sound radiator but, on the other hand, it also works in reverse, meaning that wooden plates can be easily excited into bending vibration by airborne sound.

These properties are advantageous for musical instruments, but in cases where a high sound insulation is needed (for example, door panels) it is a disadvantage, a fact which may be proved by sound insulation measurements.

For the propagation of sound within the rooms where the sound is produced, however, these effects play hardly any role, at least not as much as was originally supposed. Meyer and his colleagues[3] have experimentally studied with great care whether such a structure-borne-sound fore-runner and its radiation can be detected. The experiments were carried out in two concert halls that were richly decorated with wood but unfortunately were destroyed during World War II: namely, the world famous Leipzig Gewandhaus and the large hall in the Rheinpark at Köln-Deutz, for which E. Michel was the acoustical planner.

The result was unequivocally negative. The thin plywood panelling in Köln-Deutz had no immediate contact with the stage and was proven to be a poor sound radiator. In Leipzig, an immediate blow to the stage could be detected in the parquet floor that supported the stage, but its radiation was so weak that this sound path to the listener was meaningless for his ears.

Furthermore, the natural law which says that the first wavefront determines the perceived direction of arrival, as mentioned in Section III.1.6, leads to the conclusion that the supposed structure-borne-sound fore-runner in a wood-panelled hall obviously plays no role; otherwise we would hear the sound coming from the floor or the walls and not from the stage.

Even if experiments disprove the value of faster sound propagation in a hall via wood-panelled walls, floor and ceiling, one might still think perhaps that wood has other properties important for room acoustics which have so far not been taken into account by our calculations.

Violins, for example, are said to be highly influenced by the inner damping, for which their age might be of great importance.

It is certainly true that the sound absorption of a vibrating plate depends on its inner damping, especially if there are no other causes of damping such as absorbing material in the air cushion. However, it must

[3] Meyer, E. and Cremer L., *Z. f. techn. Phys.*, **14** (1933) 500. G. Buchmann and L. Keidel also participated in these experiments.

still be stated that the same energy losses that we observe in the case of wood can also be found in other less valuable construction materials, for example gypsum plates. Moreover, even plates with intrinsically low losses, like aluminum or steel plates, can be adjusted with respect to their damping by the use of added damping layers.

The only place in a concert hall where the special radiation ability of wood is important is the stage. But even if sound energy is radiated directly from the stage (which can happen with the cellos and contrabasses due to their supporting pegs) it is certainly very low.[4] Nevertheless, it can still be supposed that the musicians feel, and even want to feel, the vibration of the stage. Therefore the acoustical consultant is well advised to heed these wishes, especially since like a physician he can say to himself 'at least it won't hurt'. However, cases are known where an originally vibrating stage had to be replaced by a rigid one because the bass sound was too weak.

The reason for the lengthy discussion here as to whether wood should be used is the fact that wood has one great disadvantage: namely, it is flammable, especially if it is mounted in front of an air cushion. And because the ornamental wood that is visible is always a thin veneer, there is no difference from an architectural point of view whether this veneer is glued to plywood, pressed wood board, non-flammable asbestos cement plate, plaster, or something similar. The acoustical consultant can give no objective argument against the replacement of wood by another material with the same mass, bending stiffness and loss factor.

As with perforated plates, the maximum of the absorption coefficient is determined by the correct adjustment of the energy losses. In general, the losses at the supporting edges and inside the material are not sufficient; thus the absorption coefficient increases significantly if additional sound-absorbing material is placed in the air cushion behind the panel. As Meyer[5] has shown, a remarkable absorption is achieved even if the air cushion is not completely filled, but only its boundaries are treated with sound-absorbing mats. This is evidence for strong air vibrations tangential to the plate surface in this 'drawer-like' air space (see Fig. 10.12, Section IV. 10.8).

Without absorbing material filling the air space, the lowest eigenfrequencies of this 'tangential motion', namely,

$$f_x = c/(2L_x); \quad f_y = c/(2L_y) \tag{6.43}$$

[4] Cremer, L., *Physik der Geige*, S. Hirzel, Stuttgart, 1981.

[5] Meyer, E., *Elektr. Nachr. Techn.*, **13** (1936) 95.

can be observed in the frequency dependence of the absorption coefficient.[6]

Filling the backspace with sound-absorbing material is not always an advantage. Not only can the absorption coefficient become too large, but also narrowing the absorbed frequency range by the lateral decoupling can be a disadvantage. Adding or removing sound-absorbing material afterwards is one of the simplest ways to correct the frequency dependence of the reverberation time.

The naturally narrow-band absorption of most low-frequency absorbers must be evened out by using several differently tuned elements in the same room. But one can also get a broadband low-frequency absorber by putting several plates and air cushions in front of one another. Figure 6.15 shows an arrangement of oilcloths and air cushions constructed by Meyer.[5]

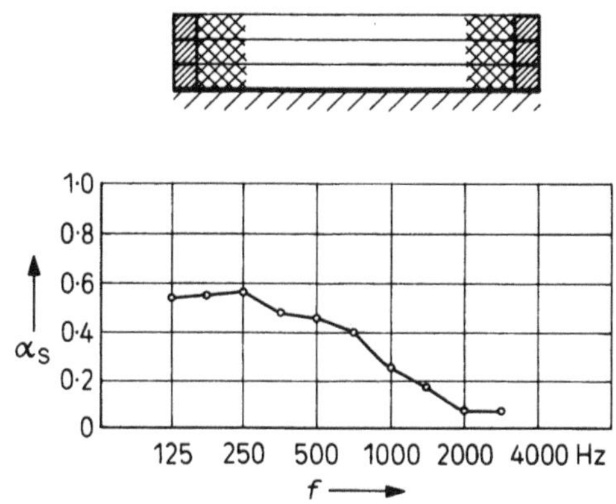

Fig. 6.15. Room-acoustical iterative network, consisting of layers of oilcloth separated by air cushions. (After E. Meyer.[5])

The frequency dependence of the absorption coefficient was measured in a reverberant room, as shown in the same figure. Such a system built by Rschevkin[7] from perforated plates is shown in Fig. 9.9, Section IV. 9.3. In both cases, an analog electrical circuit consisting of inductances and capacitances, the so-called iterative network, supplied the initial idea for such a development.

The problem of achieving a certain frequency dependence, by using circuit elements systematically put together and properly dimensioned, was first solved in the field of electrotechnics. This pioneer work can obviously be adopted in the field of room acoustics. One reason that the

[6] Schroder, F. K., *Acustica*, **3** (1953) 54.
[7] Rschevkin, S. N., *Hochfrequ. u. Elektroak.*, **67** (1959) 128.

systematic construction of combined sound absorption systems has not been further developed may be that wall coverings must also serve other purposes, such as architecture, which—like every art—must always seek new ways of design.

II.6.6 Sound-absorptive Objects

The last and most difficult category of sound absorbers to be dealt with is the individual object with arbitrary shape, where no absorption coefficient related to the area can be defined.

The influence of the distance between the objects will be studied by taking an example that can be calculated, namely air resonators, which we have previously treated in the form of perforated plates in front of an air cushion. We will generalize the picture only insofar as we look at separate air cushions for every perforation and assume that the air cushion volume can take any arbitrary shape (see Fig. 9.13, Section IV.9.5 Volume 2). Such bottle-like vibration systems, where the air inside the bottle-neck acts as a mass and the remaining volume of the bottle acts as a spring, are called Helmholtz resonators because Helmholtz was the first to use such systems for analysing sound signals.

In room acoustics today they are occasionally used as low-frequency absorbers, whereby hollows in a brick wall serve as the elastic volumes and the necks are provided by perforated bricks.[1] Assuming that the distance e between the resonators is small compared with the wavelength, this problem can be handled in the same way as the perforated plates described in Section II.6.4. In place of eqn. (6.24) for the resonance frequency, we have (see Section IV.9.5, eqns. (9.88a), (9.88b) and (9.90)):

$$f_0 = \frac{c}{2\pi}\sqrt{\frac{\pi a^2}{(l_0 + 1{\cdot}6a)\,V}} \tag{6.44}$$

where a is the radius and l_0 is the length of the bottle-neck, and V is the bottle volume. Instead of the distance d in eqns. (6.24), 6.25) and (6.28), we use the term V/e^2. Thus, for the frequency dependence of the absorption coefficient, according to eqn. (6.25), we get:

$$\alpha_{(0)} = \frac{4R}{\rho c} \Bigg/ \left[\left(1 + \frac{R}{\rho c}\right)^2 + \left(\frac{ce^2}{2\pi f_0 V}\right)^2 \left(\frac{f_0}{f} - \frac{f}{f_0}\right)^2\right] \tag{6.45}$$

[1] See, for instance, Pedersen, P. O., *Ingenirvidenskap. Skrifter*, (1940) No. 5, p. 40 ff.

and for the half-power bandwidth, according to (6.28):

$$\Delta f_H = \left(1 + \frac{R}{\rho c}\right) 2\pi f_0^2 V/(ce^2) \tag{6.46}$$

The flow resistance R can be found from eqn. (6.33). In this case there is an important dependence of the flow resistance on the distance e, because R increases with the area ratio $e^2/\pi a^2$; therefore, we can write

$$R = R'(e^2/\pi a^2) \tag{6.47}$$

where R' depends only on the dimensions of the bottle-neck. This means that, while the resonance frequency f_0 is independent of the distance e, the maximum absorption coefficient α_{max} and the half-power bandwidth Δf_H are very much dependent on this distance. The question of how this works out in the end depends on whether the flow resistance concerned is larger or smaller than ρc.

Now, for such a wall, equipped with resonators at a certain spacing e, we may divide the equivalent absorption area A by the number of resonators n, just as we did previously with the wall area S. In this case we get the increment in equivalent absorption area per resonator:

$$\delta A = \frac{A}{n} = \frac{A\,e^2}{S} = \alpha e^2 \tag{6.48}$$

In the same way, one could relate the equivalent absorptive area of other systems to the number of individual elements, for example, acoustical tiles, which often are installed as single tiles with a size of $e^2 = 0{\cdot}25\,\mathrm{m}^2$. In this case the dependency on the size of the individual tile area is contained not in α, but, of course, in δA.

If we now gradually increase the distance between the individual resonators, we depart more and more from the range of validity of the equations used for α until now, in which the appearance of e indicates the mutual influence of the resonators, and we approach the case of the single resonator, in which the value of δA is not at all influenced by the presence of other resonators.

This is the specific case in which an increase of the absorption area can only be expressed by δA because a meaningful relation to the wall area is no longer possible.

If we were to divide δA by the narrow cross-section of the bottle-neck (πa^2), we would find apparent α-values far above 1 at the resonance frequency, as we already did in eqn. (6.45), because the energy losses are not confined to the small area of the incident wave corresponding to the neck opening.

Fortunately, the case of single resonators in a rigid wall can not only be treated very exactly, as is shown in Section IV.9.5, but furthermore, if $\lambda \gg a$ (which can be assumed to hold in all practical cases), the analysis leads to the same simple equations that we found for the periodic resonator arrangement with $e \ll \lambda$. First of all there is no change in the resonance frequency of the resonator, which we could have guessed from the fact that e does not appear in eqn. (6.44). Furthermore the frequency dependence of δA is bell-shaped like that of α in eqn. (6.27), namely:[2]

$$\delta A = (\delta A)_{\max} \Bigg/ \left[1 + \frac{f_0{}^2}{(\Delta f_H)^2}\left(\frac{f}{f_0} - \frac{f_0}{f}\right)^2\right] \tag{6.49}$$

In this equation one must use the value for f_0 given by eqn. (6.44). But instead of the earlier expression for $(\delta A)_{\max}$, following from eqns. (6.26), (6.27) and (6.48):

$$(\delta A)_{\max} = e^2 \alpha_{\max} = \frac{4e^4 R'}{\pi a^2 \rho c} \Bigg/ \left[1 + \frac{R' e^2}{\pi a^2 \rho c}\right]^2$$

$$= \frac{4\pi a^2 R'}{\rho c} \Bigg/ \left[\frac{\pi a^2}{e^2} + \frac{R'}{\rho c}\right]^2$$

we now have to introduce in eqn. (6.49) the value:

$$(\delta A)_{\max} = \frac{4\pi a^2 R'}{\rho c} \Bigg/ \left[\frac{2\pi^2 a^2}{\lambda^2} + \frac{R'}{\rho c}\right]^2 \tag{6.50}$$

That is, we have to replace e^2 by $\lambda^2/2\pi$. This results from the fact that the reflected wave is no longer plane but spherical. Now the frequency dependence in eqn. (6.50) is, strictly speaking, different from that discussed before according to eqn. (6.26). But $(\delta A)_{\max}$ changes so little in the resonance range that it is enough to use for λ the wavelength λ_0 corresponding to the resonance frequency f_0. Thus, eqn. (6.50) can also be written as:

$$(\delta A)_{\max} = \frac{2}{\pi} \lambda_0^2 \Bigg/ \left[\sqrt{\frac{R' \lambda_0^2}{2\pi^2 a^2 \rho c}} + \sqrt{\frac{2\pi^2 a^2 \rho c}{R' \lambda_0^2}}\right]^2 \tag{6.50a}$$

[2]Ingard, U., *J. Acoust. Soc. Am.*, **25** (1953) 1037; in this case the resonator was in a free wave field but not embedded in the wall.

This value is largest for 'matching' of the flow resistance, which occurs when

$$R' = \frac{2\pi^2 a^2 \rho c}{\lambda_0^2} \tag{6.51}$$

and which means, in a physical sense, that the same amount of energy is transformed into heat in the resonator as is diffusely scattered by it. The denominator in eqn. (6.50) then takes the value 4, and $(\delta A)_{max}$ takes the value:

$$\left[(\delta A)_{max}\right]_{max} = \frac{\lambda_0^2}{2\pi} \tag{6.52}$$

meaning that the resonator absorbs the same amount of sound power as the incident wave delivers to an area of size $\lambda_0^2/2\pi$.

One says in this case that the resonator at its resonance frequency has an 'effective cross-section' of

$$q = \frac{\lambda_0^2}{2\pi} \tag{6.53a}$$

and an 'effective radius' of

$$\sqrt{q/\pi} = 0{\cdot}225\,\lambda_0 \tag{6.53b}$$

Therefore resonators must be spaced at least twice as far apart as the distance given by eqn. (6.53b) if they are to be independent of one another. For a resonator with a resonance frequency of 100 Hz, this effective cross-section corresponds to an area of $1{\cdot}84\,m^2$, independent of the bottle-neck cross-section. (In the case of resonators freely installed (not in a rigid wall) this cross-section decreases to half.)

The surprisingly large size of the effective absorptive area, on the other hand, corresponds to a very small half-power bandwidth, which in this case is given by

$$\Delta f_H = 8\pi^2 cV/\lambda_0^4 = 8\pi^2 V f_0^4/c^3 \tag{6.54}$$

If a resonator tuned to 100 Hz requires only a volume of $5000\,cm^3$, then the half-power bandwidth is only 1 Hz! The single resonator must, therefore, have either very low losses and therefore be very sharply tuned or it yields only small values of δA.

The small half-power bandwidth of a weakly damped resonator has still another significant characteristic, which is explained in detail in

Section IV. 9.5: the resonator has an intrinsic reverberation of its own, with a reverberation time

$$T = 2{\cdot}2/(\Delta f_H) \tag{6.55}$$

In the example given, $T = 2{\cdot}2$ s. The presence of such weakly damped resonators, therefore, may lead to curved reverberation decays, as we have found before in the case of coupled rooms.

The results of the theory for an individual resonator have been compared to arrays of periodically distributed resonators mainly because it can be seen clearly that relating the absorption to the area is preferable if the lateral periodicity is small compared to the wavelength; and relating the absorption to individual objects is preferable if they are distant from each other by half a wavelength or more.

For the individual objects that are of chief interest from a practical point of view, namely people and seats, and for the interesting wavelengths between 3 m and 8 cm that are relevant in room acoustics, both of these limiting cases can occur with the usual distances between seats: namely, 50–60 cm (19·7–23·6 in.) sideways and 80–100 cm (31·5–39·4 in.) behind one another. At low frequencies the seating represents more the case of a 'homogeneous sound-absorbing area'; at high frequencies the seats represent single objects. Therefore, both possible ways of reckoning the equivalent absorbing area can be applied. As mentioned before, the relation of the absorption to the number of objects can be formally used for an arrangement of objects in an area or array; but the relation to the area can also be formally applied if a periodic structure is extended over such a large area that the special situation at the boundaries need not be considered. It does not require geometrical reflection, which can be expected only from a homogeneous absorbing area; but it may even be applied if the periodic structure is very 'rough', like that of the seats without audience, and even more so with audience.

But it must be emphasized once more that a measurement of the equivalent absorbing area, whether related to a single object or to the area, characterizes not only the type of seat but also their separation. Moreover, the influence of the boundary on the result becomes larger and larger the smaller the number of objects. The latter influence always leads to an increase of the equivalent absorbing area, which is clearly due to the fact that the mutual shadowing of the single objects within the absorbing area does not occur at the boundaries.

Kuhl[3] has therefore recommended that one should take this influence

[3] Kuhl, W., Unpublished note of the Working Committee '*Bauakustisches Messen*'.

of the boundary into account, for measurements of only a few objects in the reverberation room, by arranging the objects in front of a wall or corner in such a way that the walls act as a mirror; and that one should cover the free boundaries by wooden panels or something similar.

For small rooms like announce studios, conference rooms and the like, it may occasionally be of interest to know how much a single chair or a single person absorbs. In this case it is useful to measure several objects in the reverberation room; they must not, however, be placed close to each other but should be separated by a distance of possibly several meters. This case of single objects spaced well away from one another is the only case where the observed change in absorbing area δA can only be related to the number of objects.

Since in large auditoriums large areas are covered with periodically arranged objects (the seats or the audience), we will limit the following discussion to that situation.

We begin with the seats. It is clear that wooden chairs represent only a small equivalent absorption area; according to the reverberation room measurements of Meyer and Jordan[4] it even makes a difference whether they have open or solid seat backs.

Chairs with open backs:

$$\delta A = 0{\cdot}01 \quad 0{\cdot}02 \quad 0{\cdot}03 \quad 0{\cdot}03 \quad 0{\cdot}07 \quad 0{\cdot}03\,\mathrm{m}^2 \tag{h}$$

Chairs with solid backs:

$$\delta A = 0{\cdot}04 \quad 0{\cdot}04 \quad 0{\cdot}04 \quad 0{\cdot}05 \quad 0{\cdot}09 \quad 0{\cdot}06\,\mathrm{m}^2 \tag{i}$$

Upholstered chairs:

$$\delta A = 0{\cdot}05 \quad 0{\cdot}12 \quad 0{\cdot}22 \quad 0{\cdot}31 \quad 0{\cdot}43 \quad 0{\cdot}38\,\mathrm{m}^2 \tag{j}$$

The values for wooden chairs are remarkably low if one takes into account that only nine objects were measured in the reverberation room. Because of this it can be understood that the reverberation time of a room with wooden chairs changes greatly with the presence of an audience; most concert halls of the last century belonged in this category.

This striking difference in acoustical conditions between rehearsal and performance is surely not desirable; this is especially true for performances that will be recorded or broadcast, where the microphone positions and other adjustments must be chosen during the rehearsal.

[4]Meyer, E. and Jordan, W., *ENT*, **12** (1935) 213.

Therefore it is not only a convenience for the audience but also an acoustical advantage that today's concert halls are generally furnished with upholstered seats. Meyer and Jordan found for such upholstered seats the absorption area given under data set (j), which has the typical frequency dependence of a porous sound absorber.

It is astonishing to the acoustician that musicians, with their excellent hearing, were able to rehearse at all under the reverberant conditions of a concert hall with wooden seats. They apparently adjusted themselves to this difference between the occupied and unoccupied hall to such an extent that they felt uncomfortable during rehearsals in the first concert halls furnished with upholstered seats.

One famous conductor even stated that if rehearsals without an audience go well, it is a sign for him that there is something acoustically 'wrong' with the hall. With respect to getting accustomed to concert halls with upholstered seats, another famous musician stated, even more paradoxically, that the concert hall sounded more reverberant to him under occupied conditions than unoccupied.

It may well be that a certain acoustical difference between rehearsal and performance is desirable, after all. Certainly, excessive upholstery must be avoided, which could lead to too low reverberation in the occupied hall. In fact, it is desirable that the upholstered portion of a seat is sufficiently covered by the sitting person that the remaining area of the seat does not absorb more sound than a wooden chair. Accordingly, it is recommended that the rear side of the seat back be made of a rigid non-porous plate. Similarly, the upholstery should not be visible from the stage, looking into an occupied auditorium. Therefore the seats should have, instead of complete upholstery on the seat-back, only a back cushion. From an architectural point of view it may be attractive if a colorful audience is set into uniformly upholstered seats; but acoustically this is almost a technical mistake.

On the other hand, with unoccupied folding seats, where the seat cushion is completely covered and the back is partly covered, the visible underside of the seat must be made sound absorptive. This may be achieved if the sound-absorbing upholstery of the underside is covered only by a porous fabric or a highly perforated plate.

These few hints show that short descriptions, such as 'wooden chair', 'upholstered seat', etc., even when supplemented with additional comments like 'flat cushions', 'folding seat with perforations on the underside', and so on, are not sufficient to define a chair adequately. This is because the number of such elements may reach a thousand or more in one room, and thus the seating is the main thing that determines the reverberation time of the empty hall. As a matter of fact, in order to evaluate properly the measured δA-values for concert hall seats, one must have a drawing of the seat construction, and the number and the spacing of the test objects, as well as the arrangement in the test room and finally its size.

On the other hand, one can predetermine the acoustical effect of seats in large halls by testing a smaller number of seats in a moderately large reverberation room. Measurements of the upholstered seats for the Herkulessaal in Munich were first carried out by the authors in a $75\,\mathrm{m}^3$ reverberant room with three seats, then in an empty $473\,\mathrm{m}^3$ foyer using 56 seats, and finally in the concert hall equipped with all 1160 seats.[5] We found the following results:[6]

3 seats in the reverberation room ($75\,\mathrm{m}^3$):

$$\delta A = 0{\cdot}19 \quad 0{\cdot}30 \quad 0{\cdot}39 \quad 0{\cdot}39 \quad 0{\cdot}32 \quad -\mathrm{m}^2 \tag{l}$$

56 seats in an empty foyer ($473\,\mathrm{m}^3$):

$$\delta A = 0{\cdot}21 \quad 0{\cdot}30 \quad 0{\cdot}30 \quad 0{\cdot}30 \quad 0{\cdot}30 \quad 0{\cdot}35\,\mathrm{m}^2 \tag{m}$$

1160 seats in the Herkulessaal ($14\,000\,\mathrm{m}^3$):

$$\delta A = 0{\cdot}21 \quad 0{\cdot}30 \quad 0{\cdot}32 \quad 0{\cdot}28 \quad 0{\cdot}31 \quad 0{\cdot}33\,\mathrm{m}^2 \tag{n}$$

The latter two measurement conditions yielded the same results as far as measurement accuracy is concerned, while the measurement of the three seats in the reverberation room gave results that were too high, as was to be expected. It may be assumed that these differences would be significantly reduced if at least 12–15 chairs were brought into the reverberation room and the precautionary measures mentioned above were taken.

It would be desirable if at least the larger seat manufacturing companies were prepared to make such measurements and to add the test results to their sales catalogues. Otherwise, as in the case of custom designs that are usually required for special banquet rooms, there is nothing to do but to insist upon early design of the seat construction and pilot production of samples. These samples must be measured in the

[5] Owing to a subsequent reconstruction we had the rare opportunity of measuring the hall before and after the reinstallation of the seats, without changing anything else.

[6] In the meantime the seats have been renewed.

reverberation room because the choice of the seat has a great influence on the rest of the work to be done by the acoustician in designing the walls and ceiling.

According to the discussion in Section II. 6.4 concerning the covering of porous materials with membranes, it is clear that seats upholstered with leather or impermeable plastic have, depending on the thickness of the upholstery, greater absorption at medium frequencies and less at high frequencies. This means that the frequency dependence is quite different from that of the clothing of the audience, and thus a certain difference of sound color is to be expected between the occupied and unoccupied hall.

On the other hand, one can lessen the sound-absorbing effect of excessive seatback upholstery, such as is often found in the balconies and upper circles of theaters because of the very steep rake, by putting a non-porous plastic membrane between the visible textile cover and the inner upholstery.

Upholstered seats cannot always be used, partly for reasons of wear and tear, partly for reasons of ease in cleaning. In any case this is true for most classrooms, lecture rooms and churches. In these cases there is the possibility of installing layers of fibrous material behind perforated plates (holes or slits), for example, underneath the seat of a folding chair, in the backs of the chairs, or even underneath desks; Fig. 6.16 gives some examples.

Fig. 6.16. Examples of the installation of porous sound-absorbing material behind perforated plates in non-upholstered seats.

As has been emphasized before, the influence of the seats on a room is so great because it enters the calculation of the overall equivalent absorption area with large areas or great numbers. Therefore there is the danger that special peaks in the frequency dependence of the absorption area may lead to valleys in the frequency response of the reverberation time. Such absorptive peaks could be caused, for example, by the uniform spacing between the seat rows. The broad maximum in absorption between 250 and 500 Hz in data sets (l) and (n) was attributed partly to the distance between the rows and partly to the height of the seats (see Section I. 6.3).

For this reason it is useful, especially if there is no opportunity for preliminary measurements, to provide certain invisible variations in the seats, for example, by changing the perforations in the bottoms of the folding seats, or by using different distances between the seat rows in the various audience areas, or by putting a non-porous membrane behind the textile covers in a certain number of seats. In this respect, because of the differences in the size and the clothing of the audience members, occupied seats show more variation than unoccupied seats. Also the greater 'roughness' of the 'audience surface' is acoustically more favorable than the monotonous grid of the seat rows, which may possibly give a certain selective quality to the room.

Finally, of course, the increase of the equivalent absorption area caused by the audience is more difficult to determine than that of the seats. On the one hand, the exaggeration of the absorption values measured in the reverberant room is presumably even greater, and, on the other hand, a 'full scale test' with well-defined sound signals is rarely possible because it would unduly try the patience of the audience.

Meyer and Jordan, therefore, employed the orchestra itself as the sound source during a concert performance, and measured sudden cut-offs in the music, using electrical filters in order to determine the frequency dependence of the reverberation time (see Fig. 1.1, Section II. 1.2). The hall in question was the former Berlin Philharmonie, which was destroyed during World War II. By comparing these results with the reverberation time of the empty hall they got the absorption areas per person shown as the solid line in Fig. 6.17, left. Above the solid line, a dotted line represents the corresponding values measured during a concert with a half-empty hall.

It would hardly have been possible to document this interesting condition with an orchestra as famous as the Berlin Philharmonie had it not been for an exceptional situation. Wilhelm Furtwängler had retired as the conductor of these concerts because he had been forbidden to perform such 'degenerate music' as (for example) Hindemith's *Mathis der Maler*. An attempt was made to find other conductors to replace him, but there were none with the same fame and importance. As a result, half of the subscription audience stayed away from the concert as a demonstration of protest. Only one fact remained nearly unchanged, to everyone's great surprise: the reverberation time.

After what we have said above about the difference between objects

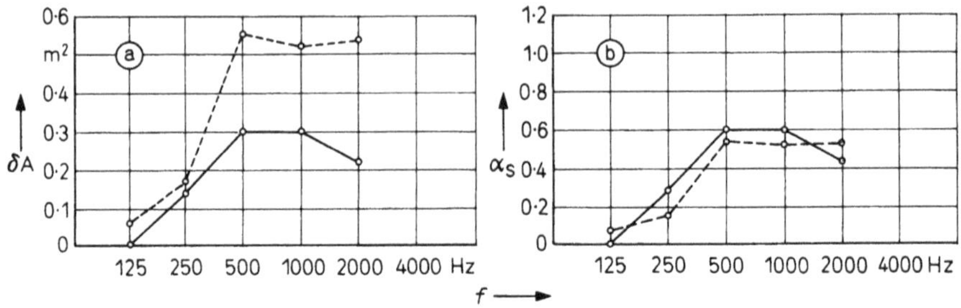

Fig. 6.17. Increase of equivalent absorption area for persons seated on wooden chairs. Left: per seat; right: per unit of occupied area. ——— Completely occupied hall; - - - half empty hall. (After Meyer, E. and Jordan, W., *Elektr. Nachr.-Techn.*, **12** (1935) 217.)

fully exposed to sound and objects that shadow each other due to their close spacing, it is understandable that the unoccupied seats, which were scattered fairly evenly throughout the hall, exhibited the same equivalent absorption area as the occupied seats.

If one divides the total measured change in equivalent absorption area, not by the number of persons present but by the entire area of the audience, one gets the apparent absorption coefficients shown in Fig. 6.17, right, for both cases. They differ so little that they appear to be measurements made under the same conditions, and therefore they may be averaged.[7] This leads to the following values for the apparent absorption coefficients:

$$\alpha_S = 0{\cdot}03 \quad 0{\cdot}20 \quad 0{\cdot}50 \quad 0{\cdot}53 \quad 0{\cdot}46 \quad — \qquad \text{(o)}$$

The corresponding absorption area per seat, taking into account an area of $0{\cdot}55\,\mathrm{m}^2$ per seat, is:

$$\delta A = 0{\cdot}02 \quad 0{\cdot}11 \quad 0{\cdot}28 \quad 0{\cdot}29 \quad 0{\cdot}25 \quad —\mathrm{m}^2 \qquad \text{(p)}$$

These values, however, were significantly lower than those measured by Sabine in a reverberant room:

$$\delta A = 0{\cdot}14 \quad 0{\cdot}23 \quad 0{\cdot}44 \quad 0{\cdot}57 \quad 0{\cdot}82 \quad —\mathrm{m}^2 \qquad \text{(q)}$$

[7] The usefulness of the relation between the equivalent absorption area and the area covered by the audience and musicians was described again later by Beranek, in a study of the reverberation time of many concert halls. (Beranek, L. L., *Proc. 3rd. ICA, Stuttgart, 1959*, Vol. II, Elsevier, Amsterdam, 1961.)

But for densely occupied rows of audience sitting on wooden chairs, Kuhl[3] reported significantly larger values:

$$\delta A = 0{\cdot}17 \quad 0{\cdot}28 \quad 0{\cdot}40 \quad 0{\cdot}44 \quad 0{\cdot}45 \quad 0{\cdot}45\,\mathrm{m}^2 \qquad \text{(r)}$$

The large differences between these data, however, may be caused not only by the differing occupancy densities; the more or less uniform sound distribution in the measurement rooms also plays an essential role. In rooms with an uneven sound distribution, such as rectangular rooms without diffusely reflecting walls, one always gets lower absorption coefficients than in highly diffuse rooms.

The reason for this is the persistence of single eigen-frequencies that are not much influenced by the absorption of the audience. For example, Kuttruff[8] reports an absorption area between 0·20 and 0·43 m^2 per seat at 1 kHz for an audience sitting in upholstered seats.

It is clear that for persons sitting in upholstered seats there is a smaller increase in absorption area between the empty and occupied conditions. In the beginning, one believed (or, better said, one hoped!) that the absorption area of highly upholstered seats would not be changed at all by the persons sitting in them. This has proven to be an error, even in the case of very comfortably furnished theaters. No room is known which in the occupied condition does not exhibit a lower reverberation time than when it is empty. The magnitude of the difference in each individual case can be clarified only by pre-tests in a reverberation room. All calculations of the reverberation time using the difference measured in other rooms are questionable for the reasons mentioned above. Nevertheless, the differences in absorption areas which were determined by comparison measurements in the occupied and unoccupied Herkules-saal, with its rather expensive seats, are reported here as an illustrative example:

$$\delta A = 0{\cdot}0 \quad 0{\cdot}05 \quad 0{\cdot}18 \quad 0{\cdot}26 \quad 0{\cdot}28 \quad 0{\cdot}23\,\mathrm{m}^2 \qquad \text{(s)}$$

Below 200 Hz no increase could be found. Occasionally even negative δA-values were reported, presumably because certain eigen-frequencies of the seats, or even of the floor, were suppressed by the weight of the seated persons. It is possible, as well, that the measurements in the occupied room, using orchestra cut-offs, were influenced by the kettledrum or

[8] Kuttruff, H., *Room Acoustics*, Applied Science Publishers Ltd, London, 1979, p. 150. (The values cited there for the Berlin Philharmonie are obviously based on an error.)

some other reverberating instrument, leading to an apparent reverberation that is longer than that of the room.

For a rough precalculation of the reverberation times of a room, it is here recommended (after Beranek[9]) that the following values of absorption coefficient should be used for the occupied audience area:

$$\alpha_S = 0{\cdot}39 \quad 0{\cdot}57 \quad 0{\cdot}80 \quad 0{\cdot}94 \quad 0{\cdot}92 \quad 0{\cdot}87 \qquad \text{(t)}$$

These values are pretty much independent of the kind of upholstered seats and the density of occupancy.

Only in one case does the audience occupancy not increase the equivalent absorption area, namely, if coupled rooms like loges or deep under-balcony spaces are equipped with upholstered seats. In this case, as described in Chapter II.3, even without audience, the coupling area (i.e., the opening into the loges or into the under-balcony space) can be regarded as totally absorbent. It has no significance for the main hall whether these coupled rooms are occupied or not; in no case is sound reflected back from this area into the hall.

The large opera houses built in the 19th century, for example, La Scala in Milan, could have such long reverberation times in the main hall, in spite of a large audience, only because a large part of that audience was seated in such coupled rooms, partly with very poor sight conditions. The more uniformly the audience is exposed to the sound, the more comfortably they are seated, and the better their view, in short, the more 'social' a room is, the more difficult it becomes for the acoustician to achieve the relatively long reverberation time desired for concert halls. As a matter of fact one must wonder how the audience, seated in the side galleries and standing at the rear of the main floor, bears up with the deficiencies in sightlines and comfort in large old-fashioned opera houses, if only 'it sounds good'—and even this virtue is questionable in the rear of a balcony. In the case of a new theater with comparable conditions, there would be a great deal of complaint!

The fact that persons and objects absorb more sound the greater the distance between them can finally be seen very clearly in the case of orchestra musicians seated on the stage. Because their separation is about twice that of the audience, they represent nearly twice the absorption area. Kuhl recommends calculation with the following values per orchestra player with instrument:

$$\delta A = 0{\cdot}60 \quad 0{\cdot}95 \quad 1{\cdot}06 \quad 1{\cdot}08 \quad 1{\cdot}08 \quad 1{\cdot}08\,\mathrm{m}^2 \qquad \text{(u)}$$

[9] Beranek, L. L., *J. Acoust. Soc. Am.*, **45** (1969) 13.

Because no seats, and especially no heavily upholstered seats, remain on the stage when only a few musicians are playing (for example, when a string quartet performs), the 'occupancy' of the stage has a much larger influence on the reverberation time than that of the upholstered chairs with more or fewer spectators. As a matter of fact, one should always describe what was present on the stage when stating the reverberation time of an 'occupied' room.

This is especially true of compositions involving a large chorus, where often in addition to 120 orchestra players there are another 250–300 members of the chorus on the stage. The latter, however, are sitting (or standing) at least as close together as the audience and therefore must be considered as contributing a comparable absorption area. Their presence, due to their large number, represents a large change in the total equivalent absorption area unless the chorus places are occupied by audience when the chorus is not there; this is the case, for example, in the Concertgebouw in Amsterdam, the New Berlin Philharmonie, the Opera House in Sydney, the Town Hall in Christchurch, New Zealand, the new Davies Symphony Hall in San Francisco, the new Victorian Arts Centre concert hall in Melbourne, and Roy Thomson Hall in Toronto.

Unfortunately, this decrease in reverberation time due to the chorus has its degrading effect upon precisely those musical works that one is normally used to hearing in large churches and which were composed for the long reverberation time of such churches. One thus naturally prefers for these works a longer reverberation time than for chamber music, which usually encounters the opposite difficulty, being confronted with an almost-empty, over-reverberant stage. Therefore, it is useful to provide certain visual and acoustical capabilities for adjustment, by the use of curtains, sliding partitions, screens, and the like.[10]

II.6.7 Example of a Reverberation Time Calculation

If the absorption coefficients or the equivalent absorption areas of the boundaries of the room are known, it is not difficult to calculate the reverberation time of a room as a function of frequency. Because

[10] Variable acoustical banners are provided in the new Davies Symphony Hall, San Francisco, for a dual purpose: they permit the rehearsal condition to be adjusted to match the performance condition, and they allow the reverberation time to be changed to be suitable for different kinds of music. The available range of mid-frequency reverberation time is from 1·9 to 3·1 s in the unoccupied hall and from 1·6 to 2·2 s in the occupied hall. A value around 2 s is usually used.

absorption coefficients are given as the Sabine coefficients α_S it is sensible to use the simple Sabine reverberation equation, (1.35). The assumptions on which this is based are almost completely fulfilled for concert halls, theaters, conference rooms, and so on.

If one uses α_S-values in the Eyring equation, (2.26), one gets lower reverberation times, mostly due to the fact that the Sabine coefficient α_S is always larger than the energy absorption coefficient α. Because, in reverberation time calculations, one usually does not take into account *all* of the absorbing areas, one gets reverberation times which are a little too long.

The starting point of a reverberation calculation is usually an acoustical goal, the desired reverberation time. As will be shown in detail in Section III. 3.3, this depends on the purpose of the hall, that is, whether high speech intelligibility is required or whether the room is to be used for musical performances. High speech intelligibility requires a short reverberation time, below 1 s; music performances, depending on the musical style, require reverberation times up to 3 s. Because a person in an upholstered chair in the audience area represents an absorption area of about 0·6 m^2, a given reverberation time T requires a volume per seat $V_p(m^3)$ (see eqn. (1.40a)) of at least

$$V_P \approx 4T \tag{6.56}$$

To achieve a reverberation time of 2 s, a hall volume of approximately 8 m^3 per seat is necessary, even if the other boundaries of the hall are totally reflecting. For a reverberation time of 1 s, a volume of 4 m^3 per seat will suffice.

As an example the reverberation time calculation for the Herkulessaal in Munich, with a volume of about 14 000 m^3, will be briefly described. This concert hall provides seating for 1100 persons as well as a very large orchestra and chorus. Grouping together those surfaces that have similar acoustical effect, we get the following picture, taking into account only those areas that have a significant influence on the total absorption:[1]

[1] In this respect, the following kind of reverberation calculation would not be suitable for determining the reverberation time of an empty hall without seats. One would get a reverberation time that is much too long, because a number of minor absorbing elements are neglected (such as ventilation openings, cracks under doors, etc.) which are negligible in comparison with the absorption of the seats or the audience, but might be dominant in a hall without seats.

Plastered walls, 1000 m^2; absorption coefficient in accordance with data set (c):[2]

A	10	10	20	20	30	40	m^2

Vibrating surfaces made of gypsum plasterboard, 5 cm in front of a rigid wall, the air cushion filled with mineral wool; hollow door panels filled with mineral wool; total area = 600 m^2; α_S measured in the reverberation room:

α_S	0·37	0·20	0·15	0·08	0·06	0·08	
A	222	120	90	48	36	48	m^2

Tapestries, 5 cm from a rigid wall, and tapestries in front of the 'second balcony';[3] total area = 520 m^2; α_S measured in the reverberation room:

α_S	0·17	0·40	0·70	0·86	0·84	0·82	
A	88	208	364	447	437	426	m^2

Wooden coffered ceiling, according to Fig. 6.14, 1000 m^2; measured in the reverberation room:

A	250	320	200	90	40	40	m^2

Floor, floating parquet floor, plus hollow wooden stage, total area 1300 m^2; α_S according to measurements of thick wood-chip board in front of an air cushion:

α_S	0·11	0·07	0·03	0·01	0·01	0·02	
A	143	91	39	13	13	26	m^2

Seats, 1100 upholstered seats, with thick seat upholstery and moderately thick back upholstery, as well as fabric-covered rear of the seatback; δA according to data set (m):

A	231	330	330	330	330	385	m^2

Audience, 1100, seated on upholstered chairs; δA according to data set (s):

A	0·0	55	198	286	308	253	m^2

[2] The plastered part of the wall surface is included here only in order to give a complete accounting of the room surfaces.

[3] The second balcony has not yet been completed. Instead, the opening to the area where this balcony would be installed is closed by a plaster wall in front of which hangs a tapestry. The absorption now presented to the hall perhaps resembles that of the future balcony.

Musicians, large orchestra with 100 members; δA according to data set (u):

$$A \quad 60 \quad 95 \quad 106 \quad 108 \quad 108 \quad 108 \quad \mathrm{m}^2$$

Chorus, 200 members sitting on wooden chairs; δA according to data set (r):

$$A \quad 34 \quad 56 \quad 80 \quad 88 \quad 90 \quad 90 \quad \mathrm{m}^2$$

Losses during sound propagation with a room volume $V = 14\,000\,\mathrm{m}^3$ and a relative humidity $\phi = 70\%$; calculated according to eqn. (1.45a):

$$A \quad 0 \quad 1 \quad 3 \quad 14 \quad 54 \quad 218 \quad \mathrm{m}^2$$

From the sum of all the equivalent absorption areas at the various frequencies and the Sabine reverberation equation, one gets the reverberation times that can be expected in the fully occupied hall with large orchestra and chorus as follows:

$$T \quad 2{\cdot}2 \quad 1{\cdot}8 \quad 1{\cdot}6 \quad 1{\cdot}6 \quad 1{\cdot}6 \quad 1{\cdot}4 \quad \mathrm{s}$$

For comparison we measured:

$$T \quad 1{\cdot}8 \quad 1{\cdot}6 \quad 1{\cdot}5 \quad 1{\cdot}5 \quad 1{\cdot}4 \quad 1{\cdot}3 \quad \mathrm{s}$$

If we omit the equivalent absorption area contributed by the audience, the orchestra and the chorus, we get the reverberation times which can be expected for the empty hall:

$$T \quad 2{\cdot}4 \quad 2{\cdot}1 \quad 2{\cdot}2 \quad 2{\cdot}4 \quad 2{\cdot}4 \quad 1{\cdot}9 \quad \mathrm{s}$$

We measured in the empty room:

$$T \quad 2{\cdot}5 \quad 2{\cdot}2 \quad 2{\cdot}2 \quad 2{\cdot}3 \quad 2{\cdot}3 \quad 2{\cdot}0 \quad \mathrm{s}$$

As one can see, the accuracy of the calculation for the reverberation time of the empty hall is somewhat better than that for the fully occupied hall. This is a consequence of the fact that the absorption of the musicians and the chorus depends very much on the distance between the individual members and this cannot be accurately estimated.

For an orchestra of normal size, the reverberation time of the Herkulessaal lies between the extreme values given above, namely, around 2 s at mid-frequencies.

Chapter II.7

The Fine Structure of Reverberation and the Room-acoustics Criteria Based Thereon

II.7.1 Early and Late Sound Decay

After W. C. Sabine's fundamental investigations, the reverberation time as a function of frequency was, for many years, regarded as the only necessary parameter for describing room-acoustical conditions. However, it gradually became evident that different rooms with the same reverberation time were subjectively judged to be very different; indeed, the subjective impression changed even from location to location within the same auditorium.

For this reason additional criteria have been developed over the years, which nowadays are sometimes regarded as more important than the reverberation time, especially by their creators.

These criteria are based on the deviations from the ideal exponential decay envisioned by Sabine. Although they can be treated in part by statistical methods (as, for example, in Chapters II.2 and II.3), they generally go beyond the proper limits of statistical room acoustics, as we have seen with the direct and first-reflected sounds in Fig. 2.1.

But since it is advantageous to build upon the statistical representation and to refer back to statistical criteria for an evaluation of the new criteria based on the fine structure of the reverberation time, we will treat them at the end of this Part. Thus, we will clearly see the limits to the validity of statistical room acoustics.

If we exclude the cases of Section II. 2.3, where the same room surfaces are met again and again with the same angles of incidence, and assume, instead, that we are always dealing with at least partially diffuse reflections, then it happens that the assumptions underlying the statistical room-acoustics theory are increasingly well fulfilled as the reverberation process goes on. Sabine's own subjective observations of the duration of the reverberation correspond to this late phase of the decay process.

Nevertheless, we have already demonstrated by statistical analysis (see eqn. (2.46) and (2.48)) that we can expect, during the early part of the decay, a different slope of the level versus time curve depending on whether the excitation is steady-state or impulsive.

Now, it is seldom that a listener has the opportunity to hear the final phase of the sound decay process; one such example was given in Fig. 1.1, which shows the reverberation record for a few bars of Beethoven's Coriolanus Overture. But usually the later part of a reverberant decay excited by a specific impulse in running speech or music is already masked by subsequent signals once it has dropped by about 10 dB.

Thus, we find it plausible that Atal *et al.*[1] have found that test persons, asked to compare running speech and music signals recorded in concert halls against similar but anechoic signals modified by the addition of artificial reverberation with various decay rates, adjusted the latter signals to match the *early* decay of the concert hall signals. They therefore recommended the use of the 'Initial Reverberation Time' T_I as a room-acoustics criterion, instead of the traditional reverberation time T. T_I is also defined in terms of the time required for a 60 dB decay, but it is reckoned from a straight line fitted to the slope observed during the first 160 ms, or, alternatively, the first 15 dB, of the actual decay. Since the latter definition of T_I was recommended for concert halls, it is this definition that has been retained here (see Section III. 2.1).

This recommendation became practically useful, however, only with the development (also by Schroeder, see Section II.1.3) of the 'backwards integration' method of recording the reverberation time in a hall. The extreme fluctuations that typically follow the cutoff of steady-state excitation are so great as to make impossible an accurate assessment of the early part of the reverberation decay, using the classical methods. On the other hand, if we use the backward integration of impulse excitation to determine the reverberation curve, the value of T_I clearly defined by this curve corresponds

[1] Atal, B. S., Schroeder, M. R., and Sessler, G. M., *Proc. 5th. ICA, Liège, 7–14 Sept., 1965*, Paper G-32.

to steady-state excitation, not to impulse excitation (which would exhibit a steeper slope).

Kürer and Kurze have shown,[2] using different source and receiver locations in the Berlin Philharmonie, that the initial slope given by T_I depends on the location in the hall, in contrast to the ideal Sabine reverberation which is location-independent. They also discovered that T_I may even exceed T (evaluated from the usual level range between -5 and -35 dB on the decay curve). Thus, we may encounter not only 'sagging' curves of level versus time, but also 'ballooned' curves. Corresponding contrasts in the decay curve shape following impulsive excitation are given by Schultz[3] for Boston Symphony Hall, Clowes Hall, Indianapolis, and Philharmonic Hall, New York (original version).

Kürer and Kurze further found it expedient to extend the level range used for defining the initial slope of the decay to the region between 0 and -20 dB. Since they used the word *Anfangsnachhallzeit* to refer to the 'initial reverberation time', other authors have adopted the symbol T_A to designate the reverberation time based on the decay slope in the range from 0 to -20 dB, reserving the symbol T_I to designate the similar quantity based on the range from 0 to -15 dB. (We observe, again and again, that the names given to room-acoustics criteria are not always very ingenious!)

Jordan,[4] on the other hand, felt justified in attending only to the restricted decay range between 0 and -10 dB; he called the corresponding reverberation time Early Decay Time, which he abbreviated *EDT*. (Here we prefer to use, instead, the symbol T_E to emphasize the close relation to T_I and T_A.) (Linguistically, and contrary to the tendency apparent in the nomenclature described above, it would seem more appropriate to regard a quantity described by 'early' as referring to a later portion of the decay process than a quantity described by 'initial'.)

Measuring the reverberation time with backwards integration of an impulse allows such a shortening of the evaluation range for early reverberation time without loss of significance, especially if the slope is evaluated by a computer and not 'by hand'. Since T_E can depart even farther from T than T_I or T_A, the tendency nowadays is to give preference to T_E.

---

[2] Kürer, R. and Kurze, U., *Acustica*, **19** (1967/68) 313.

[3] Schultz, T. J., *IEEE Spectrum*, June (1965) 56–67; see Fig. 6.

[4] Jordan, V., *Forty-seventh AES convention, Copenhagen, 1974.*

In this connection, certain limitations of measurement may play a role. At low frequencies, the pistol shots sometimes used in measuring reverberation time do not contain enough energy to produce sound levels more than about 35 dB above the background noise, so that, at best, there is only a limited decay range available for evaluation. But, with Chaucer, we 'hold it wise... to make a virtue of necessity'; although of necessity we cannot excite a long decay, we find it possible, nevertheless, to define criteria from the curtailed reverberation that are subjectively more meaningful than the '60 dB reverberation time'!

Jordan uses the dependence of T_E on location in the auditorium as the basis for introducing a further criterion. His experience with lack of communication between musicians on the stage of the Radiohus, Copenhagen, led him to install overhead sound-reflecting panels. Whereas the values of T_E in the rest of the auditorium do not change with such an installation, the values of T_E on the stage become shorter. Before the installation, it was found that T_E (stage) > T_E (hall); but afterwards, T_E (stage) < T_E (hall). Jordan thereupon recommends the ratio, which he calls the 'Inversion Index':

$$II = \frac{T_E(\text{hall})}{T_E(\text{stage})} \tag{7.1}$$

as a further criterion of concert hall quality, with the requirement that $II > 1$. He confirmed in later consulting practice that this condition was fulfilled in halls where the communication between the musicians was satisfactory.[5]

II.7.2 Characteristics and Representation of the Impulse Response

Until now, we have characterized the difference between the early and late sound decay in terms of a reverberation process composed of a sum of exponential decays with differing damping constants, as in eqn. (2.46). But the very knowledge that the *best* representation following steady-

[5] In his investigation at Radiohus, Copenhagen, mentioned above, Jordan already applied the principle of his 'inversion index' in a 'time criterion' for room acoustics which he had developed earlier; for systematic reasons, we postpone discussion of that criterion until eqn. (7.13).

state excitation is achieved by integrating an impulse makes the description in eqn. (2.46) seem rather doubtful.

The impulse response of a room is composed of numerous single impulses, starting with the direct sound and continuing with discrete and separate reflections from the various room surfaces, as we demonstrated schematically in Fig. 2.1. The integration of this response results, at the beginning, not in a straight line or even a smooth curve, but in a series of marked steps, rather like a stair. (This behavior was well-known to those who favored the evaluation of an 'initial reverberation time'; they even thought it reasonable to evaluate the mean slope of such a stair-like response as a means of characterizing the acoustics of an open-air theater.)

On the other hand, other criteria had already been developed on the basis of the integration of an impulse response, before it was used to represent the slope of the onset or the decay under steady-state excitation. From the beginning, these criteria took into account the difference in structure between the early and the late parts of an impulse response.

But before we enter into a discussion of those criteria it is appropriate to consider in more detail the characteristics of impulse responses and the possible kinds of representation.

The first consideration is the choice of the impulse source. The handiest of these is a pistol, such as is used for the starting signal at races, and which we used in the examples of echograms discussed in Section I. 4.3. (It may be mentioned again that we will use the expression 'echogram' for an impulse response not only when we are interested in the diagnosis of disturbing echoes but also when we are interested generally in the distribution of reflections over a period of time. The alternative expression 'reflectogram', which also includes the direct sound, may be more appropriate for suggesting the objective nature of the record, but it is seldom used in practice.)

The course of pressure versus time for the sound in the far field of a pistol shot consists of a brief over-pressure phase, followed by a longer and lower under-pressure phase, as shown by Gottlob[1] in Fig. 7.1. The same is true of the spark impulses often used in small-scale acoustical model tests.

The corresponding spectrum exhibits a maximum, with skirts sloping

[1] Gottlob, D., Dissertation, Göttingen, 1973.

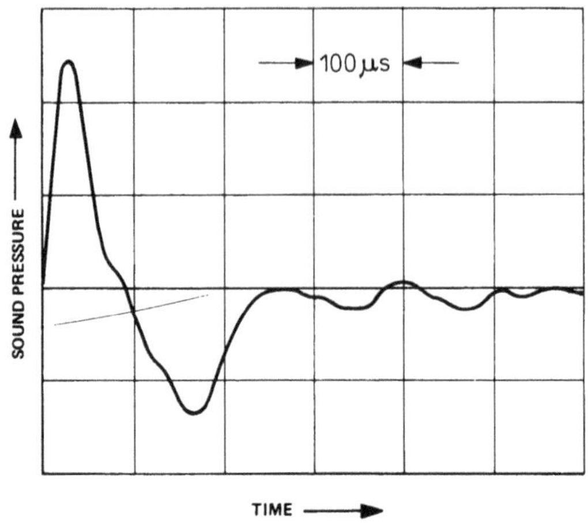

Fig. 7.1. Time dependence of the sound pressure in the far field of a spark. (After Gottlob.[1])

off to both higher and lower frequencies.[2] Because the physical acoustical processes are non-linear in differing degrees in the initial part of an explosive impulse, becoming linear as the spherical wavefront expands, the frequency of the maximum is lower, the more intense the impulse. If the frequency of the maximum in the spectrum is low, this means an impulse of longer duration and therefore a tendency for the subsequent reflections to overlap sooner than with short impulses (for which the maximum in the impulse spectrum occurs at high frequencies).

Reichardt[3] has shown, furthermore, that the spectral distribution of energy for a starting pistol corresponds approximately to that of speech. And since the pistol also has a preferential directional response along the forward axis of the barrel, it resembles speech in this respect also. The use of a pistol is therefore especially appropriate for evaluating the speech qualities of a room.

In each case, the intensity of the reflections depends on frequency, a fact that is best proved by the dependence of the reverberation time on frequency. This dependence is of special interest because it governs the change of timbre of a sound during the course of the reverberation.

[2] Cremer, L., *Akustische Zeitschrift*, **5** (1940) 46.
[3] Reichardt, W., *Schalltechnik*, **17** (1957) No. 22.

We can determine this frequency dependence, both for the impulse responses and for the criteria that depend on them, by using frequency band filters (either octave or third-octave) in the measurement equipment (post-filtering the signal). This reduces the available energy of the signal of interest, but it also correspondingly reduces the unwanted noise.

Another method makes use of loudspeakers that radiate frequency-band-filtered signals (pre-filtering the signal).

Especially distortionless[4] are the so-called 'Gauss-tone impulses' (Gauss-tones, for short). These are short pure-tone impulses whose time envelope is bell-shaped like the Gaussian distribution (see Section III.1.4, especially Fig. 1.14). Here it is helpful to adjust the width of the impulses to the filtering characteristics of the ear.

Although with such impulses we would obtain, from a room impulse, responses representative of the responses of our ears (see Section III. 1.3), experience has shown that they are not well adapted for detailed evaluation, as with echograms.

For instance, Lehmann[5] has demonstrated that such Gauss-tones can lead to rather different impulse responses (and thus also to differences in the corresponding criteria), for changes of only a few centimeters in the source or receiver position. But our ears do not detect major changes in the acoustical quality of a hall with such minor changes in location!

For the recording of echograms in large auditoriums, therefore, it appears better to work with pistol shots and to filter the responses in bands no narrower than two octaves.

Just how crucial is the frequency range of an impulse response was demonstrated by a simple laboratory experiment (see Fig. 7.2) in which Gauss-tones were used for the excitation.[6]

In this experiment, a picket fence with rather large slits was placed in front of a rigid wall (shown in the figure at the right; the dimensions are given in the lower part of the figure). The loudspeaker at the left radiated Gauss-tones, once with a carrier frequency of 500 Hz (see echogram at right, above) and again with 4000 Hz (see right, below). The microphone recorded the direct sound, the reflection from the fence, and the reflection from the wall. The oscilloscope echograms show that at 500 Hz the

[4] The Gaussian time envelope results in a Gaussian (i.e. smooth) spectral function without phase distortion, so that signals still appear as Gauss-tones even after frequency-dependent radiation and reflection.

[5] Lehmann, P., Dissertation, TU Berlin, 1976.

[6] Gerwig, H. D. and Zemke, H. J. *Acustica*, **15** (1965) 304.

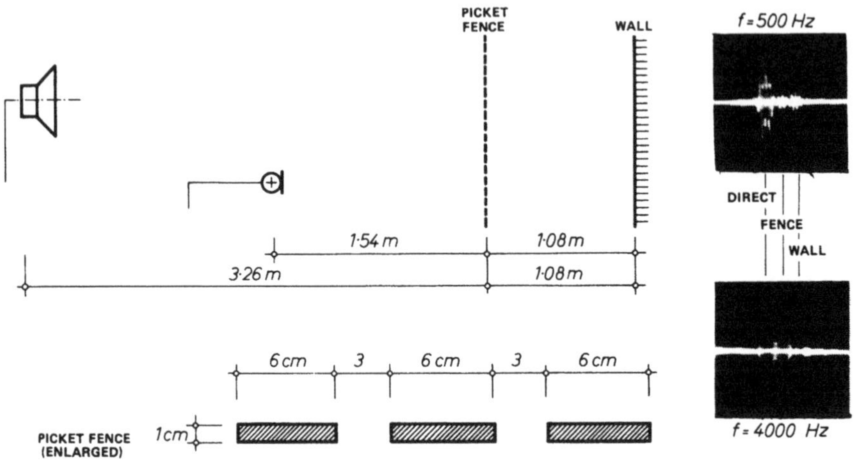

Fig. 7.2. Reflections of a Gauss-tone, with low and high carrier frequencies, from a picket fence in front of a rigid wall. (After Gerwig and Zemke.[6])

sound impulse is diffracted freely through the fence and is reflected only from the wall. At 4000 Hz, however, we observe a strong reflecion from the fence and hardly any from the wall. Such 'frequency switches' occur more or less with all non-specular wall coverings and especially with single objects as reflectors.

In Fig. 7.2 we have, as in Part I, recorded the undistorted (unfiltered) $p_{(t)}$-function at the microphone. If sharp impulses are used (with spectral maxima at high frequencies), we get oscillograms such as Schodder[7] recorded at the Royal Festival Hall (see Fig. 7.3).

One is immediately struck by the complicated structure of the reflection patterns that reach the ear. The echograms in the same row correspond to the same source location but different receiver locations; those in the same column refer to the same receiver location but different source locations.

The gain of the receiving system was adjusted to give the same peak reading for the direct sound in each case. Therefore, the later reflections in the echogram at top left (close to the source) appear much weaker than if the same gain were used for all the echograms. But even if we take this into account, it is evident that the impulse reponses for different locations in the same auditorium are extraordinarily different.

[7] Schodder, G. R., *Acustica*, **6** (1956) 445.

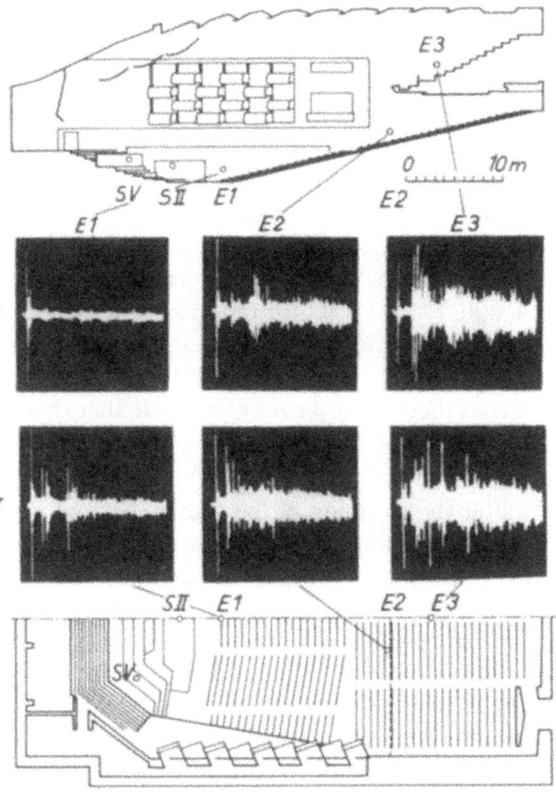

Fig. 7.3. Echograms ($p_{(t)}$) from the Royal Festival Hall. Rows: different source locations; columns: different receiver locations. (After Schodder.[7])

This holds equally for changes both in receiver location and source location.

If we were surprised at first that rooms with nearly the same reverberation time could sound quite different, now a study of the different impulse responses in the same hall makes it astonishing that one would dare to speak of *the* acoustics of a hall at all, without at least indicating where the observation was made!

The *common* reverberation time appears only in the later part of the decay, as the peaks decrease according to $e^{-2\delta t}$; and this occurs only if the reverberation time is independent of frequency. Because of this generally monotonic decrease of the peak levels, the echogram looks rather like a 'fir tree' turned on its side. Since it is generally unfavorable if this tree 'lacks some limbs' and favorable if it is 'fully branched', an experienced observer can make good (if non-quantitative) use of such fir-tree criteria.

This fir tree has the unusual quality that 'its top is never fully attained'; also, the branches (representing reflections) become more and more dense toward the top (toward later time).

We have already mentioned in Section I. 2.2, based on a discussion of the image sources in a rectangular room, that the number of reflections increases with the third power of time. Furthermore, we shall learn in Section III. 1.4, eqn. (1.15a), that for random noise (which an impulse response resembles in its late phase) the ear cannot distinguish the difference between responses having about 2000 impulses per second and those having more. From these facts, we may conclude that there is a time limit.

$$t_{st} = 2\sqrt{V}\ \text{ms} \tag{7.2}$$

(with V in m^3) after the stopping of the source, beyond which there is no longer any point in studying the fine structure of the reverberation. But long before this time, the $p_{(t)}$ function has already conveyed much more information than our ears are interested in or are able to evaluate.

Since the ear, in transforming mechanical vibrations into nerve impulses, works like a rectifier, it is reasonable to rectify the impulse responses picked up by the microphone before recording them. Mathematically, the simplest way to do this is to square the pulses. Figure 7.4 shows an example of such a representation of $p^2_{(t)}$.[5] This is equivalent to plotting the energy density, and is therefore an especially suitable presentation for statistical room acoustics.

On the other hand, plotting $p^2_{(t)}$ has the disadvantage that the smaller reflections tend to disappear, in contrast to a linear plot of $p_{(t)}$; for this reason, a $p^2_{(t)}$ plot corresponds less well with our subjective impression. As we shall

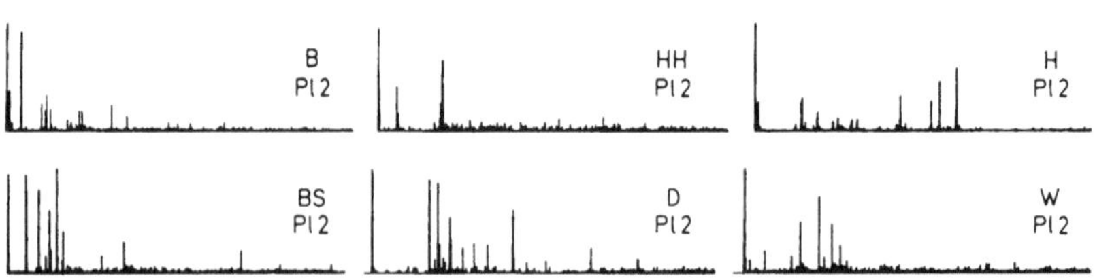

Fig. 7.4. Echograms in terms of $p^2_{(t)}$. (After Lehmann.[5]) These examples correspond to the halls shown in Section III.2.7, Fig. 2.22. Pistol shots were fired from the middle of the stage; the microphone was always at location 2.

learn in Section III. 1.2, a plot of $|p^{2/3}|$ or $|p^{1/2}|$ would correspond better to our loudness sensation.

Even more important are two other psychological effects: the so-called 'inertia' of the ear (see Section III. 1.4) and the masking of weaker reflections by stronger earlier reflections (see Section III. 1.3). Both effects act to reduce the amount of impulse information in which the ear is interested.

Two methods have been developed for taking into account the reduction in information caused by the first effect, the inertia of the ear. One (see Part III, Fig. 1.16) is to square the microphone signal ($E_{(t)}$) and pass it through a 'smoothing circuit' (a series combination of a resistor and capacitor; the voltage across the capacitor is plotted). Such a smoothing circuit simulates the inertia of the ear and yields a plot that corresponds a bit better to the hearing impression. This plot resembles a 'fir tree covered with snow', since many of the smaller branches disappear. Figure 7.5 shows two examples of the original echograms and the

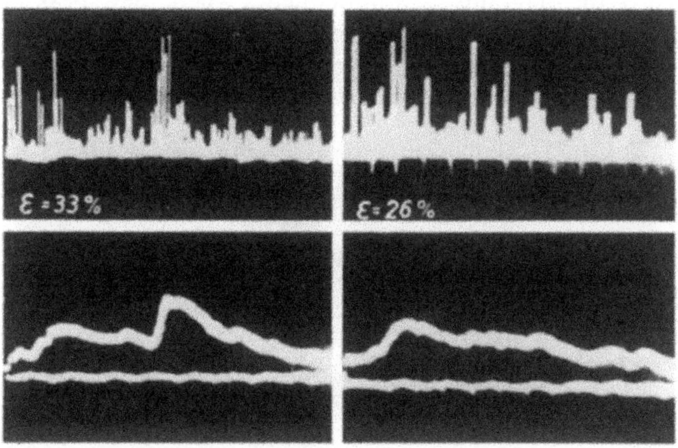

Fig. 7.5. Echograms of $p^2_{(t)}$ (above) and the same smoothed to correspond to the response of the human ear (below); left diagrams for the left ear, right diagrams for the right ear. (After Niese.[8])

corresponding 'ear-smoothed' echograms, recorded by Niese[8] with microphones in the left and right ears of an artificial head ('dummy'). In spite of the reduction of information caused by the 'ear-smoothing', there are still significant differences between the responses at the right and left ears.

The other method, developed by Schodder,[7] splits the impulse

[8] Niese, H., *Hochfrequenz u. Electro akustik*, **65** (1965) 4.

response $p^2_{(t)}$ into small time intervals of 5–10 ms between the instants $0, t_1, t_2, t_3, \ldots$ and integrates either

$$\int_0^{t_1} p^2_{(t)}\,\mathrm{d}t, \qquad \int_0^{t_2} p^2_{(t)}\,\mathrm{d}t, \ldots, \text{etc.}$$

to get a step-wise onset for steady-state excitation, or

$$\int_0^{t_1} p^2_{(t)}\,\mathrm{d}t, \qquad \int_{t_1}^{t_2} p^2_{(t)}\,\mathrm{d}t, \ldots, \text{etc.}$$

to get an impulse response with reduced information content by transforming $p^2_{(t)}$ into a fence-like structure.

Neither method has been generally applied. The same holds for all representations of impulse responses in which the results for different frequency regions are all depicted together on the same diagram.[9,10,11] For example, Fig. 7.6 (right) shows such a representation; it reminds us somewhat of the diagrams used in phonetics where they speak of 'visible speech'; by analogy, we may speak here of 'visible room response'.

Here, the impulse responses $p^2_{(t)}$ recorded in different frequency groups

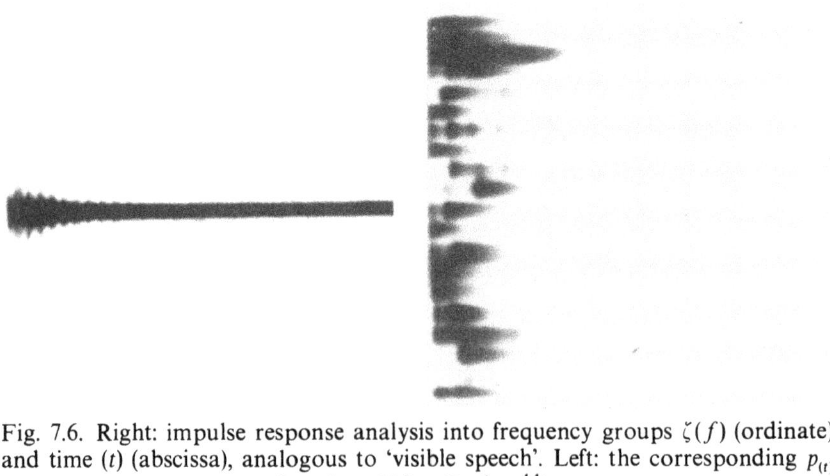

Fig. 7.6. Right: impulse response analysis into frequency groups $\zeta(f)$ (ordinate) and time (t) (abscissa), analogous to 'visible speech'. Left: the corresponding $p_{(t)}$ function. (After Rother.[11])

[9] Somerville, T. and Gilford, C. L. S., *BBC Quarterly*, **7** (1952) 41.
[10] Sacerdote, G., *Electronica e Televisione Italiana*, Nov/Dec (1953) No. 6.
[11] Rother, P., *Proc. 5th. ICA, Liège, 7–14 Sept., 1965*, Paper I–22. See also *Archiwum Akustyki, Warsaw*, **2**, 3 (1968) 221.

$\zeta(f)$ (see Section III.1.3) are transformed into brightness of light (as in sound film recording) and are plotted, one above the other, along the horizontal time axis running from left to right, so that the ordinate corresponds to a rough frequency scale.

In the apparatus developed by Rother,[11] the broadband impulse response is first stored on a magnetic drum; it rotates once per second during recording but sixteen times per second on playback and analysis, so that the entire presentation is available soon after the recording. By periodic repetition and the use of a persistent fluorescent screen, one can get a permanent picture even without a photograph. Furthermore, since each band is adjusted to the same peak-value of the direct sound, the differences due to the type of impulse are nearly eliminated.

In the recorded time interval of 1 s, we can see in Fig. 7.6 (right), from the bottom to the top, the frequency dependence of the reverberation; but we can also recognize (from left to right) the structure of a flutter echo, which can also be seen in the $p_{(t)}$ plot at the left of the figure.

Generally, in this representation mistakes in the choice of room boundary materials show up as inhomogeneities on the vertical (frequency) scale, while mistakes of room shape show up on the horizontal (time) scale.

In a similar manner, the acoustician tests for these effects when he claps his hands in a hall: he listens both for echoes (temporal inhomogeneities) and for changes of timbre during the reverberation (frequency-dependence of reverberation). But a two-dimensional representation, like that of Fig. 7.6, stores the information from this brief event clearly and permanently in black and white, so that it can be studied and discussed (like a physician with an X-ray photograph); it can even, in principle, be evaluated quantitatively.

II. 7.3 Comparison of the Direct and Reflected Sound Energy

Quite often we find in the room-acoustics literature endeavors to condense the information contained in the impulse response of a room into a single-number criterion of quality. The evident hope is to find one rating which so greatly excels all the others that it alone must be recognized as the absolutely valid room-acoustics criterion.

The reverberation times for the early decay, T_E, T_I, T_A and Sabine's reverberation time T are examples of such criteria, though they are 'single-number' only if we neglect for the moment their dependence on frequency.

But all of these criteria take for granted, in principle, both for the decay $E_{-(t)}$ and the onset $E_{+(t)}$, a continuous change of energy density. The clearest contradiction to this assumption lies in the direct sound, which introduces a step-wise change in level both in the decay and (more clearly) in the onset of sound in a room.

One group of criteria compares the direct sound, whose energy density in the steady-state falls off inversely proportional to the square of the distance from the source, against all of the later-reflected sound. For non-directional radiation, the energy density of the direct sound may be expressed by

$$E_{\mathrm{D}} = \frac{P}{4\pi c r^2} \tag{7.3}$$

where r is the distance between source and receiver. Comparing this with the statistically distributed mean (reverberant) energy density (see eqn. (2.19))

$$E_{\mathrm{St}} = \frac{4P}{cA} \tag{7.4}$$

we can define a distance r_{H} at which the energy densities of the direct and the reverberant sound are equal:[1]

$$r_{\mathrm{H}} = \frac{\sqrt{A}}{4\sqrt{\pi}} \tag{7.5}$$

or:

$$r_{\mathrm{H}} = 0{\cdot}057\sqrt{\frac{V}{T}} \tag{7.5a}$$

where r is in m, V is in m^3 and T is in s.

We can see from this formula that the limiting distance, r_{H}, within which the direct sound predominates and outside of which the reverberant sound predominates, is surprisingly small. For instance, in a living room with volume $V = 100\ \mathrm{m}^3$ and reverberation time $T = 0{\cdot}5$ s, the *Hallradius* is only 81 cm. Even in a concert hall of volume $V = 20\,000\ \mathrm{m}^3$ and reverberation time $T = 2$ s, it is no more than 5·7 m, a distance that is far smaller than a typical room dimension.

[1] The subscript H comes from the German *Hallradius*, literally 'reverberation radius', for which no English term has, so far, been widely adopted.

This would mean that only for the musicians and the conductor (and the microphones placed in their vicinity) is the direct sound *not* overpowered by the statistical (reverberant) sound.

But it must be emphasized that the ratio that defines the *Hallradius* concerns only steady-state excitation, that is, long chords. For running speech and music, the direct sound always brings the 'new signals' first, and therefore these are always clearly perceived, even at the most distant seats.

Furthermore, only a few sound sources have non-directional radiation. In general, as with speech, there is a pronounced directivity, such that the sound radiated in some directions is significantly stronger than in other directions. If we introduce the directivity coefficient γ, that is, the ratio of maximum to mean sound intensity, we get for r_H the increased value:

$$r_H = 0{\cdot}057\,\sqrt{\gamma}\,\sqrt{V/T} \tag{7.5b}$$

Since γ may take on values as high as 100, especially for directional loudspeakers, the steady-state direct sound may predominate throughout the entire audience.

The ratio

$$\frac{E_{St}}{E_D} = \left(\frac{r}{r_H}\right)^2 \tag{7.5c}$$

derived under ideal conditions in the steady state, can also be determined by integration of an impulse response, but only if we separate the direct sound from the first reflection (from the stage floor) and all subsequent reflections; we can do this by integrating only over the duration of the impulse Δt. The steady-state energy density is given by integration from 0 to ∞. Thus, we get the more general relation:

$$\frac{E_{St}}{E_D} = \int_0^\infty p^2 \mathrm{d}t \bigg/ \int_0^{\Delta t} p^2 \mathrm{d}t \tag{7.6}$$

Maxfield and Albersheim[2] proposed a variant of eqn. (7.6) as a criterion especially suitable for the placement of microphones for recording:

$$\int_{\Delta t}^\infty p^2 \mathrm{d}t \bigg/ \int_0^{\Delta t} p^2 \mathrm{d}t = \lambda \tag{7.7}$$

[2] Maxfield, I. and Albersheim, W., *J. Acoust. Soc. Am.*, **19** (1947) 71.

they called this quantity 'liveness'. (Here we have replaced their letter L for liveness by λ, primarily to avoid confusion with a level; in addition, we have chosen a lower-case Greek letter, since 'liveness' is a dimensionless quantity; finally, since it represents a ratio of energies, like an efficiency, we prefer to speak of a 'liveness coefficient', reserving the word 'liveness' for the subjective sensation.)

With respect to equations like (7.6), the notion of removing the direct sound energy from the numerator (which otherwise describes the statistical sound field) with the idea of forming a quotient of non-direct to direct sound, is sometimes realized in the literature by multiplying E_{st} in eqn. (7.4) by $(1-\alpha)$. This procedure not only makes the calculation more difficult (since, in addition to V and T, we have to find the total surface area S in order to determine the mean absorption coefficient, α), but it is also a dubious compromise between a consistent statistical theory and the actual time-distribution of reflections. The definition given by eqn. (7.7) is surely preferable.[3]

In the German literature, we also find[4] the logarithm of $1/\lambda$, under the name *Hallabstand* (literally, 'reverberation distance')[5] with the symbol H:

$$H = 10 \log (1/\lambda) \text{ dB} \tag{7.7a}$$

(The question may be raised, however, whether it is reasonable to increase the number of room-acoustics criteria by introducing both the ratio and its logarithm!)

Since our list of criteria here follows systematic rather than historical lines, we may mention two criteria in which the reference quantity, to which the statistical energy is compared, is not the direct sound energy at the observation point but the direct sound energy at a fixed, defined distance from the source. The latter is, in the case of a non-directional source, a measure of the radiated sound power. (If the source does have significant directivity, the relation can be restored by the introduction of the directivity coefficient γ.) If, in the numerator, we include the direct sound energy measured at the observation location (as in (7.6) but not in

[3] This problem was recently tackled once more by W. Reichardt (*Akustica*, **45** (1980) 238).

[4] See, for instance, Reichardt, W., *Grundlagen der technischen Akustik*, Akademische Verlag, Leipzig, 1967, p. 509.

[5] The German word *Abstand* (distance) refers here to a distance between levels in a diagram, not to a real physical distance, such as r_H. Since level differences in German are identified (see DIN 5455) by the final syllable '*mass*' in a word (for example, *Schalldämm-Mass*, which is called in English 'reduction index'), we could, by analogy, call the quantity $(-H)$ the 'liveness index'.

7.7)), we get the criterion introduced by Lehmann,[6] which he called *Stärke-grad*, or 'strength coefficient' (another γ!):

$$\gamma_{(x)} = \frac{\int_0^{\infty} p_{(x)}^2 \mathrm{d}t}{\left(\int_0^{\Delta t} p_{(rS)}^2 \mathrm{d}t\right)\left(\frac{4\pi r_S^2}{S_0}\right)} \tag{7.8}$$

With r_S in m and S_0 in m^2 this quantity defines some kind of 'room-acoustical efficiency', even though it concerns a ratio of energy densities, rather than a ratio of the corresponding powers. At least, it tells us the steady-state energy density at a special location, for a given power of the source. Moreover, this is a criterion that does not depend, as do the criteria discussed previously and hereafter, on the time dependence of the impulse response at only a particular location.

Lehmann also introduced the logarithm of the strength coefficient:

$$G = 10 \log \gamma \text{ dB} \tag{7.8a}$$

calling it the *Stärkemass*, which we may translate as the 'strength index'. This quantity gives the difference between the sound pressure level at the location of the receiver and the sound power level of the source. The same quantity was introduced earlier by Yamagushi,[7] who called it the 'stage seats transmission characteristic'.

Finally, we mention a concept based on the difference in the time-of-arrival of the direct sound and the first reflection. Kuhl[8] regarded this temporal interval as very important for the placement of microphones: the larger this time interval, the larger the recording room seems. But Beranek[9] emphasized that this first interval in the impulse response, which he called the 'initial time delay gap', is the single most important room-acoustics criterion of all: the smaller the better. He determined this quantity, which we here designate Δt_i, not on the basis of impulse responses in the hall, but (much more easily) from the architectural plans

[6] Lehmann, P., Dissertation, TU Berlin, 1976.

[7] Yamagushi, K., *J. Acoust. Soc. Am.*, **52** (1972) 1271.

[8] Kuhl, W., *Proc. 2nd. ICA, Cambridge, Massachusetts, 1957*, p. 53.

[9] Beranek, L. L., *Music, Acoustics and Architecture*, John Wiley, New York, 1962, (reprinted Krieger Publishing Co., Huntington, New York, 1979), p. 417.

of the building. Thus, the initial time delay gap can be determined in advance, during the planning of the hall.

Certainly, it would be more convincing if the criterion were determined from an echogram actually made in the finished hall. But then we would be confronted with the question of how to deal with the difference in level between the direct and first-reflected sound: what should we do if the first reflection is rather weak but a stronger one follows shortly thereafter?

A further development of this criterion, quantitatively taking into account the energy differences in the early-arriving sounds, would require a consideration of the properties of the ear, and especially the 'inertia of hearing'. The fact that small values of Δt_i are preferred is related to the fact, discussed in Section I.4.1, that the ear perceives the first reflection, together with the direct sound, as a single, reinforced signal, provided that this reflection follows the direct sound with a short enough delay.

II.7.4 Comparison of the Early and Late Portions of the Impulse Response

In Section I.4.1, we have mentioned that not only are the direct sound and the first reflection to be regarded as useful, but also all further reflections that arrive at the listener's position with a time delay not greater than $t_g = 50\,\mathrm{ms}$, which we called (with Henry) the 'limit of perceptibility'. (We shall discuss this matter in greater detail in Section III.1.4.)

On this basis, Thiele[1] proposed a criterion that compares the 'useful' sound with the total sound, a ratio that can be determined from an impulse response in the hall:

$$\int_0^{50\,\mathrm{ms}} p^2 \mathrm{d}t \bigg/ \int_0^{\infty} p^2 \mathrm{d}t = \vartheta \qquad (7.9)$$

He called this ratio *Deutlichkeit* (distinctness) and used the symbol *D*. He expected the distinctness of speech to be greater, the greater the value of *D*. In

[1] Thiele, R., *Acustica*, **3** (1953) 291.

the open air, this ratio reaches its maximum value of unity; thus, the value of D may be reported as a percentage.

Since here again the word 'distinctness' implies a subjective evaluation that could only be determined quantitatively in scaling experiments with a number of test persons; and since the ratio involved is again a ratio of energies, we propose to call this criterion the 'distinctness coefficient' and to use for the symbol the lower-case Greek ϑ.[2]

The logarithm of ϑ is not used in the literature. But Beranek and Schultz[3] introduced the modified ratio:

$$\int_{50\,\mathrm{ms}}^{\infty} p^2 \mathrm{d}t \Bigg/ \int_{0}^{50\,\mathrm{ms}} p^2 \mathrm{d}t = \frac{1-\vartheta}{\vartheta} \tag{7.10}$$

which they called the 'reverberant-to-early-sound ratio'. In their curves they plotted, as a function of frequency, the quantity ten times the common logarithm of this ratio, in dB. In Germany, this quantity is called the *Hallmass* (reverberation index):

$$R = 10 \log\left(\frac{1-\vartheta}{\vartheta}\right) \mathrm{dB} \tag{7.10a}$$

(With respect to the criteria defined in eqns. (7.7) and (7.7a), the 'liveness coefficient' and the 'liveness index', the term 'reverberation' there referred to all of the sound that follows the direct sound; here, however, we are regarding as 'reverberant' only the sound arriving later than 50 ms; Beranek and Schultz also used the word 'reverberant' in this restricted sense. This discrepancy could be avoided if we were to call R the 'late-to-early-sound index'.)

It was found rather early in studies of the limit of perceptibility, that the value of the limit depends on the character of the signal, and especially that it is greater for music than for speech. On the basis of careful observation of the shortest duration of musical notes and of the transients of musical instruments, Reichardt[4] proposed a value of $t_g = 80$ ms for the limit of perceptibility for music. He proposed to replace Thiele's concept 'distinctness' (referring to speech) with the concept

[2] In personal discussion with one of the authors, Thiele agreed to this nomenclature, particularly since it is an improvement on the awkward expression '50 ms energy-portion', which he had been considering.

[3] Beranek, L. L. and Schultz, T. J., *Acustica*, **15** (1965) 307.

[4] Reichardt, W. and Kussev, A., *Z. Elektr. Inform. u. Energietechnik. Leipzig*, **3** (2) (1972) 66.

'clearness' (referring to music), and defined a 'clearness index' by:

$$C = 10 \log \frac{\int_0^{80\,\mathrm{ms}} p^2 \mathrm{d}t}{\int_{80\,\mathrm{ms}}^{\infty} p^2 \mathrm{d}t} \ \mathrm{dB} \qquad (7.11)^5$$

(He refrained from introducing a special name for the argument of the logarithm itself.)

The criteria defined by eqns. (7.9), (7.10) and (7.11) share the disadvantage (although it seldom arises in practice) that a very slight shift in the time of arrival of a strong reflection can significantly change the value of the criterion, depending on which side of the limit of perceptibility it falls on.

Lochner and Burger[6] avoid this problem, which is created by the concept of an abrupt change from useful to non-useful sound just at the limit, by introducing a weighting factor $a_{(t)}$ that depends on the delay time t and on the relative level of the reflection (see Fig. 7.7 and Section III. 2.2). The factor decreases from 1 for the early reflections to 0 for

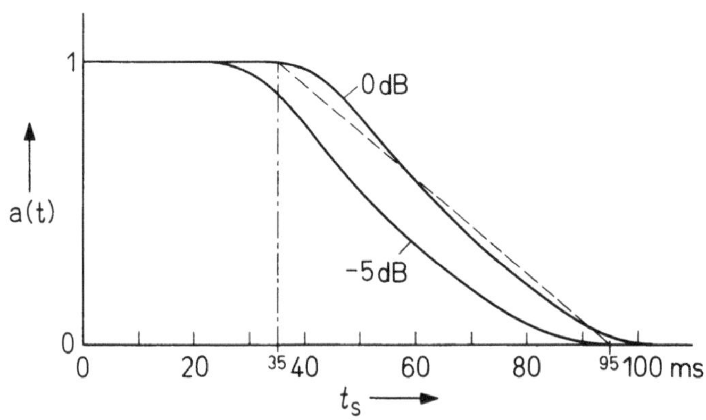

Fig. 7.7. Weighting factor $a_{(t)}$ for the useful part of sound energy during a reverberation process. (After Lochner and Burger.[6])

[5] Reichardt, W., Abdel Alim, O. and Schmidt, W., *Acustica*, **32** (1975) 126.
[6] Lochner, J. P. A. and Burger, J. F., *Acustica*, **8** (1958) 1.

reflections after 95 ms. They call the weighted ratio so defined:

$$\eta = \frac{E_{\text{useful}}}{E_{\text{detrimental}}} = \frac{\int_0^{95\,\text{ms}} p^2_{(t)}\, a_{(t)}\, \mathrm{d}t}{\int_{95\,\text{ms}}^{\infty} p^2_{(t)}\, \mathrm{d}t} \tag{7.12}$$

the 'signal to noise ratio'. (They should have added the adjective 'weighted' before 'signal'.) Since this is again a ratio of energies, we may use for the symbol a lower case Greek letter, for example η.

Another criterion that avoids a sharp limit in the time domain, but introduces one in the level domain, was proposed by Jordan;[7] he called it the 'rise time', t_r. This criterion is based on the fact that (forward) integration over the impulse response corresponds to the onset for steady-state excitation. It is defined as the instant in the impulse response at which the early and late energy are equal:

$$\int_0^{t_r} p^2 \mathrm{d}t = \int_{t_r}^{\infty} p^2 \mathrm{d}t \tag{7.13}$$

It is also the time required, after the start of steady-state excitation, for the room energy to reach a level 3 dB below the asymptotic final level, and is the criterion in terms of which Jordan first defined his 'inversion index' (*II*), mentioned in Section II.7.1, above. Here again, Reichardt[8] proposed a modification according to which the 'rise time' is defined by the time required for the level to reach -5 dB (rather than -3 dB) with respect to the final level. We shall call this quantity t'_r.

Although the criteria t_r and t'_r do not change abruptly when a strong reflection shifts a bit in time around the -3 dB (or -5 dB) level, the effect of a strong reflection is especially remarkable there.

It would be easy to augment the number of room-acoustics criteria based on the distinction between useful (early) and detrimental (late) sound, by proposing other time limits or new weighting functions. If one of the authors has added one more to the criteria already discussed, it

---

[7] Jordan, W. J., *Proc. 3rd. ICA, Stuttgart, 1959*, Vol. II, Elsevier, Amsterdam 1961, p. 922.
[8] Reichardt, W., *Z. Hochfr. u. Eleckroakustik*, **79** (1970) 121.

was with the intention of introducing a criterion that is defined (like Sabine's reverberation time) on a purely mathematical basis, free of arbitrary choices of the limiting values.

Generally, it is possible to characterize the distribution of a positive, decreasing function, such as $p^2_{(t)}$, with respect to the 'distance' t (in time) from the origin $t=0$, by means of moments of degree n, related to the total area under the curve:

$$M_n = \frac{\int\limits_0^\infty t^n p^2_{(t)}\,\mathrm{d}t}{\int\limits_0^\infty p^2_{(t)}\,\mathrm{d}t} \tag{7.14}$$

For the simplest case, $n=1$; this moment defines the distance t_s from the origin of the 'center of gravity' (*Schwerpunkt*) of the 'decay curve area'. Thus, Cremer proposed this 'center-time' (Schwerpunktzeit) as a further room-acoustics criterion:

$$t_s = \int\limits_0^\infty t p^2_{(t)}\,\mathrm{d}t \Big/ \int\limits_0^\infty p^2_{(t)}\,\mathrm{d}t \tag{7.15}$$

Kürer[9] has developed a rather simple method to measure the center-time. But, in fact, any of the criteria discussed here can be readily evaluated nowadays by analog-to-digital transformation with the help of a digital computer.

Kürer also demonstrated that the center-time criterion is appropriate for characterizing a special location in a room with respect to the 'distinctness' and 'clearness' of the sound there (see Part III, Fig. 2.5).

In the case of a purely exponential decay, all of the criteria discussed in this section are dependent on the reverberation time (and on each other). It is therefore useful in certain circumstances to give the deviations from the values corresponding to the ideal exponential case, regarded as a typical 'expected value', so that the actually measured value in a given case may be compared to the expected value, either as a quotient or as a

[9] Kürer, R., Dissertation, TU Berlin, 1972.

difference. We conclude this section with a summary of the corresponding 'expected values' for the various criteria:

T	=	0·5	1	2 s
ϑ	=	75	50	29%
R	=	−4·8	0	3·9 dB
C	=	9·1	3·1	−1·4 dB
t_r	=	25	50	100 ms
t_r'	=	14	28	56 ms
t_s	=	36	72	144 ms

II.7.5 Criteria for Annoyance Due to Echoes

For the integrals of the last section, it did not matter how the final values came about; that is, it was immaterial how the individual contributing reflections were distributed over time. Our most familiar experience showing that this is *not* always the case is the perception of echoes. Not only does a single reflection arriving within the time limit t_g not produce an echo at all, but the later-arriving reflections produce echoes only if they stand out above the general level of reverberant sound (as we have explained in Section I. 4.1).

Niese[1] tried to take into account these psychological observations by plotting the impulse response—he used $|p_{(t)}|$—together with an exponential decay (see Fig. 7.8, top: triangular peaks and solid line). He considered as 'detrimental' only those areas of the impulses that, after $t_g = 33$ ms, lie above the decay curve. The exponential decay that he used in this case was not simply the reverberation of the room, but the decay to be expected taking into account the additional time constant of the ear (see Section III. 1.4). Also, the initial value of the decay had to be evaluated in terms of an integral of the impulses arriving before 33 ms.[2] But these early impulses are not all to be regarded as fully 'useful'; instead, Niese used a weighting function based on arrival time, similar to that of Lochner and Burger.

To express the result of this rather complicated evaluation, Niese combined the 'detrimental' (*schädlich*) areas S and the 'useful' (*nützlich*)

[1] Niese, H., *Hochfrequenz u. Elektroak.*, **65** (1956) 4.
[2] Niese, H., *Acustica*, **11** (1961) 201.

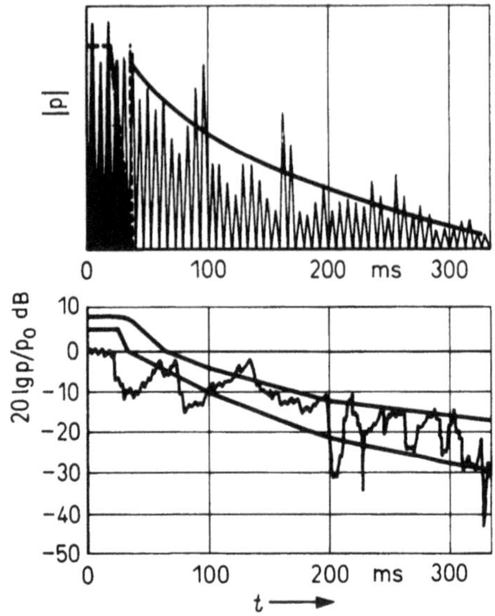

Fig. 7.8. Single-number criteria for the evaluation of echoes in an impulse response. Top: after Niese.[1] Bottom: after Bolt and Doak.[3]

areas N in the form:

$$\varepsilon = \frac{S}{N+S} \tag{7.16}$$

as a criterion for the existence of echoes; he called this quantity the 'echo coefficient'. It is an open question whether the essential idea of Niese (namely, only to count as detrimental those reflections that, after a certain limit time, exceed the reverberation curve) could be realized in a simpler way.

But more important is the fact that, in Niese's calculation, in spite of the considerable complication, the existence of a number of peaks that exceed the general decay only slightly would produce the same echo coefficient as a single strong reflection that certainly would be much more annoying.

One may ask, therefore, if the evaluation of echo annoyance is not better handled by the method of Bolt and Doak,[3] as shown in Fig. 7.8 (bottom). Here the impulse response is plotted in terms of level versus

[3] Bolt, R. H. and Doak, P. E., *J. Acoust. Soc. Am.*, **22** (1950) 507.

time, starting with $L_{(0)}=0$ dB. This curve is compared with a set of standard decay curves and is classified according to the curve that it just exceeds. (This 'tangent-fitting' principle is well-known in the evaluation of noise spectra.)

We now consider the case in which a sequence of rather weak reflections can give rise to severe annoyance, as, for example, in the flutter echoes that we discussed in Section I. 4.3. (We will come back to the subjective aspect of such phenomena in Section III. 3.4.) Here we only show how such periodic impulses (which may even be difficult to find in the impulse response, where they are mixed together with non-periodic impulses) can be filtered out and identified. For this, we have to record the so-called auto-correlation function corresponding to the impulse response $p_{(t)}$:

$$\phi_{(\tau)}=\int_0^\infty p_{(t)}p_{(t+\tau)}\,\mathrm{d}t \qquad (7.17)$$

Kuttruff[4] has proposed a method that allows this function to be evaluated with the simple means available to most room acousticians. It will be noticed that the quantity $\phi_{(0)}$ is nothing other than the integral that appears in the denominator of the 'distinctness coefficient' (see eqn. (7.9)).

If the time intervals between impulses were random, then ϕ would vanish for all τ greater than Δt, where Δt is the width of the impulses. But if the echogram contains periodic sequences of impulses with time-intervals τ_1, then ϕ would show peaks at $\tau=\tau_1$, $2\tau_1$, $3\tau_1$, ... Therefore, Kuttruff proposed to characterize the randomness of the impulse response by the ratio of $\phi_{(0)}$ and the maximum value of ϕ outside the origin:

$$\Delta=\frac{\phi_{(0)}}{\phi_{\max}(\tau\neq 0)} \qquad (7.18)$$

The larger the value of Δ, the more random the impulse response. He called this ratio the 'temporal diffusion', referring to the fact that a flutter echo can be produced only by parallel (highly *non*-diffuse) walls producing specular reflections, but not by diffuse-reflecting walls.

This criterion is useful not only for the diagnosis of such defects as flutter echoes, but also because it gives a general measure of the

[4] Kuttruff, H., *Acustica*, **16** (1965/66) 166.

randomness of a sequence of reflections. Assuming that any regular sequence of reflections is undesirable for hearing, because it may divert the listener's attention from the signal of interest, Holtzmark[5] proposed that one should make an auto-correlation analysis according to eqn. (7.17) for every echogram, or a spectral analysis of its Fourier-transform, the so-called power spectrum. For this analysis, he recommended multiplying the echogram by $e^{2\delta t}$ in order to compensate for the generally decreasing amplitude during the decay.

Because, so far, the only significant investigations have been those of Holtzmark and Kuttruff, it is still possible that the auto-correlation analysis of the impulse response will become more important in future studies of room acoustics.

II.7.6 Criteria for Directional Distribution of Sound

If room-acoustics evaluations required only the analysis of reverberation or of a single impulse response as a function of frequency and time, then a single-channel recording on a disk or tape would be entirely satisfactory, with playback over a single loudspeaker or through 'mono' earphones. But that is no way to convey the impression of spaciousness that one gets when listening in a concert hall. Our ears are sensitive not only to the time dependence of sound reflections and to the change of timbre during reverberation, but also to the direction of arrival of the reflections.

We are not referring here to our ability to localize the various instruments on the stage; this ability to spot the location of a sound source with a precision of about 3° is based on the 'law of the first wavefront' and it works equally well in all rooms, so long as the direct sound is responsible for that first wavefront (a condition that is frequently violated by an inept installation of a sound-reinforcement system!). (See Sections I. 5.2 and III. 1.6.) We mean, instead, the additional spatial impression of the room that depends on the later-arriving reflections, which convey a sense of the dimensions and the shape of the room.

The distribution of sound reflections in different directions not only differs on the average from room to room, but also from place to place within the same room. It is worthwhile, therefore, first to determine what

[5] Holtzmark, J., *Proc. 3rd. ICA, Stuttgart, 1959*, Vol. II, Elsevier, Amsterdam, 1961, p. 905.

kinds of differences can occur; this question was first studied by Meyer and his collaborators.

Thiele,[1] in his fundamental experiments studied this problem under steady-state conditions, using a warble tone of at least 2000 ± 1000 Hz. In this frequency range, a microphone placed at the focus of a parabolic reflector approximates the subjective directional characteristics of a listener (see Section III. 1.5).

In order to represent the measured distribution of sound direction, Thiele first used a continuous recording from a rotating receiver system. Fig. 7.9 shows the directional distribution of the intensities in the

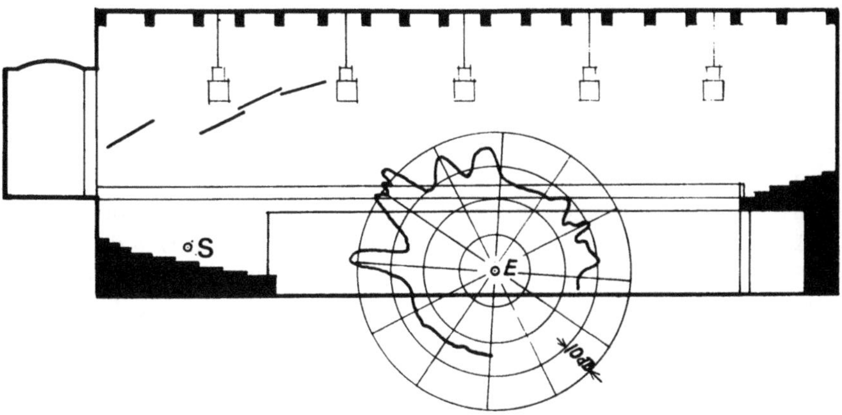

Fig. 7.9. Directional distribution of the sound intensity at location E on the main floor of the Herkulessaal in Munich. Source at S. (After Thiele.[1])

longitudinal section for a mean frequency of 4000 Hz, measured at a location in the middle of the main floor of the Herkulessaal in Munich. This polar diagram shows the respective maxima corresponding to the direct sound coming from location S, the reflected sound from the plexiglas panels above the stage, and from the coffered ceiling (see I. 5.3).[2]

Thiele proposed to condense the information from such a directional distribution into a single-number criterion called 'directional diffusion' with the symbol *d*. Since it is again a quantity with dimensions of unity, we will use the Greek letter Θ, instead. For this criterion Thiele first determined an auxiliary quantity, the sum of the absolute differences

[1] Thiele, R., *Acustica*, **3** (1953) 291.
[2] Cremer, L., *Schalltechnik*, **13** (1953) No. 5.

between the incoming intensity J over all room directions within the solid angle of interest Ω and the average incoming intensity $\bar{J}$; he normalized this sum to the average intensity $\bar{J}$:

$$\mu = \frac{1}{\Omega \bar{J}} \int_{\Omega} |J - \bar{J}| \, d\Omega \tag{7.19}$$

This quantity is zero for a non-directional, isotropic sound field, and has its maximum value μ_0 in open air. But since μ_0 depends on the directivity of the receiver, Thiele tried to eliminate this dependence by dividing μ by μ_0; and then, in order to get a quantity that increases with the degree of diffusion, he subtracted this ratio from 1. Thus he gets:

$$\Theta = 1 - (\mu/\mu_0) \tag{7.20}$$

But even here, the directivity of the receiver enters the picture, because a less directive receiver would round off the peaks and present a more smoothed directional distribution than a highly directional receiver. (Furthermore, it would appear mathematically more appropriate to make use of the variance of J, instead of the average of $|J - \bar{J}|$.)

The chief drawback to this criterion, as with all the 'averaging' criteria, is that eqn. (7.20) may yield the same value for a single strong directional peak as for several small peaks and valleys that are distributed equally over the solid angle and that have no subjective significance whatever.

However, until we learn more about our differential thresholds for directional impressions in a room, we cannot hope to replace Thiele's single-number characterization of the directional distribution with a criterion that is subjectively more adequate.

In every single-number representation, we lose information about the details of the distribution. Therefore, Meyer and Thiele[3] later proposed an entirely different kind of representation, according to which the entire solid angle is subdivided into steps of 10° in both directions. Since these steps corresponded to the directivity of the receiver, the information content was not diminished in their representation. The discrete measured fraction of power in each element of solid angle was represented by a rod stuck into a corresponding hole in a sphere, the length of the rod being proportional to the measured power. The resulting 'hedgehog', with its 200 quills, represented quite graphically the directional distri-

[3] Meyer, E. and Thiele, R., *Acustica*, **6** (1956) 425.

bution of incoming intensity, and this information could be preserved by means of photographs from above and from the side.

Fig. 7.10 shows photographs corresponding to three locations in the Beethoven Hall of the Liederhalle, in Stuttgart,[4] taken from the work of Junius.[5] The six hedgehogs shown above the longitudinal section correspond to the directional distributions of the early- and late-arriving

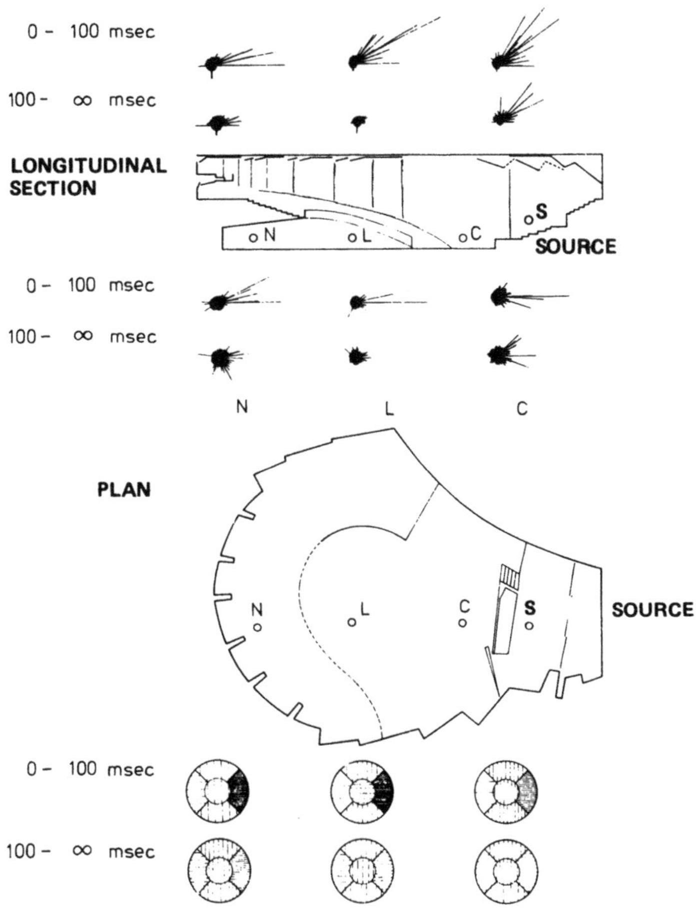

Fig. 7.10. Directional 'hedgehogs', for early- and late-time arrival, at three locations in the Beethoven Hall of the Liederhalle, in Stuttgart. (After Junius.[5])

[4] Cremer, L., Keidel, L. and Müller, H. A., *Acustica*, **6** (1956) 466.
[5] Junius, W., *Acustica*, **9** (1959) 289.

sound at the three seats (hedgehogs seen from the side). Another six, shown above the plan view, show the corresponding distributions in the horizontal plane (hedgehogs seen from above). In the upper group of hedgehogs, the rods directed toward the upper right represent sound reflections from the part of the ceiling and the reflectors above the stage.

In this figure, we see also something about the time-dependence of the directional distribution, based on earlier measurements by Meyer and Burgtorf.[6] For these studies, instead of a steady-state warble-tone, they used impulses for the sound source; the resulting incoming impulse responses were integrated over a longer time. The upper three hedgehogs in each group correspond to the time interval between 0 and 100 ms, the lower three to the interval from 100 to ∞ ms. As was to be expected, the directional distribution is more uniform in the later time interval (the hedgehog curls himself into a ball, so to speak, as the reverberation progresses).

These measurements make it clear that the assumptions underlying the statistical theory are very poorly fulfilled in a large hall, even in the later time period, and not at all in the early period. The latter is easy to understand since, during the first 100 ms, the direct sound from the stage has not yet reached the rear wall of a room of this size. It is not surprising, therefore, that, even in the time period beginning with 100 ms, the directional distribution is still far from uniform. (Since the volume of the Beethoven Hall is about 16 000 m^3, we should not expect a statistical field, according to eqn. (7.2), before $t_{st} = 253$ ms.)

Since an extension of the hedgehog representation to include its dependence on time would present more information than can be easily assimilated, Junius tried to condense the directional distribution given by the complete hedgehog, with its 200 quills, into a simpler form, over five principal directions: front, rear, left, right and above. These were believed to correspond to the most important directional sensations of a listener in a hall.

Such a condensation is shown in the lower part of Fig. 7.10, for both the early and later time intervals. Here, instead of representing the amount of incoming power from each direction by the length of the quill, it is presented by the darkness of the shading.

The need for highly directional receivers in these studies of directional distribution means not only that the results obtained (and therefore the

[6] Meyer, E. and Burgtorf, W., *Acustica*, **7** (1957) 525.

criteria derived from them) are frequency dependent, but also that the studies are expensive. Above all, the directional receiver takes up much more space than the human head, which derives its directional information by comparison of two signals, from the right and left ears.

We have already seen in Fig. 7.5 that these two signals are very different, even after their information content is restricted by rectifying and smoothing. It has been the hope of room acousticians, therefore, that a comparison of the signals from the right and left ears of a dummy head, and, in particular, some kind of transformation of the corresponding impulse responses, would lead to a directional criterion that matches our subjective hearing impression.

Up to this point, all of the proposed criteria have been based on an immediate comparison of the left-ear and right-ear sound pressures, $p_{l(t)}$ and $p_{r(t)}$, such that these signals are directly multiplied and integrated over various observation times. For example, Danilenko[7] defined a 'binaural distinctness coefficient':

$$\vartheta_{\mathrm{B}}=\int_0^{t_g} p_{l(t)}p_{r(t)}\,\mathrm{d}t \Big/ \int_0^{\infty} p_{l(t)}p_{r(t)}\,\mathrm{d}t \tag{7.21}$$

where he called t_g the 'limit time' (g from *G̲renze* = limit); this reverts to Thiele's 'distinctness coefficient' (eqn. (7.9)) if the left and right pressure signals are the same, as they would be for sounds arriving in the median plane of the head. The more the sound arrives from the sides, and the higher the frequency, the greater are the differences between the left and right ear signals, a condition that leads to smaller values for the integrals. Indeed, since the right and left ear signals can sometimes be out of phase, the integrals can even assume negative values, though this is unlikely in practice because of the typical preponderance of the direct sound and early reflections, which arrive more or less frontally. In each case we must expect the results to be strongly influenced by random fluctuations. Since these affect both the numerator and the denominator, this criterion is subject to much greater scattering than the simple 'distinctness coefficient'; indeed, the scatter in this criterion may exceed the variation that it attempts to describe.

Kürer[8] has extended the concept of 'center-time', defined for a

[7] Danilenko, L., Dissertation, TH Aachen, 1968.
[8] Kürer, R., Dissertation, TU Berlin, 1972.

monophonic signal in eqn. (7.15), to a binaural form:

$$t_{sB} = \int_0^\infty t\,|p_{l(t)}p_{r(t)}|\,\mathrm{d}t \Big/ \int_0^\infty |p_{l(t)}p_{r(t)}|\,\mathrm{d}t \tag{7.22}$$

Since this formula uses the absolute values of the products of the pressures, it avoids negative values and thus reduces the fluctuations in the numerator and denominator. But this criterion has the disadvantage that the denominator changes with the angle of incidence.

It would seem more suitable to choose, instead, for the denominator, as is customary for the normalization of the so-called 'cross-correlation function'

$$p_{lr} = \int_0^\infty p_{l(t)}p_{r(t+\tau)}\,\mathrm{d}t \tag{7.23}$$

the geometric mean of the integrals

$$\int_0^\infty p_{l(t)}^2\,\mathrm{d}t, \qquad \int_0^\infty p_{r(t)}^2\,\mathrm{d}t$$

such as we have already encountered in the denominator of eqn. (7.9) for the 'distinctness coefficient'.

In this case, we get the interaural cross-correlation function:

$$\kappa_{(\tau)} = \frac{\int_0^\infty p_{l(t)}p_{r(t+\tau)}\,\mathrm{d}t}{\left[\int_0^\infty p_{l(t)}^2\,\mathrm{d}t \int_0^\infty p_{r(t)}^2\,\mathrm{d}t\right]^{1/2}} \tag{7.24}$$

The numerator in this equation can still take on both positive and negative values. But it is only the absolute value $|\kappa|$ that is of interest, and since this again is a ratio of energies, it is reasonable to call $|\kappa|$ the 'interaural correlation coefficient'.[9]

[9] Cremer, L., *Acustica*, **35** (1976) 215.

Damaske[10] recommended the *maximum* of $|\kappa_{(\tau)}|$ as a room-acoustical criterion that characterizes the 'coherence' of the two ear signals. (See Section IV. 13.5, Volume 2, for a discussion of the distinction between correlation and coherence.) Here we observe only that the designation 'maximal interaural *correlation* coefficient' would be more suitable since we are interested more in a criterion that distinguishes between reflections incident from the front and the sides, rather than in whether a more or less statistically diffuse sound field is achieved (the question that interested Damaske).

We can practically exclude the latter possibility by restricting the integration time to, say, $t_g = 50$ ms. During this time, we may expect single, mutually coherent sound waves incident from certain directions. Keet[11] adopted the limited integration time t_g setting $\tau = 0$, in eqn. (7.24), to get the criterion:

$$[\kappa_{(0)}]_{t_g} = \frac{\int_0^{t_g} p_l p_r \,\mathrm{d}t}{\left(\int_0^{t_g} p_l^2 \,\mathrm{d}t \int_0^{t_g} p_r^2 \,\mathrm{d}t\right)^{1/2}} \tag{7.25}$$

which he called the 'short time correlation coefficient'.

All of the impulses contribute in the same way to the denominator; but they contribute more to the numerator the more they are incident in or near the median plane; therefore, the value of $\kappa_{(0)}$ will be greater for conditions where sound energy in the median plane is predominant.

Gottlob[12] combined the short time integration of Keet with Damaske's proposal to use the maximum value of $|\kappa|$, but restricted τ to values less than 1 ms only. Since the frontally incident direct sound governs the maximum value, however, the result is nearly the same as $\kappa_{(0)}$. But the evaluation of the maximum allows us to compensate for an oblique position of the dummy head with respect to the direct sound. Unfortunately, κ_{max} can also be dominated by a strong oblique reflection, especially if the direct sound is attenuated (see Fig. 7.11), for example, by

[10] Damaske, P., *Acustica*, **19** (1967/68) 199.
[11] Keet, W. de V., Dissertation, Capetown, 1969.
[12] Gottlob, D., Dissertation, Göttingen, 1973.

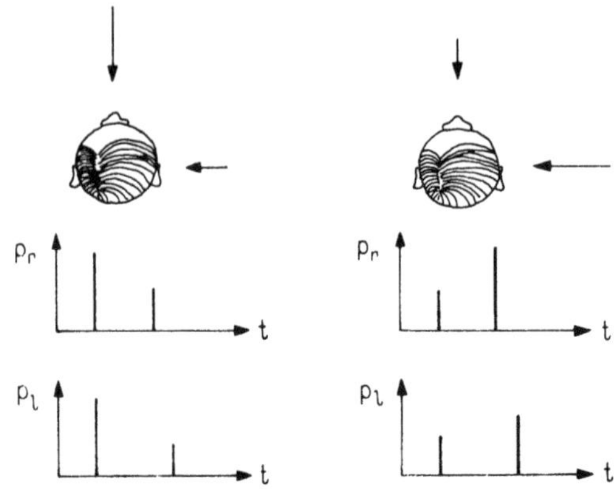

Fig. 7.11. Sketch showing the different evaluations of κ according to eqn. (7.24). Left: the direct frontal impulse predominates. Right: the reflected lateral impulse predominates.

passing over the audience at grazing incidence. Thus, the correlation between κ_{max} and the distribution of directions is still not unequivocal.

It should be mentioned here that $\tau \approx 0{\cdot}5$ ms is the greatest time delay to be expected between the two ears, even if the sound is incident from 90° to the side. We shall discuss the relation of this rather complicated criterion to subjective judgments in Section III. 3.1; here, we are entering upon matters that have not yet been fully resolved, but are still under development in discussions among room acousticians.

We can, however, state that, like the directional characteristics of the ear, all binaural criteria depend on frequency and, therefore, on the kind of impulse used. For this reason, test signals must either be specified as to their frequency content or restricted to certain especially chosen examples.

In the hope of deriving criteria that combine the requirements for desirable distributions of reflections in both time and direction of arrival, Reichardt and Lehmann[13] first combined these aspects in their 'room impression index' (*Raumeindrucksmass*):

$$R = 10 \log\left\{\left(\int_{25\,\mathrm{ms}}^{\infty} p_k^2\,dt - \int_{25\,\mathrm{ms}}^{80\,\mathrm{ms}} p_r^2\,dt\right) \Bigg/ \left(\int_{0}^{25\,\mathrm{ms}} p_k^2\,dt + \int_{25\,\mathrm{ms}}^{80\,\mathrm{ms}} p_r^2\,dt\right)\right\} \mathrm{dB} \qquad (7.26)$$

[13] Reichardt, W. and Lehmann, U., *Acustica*, **40** (1978) 277; see also Reichardt, W., *Gute Akustik—aber wie?*, VEB Verlag Technik, Berlin, 1979.

Here p_k is the sound pressure recorded with an omnidirectional (*Kugel*) microphone and p_r is the sound pressure recorded with a 'frontally directed' microphone (*Richtmikrofon*), defined by the requirement that it responds equally to sounds arriving within $\pm 40°$ of dead ahead and not at all to sounds arriving from directions outside that range; this requirement can scarcely be met even in a narrow frequency range and by no means at all frequencies of interest.

In this respect, the recommendation of Jordan,[14] to use a pressure-gradient microphone (receiver of first order) for the second receiver, is better defined and, moreover, is realizable over a wide band of frequencies. Jordan orients the directional microphone 'against' the frontal sound, so that he gives preference to the lateral sound and ignores the sound arriving in the median plane. With this combination of microphones, he chooses for a criterion the ratio of the integrals of the squares of the recorded early sound pressures, calling it the 'Lateral Efficiency'

$$LE = \int_{25\,\mathrm{ms}}^{80\,\mathrm{ms}} p_1^2\,\mathrm{d}t \Bigg/ \int_{0}^{80\,\mathrm{ms}} p_0^2\,\mathrm{d}t \qquad (7.27)$$

Here, p_0 is the sound pressure recorded with the omnidirectional (zero-order) microphone and p_1 is the sound pressure recorded with the pressure gradient (first-order) microphone.

Here again we are dealing with developments that are still under discussion.

[14] Jordan, V., *Acoustical Design of Concert Halls and Theatres*, Applied Science Publishers Ltd, London, 1980, pp. 159 and 191.

Part III:
Psychological Room Acoustics

Chapter III. 1

General Results of Psychoacoustics

III. 1.1 Psychoacoustical Research Methods

The proper aim of the room acoustics consultant, no doubt, is that the users of the auditorium, whether they be performers or audience, should be satisfied with the acoustics.[1] But what do we mean when we say that a room has good acoustics?

If we are concerned with a lecture room or a theater the question is easily answered: the acoustics are good if one can readily understand what is spoken. One can even measure this quality objectively with a bit of effort, in terms of the percent word intelligibility (or articulation index) (see Section III. 2.2).

The situation is altogether different when it comes to musical performances. In a concert hall almost everyone feels entitled to be dissatisfied with the 'acoustics'. No matter that the measured reverberation time is optimum and that the 'early reflections' are faultless: for the listener, his ear is the final judge and this is his right.

Moreover, if the judge is a well-known musician or music critic, his assessment of the acoustics will usually be accepted as valid even though it may not be based upon observations that others could reproduce. No acoustical consultant would dare to challenge the auditory evaluation of a music critic or to suggest that his views be put to an objective test!

[1] In German, the word *Akustik* meant originally the part of physical science that deals with sound. In English the word has that meaning, too; but we also speak of the 'acoustics' of a concert hall, meaning 'the way it sounds'. For the latter concept Cremer uses the German word proposed by Michel: *Hörsamkeit*, literally 'hearing-ness', but often translated generally as 'acoustical properties'. (See Michel, E., *Hörsamkeit grosser Räume*, Viewig, Braunschweig, 1921.) We will stick with 'acoustics' for both meanings.

Nevertheless it is generally true that subjective judgments vary greatly, even for the same person. This holds even for purely psychoacoustical matters such as equal loudness judgments into which no extra-acoustical influences or prejudices enter.

It is, of course, always possible, assuming that the individual result can be quantified by a number x_k, to calculate an average value of the differing subjective results according to the formula:

$$\bar{x} = \frac{1}{n} \sum_{k=1}^{n} x_k \tag{1.1}$$

Furthermore it is possible to compare the mean values obtained under different conditions, for example, in two different rooms. But such a comparison becomes meaningless if the scatter in the individual judgments, as characterized by the standard deviation:

$$\sigma = \sqrt{\frac{\Sigma (x_k - \bar{x})^2}{n-1}} \tag{1.2}$$

greatly exceeds the difference between the average values.

However, even when the difference in mean values exceeds the standard deviations (assumed to be equal in both cases), this could be the accidental result of the samples chosen for comparison, particularly if the samples are small in number. Mathematical statistics requires refutation of the so-called 'null hypothesis' (that the mean values are *not* different, but are the same) if only the number of test results under both conditions is sufficiently increased. We can regard the resulting difference as 'significant' only if the probability of the null-hypothesis is smaller than 5% and as 'very significant' only if it is smaller than 1%.

Unfortunately the relationships between this probability, the differences between the mean values, the standard deviations, and especially the numbers of the test results are not expressible in formulas: instead, tables are required. Furthermore, the tests for significance themselves depend on the special problem at hand. The reader is referred to the literature concerning mathematical statistics.[2]

The upshot, again and again, is that subjective judgments in room

[2] Arkin, H. and Colton, R. R., *Statistical Methods*, 5th. edn., Barnes and Noble Books, New York, 1972; Meyer, S. L., *Data Analysis for Scientists and Engineers*, Wiley and Sons, New York, 1975; Davenport, W. B., Jr. and Root, W. L., *An Introduction to the Theory of Random Signals and Noise*, McGraw-Hill, New York, 1958.

acoustics do not allow significant conclusions. Unassailable psychoacoustical conclusions require, in addition to a large number of test persons, further conditions that are hardly ever fulfilled in practice. For example, only one thing should be changed at a time if one is to evaluate accurately the effect of that change. Instead, comparisons of different concert halls are often made with different musical compositions and with different performers. Even if one wishes to compare the sound at different seats in the same hall, the musical signal is always changing (sometimes strings only, sometimes full orchestra with brass and percussion, sometimes pianissimo, sometimes fortissimo); the acoustical sample may differ more than, say, the 'acoustics' between the main floor and the balcony or between center and side seats in the hall.

It is even more difficult to fulfil the requirement that the observer not be aware of what experimental condition has been changed during his observations; in fact, this requirement is impossible to meet for *in situ* listening judgments. It can, however, be respected to some degree to the extent that we are able to record faithfully the sounds in question (on disks or tape, or in computers) and to reproduce (without identification) the recorded sound in a 'neutral' place by means of loudspeakers or earphones.

Under such conditions it is possible to test the repeatability of a person's judgment by presenting the same situation or the same comparison several times without his knowledge. By contrast, when a person knows what room he is in, it is probable that his acoustical judgment will be influenced by this knowledge, albeit unconsciously. Even at first hearing, one's acoustical evaluation may be biased by knowing the hall's reputation.

And this brings us to another important point: valid acoustical judgments must be made without reference to the opinions of other people. Thus, ideally there should be no discussions between persons during their listening evaluations; but it is quite impossible to avoid such comparisons during the first rehearsals in a new hall.

Furthermore, for valid judgments the situations to be compared should follow one another quickly in time; but again, this is practically impossible except with the playback of recorded examples. Comparisons, live, between different halls entail delays of (at least) several hours, more likely several days.

In fact, comparisons are frequently made of concert listening experiences separated by years! In such cases the musical judge may legitimately say that he remembers that he found the experience especially beautiful or especially disappointing. His evaluation gains weight only if it is written down at the time, for example, on a detailed questionnaire.

In most cases, however, earlier musical judgments are stated without reference to notes and are subject to the uncertainties of human memory.

Anyone who has been obliged for years to listen to such judgments—made under psychologically questionable conditions—can observe some inconsistency in the responses. For instance, at the first rehearsals in a new hall the artists tend to react either very cautiously or with extreme judgments ('fantastic' or 'disaster'). It is very important in such cases to take into account the hall in which they are accustomed to play.[3]

Just after the opening of a new hall, which usually entails a spectacular artistic event (or at least a social celebration), there is usually a 'happy ending' mood; this may indeed be a good time for the acoustics consultant to gather favorable acoustical opinions of the hall!

During the next phase, however, one must expect increasingly critical opinions as the number of people grows who hear the hall for the first time. And as with all subjective judgments, the discontented critics get more attention than the contented ones. Whoever is critical of the hall is at least interesting; whoever praises the hall risks the accusation of insufficient listening experience or of poor acoustical judgment.

The first two seasons are the proper time for evaluating the acoustics of the hall, for then, with many different musical compositions and performing groups, there will be a gradual settling of the judgment of the musical quality. In this period it may even happen that people who previously criticized the acoustics now declare that the sound has improved.

And indeed this is physically possible (at least in principle!): the seat upholstery may change through use, the paint may become porous, etc. But so far there have been no systematic objective investigations of such effects that could account for a change of critical mood. Instead, the acoustics consultant is confronted again and again with the physically untenable notion that a concert hall attains its fine acoustical quality in the course of time as a consequence of the vibrations excited in the structure by the music played there: the better the music, the better the hall will become! The implication is that halls improve with use as do old violins.

---

[3] See the discussions about London's Royal Festival Hall just after its opening; the musicians who had played earlier in the Royal Albert Hall were irritated by the acoustical clarity of the new hall. *Royal Inst. of Brit. Arch. Journal*, **59** (1951) 47.

Now it is by no means proved whether old violins acquire their fine quality as a result of being often played, or by whom, but it is surely more probable that intense mechanical excitation could influence the thin wood of a violin than that airborne sound in a concert hall could modify its massive structural elements! There is no doubt that musicians become aware of the peculiarities of either an instrument or a concert hall and in time will come to play better in each case. Also it is conceivable that the listeners become better acquainted with a room. At the very least we may conclude that a definitive judgment on the acoustical quality of a room becomes possible only when this 'asymptotic state' is at long last attained.

If the hall continues to please during this period, critical voices tend to be silenced as more beautiful performances take place and as one successful season follows another. (At this time the earlier critics may be reproached for lack of acoustical judgment!)

Even among the acoustical judges, however, there will be systematic differences; it is understandable that those who were involved in the conception and design of the project tend toward a more favorable opinion. Moreover, the inhabitants of the community may take pride in their hall, whereas listeners from out of town may remember their own hall more favorably. On the other hand, someone who has travelled far to hear the new hall may be inclined to overvalue the listening experience that he has travelled so far to enjoy.

With respect to rooms used for musical performances, the judgments of musicians are of special interest, particularly the professional players. Unfortunately, fruitful discussion between such musicians and acousticians is especially difficult. The acoustician must learn not to mind the fact that musicians often combine their (valid) aesthetic musical judgments with architectural (physical) explanations (or even proposals for changes in the hall) that are manifestly wrong. Understanding is rendered still more difficult by the fact that musicians and acousticians speak different languages and it may not be possible to translate music into physics. For instance, a statement such as 'the sound is not "responsive" enough' could mean either that the reverberation time is too great or too small. Furthermore, musicians tend to exaggerate: their scale of quality seems not to include continuous small gradations from 'excellent' to 'very good' to 'fair' and so on; the hall is either marvellous or a disaster.

Nevertheless, it is both important and interesting for the acoustician to try to answer the question whether the judgments of the musicians can be correlated with physical statements or even measurements in the hall.

Now, before we discuss the accepted criteria for room acoustics in a hall, we must consider some basic psychological phenomena. In this discussion we will become acquainted with the different levels of difficulty in psychoacoustical tests.

The simplest test concerns the question of whether an acoustical signal can be perceived at all: we may call this the 'threshold of perception'.

Related to this is the question of whether two signals are different (the 'difference limen'); if we restrict the question to some special quality of the signal, the perception of differences requires more ability on the part of the listener.

Again a step more difficult is the question of whether two signals are *equal* in some special respect, such as 'loudness'. In this case there are two possible approaches. The simple method is based on the possibility that the test subject can control the signals himself. He may begin with the situation where, without doubt, signal A_1 is 'more' (in respect to some quality) than signal A_2; then he decreases A_1 until he is certain that it is less than A_2, and so on. This method is called 'alternating approximation'. (It would also be possible for A_1 to approach A_2 from one side; but this method is less satisfactory because it is not certain whether approaching from the other side would result in the same final value.)

The second method requires more time but it is regarded as more reliable by psychologists. In this case the test person hears different constant-stimulus signals, $A_1, A_2 \ldots A_n$, as well as one reference stimulus, A_0, for comparison. He must decide in each case whether A_n is greater or less than A_0.

The most difficult task is the *quantification* of sensations. So long as we are concerned only with 'scaling' in gross steps (say, between 'extraordinary' and 'insufficient'), the task may be easy, even easier than to establish a difference threshold. But it is difficult to decide when a sensation has been doubled or halved. Some psychologists even doubt that such judgments are possible at all.

III.1.2 Loudness

The threshold of hearing for pure tones in the frequency region of 1000 to 4000 Hz lies at the limit of what is physically and physiologically just possible. At higher, and especially at lower, frequencies the corresponding threshold sound pressure increases, as shown by the lower curve in Fig. 1.1.

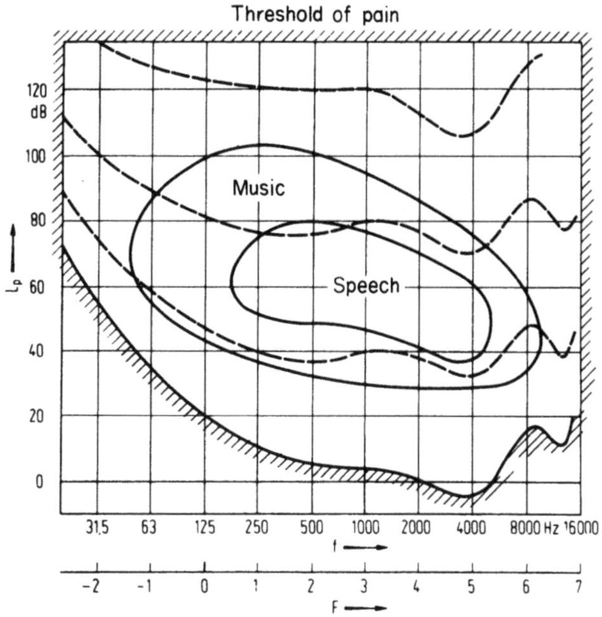

Fig. 1.1. Curves of equal loudness level.

In this figure the ordinate represents the sound pressure level, defined by

$$L_p = 20 \log p/p_0 \qquad \mathrm{dB}$$
$$(p_0 = 2 \times 10^{-5}\ \mathrm{Pa}) \tag{1.3}$$

it is plotted on a logarithmic frequency scale. (The F scale on the abscissa is the Frequency Index, defined below in eqn. 1.10.)

The hearing threshold depends on the angle of incidence of the incoming sound; the plotted sound pressure is that of the undisturbed wave. The curves of Fig. 1.1 pertain to frontal incidence in a free sound field. The fluctuations in the curves at frequencies above 2000 Hz arise from the disturbance of the sound field by the head, whose width becomes comparable with the wavelength at that frequency.

Since every sound receiver exhibits frequency dependence as well as upper and lower frequency limits, we must expect similar behavior in the ear, that most complicated and most efficient of receivers.

In this case the upper limit is sharply defined and is almost independent of the sound pressure; for young people it lies in the vicinity of 16 000 Hz but it becomes lower with increasing age.

The low-frequency limit of audibility is more difficult to define because it is very difficult to produce pure tones in the ear at low frequencies. The motion of the mechanical elements of the ear at low frequencies becomes so great as to introduce non-linear distortion, leading to the production of partial tones of higher order. These partial tones of higher frequency, rather than the intended low-frequency pure tone, may establish the threshold of audibility.

The reason for limiting the frequency scale to 20 Hz, at the left in Fig. 1.1, is that just below this frequency is the 'flicker limit' characteristic of all our sense organs. Above that limit, sequences of acoustic impulses blend into one another to produce the perception of a steady sound; but below the limit we tend to hear the individual impulses separately.

The upper limit of sound pressure level in Fig. 1.1 is also arbitrarily chosen. This does not mean that higher sound pressure levels cannot exist; unfortunately they can. But the designation 'threshold of pain' indicates that in the neighborhood of the indicated levels a different kind of aural perception occurs: a little feeling of pain in the ears. Sudden mechanical damage to the ear occurs only at much higher levels; but permanent hearing loss and other harm can occur with exposure to sound at a level below the pain threshold, if it is sufficiently prolonged.

If we consider that the range between the threshold of audibility and the threshold of pain spans about 130 decibels, corresponding to a ratio of sound intensities of $10^{13}:1$, we begin to comprehend the astonishing capability of human hearing. It can cope, on the one hand, with peaks of sound so loud as to produce pain but it can also detect, on the other hand, the very quietest of sounds, which must originally have served as a warning of approaching danger.

Furthermore, the ten-octave span in frequency of audible sounds is extraordinary. By contrast, the range of visible wavelengths of light covers only about one octave.

In Fig. 1.1, the area enclosed by hatched lines is the so-called audible region. Man uses only a small part of this region for speech; even for music he does not require the entire region. (We note that the upper portion of the area designated 'music' in Fig. 1.1 applies only to fortissimo passages for large orchestra with chorus. The very high levels sometimes reached by rock bands, with the help of electroacoustical amplification, are not included in the figure because they are actually harmful to the ears and should be prohibited.)

We must pay special attention to the sound that lies below the areas corresponding to speech and music in Fig. 1.1 because, particularly for levels around 20 to 25 dB, these often occur as disturbing background noise in auditoriums. Usually the noise generated by a large audience exceeds those levels; but from time to time in a performance there come moments when the audience literally holds its breath, in thrall to the music. In these precious moments even a noise of only 20 dB can be perceived as disturbing. It is for this reason that effective noise control plays such an important role in the design of theaters and concert halls.

This is especially true if the auditorium is to be used for recording. The listener, sitting in front of his loudspeakers at home, is deprived of certain acoustical cues that, if he were sitting in the auditorium, would help him distinguish between the music of interest and the unwanted noise. In addition, during quiet passages he (or the sound engineer) sometimes turns up the gain so that he can hear the music better, but this also increases the recorded background noise that was unintentionally picked up from the auditorium.

Figure 1.1 also shows three dashed curves, corresponding to the levels of pure tones at various frequencies that are perceived as being equally as loud as a 1000 Hz tone at 40, 80 and 120 dB, respectively. If the ear were a linear receiver, these curves would lie parallel to the threshold curve which corresponds to zero loudness level. As a matter of fact, surprisingly, this actually is nearly so for frequencies above about 1000 Hz, despite the enormous differences in sound pressure ratios. Below 500 Hz there is a rapid increase of the threshold curve toward low frequencies, but this disappears for high sound levels. This 'compression' of the contours of equal loudness level at low frequencies has important consequences for room acoustics.

McNair[1] noted that, as a result of this compression, the timbre of a broad-band signal must change subjectively as the loudness decreases during a reverberant decay. Since he regarded such a change as undesirable, he concluded that one should provide for an increase in reverberation time toward low frequencies in auditoriums (about 50% longer at 125 Hz than at 500 Hz).

A similar consideration concerns the perceived spectral balance during steady excitation, as with sustained chords. For this condition we found

[1] McNair, W. A., *J. Acoust. Soc. Am.*, **1** (1930) 242.

(in Chapter II.1) that the energy density E in the room is given by

$$E = P/4cA = 1{\cdot}53\,PT/cV \quad (1.4)$$

where P is the source power in watts, c is the speed of sound in $\mathrm{m\,s^{-1}}$, and V is the room volume in $\mathrm{m^3}$. Since the available power is always limited, it may happen in large halls that the energy density and the corresponding sound pressure level are too low, a problem that cannot be overcome by increasing the reverberation time with increasing room volume. The loss of sound level is subjectively more noticeable at low than at high frequencies, however; moreover, we are much more sensitive to changes in timbre than to changes in overall sound level. Therefore it is desirable, particularly in large rooms, that the reverberation time should increase toward the low frequencies. This is especially true in large churches where the low-frequency flue pipes of the organ are inherently less powerful than the high-frequency pipes.

On the other hand, it may happen in small rooms that an orchestra accustomed to playing in large halls will complain of too much low-frequency sound.

In fact, Békésy[2] found that musicians, who were asked to adjust the variable sound-absorbing elements in small studios of 180 to 4000 $\mathrm{m^3}$ to achieve the best tonal balance, chose a reverberation time that was practically independent of frequency (see Fig. 1.2).

In Figure 1.1 we have effectively plotted the sound pressure on a logarithmic scale (by plotting the sound pressure *level* on a linear scale);

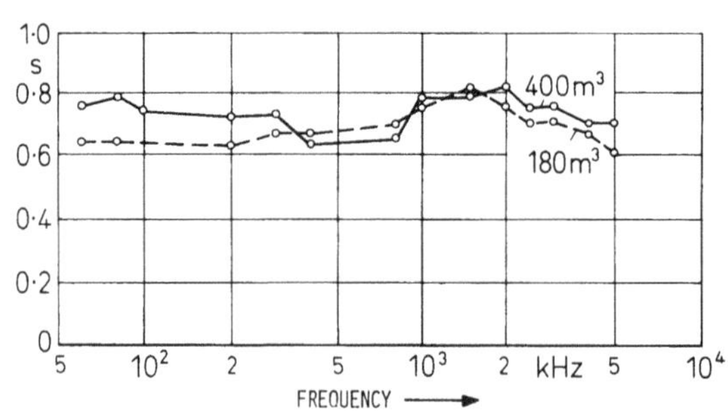

Fig. 1.2. The preferred frequency dependence of reverberation time in small studios, adjusted by musicians with the aid of variable sound-absorptive elements. (After Békésy.[2])

[2] Békésy, G. v., *Ann. Phys. (Leipzig)*, *V*, **19** (1934) 665.

the primary reason for this, nowadays, is to accommodate the enormous range of levels of practical interest.

But it was assumed originally that a logarithmic evaluation of the sound pressure would correspond to the subjectively perceived loudness. When it became evident that such an evaluation led to different scales for different frequencies, an alternative approach was adopted. Each sound was directly compared with a pure tone at 1000 Hz, whose level was adjusted until it sounded just as loud as the test sound. The sound pressure level of the 1000 Hz tone, when it was so adjusted, was assigned as the 'loudness level' L_s of the test sound. In order to help distinguish the loudness level from the sound pressure level, a new unit was adopted for loudness level, called the phon.

$$L_s = [L_p \text{ (1000 Hz)/dB}] \quad \text{phon} \tag{1.5}$$

Since this unit has found its way into regulatory legislation, we are obliged to use this definition for characterizing noises, although we have recognized in the meantime that a true subjective evaluation of loudness is by no means given by the loudness level.

Fletcher was the first to raise the question of a subjectively adequate scale of loudness, basing his work on the plausible assumption that the loudness of a sound heard with both ears should be twice as great as when heard with only one ear, since twice the number of nerves would be excited. Thus, he increased the level of a sound heard with only one ear until it was judged to be equally as loud as the original sound heard at a lower level with two ears; the difference in loudness levels corresponded to doubling the subjective loudness.

Although many psychologists today doubt that Fletcher's hypothesis is correct, their later researches have produced nearly the same results. Apparently the ear executes a spectral analysis and then evaluates the low-frequency and high-frequency components separately, with respect to loudness. Thus, loudness may also be doubled by adding together equally loud signals belonging to distant parts of the frequency spectrum.

A quite different approach, the estimation of loudness fractions, also resulted in the same loudness scale: an increase of 10 phons in loudness level always corresponds approximately to a doubling of the loudness.

It was found desirable to introduce another new unit for loudness, called the 'sone'; it is defined by the loudness of a sound whose loudness level is 40 phons.

Thus, the relation between the loudness S (in sones) and the loudness level L_s (in phons), for pure tones above 40 phons, is given by

$$S = 2^{(L_s - 40)10} \tag{1.6}$$

Above 1000 Hz, at which frequency the loudness level and the sound pressure level are equal, we may also express L_s by the rms sound pressure $\tilde{p}$, replacing L_s (in phons) by L_p (in dB). We thus get:

$$S = \left(\frac{\tilde{p}}{\tilde{p}_1}\right)^{0\cdot 6} \approx \left(\frac{\tilde{p}}{\tilde{p}_1}\right)^{2/3} \quad \text{sones} \tag{1.7}$$

in which $\tilde{p}_1$ is the rms sound pressure corresponding to 40 phons, i.e. 2×10^{-3} Pa. As for the intensities, we get the corresponding relation:

$$S = (J/J_1)^{1/3} \quad \text{sones} \tag{1.7a}$$

where J_1 is the intensity 10^{-8} W m^{-2} corresponding to 1 sone.

Zwicker and Feldtkeller[3] found that the exponent k for the intensities ($2k$ for the sound pressures) depends upon the nature of the signal. For broad-band noises, which are mathematically more similar to running speech and music than are pure tones, a lower value of the exponent ($k = 1/4$) is better:

$$S = (J/J_1)^{1/4} = \left(\frac{\tilde{p}}{\tilde{p}_1}\right)^{1/2} \quad \text{sones} \tag{1.7b}$$

The power laws with exponents $k < 1$ have one tendency in common with the logarithmic function, that of 'saturation': the higher the sound pressure, the smaller the increment of sensation for equal increments of excitation. From eqns. (1.3) and (1.5) we can derive:

$$\frac{dL_s}{d\tilde{p}} = \frac{(20 \log e)\, \tilde{p}_0}{\tilde{p}} \sim \frac{1}{\tilde{p}} \tag{1.8a}$$

and from eqns. (1.7) we get:

$$\frac{dS}{d\tilde{p}} = 2k\tilde{p}^{(2k-1)}/\tilde{p}_1^{2k} \sim \frac{1}{\tilde{p}^{(1-2k)}} \tag{1.8b}$$

Again it turns out that an echogram, with pressure p plotted against time, does not correspond to the evaluation by the ear. Whereas the

[3] Zwicker, E. and Feldtkeller, R., *Das Ohr als Nachrichtenempfänger*, Hirzel, Stuttgart, 1967, p. 130.

exponents k (in eqns. (1.7) to (1.7b) were determined with steady sound signals, a transformation from a p-scale to a $\sqrt{p}$-scale will yield a more ear-like evaluation for impulse responses; it enhances the weaker reflections relative to the stronger ones, corresponding better with our subjective aural perception.

In one respect, the power laws differ fundamentally from the logarithmic function: in the latter case for each recording of a level the appearance depends on the mean level. This is not true for the power law, for which the echograms remain similar at all levels.

Therefore, the exponential decay of sound pressure results in an exponential decay of the loudness, to the extent that the exponent k can be regarded as constant during the decay:

$$S = S_0 e^{-2k\delta t} \tag{1.9}$$

In this equation the decay of loudness appears with a reduced decay coefficient, $k\delta$. Thus one can say that the subjective reverberation time, so defined, is three times as long as the objective reverberation time for pure tones and narrow-band noises, and four times as long for broad-band noises. (This does not mean, however, that we hear a two-second reverberation as though it were three or four times longer, because we are already accustomed to express our subjective sensation of a sound decay by the objective reverberation time.)

Now, the power laws are valid only above 40 phons (i.e. 1 sone). Below this limit, as stated by Fletcher, the loudness increases more rapidly with increasing sound pressure (or loudness level), see Fig. 1.3. (On account of the large differences in sound pressure, it is again appropriate to plot

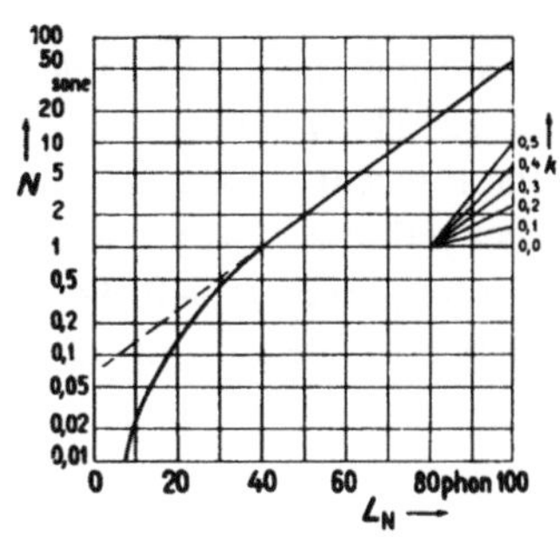

Fig. 1.3. Dependence of the loudness on the loudness level. (Adapted from Zwicker and Feldtkeller,[3] Fig. 49.2.)

both sound pressure and loudness on a logarithmic scale; on a linear scale, changes in the exponent k would hardly be noticeable.) The consequence is that echograms, when plotted in a subjectively accurate way, depend on the loudness of the source and are not similar for different excitations. This is also one of the reasons that room-acoustics impressions depend on the loudness of the sound. For music played *piano*, the perceived direction of the incident sound coincides with the extent of the sound sources; in *fortissimo*, however, we become aware of the room because of sound reflected from the walls and ceiling (see Section III. 2.5). Note, however, that another phenomenon comes into play here which complicates the matter: when instruments are played loudly, they produce sounds richer in overtones.[4]

III.1.3 Balance and the Perception of Timbre

In practically all frequency plots in acoustics, it is the logarithm of frequency that is plotted, as in Fig. 1.1. This is justified, in the first place, by the extraordinary range of audible frequencies. But also the logarithmic scale corresponds to the musical scale of tones separated by equal 'half-tone' intervals (despite some inconsistencies in the vertical notation of the musical staff D to E is a whole step, while E to F is only a half step); the frequency of each tone is higher by a factor of $2^{1/12}$ (about 6·0%) than that of the next lower half-tone.

For the abscissa in Fig. 1.1, it is also possible to define a Frequency Index[1] F plotted linearly in the figure and defined by

$$F = \log_2 (f/125) \quad \text{oct.} \tag{1.10}$$

We call this an 'index' because it is a logarithmic ratio of frequencies with a reference frequency of 125 Hz; all tones, which in musical usage are named with a letter of the musical scale and a subscript n, correspond to a value of F whose integer value is always the subscript n. For example, the tuning note a_1, whose frequency is 440 Hz, corresponds to $F = 1{\cdot}82$.[2]

It would also have been possible to represent the abscissa of Fig. 1.1

---

[4] Wettschureck, R., Dissertation, TU Berlin, 1976.

[1] DIN 13320, June 1979.

[2] The American Standard Pitch, adopted in 1936, uses a different set of designations for tones, with capital letters instead of lower case and different subscripts: for example, the tuning note (440 Hz) is named A_4.

with a piano keyboard, where the equally spaced keys would (at least approximately) represent equal musical intervals.

The octave relationships represented in the musical designations a_0, a_1, a_2, ... may have a physiological basis, not only because of the consonance that appears when two identical melodies are played an octave apart (which sound rough if not performed with absolute accuracy) but also on account of their apparent similarity if they are played one after the other.

Let us assume that the filtering process in the ear corresponds to filters of equal relative width $\Delta f/f$, for instance, one-third octaves. Then the analysis of a sequence of Dirac impulses with a low fundamental frequency will display the low partial components only as separate sinusoidal tones (see Fig. 1.4). In the higher third-octave bands, several such tones fall within a single filter bandwidth and amplitude-modulated sounds appear, of which the envelope has the period of the fundamental frequency. If we listen to such a tone combination, we hear a pitch that is independent of the carrier frequency (i.e. the frequency of the filter). Schouten[3] has called this pitch the 'pitch of the residue'. It has been physiologically proved that the nerve impulses that leave the inner ear after its filtering contain this period.

Therefore, it is possible, though not yet proven, that the difference limens between neighboring frequencies are determined not only by places of maximal excitation in the inner ear but also by differences in the periods of the nerve impulses.[4] It is certain, however, that the difference limens of pitch at high frequencies (above about 500 Hz) exhibit a constant relative width; but they change more and more, toward low frequencies, to constant absolute widths. If we plot the summation of the (approximately) 600 directly perceivable pitch differences, thus defining a magnitude of pitch, against the log-frequency of the test tones (at a level of 60 dB), we get the curve of Fig. 1.5.[5] The same curve results when people are asked to double or to halve the sensation of pitch.

Stevens *et al.*[6] defined a unit for this pitch scale by arbitrarily assigning 1000 units of pitch to correspond to a frequency of 1000 Hz; the unit is

[3] Schouten, J. F., *Proc. K. Ned. Akad. Wet.*, **41** (1938) 1083; **43** (1940) 356 and 991.

[4] A simple model that produces such a two-fold pitch analysis was developed by L. Schreiber (Dissertation, TU Berlin, 1962).

[5] Stevens, S. S. and Davis, H., *Hearing*, Wiley, New York, 1938, p. 81.

[6] Stevens, S. S., Volkman, J. and Newman, E. B., *J. Acoust. Soc. Am.*, **8** (1937) 185.

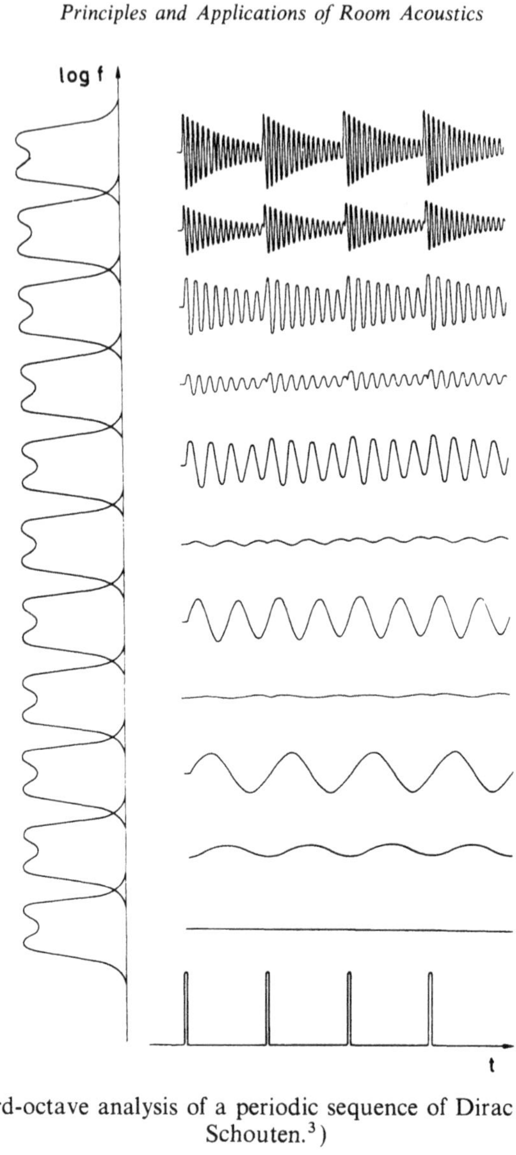

Fig. 1.4. Third-octave analysis of a periodic sequence of Dirac impulses. (After Schouten.[3])

called the 'mel', from the word 'melody'. This relation is doubtful, however, because for a melody played once in descant and once in bass, the intervals (i.e. the differences on the abscissa of Fig. 1.5) remain the same, but not those on the ordinate.

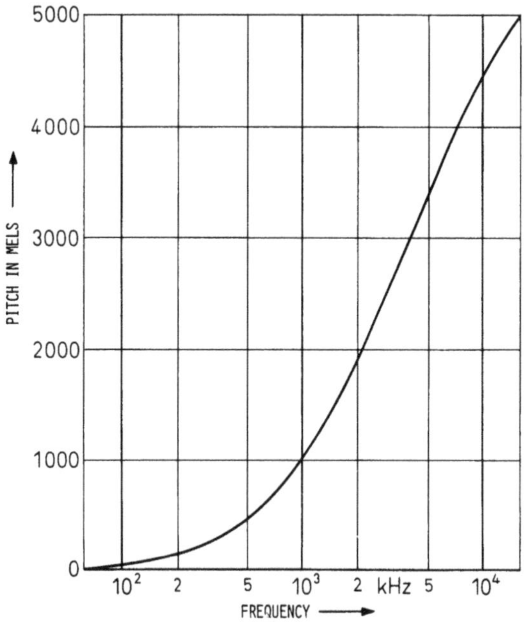

Fig. 1.5. Pitch in mels as a function of frequency.

A third method exists which results in the same summation of difference limens; it is especially important for consideration of the loudness of complex signals. Zwicker and Feldtkeller[7] compared the loudnesses of narrow-band random noises, holding the rms sound pressure constant but increasing the bandwidth in steps. By this means they discovered the astonishing fact (see Fig. 1.6) that for small bandwidths Δf all of the signals were perceived as equally loud, whatever the bandwidth. But when a certain limiting bandwidth Δf_g was exceeded (which does not depend upon the level), the perceived loudness increases with increasing bandwidth. This may be taken as an indication of the physiological fact that now the nerves in a neighboring filtering region are being excited; the reason is that they respond, according to the saturation tendency, in the steeper part of the $S_{(J)}$-function of eqn. (1.7b), so that their contributions accrue with greater increments.

The widths of the 'frequency groups' determined in this manner change

[7] Zwicker, E. and Feldtkeller, R., *Das Ohr als Nachrichtenempfänger*, Hirzel, Stuttgart, 1967, p. 123.

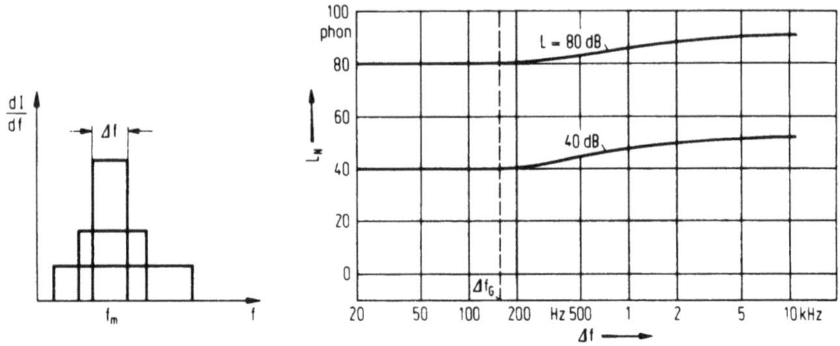

Fig. 1.6. Determination of the 'width of frequency groups', Δf_g. (After Zwicker and Feldtkeller.[7])

in the same way with center-frequency as the difference limens for pure tones presented one after the other, and they span rather precisely 100 mels.[8] For calculation of the loudness of broadband signals, Zwicker and Feldtkeller subdivide the audible frequency region (20–16 000 Hz) into 24 groups.

Now, having found a means of plotting frequency and (in the last section) a means of plotting sound pressure that correspond to the behavior of the human ear, we may be tempted to suppose that plotting the partial loudnesses, in their respective frequency groups, would yield an adequate representation of the subjectively perceived spectrum and that the total plotted loudness area would adequately represent the overall loudness of the signal.

But this would overlook another essential aspect of the hearing process, namely, the 'masking' of one tone by another, or by a narrow band of noise. In this connection, the level of the masking signal need be only about 6 dB higher than that of the masked signal in order to be effective.

This masking effect is so well known to everyone from everyday experience that it is astonishing how seldom laymen (including musicians!) take this phenomenon into account in making judgments about room acoustics. Remarks such as: 'There are places in this hall where the

[8] Zwicker and Feldtkeller's presentation of the mel scale differs from the original scale shown in Fig. 1.5. They proposed for this 'group number' ζ a new unit, Bark, in honor of H. Barkhausen.

violins cannot be heard at all' may be understood only by assuming that the listener found himself at such places at times when the orchestra was playing in fortissimo with full brass, times when the violins would never be heard in *any* location no matter how loud they were playing.

When we consider that a single trombone can radiate 6 watts of sound power while a violin can produce at most about 6 milliwatts,[9] we see that a section of 16 violins at its loudest produces a sound level 18 dB below that of one trombone; it is therefore hardly surprising that the violin section often cannot be heard above the brass choir!

A room-acoustics consultant, in the design of a concert hall, can change the balance between instrumental groups by at most about 6 dB, providing nearby reflecting surfaces for the weaker instruments; he can succeed only in assuring that the masking, which is always possible, happens less often.

Much greater level differences can be achieved by the musicians themselves, playing at different dynamics under the guidance of the conductor.

But now, in the context of loudness balance, we see what an extraordinary location the orchestra conductor occupies. He stands very close to the string section while the brasses are relatively far away. A listener at the back of the hall may sometimes be very surprised to see the conductor urge the violins to play more softly with his left hand, while, with his right hand, he encourages the brasses to play more loudly—although for that listener the strings are already masked.

'A classic example occurs in the first movement of Beethoven's Ninth Symphony. When the strings play the main theme in pianissimo at the beginning, they can be readily heard—provided the audience has quieted down—even though the horns are holding a long chord. But when they repeat the passage in fortissimo at bar 301, it is much more difficult to hear them because now the winds and brass, including trumpets, play the chords in fortissimo as well.

In striking contrast, in the context of auditory masking, are the pins dropped on the stage in antique theaters in order to demonstrate to the astonished tourists the admirable acoustics of the building. In fact, all that is demonstrated is the 'non-masking' associated with the extraordinary quietness that still exists at places far from the noise of traffic.

Auditory masking occurs not only in those parts of the frequency

[9] Sivian, L. J., Dunn, H. K. and White, D. S., *J. Acoust. Soc. Am.*, **2** (1931) 330.

spectrum where there are components of the masking sound. Figure 1.7 shows as a function of frequency the level of a test tone L_T that can just be heard when a narrowband noise at 1000 Hz is simultaneously sounded, the spectral width of which is still within the corresponding frequency group. This level is called the 'masking threshold'; it appears in the figure as a peak rising above the normal hearing threshold, which is broader the higher the level of the masking sound. (For this reason, as we have seen in Fig. 1.1, the ear cannot be represented by linear networks.) The peak has a steep slope toward low frequencies but a rather mild slope toward high frequencies, where several frequency groups are always covered. One can conclude that the same thing happens for the regions of excited nerves in the inner ear.

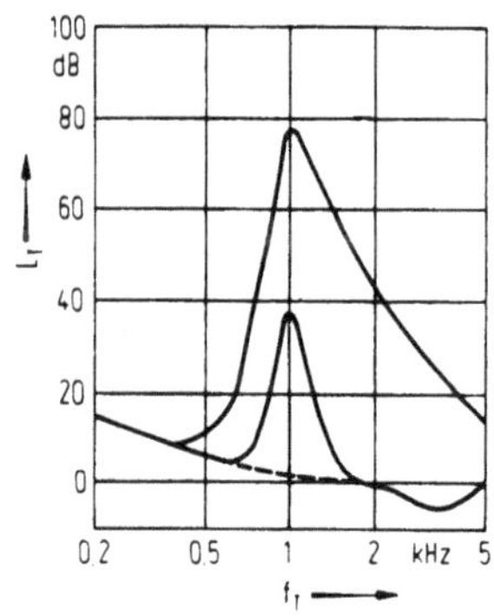

Fig. 1.7. Masking of a tone by narrowband noise. (After Zwicker and Feldtkeller.[7])

Zwicker and Feldtkeller[10] took into account this broadening of the excitation and the masking in a simple but adequate way in their development of the 'loudness-group-spectra' diagrams, adding to the right of each frequency group (in which the loudness component for that group is represented by a horizontal line) a downward-sloping line, or scarp, representing the upward spread of masking. If the partial loudness of a higher frequency group lies below this scarp, it does not contribute to the total loudness: it is totally masked. The slight broadening toward lower frequencies is not taken into account as being relatively unimportant.

[10] Zwicker and Feldtkeller,[7] p. 184 ff.

Figure 1.8 shows the loudness-group-spectra constructed in this way for two 1000 Hz tones at different levels.

The broadening of the loudness area with increasing level makes it understandable that the loudness of a sinusoidal tone increases with increasing level more rapidly than that of a broadband noise. The increase in loudness of the latter may be estimated from the third-octave levels that are marked at some of the horizontal lines in the diagram.

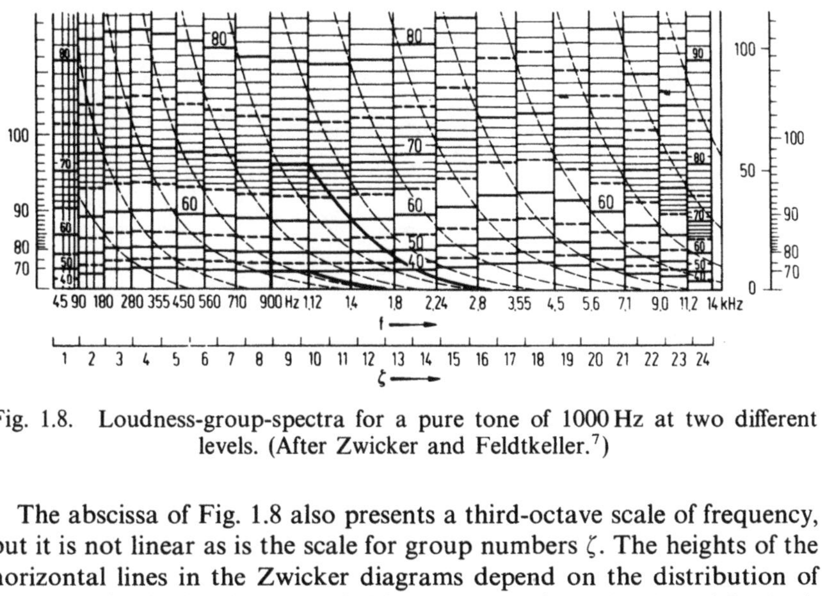

Fig. 1.8. Loudness-group-spectra for a pure tone of 1000 Hz at two different levels. (After Zwicker and Feldtkeller.[7])

The abscissa of Fig. 1.8 also presents a third-octave scale of frequency, but it is not linear as is the scale for group numbers ζ. The heights of the horizontal lines in the Zwicker diagrams depend on the distribution of directions in the incident sound. Diagrams have been developed for both frontal and isotropically diffuse sound fields.

Zwicker and Feldtkeller[11] have also developed an apparatus that records the loudness-group-spectra automatically. For this purpose, the scarps are produced by a decay process during the transport of the recording paper. Figure 1.9 shows a diagram of the noise of a planing machine recorded by this method.

The room-acoustical question as to how much the loudnesses differ at different locations in a concert hall, or at comparable locations in different halls, can be decided exactly only by comparison of the corresponding areas in such loudness diagrams. This would be even more

[11] Zwicker and Feldtkeller,[7] p. 200.

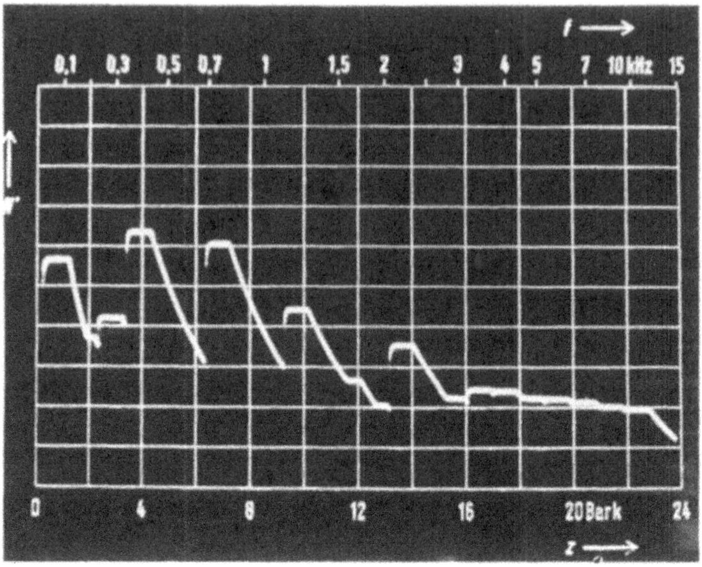

Fig. 1.9. Loudness-group-spectrum of the noise of a planing machine, recorded by automatic apparatus. (After Zwicker and Feldtkeller.[7])

true for a comparison of the impressions of timbre, for which we may take the loudness-group-diagrams as a representation of 'subjective spectra'. The loudness-group-spectra would make clear the extraordinary dependence of all room-acoustical judgements upon the nature of the acoustical signal in each case.

III. 1.4 Limit of Perceptibility and the Inertia of the Ear

Steady-state signals do not occur in speech and only seldom in music, even if we consider that a duration of one second is long enough for all field- and ear-conditioned transients to have died away.

The duration of speech elements may be derived from the fact that we speak about five syllables in a second; each syllable may consist of a vowel preceded and followed by consonants, the transitions to and from the vowels being perceived as characteristic of the speech. Thus, there are about 15 distinguishable speech elements per second, corresponding to a duration of about 70 ms per speech element.

With respect to music, Reichardt and Kussev[1] found, by recording and

[1] Reichardt, W. and Kussev, A., *Z. elektr. Inform. u. Energietechnik, Leipzig*, **3** (2) (1972) 66.

analysing examples of music from different periods, that the shortest duration for notes in string passages is about 73 ms. But since both in speech and music one can perceive certain details within these time durations, we can assume that the ear is capable of discriminating even shorter time intervals.

On the other hand, in Section III. 1.2 we were led to 20 Hz, the flicker limit, as the frequency below which (as the frequency decreases) the sequence of individual impulses begins to be recognized as such, and above which (as the frequency increases) the impulses blend together into a steady tone. (Everybody recognizes this phenomenon in the sound of a motorcycle starting up.)

Thus, we may say that the ear can distinguish sound impulses that are separated in time by:

$$\Delta t_{\min} = 1/20\,\text{Hz} = 50\,\text{ms} \qquad (1.11)$$

The same time limit appears for the perceptibility of a reflection following the direct sound.

If two people, walking and talking in a reflection-free field, approach the wall of a large building, they become aware at a certain distance that they are receiving reflections of their speech from the wall. They may perceive single words as echoes. This impression becomes stronger upon closer approach, since the reflected sound pressure increases. But when the distance becomes less than 17 m, corresponding to a transit time of 100 ms for the path to and from the wall, this impression tends more and more to disappear, even though the reflected sound pressure is still increasing. The reflected sound finally blends together with the direct sound at about half this distance (i.e. for a delay of about 50 ms).

We have already introduced this limit in Sections I. 4.1 and I. 5.1 as the 'limit of perceptibility' and used it there as a criterion for identifying 'useful' sound reflections.

This 'threshold' is by no means as sharp as that for masking; moreover, it is obvious that it will depend upon the nature (i.e. the shortness) of the signal as well as the direction of incidence of the reflection on the ear.[2] In any case, such a threshold of time delay between the direct sound and the first reflection exists under all conditions.

If our strollers approach still closer to the wall, they reach a point where they are no longer aware of the reflection at all, because it no longer has any disturbing effect on the intelligibility of their speech.

[2] Stumpp, H., *Beihefte zum Gesundheitsingenieur, Series II*, (1936) No. 17.

Nevertheless, if the reflection were suddenly removed, its absence would be immediately noticed; in other words, the reflection was still being perceived, but not as an echo separate from the direct sound.

The effect, described above, of walkers approaching a large reflecting surface becomes even more pronounced if they maintain a substantial lateral distance between each other, because then the reflected sound is not so weak in comparison with the direct sound.

With the help of a magnetic tape recorder, and playback of both the signals by means of loudspeakers, it is possible to present this phenomenon to many test persons at the same time. Such tests were first carried out by Haas,[3] at the suggestion of Meyer; 80 (!) test subjects were used, in a lecture room having a reverberation time of 0·8 s.

In order to be able to reduce the time delay to zero, he used two sets of erase/record/playback heads; the difference between the 'time separations' of the record and playback heads is the effective time delay for the experiment. It can be varied (and can even be made negative) by changing the location of one set of the record and playback heads.

Figure 1.10 shows the percentage of test subjects who felt 'just disturbed' by a reflection of running speech as loud as the direct sound, plotted as a function of the time delay. The scatter in the test results makes it doubtful that one can establish a threshold with confidence. But since the curve that can be drawn through this 'galaxy' corresponds to the cumulative distribution of a Gaussian population, we can regard the

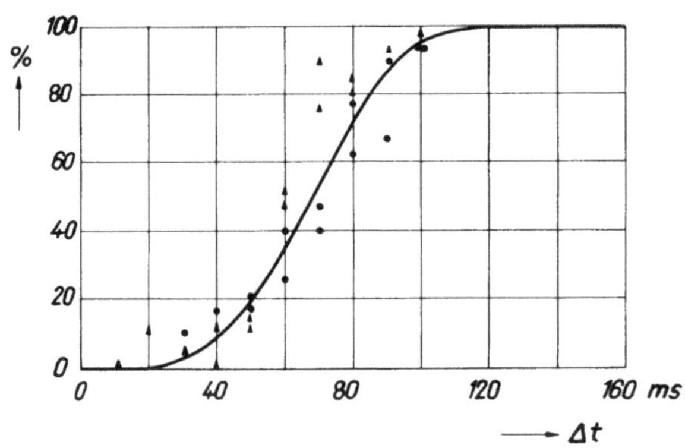

Fig. 1.10. Percentage of test persons who felt 'just disturbed' by an equally loud reflection with various time delays, Δt. (After Haas.[3])

[3] Haas, H., *Acustica*, **1** (1951) 49; see also Section III. 1.6.

deviations from the curve as random. Therefore it makes sense to regard the value of Δt for 50% response on this curve as a reasonable mean value: it is 68 ms. The fact that this value is greater than 50 ms may be due to the additional reflections in the closed test room. A repetition of the test in the open air led to a mean time delay of only 44 ms.

The dependence of the disturbance threshold on the nature of the signal can be illustrated by the fact that the results reported above were determined with a speech speed of 5·3 syllables per second. If the speed was increased to 7·4 per second, the 50% response occurred at a delay of only 40 ms; and at 3·5 syllables per second it occurred for 92 ms delay. (Before the introduction of electronic sound systems into churches, the preacher often felt himself forced to use an unnaturally slow manner of speaking in order to compensate for the long reverberation. The resulting impression of pathos was only a secondary consequence.)

As expected, the limit of perceptibility also depends on the level difference between the direct and reflected sounds. Decker,[4] who was interested in echoes on transmission lines, found that the permissible time delay Δt increases by 50% if the level of the echo is decreased to 3 dB below the primary sound signal. Haas also found in his lecture room experiments that attenuating the echo by 3 dB raised the time delay to 108 ms; with the echo level 10 dB below the direct sound there was no disturbance at all.

The results obtained by Muncey *et al.*[5] are somewhat different; but they carried out their experiments in a highly damped room having a reverberation time of 0·2 s. Figure 1.11 shows the dependence of ΔL on

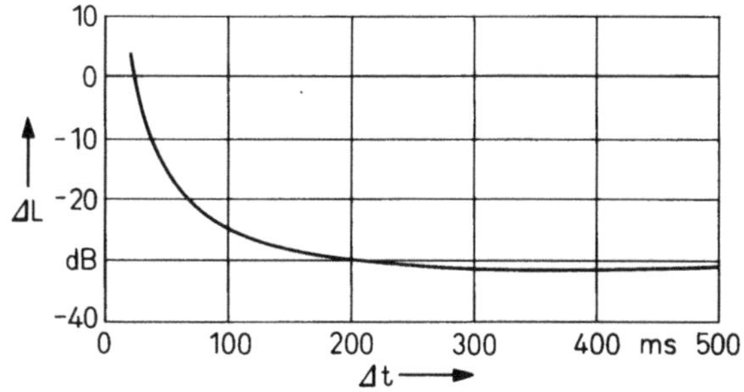

Fig. 1.11. Dependence of the level difference between direct and reflected sound on the time delay when 50% of the test subjects felt disturbed. (After Muncey *et al.*[5])

[4] Decker, H., *Elektr. Nachr. Techn.*, **8** (1931) 524.
[5] Muncey, R. W., Nickson, A. F. B. and Dubout, P., *Acustica*, **3** (1953) 168.

Δt, when 50% of the test subjects felt disturbed by the reflection. Here the independence of Δt on level difference does not appear until the echo is 30 dB below the direct sound; the 0 dB point corresponds to a time delay of only 25 ms. The effect of a limit of perceptibility was thus also confirmed by them.

If we now ask for the physical and physiological facts that could underlie the limit of perceptibility, we can begin by excluding wave interference between the direct and reflected sounds (see Section IV. 2.1, Volume 2); for that effect the 8·5 m distance to the reflecting wall (corresponding to $\Delta t = 50$ ms) is much too large for the wavelengths of interest. In order to get perceptible changes in timbre due to wave interference we must approach to less than 1 m from the wall. This kind of disturbance occurs below, not above, a limiting distance, and thus we may say that a change in timbre due to interference effects sets a lower limit on the time delay for useful sound.

Between these two limits, however, the reflected sound offers the advantage (for equal strength of the direct and reflected sounds) of increasing the loudness just as if the direct sound intensity were doubled. The energies of the two waves add. (This energy summation demonstrates that the phases play no role, a condition that we always assume in statistical room acoustics.)

The energy addition of two brief actions close together in time is often compared to a pendulum starting from rest: if pushed twice by excitations following one another quickly, it reaches the same displacement as if it were pushed only once by an excitation with twice the impulse:

$$\int_{\Delta t} F \, \mathrm{d}t = m v_0 \tag{1.12}$$

(F = force; m = mass; v_0 = initial velocity of the pendulum.)

Since this procedure is customarily used to measure the recoil of a rifle mounted on a pendulum, an apparatus used to carry out such an integration is called 'ballistic'. For the procedure to be valid, it is necessary that during Δt the deflection of the pendulum is small compared with the measured maximum deflection, and this requires that the ballistic measuring instrument must have a sufficiently large mass or, more generally, a sufficiently large inertia.

Although this analogy fails from the very beginning because of the fact that the ear integrates energies, not forces, nevertheless one often finds in the literature the assertion that the ear acts as a ballistic instrument. At least it is usual to speak of the 'inertia of the ear' to characterize its ability to integrate intensities over short time intervals.

By 1929, Békésy[6] had already found that the loudness level of short tones with rectangular envelopes increases with increasing duration Δt up to about 200ms, at which point the loudness level reaches that for a steady tone (see Fig. 1.12). Beyond this duration the loudness level

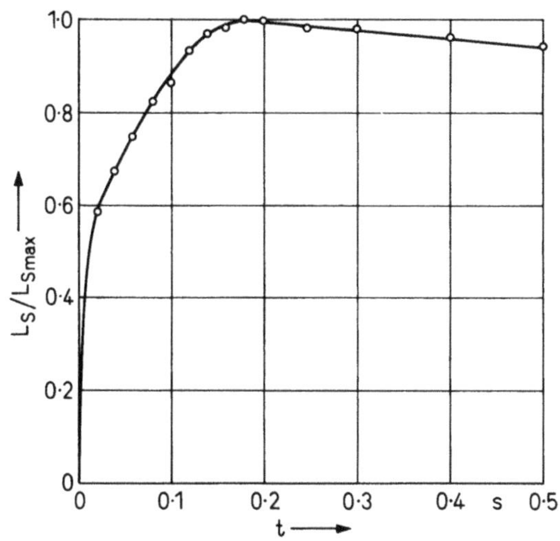

Fig. 1.12. Increase of the loudness level with duration of short tones. (After Békésy.[6])

decreases slowly, an effect that Békésy called 'fatigue'. If we neglect this latter effect, which is of no interest in the present context, we can compare the initial increase of loudness level and the asymptotic approach to the steady-state response with the charging of a capacitance C through a resistance R. The latter case is characterized by a relaxation time $\tau = RC$, where the voltage U across the capacitance increases according to the relation

$$U = U_{\infty}(1 - e^{-t/\tau}) \tag{1.13}$$

[6] Békésy, G. v., *Phys. Z.*, **30** (1929) 118.

A ballistic system, such as a pendulum at rest, would not approach a final steady value but would oscillate instead.

Munson,[7] who repeated this experiment with improved measurement techniques at very different frequencies and loudness levels, found the same increase of the loudness level.

A complementary experiment concerns the subjective decay of an abruptly interrupted tone. For this test Steudel[8] took advantage of the masking effect. After the interruption of the tone, he observed the time until a masked steady tone could he heard, the level of which he varied step-wise. By this means he found straight lines of $\Delta L_{(t)}$, which determine a reverberation time of 300 to 400 ms for the ear (see Fig. 1.13). The

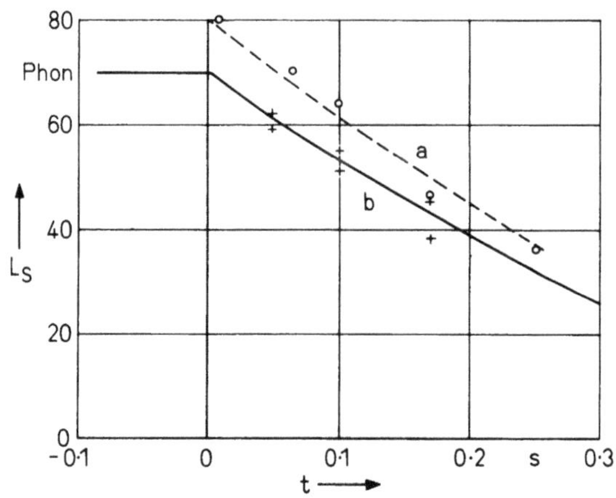

Fig. 1.13. Decay of the subjective loudness level of a pure tone after its interruption. (After Steudel.[8])

corresponding relaxation time for the electrical analog of eqn. (1.13) would be 22 to 28 ms.

Niese,[9] too, found in later investigations that the relaxation time of the ear is approximately 23 ms.

When we find mention of relaxation times (or time constants) in the

[7] Munson, W. A., *J. Acoust. Soc. Am.*, **19** (1947) 584.

[8] Steudel, U., *Z. Hochfr. u. Elektroak.*, **41** (1933) 116.

[9] Niese, H., *Z. Hochfr. u. Elektroak.*, **65** (1956) 4; **66** (1958) 115; **68** (1959) 26; **70** (1961) 5.

literature, we must be careful to distinguish whether they refer to the energy or the pressure of the sound. In the latter case, for which the value of τ is doubled, the reciprocal of this quantity corresponds to the decay coefficient introduced in Part II. (Also, τ could be related to the loudness, i.e. to $1/k\delta$ in eqn. (1.9)).

Steudel was also the first to investigate the loudness of 'one-sided' sound pressure impulses (clicks), which can be presented to the ears only with headphones.

Since the ear analyses all signals by dividing them into frequency groups, and since we cannot expect all these frequency groups to have the same relaxation time, it is better to use test signals whose spectral components all lie within a single frequency group. This is especially successful with tone impulses having an envelope shaped like a Gaussian distribution, the so-called Gauss-tones; we have already used Gauss-tones for the production of frequency-dependent impulse responses (echograms, see Section II. 7.2), and have noted the advantage that their spectral distribution is also Gaussian. If the 'duration' Δt is defined so that the dependence on time is given by

$$p_{(t)}=\hat{p}e^{-\pi[(t-t_0)/\Delta t]^2}\cos\left[\omega(t-t_0)\right] \tag{1.14a}$$

where Δt is the time interval in which p exceeds 0·456 $\hat{p}$, then the spectral function is given by

$$\check{p}_{(f)}=\check{p}e^{-\pi[(f-f_0)/\Delta f]^2} \tag{1.14b}$$

where f_0 is the carrier frequency and $\Delta f=1/\Delta t$.

Schwarze[10] adjusted the loudness levels of Gauss-tones having different Δt to match those of Gauss-tones for which Δf is the bandwidth of the corresponding frequency group. Figure 1.14 shows the differences in peak levels of equally loud 1000 Hz Gauss-tones plotted against Δt, as determined with six test persons. The points at the right boundary correspond to steady tones.

The decrease of the straight line at 3 dB per doubling of Δt corresponds to the linear region of eqn. (1.13), i.e. to the total energy at the ear, which for a Gauss-tone is proportional to:

$$\int_{-\infty}^{+\infty} p^2\mathrm{d}t=\frac{\hat{p}^2\Delta t}{2\sqrt{2}} \tag{1.14c}$$

[10] Schwarze, D., Dissertation, TU Berlin, 1963.

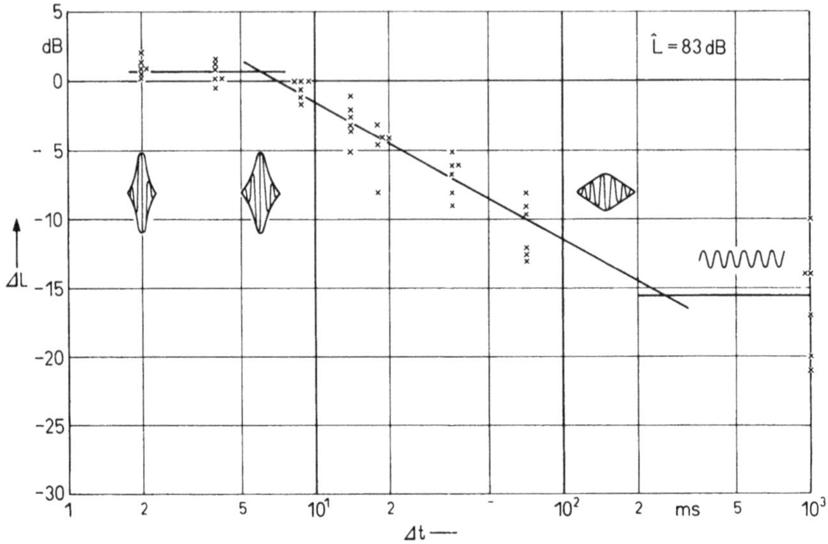

Fig. 1.14. Comparison of the loudness levels of various Gauss-tones ($f_0 = 1000$ Hz) with different durations. (After Schwarze.[10])

This decrease, which was found by other authors using different impulses, proves that the relaxation scheme of eqn. (1.13) must be applied to J and not to $S = J^k$.

The asymptotic transition to the steady-state value does not appear when we plot Δt on a logarithmic scale. But the extrapolation in Fig. 1.14 of the falling straight line to the horizontal line, which corresponds to the mean of the steady-state values, shows that this value is approached only for durations greater than 250 ms. Schwarze found the same points of intersection also for frequencies of 250 and 4000 Hz. If we were to consider these durations as characteristic of the relaxation time of the ear, it would mean a much larger value than those reported above.

Port[11] made similar comparisons using noise impulses whose envelopes were rectangular and whose frequency components lay within one frequency group; he found points of intersection between the falling line and the steady-state value at a duration of 70 ms. He doubts, however, that one can determine the ear's relaxation time from this time

[11] Port, E., *Acustica*, **13** (1963) 212.

limit. At the least we may have to take into account that the 'intersection duration' is influenced by the different envelopes of the test signals.

Furthermore Schwarze found, in investigations similar to those of Steudel, a shorter self-reverberation of the ear. For this purpose he presented one 'frequency-group-wide' Gauss-tone with constant peak level and followed it with a second such Gauss-tone after a delay which he varied between 10 and 100 ms. The peak level of the second tone was adjusted by the test subjects so that the second tone could barely be recognized as a separate impulse. He called this level the 'merging limit' (*Verschmelzungsschwelle*).

Although this level certainly must exceed the masking threshold level, the falloff of the 'merging limit' level with increasing separation of the Gauss-tones t (as shown in Fig. 1.15) permits a determination of the ear's own reverberation time (or relaxation time). If we lay a straight line along the drooping curves of Fig. 1.15, we find the ear's own reverberation time to be about 240 ms, corresponding to $\tau = 17$ ms.

According to Reichardt,[12] the large differences measured by different researchers for the ear's relaxation time may in fact result from actually existing large differences between individual test persons.

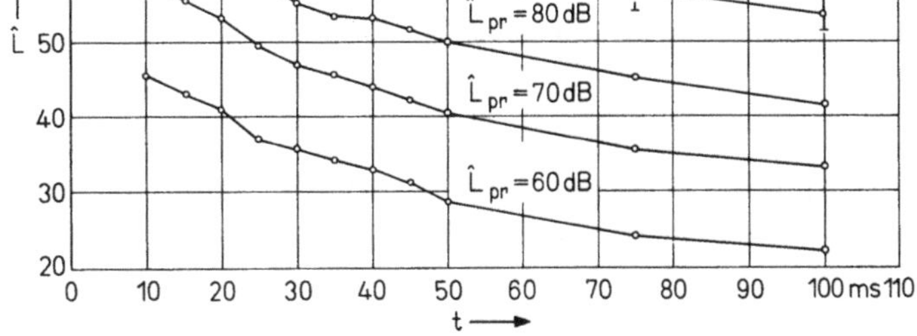

Fig. 1.15. Level difference of the 'merging limit' for two frequency-group-wide Gauss-tones, plotted against their time separation. (After Schwarze.[10])

[12] Reichardt, W., *Acustica*, **15** (1965) 345.

Although the correct determination of the relaxation time of the ear is for this reason rather uncertain (or, expressed more generally, although the model of the charging capacitance may be too simple to represent adequately the far more complicated processes in the ear), we may still regard this model as suitable for representing the observed integration of short impulses of excitation and the transition to a steady-state response.

If we have a sound field situation such as is represented in the upper part of Fig. 1.16, where a receiver is placed between a sound source and a

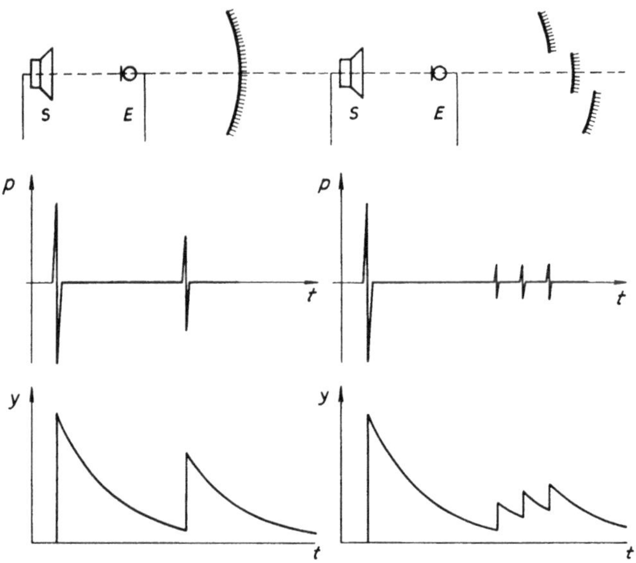

Fig. 1.16. The $p_{(t)}$-dependence (middle) and the 'ear-imitative' function (bottom) for a single reflection (left) and three smaller reflections (right).

concave reflecting wall, the objective $p_{(t)}$-representation of a single-period tone impulse (shown in the middle, left) will hardly be adequate to represent the subjective sensation; instead we will need the relaxation processes excited by the rectified (squared) impulse, as shown in the lower part.

The shorter the delay between the direct sound and the reflection, the shallower the trough between the responses to these impulses and the stronger the impression of a single-peaked response.

But another aspect is also very important in the evaluation of echograms, according to experience. In Fig. 1.16, right, the reflection has been broken up into three smaller ones separated in time. One might assume that this change would transform the strong echo into a harmless one, as suggested by the $p_{(t)}$ plot in the middle. But within the 'integration circuit' of the ear there is a buildup to a value higher than that of the individual impulses, as shown in the ear-imitative representation, below, right. This integration effect becomes stronger the closer the individual impulses follow one another.

That this representation corresponds better to the aural impression was demonstrated by a model experiment carried out in the anechoic room of the Institute for Technical Acoustics of the Technical University (ITA), Berlin. A comparison was set up in which these two $p_{(t)}$-time sequences were presented for evaluation by means of speed-transformed magnetic tape recordings. The two echo sequences, left and right in the figure, were judged to be almost equally loud.

Also, Meyer and Schodder[13] found that the interpolation of a new reflection in the gap between the direct sound and an existing reflection improves the 'quality of speech transmission'; their observation becomes understandable in the light of this relaxation process.

We now return to the increase of loudness by a single reflection, which we will choose to be equally strong. Schwarze[10] has investigated this process with the help of Gauss-tones. The test person had to compare a single Gauss-tone of given peak level with a pair of Gauss-tones whose time-separation δt was changed; his task was to adjust the (equal) peak levels of the pair so that they sounded equally as loud as the single given Gauss-tone. Figure 1.17 shows the results in terms of ΔL as a function of δt, for different peak levels of the single Gauss-tone with a carrier-frequency of 1000 Hz.

Since the duration Δt of the Gauss-tones was adjusted so that the frequency bandwidth corresponded to that of the frequency group, i.e. $\Delta t = 6$ ms, the duration of the Gauss-tones in the pair was always smaller than the time separation δt between the pair of Gauss-tones. For $\delta t = 10$ ms, for example, ΔL was found to be about -3 dB, corresponding to an ideal energy summation. As Fig. 1.15 leads us to expect, this summation effect decreases significantly at $\delta t = 50$ ms and nearly vanishes at 100 ms.

As Haas had already discovered using speech, Schwarze, using carrier

[13] Meyer, E. and Schodder, G. R., *Göttinger Nachrichten, Math. Phys. Kl.*, IIa (1962) No. 6.

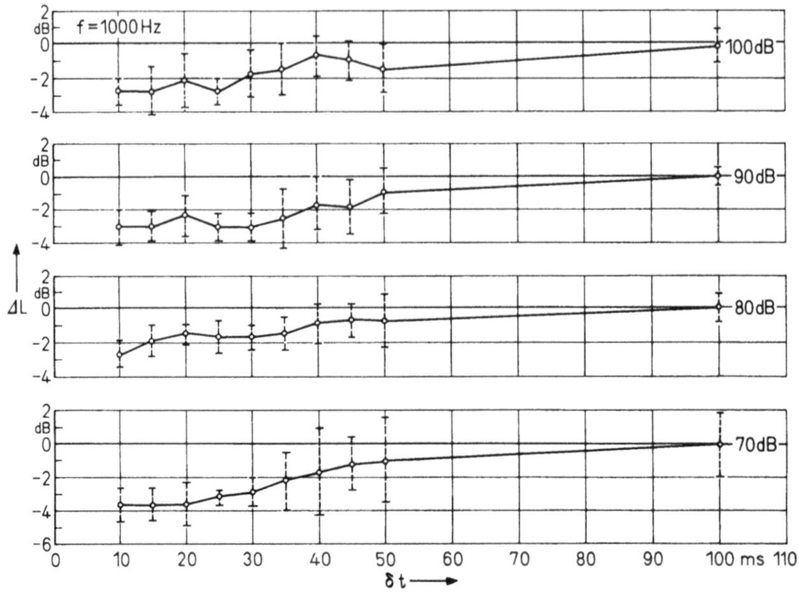

Fig. 1.17. Comparison of the loudness level of a single given Gauss-tone with that of a pair of equal Gauss-tones, where their time separation δt was varied and their peak level was adjusted for equal loudness with the single tone. (After Schwarze.[10])

frequencies of 250 and 4000 Hz, never found summations that exceeded that of simple energy summation. This fact must be emphasized because other authors have reported greater increments for the addition of delayed signals. According to Aigner and Strutt,[14] for this effect to occur a tone color difference between the two signals was required. According to Lübcke,[15] the time delay alone is sufficient to achieve the effect without a difference in tone color.

Meyer and Schodder[13] observed such additional increments of loudness at loudness levels above 65 phon, but never more than 2 phons at most. In any case this effect is not important enough to be seriously considered in room-acoustics matters.

But reflections with short time delay may be very useful for other reasons: for better recognition of signals, for compensation of the direc-

[14] Aigner, F. and Strutt, M. I. O., *Z. techn. Phys.*, **15** (1934) 355.
[15] Lübcke, E., *Z. techn. Phys.*, **16** (1935) 77.

tional characteristics of the sound source which may degrade the direct sound at certain frequencies, and many others as well.

Another threshold should be mentioned that also concerns the so-called inertia of the ear. If sound reflections follow one another at irregular intervals, as is usually the case in rooms, and if we compensate for the normal decay of sound with an artificial, exponentially increasing amplification, we generate a random noise. If the source signal is a Dirac function and so also are all the reflections, we call this noise a 'white' noise because, as in white light, the energy is then equally distributed over all equal frequency intervals Δf.

Communications theory also makes use of this analogy when spike impulses follow each other for the most part at such great intervals that they may be perceived as single impulses, like the first few raindrops on a window.

As the impulse sequence becomes more dense, we speak at first of a pattering, as of light rain, and then we continue to notice changes in the subjective quality of the sound as the impulses become more and more frequent and their mean separation becomes shorter and shorter. The question is whether there exists a threshold of the mean impulse density (number of impulses per unit of time) $\bar{m}_{max}$ such that a further increase in the mean density $\bar{m}$ is no longer discernible by ear. Schreiber[16] has investigated this problem theoretically and experimentally; he found in both cases the threshold value:

$$\bar{m}_{max} = 2800\,s^{-1} \tag{1.15}$$

With a spectral roll-off toward high frequencies, as usually happens in room acoustics, this threshold density decreases; for example, for low-pass filtering at 2000 Hz the threshold drops to

$$\bar{m}_{max} = 2000\,s^{-1} \tag{1.15a}$$

Here the mean separation of the impulses is 0·5 ms, only about 1/100 of the limit of perceptibility; but this is not surprising because now, instead of dealing with only two impulses, the ear can discriminate on the basis of a sequence of impulses with longer time separations.

Finally we return to Fig. 1.14. The regime of the sloping straight line, where the total energy of the Gauss-tone governs the loudness level, is bounded (according to Schwarze's investigations) at very short values of

---

[16] Schreiber, L., *Frequenz*, **14** (1960) 399.

Δt, namely, where $\Delta f = 1/\Delta t$ becomes equal to the frequency-group width Δf_g. Since the loudness increases as soon as the spectrum of the signal exceeds this bandwidth (for equal effective sound pressures, see Fig. 1.6), the loudness of a shorter Gauss-tone will be greater than that corresponding to its total energy. In Fig. 1.14 this means a displacement of the plotted peak levels to lower values, as shown by the levelling off of the curve to a horizontal line for low values of Δt. The fact that Schwarze found nearly equal peak levels for all durations Δt less than $1/\Delta f_g$ can be expressed by the simple statement that the ear evaluates the loudness in this regime of short Δt according to the peak level, independent of the duration of the impulses.

III.1.5 Perception of Direction

All the problems discussed so far have concerned the reception of a single sound signal in one ear, or the same signal in both ears, as when the same signal excites both earphones or when sound is incident in the median plane in an open field. From the standpoint of reproduction technique we may think of these as 'one-channel' (or monaural) problems. Even with this restriction, different directions in the median plane (front, behind, above or below) can be to some extent distinguished; but most sensation of direction vanishes when we close one ear.

A comparison of the left-ear and right-ear signals (made possible by inter-connections in the nerve paths) is essential for the perception of direction. This is easy to understand if we consider the special case of sound incident in the horizontal plane.

For wavelengths that are comparable with the dimensions of the human head or smaller, we get for lateral incidence an increase of sound pressure at the near ear because of the reflection at the hard skull, whereas the opposite ear is shadowed by the head. Although, as shown in Fig. 1.18, the ears are not placed precisely laterally but instead lie a bit behind the middle, and also the shape of the head is not symmetrical with respect to the transverse plane, the level difference between the left and right ears is still a useful clue to the azimuthal incidence angle ϕ.

The perception of direction is also based on the difference in time of

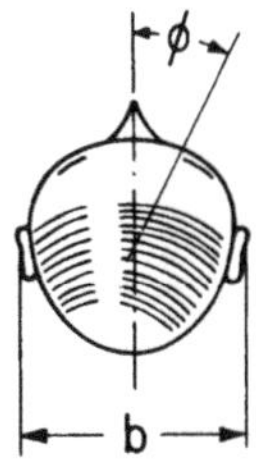

Fig. 1.18. Diagrammatic sketch for perception of direction, in terms of the azimuthal angle ϕ in the horizontal plane.

arrival of the sound at the two ears; this is—geometrically simplified—given for an ear separation distance b by

$$\Delta t \approx \frac{b \sin \phi}{c} \tag{1.16}$$

Since it is possible, for frontally incident sounds ($\phi \approx 0$), to distinguish differences in incidence angle as small as 1°, the nerve system must be able to distinguish (for $b = 18\,\text{cm}$) time differences as small as $10^{-5}\,\text{s}$ between the left-ear and right-ear signals.

For pure tones this is possible only if the different individual periods are still recognizable in the pattern of nerve impulses at the location where the nerve paths from the two ears come together. This implies that at higher frequencies (above 1000 Hz) where the level difference between ears becomes more pronounced, the time difference between the signals from the two ears loses its importance. The temporal evaluation of the nerve impulses is limited, since the triggering of nerve impulses is subject to unavoidable fluctuations. Moreover, if the phase difference between the signals is of interest (that is, the ratio of the delay-time to the period), this quantity is determined with certainty only at lower frequencies.

As Boerger[1] has shown, for Gauss-tones with frequency-group bandwidth, the carrier frequency pattern and its phases govern the perception of direction at frequencies below 1000 Hz. But for all impulse signals above 1000 Hz, including Gauss-tones, the time delay, in addition to the level difference, plays a decisive role, as Boerger also proved. Here it is the

[1] Boerger, G., Dissertation, TU Berlin, 1965.

envelope, especially the ascending slope, that determines the nerve impulses that are to be compared.

For experiments on the perception of direction, Gauss-tones again offer the advantage of excitation with restricted time and frequency regions. Both are important in resolving the question of how many directions can be perceived simultaneously. In a continuous presentation of several signals, the test person can pay attention to the different sources one after the other. With Gauss-tones simultaneously radiated from different sources this possibility disappears.

Whereas Békésy[2] assumed that, within an interval of 0·8 s, man can concentrate on only a single effective locale (*Präsenzraum*), Franssen[3] assumed that the ear analyses the sound pressures into different frequency groups not only for spectral evaluation but also for the perception of direction.

Boerger demonstrated, in his investigations with Gauss-tones, that the ear can simultaneously determine only two directions, and then only when one signal belongs to the frequency region below 1000 Hz (where the carrier frequency governs the nerve impulse pattern) and the other signal belongs to the frequency region above 1500 Hz (where the envelope of the impulse is decisive). He obtained this result in an experiment in which he presented the two Gauss-tones to the test persons from two loudspeakers placed at $+15^\circ$ and -15° with respect to the frontal (0°) direction. The subjects were asked to decide which loudspeaker radiated the high-frequency sound and which radiated the low-frequency sound. Figure 1.19 shows the percentage of correct responses plotted against one carrier frequency, for two values of the other carrier frequency: 500 Hz and 6000 Hz. Only the portions of the curves above the 50% correct responses contribute to proof of the ability to perceive two directions at once.

If the Gauss-tones were radiated with a time delay of at least 10 ms, the two directions could always be heard, even for equal carrier frequencies. (For shorter Δt no reliable results could be obtained.) But in this case the (variable) angles of the loudspeakers with respect to the frontal direction had to be as much as 35° for the two directions to be distinguished. These angles decreased as the carrier frequencies became increasingly different (either lower or higher) and as the time delay increased, as shown by

[2] Békésy, G. V., *Phys. Z.*, **31** (1930) 824 and 857.
[3] Franssen, N. V., Dissertation, Delft, 1960.

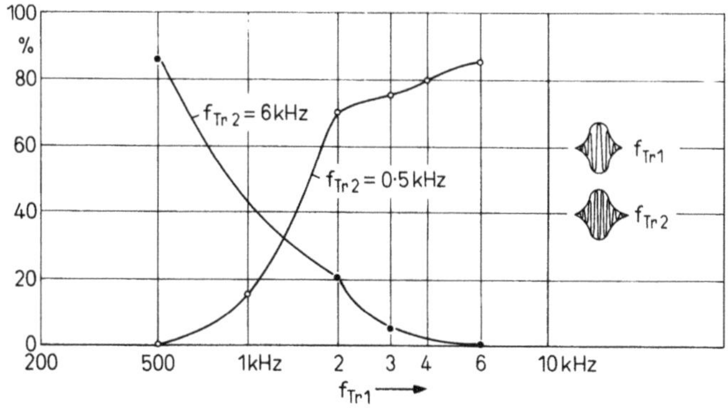

Fig. 1.19. Percentage of correct judgments concerning the $\pm 15°$ directions of two simultaneously radiated Gauss-tones, plotted as a function of one carrier frequency for two values of the other carrier frequency: 500 Hz and 6000 Hz. (After Boerger.[1])

the results summarized in Fig. 1.20. The distribution of the impulse spectra in frequency groups, which is so important in the evaluation of the loudness and tonal color, is not very important for the perception of directional differences.

All of the directional criteria introduced up to this point have the disadvantage of being ambiguous. The reason is that the same time delays and the same sound level differences can occur for oblique incidence either from in front or from behind (though the actual incidence angles may be slightly different because of the asymmetry of the ear placement on the head).

But a slight turning of the head about the vertical axis changes both the time delay and the level difference in opposite senses, depending on whether the sound arrives from the front or from the rear. It was for a long time supposed, therefore, that involuntary and unconscious head motions of this kind helped a person to distinguish between forward sounds and those from the rear.

In the tests with Gauss-tones, however, there was not enough time for the subject to turn his head or to evaluate the corresponding effect. Nevertheless, when the sounds were incident from the forward region there was never any doubt for the test persons that this was the case.

There is another problem that remains unsolved. Time delays and level differences between the ears can occur not only for sounds arriving in the horizontal plane but also for sounds from arbitrary directions around the room, except those arriving in the median plane. Even in this case, however, different arrival directions can be readily distinguished and

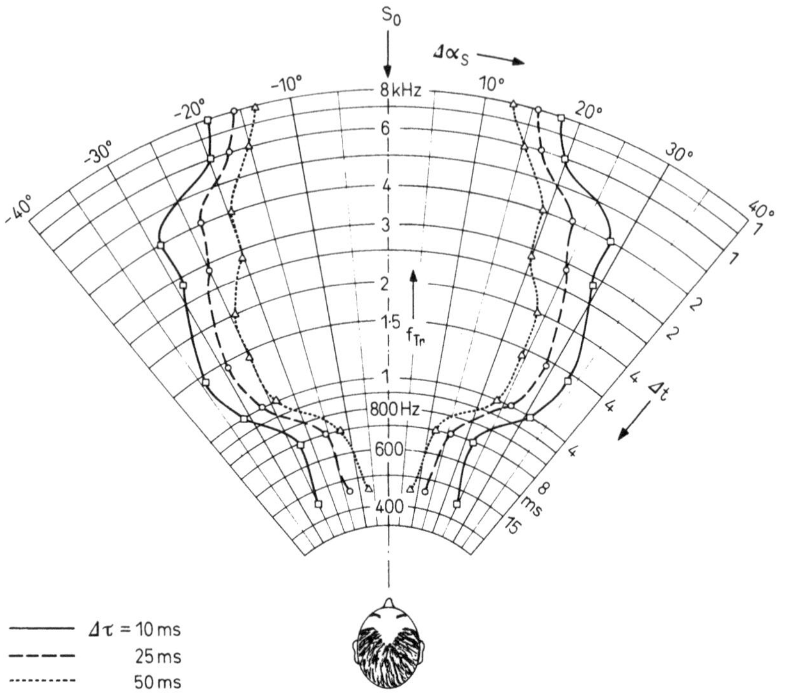

Fig. 1.20. Thresholds of localization for two Gauss-tones with spectral width $\Delta f = 1/\Delta t$ of the respective frequency groups, and with time differences of 10, 25 and 50 ms.

especially there is no doubt as to whether the sounds arrive from in front or from behind.

It is sometimes supposed that the external part of the ear (i.e. the pinna or auricle) acts to enhance the high-frequency sounds arriving from the front and to weaken those from the rear and that the listener can therefore determine, from the resulting tonal distortion of a familiar sound at sufficiently high frequencies, whether the sound comes from in front or behind.

This view was given support by the observation of Kietz[4] that the ability to distinguish between sounds from in front and from behind disappears when the ears are fitted with small tubes that extend far enough beyond the pinna; this ability is regained, at least partially, when

[4] Kietz, H., *Acustica*, **3** (1953) 73.

the tubes are fitted with artificial outer ears at the free ends. The restoration of directional perception was incomplete because the actual external ears on the head lead to field changes quite different from those of the artificial outer ears in 'free' space.

It is also interesting that the front/rear distinction is lost if the tubes are fitted with forward-directed horns at the outer ends, i.e. with *more* directional reception than the natural external ear can provide. Kietz concluded that our experience is based on the existing differences between sounds arriving from in front and from behind. He even suggests that if our external ear structure were changed, we would, in time, learn from daily observation which sounds come from in front and which from the rear. This accords with the fact that, as a child grows older, the size and shape of his head changes significantly; nevertheless, every child is capable of distinguishing between sounds from in front and behind.

The most careful and interesting investigation of the question of directional hearing in the median plane, and not merely of front/rear discrimination, has been carried out by Blauert.[5] He measured with probe microphones at the entrances of the outer ears the level differences between sound waves incident from the front ($\phi = 0°$) and those from directly behind ($\phi = 180°$); Fig. 1.21 shows the results.

At low frequencies where the head is small compared with the

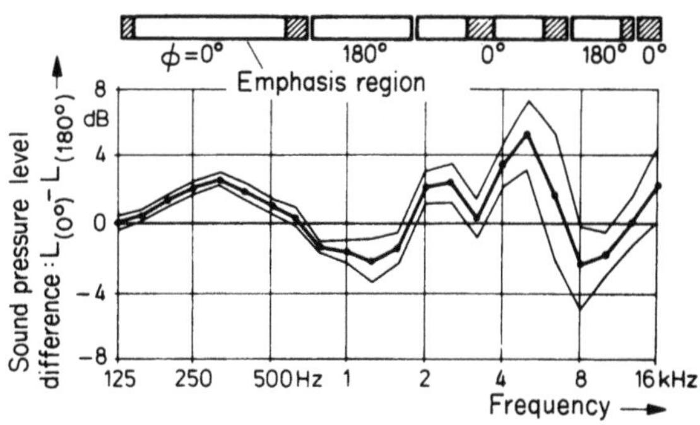

Fig. 1.21. Sound pressure level difference at the eardrum between sound waves incident from the front ($\phi = 0°$) and from the rear ($\phi = 180°$), plotted against frequency. (After Blauert.[5])

[5] Blauert, J., Dissertation, TH Aachen, 1969; see also his book, *Räumliches Hören*, Hirzel, Stuttgart, 1974. It is the most extensive treatment to date of the problems of perception of direction, and is recommended to readers who wish to know more about the subject. English translation from MIT Press, Cambridge, Mass., 1982.

wavelength, no differences occur; but by 250 Hz the frontal sound predominates, whereas around 1000 Hz the rear sound predominates. This alternation continues up to the highest frequencies. Thus the tendency is by no means monotonic, as one might at first assume; the disturbance of the incoming sound waves by the head is much too complicated for that. Blauert even speaks of the 'comb-filter' effect of the head upon the sound field, but this is a bit exaggerated.

Because of the alternating tendency for front/rear dominance shown in Fig. 1.21, it is likely that in any broadband signal one would have enough tonal distortion clues to be able to distinguish between sound arrival from the front or the rear.

Indeed, Blauert was able to show that one does not even need a broadband signal for directional impressions in the median plane. If the test person is seated in front of a single loudspeaker in an anechoic room and a one-third-octave-band noise is presented with a mid-frequency that slowly progresses from low to high frequencies, he perceives the sound as arriving frontally until about 600 Hz; but then this impression gradually reverses, until at 1000 Hz the sound appears to arrive from behind—even though the test person is quite aware that the sound comes from the loudspeaker in front of him. At 2000 Hz the impression is again frontal; then the direction of incidence moves again, this time in a large arc over the head, toward the rear. The same directional impressions occur when the real sound is incident from above, from behind, or with the same phase from both sides. Figure 1.22 shows the relative frequency with which the test persons, with 95% certainty, give the answer 'behind',

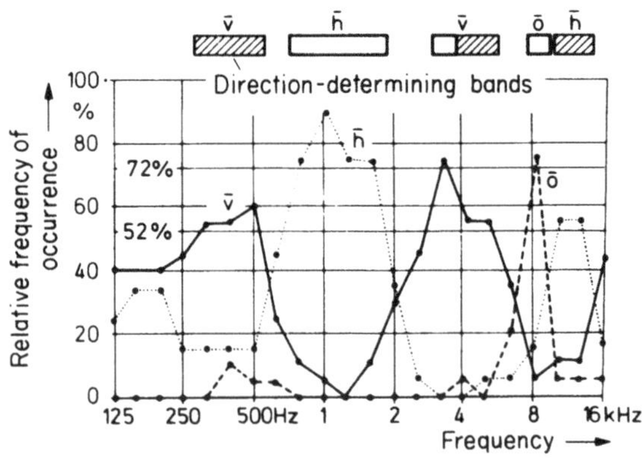

Fig. 1.22. Percentages of test persons who give, with 95% certainty, the answers 'behind' (h = *hinten*), 'above' (o = *oben*), or 'in front' (v = *vorn*) more frequently than the other two answers together. (After Blauert.[5])

'above', or 'in front' more often than the other two answers together. Blauert calls the frequency bands identified with the three directions at the top of the figure the 'direction-determining bands'. A comparison of the 'front' and 'behind' curves of Fig. 1.22 with Fig. 1.21 allows the conclusion that the ear associates with each direction a frequency region that is enhanced for any broadband signal.

We leave open the question of whether this is the result of a learning process or whether there are physiological reasons.

There is no doubt, however, that localization is easier for the listener if he is familiar with the signal. Plenge and Brunschen[6] proved this by comparing the judgments of perceived directions (only five directions were used), first with familiar speakers and then with unknown speakers, as shown in Fig. 1.23.

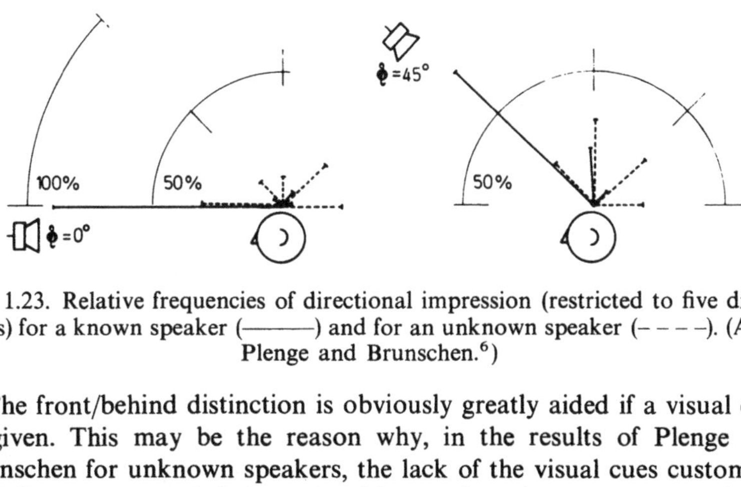

Fig. 1.23. Relative frequencies of directional impression (restricted to five directions) for a known speaker (———) and for an unknown speaker (- - - -). (After Plenge and Brunschen.[6])

The front/behind distinction is obviously greatly aided if a visual clue is given. This may be the reason why, in the results of Plenge and Brunschen for unknown speakers, the lack of the visual cues customary in daily encounters led the test subjects, when they were uncertain, to locate the sound source behind them.

For room-acoustical applications, it is important that the different directions can be correctly perceived, even in the median plane.

III. 1.6 The 'Law of the First Wavefront'

While we are enumerating the possible modes of perception of direction, we must not fail to mention those cases in which such perception does not occur.

[6] Plenge, G. and Brunschen, G., 7th. Int. Cong. on Acoustics, Budapest, 1971, 19 H 10.

For room acoustics, a phenomenon that we have already mentioned in Section I.2.1 is of special importance: namely, if there are several loudspeakers distributed about a space, all operating in phase, then as we walk about from one loudspeaker to another we always perceive the nearest source to be the only source operating. In connection with the mirror principle, we have pointed out that in such cases the sound field would be the same if each distant loudspeaker were replaced by a perfectly reflecting wall, midway between the locations of the near and distant loudspeaker and perpendicular to the line between them.

Since the common criterion for the perceived location of the source in both cases is the incidence direction of the wave whose front arrives first, we called this phenomenon (in Section I.5.2) the 'law of the first wavefront'.

This reference to an objective sound field still leaves open the question whether it is an innate or a learned ability to ignore the directions of later sound reflections in a room that could carry no warning of danger.

Wallach *et al.*[1] called this phenomenon, which they found can be observed even with headphones, the 'precedence effect'. Their use of the word 'effect' implies that they are more inclined to assume a physiological process behind the phenomenon rather than only 'an educated faculty of attention'.

Haas had the opportunity, with the magnetic tape recording arrangement mentioned above (Section III.1.4),[2] to investigate this phenomenon quantitatively for different combinations of level difference and time delay of the primary and secondary signals. The question that was asked of the test persons did not refer to the perception of direction at all; instead they were asked to decide the combinations of ΔL and Δt for which they judged both signals to be equally loud. His astonishing results are shown in Fig. 1.24: in order to achieve equal loudness, with time delays between 10 and 25 ms, the level of the secondary signal had to be about 10 dB higher than that of the primary signal.

But if the test persons are asked to adjust the strength of an echo until it just becomes imperceptible, as was done in the experiment of Meyer

---

[1] Wallach, H., Newman, E. B. and Rosenzweig, M. R., *Am. J. Psychology* **LXII** (1949) 315. The authors also refer to previous investigations by G. v. Békésy (*Phys. Z.*, **31** (1930) 824 and 857), who regarded the phenomenon as a process of masking; and to the work of Langmuir *et al.* (OSRD Report 4079, June 1944), who observed the phenomenon in sound localization under water. For further information, the reader is referred to Gardner, M. B., *J. Acoust. Soc. Am.*, **43** (1968) 1234; and to Cremer, L., *J. Audio Eng. Soc.*, **25** (1977) 420.

[2] Haas, H., *Acustica*, **1** (1951) 49.

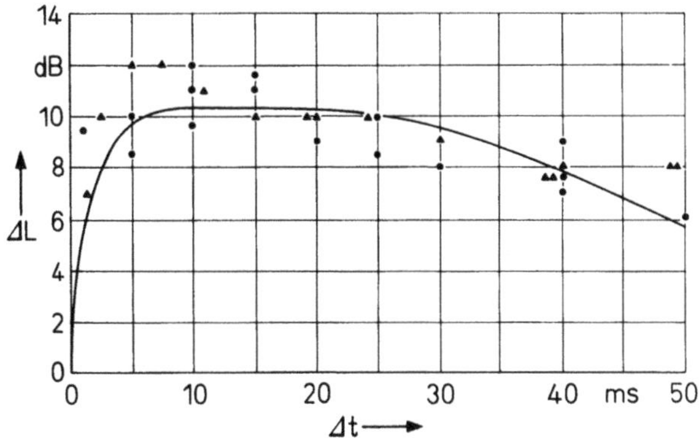

Fig. 1.24. Dependence of the level difference between secondary and primary sound signals upon the time delay, when they are judged to be equally loud. (After Haas.[2])

and Schodder,[3] then for time delays between about 10 and 25 ms the secondary signal permits a level only about 4 dB higher than that of the primary signal (see Fig. 1.25, solid line).

Yet another curve of $\Delta L - \Delta t$ was found by Lochner and Burger[4] (see Fig. 1.25, dashed line), who asked the test persons to respond when they perceived the 'echo' loudspeaker as a second source, thus indirectly introducing the notion of direction into the test. A comparison of the two curves of Fig. 1.25 demonstrates the importance of the exact formulation of the question in psychological investigations.

In this connection we may also refer to the results of Boerger, already presented in Fig. 1.20. They included equal Gauss-tones that could be distinguished with respect to their directions for time delays of only 10 ms. This demonstrates that the discriminatory ability of an observer depends to a great extent on what he is asked to concentrate upon.

Here the choice of signal certainly plays an important role. The results shown in Figs. 1.24 and 1.25 concern running speech only.

All this seems to make it more probable that the law of the first wavefront is based on a learned ability to disregard unimportant information, particularly since speech is a signal for which even a child can rely upon visual cues for the location of the source.

---

[3] Meyer, E. and Schodder, G. R., *Göttinger Nachrichten, Math. Phys. Kl*, IIa (1962) No. 6; the expression *verdeckt* (masked) used by the authors referred only to the directional impression.

[4] Lochner, J. P. A. and Burger, J. F., *Acustica*, **8** (1958) 1.

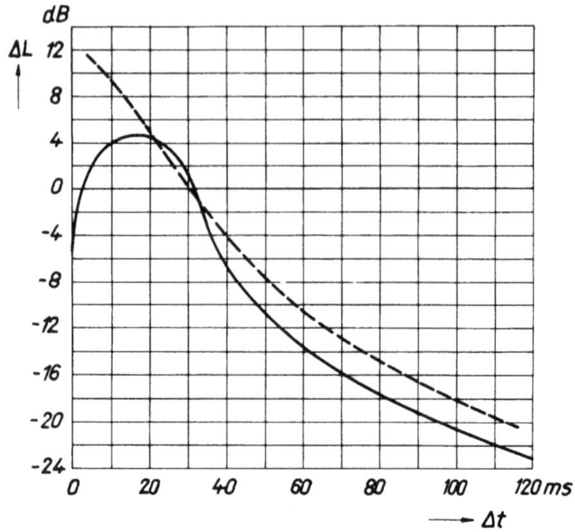

Fig. 1.25. Dependence of the level difference between the secondary and the primary sound source upon the time delay: (a) if the 'echo' is not perceived at all (———) (after Meyer and Schodder[3]); (b) if the 'echo' loudspeaker is just perceived as a second source (– – –) (after Lochner and Burger[4]).

III.1.7 Localization Inside the Head and 'Subjective Diffuseness'

The most striking example of mistaking the direction of incident sound occurs when the listener localizes the sound source inside the head.[1] We must point out here that in psychoacoustics, the term 'localization' embraces both the notions of direction and distance of the source. If we omit consideration here of the distance of a source in free space, the reason is that in rooms the perception of distance is given primarily by the ratio of the direct sound (including reflections with very short time delays) to the later-arriving sound; thus the criteria for free-field distance judgments play no significant role.

Localization inside the head was first observed in the early attempts to get a stereophonic impression by playing back over headphones the two

[1] A comprehensive account of all the cases of localization of sound inside the head is given by Plenge, G., *Über das Problem der interkranialen Ortung von Schallquellen bei der akustischen Wahrnehmung des Menschen*, Habilitationsschrift, TU Berlin, 1973 (an unpublished paper written to qualify the author for a lectureship).

signals picked up by microphones separated by a typical ear distance. In this context a stereophonic impression means that the sounds from various sources are supposed to be heard as if they were coming from their correct directions in front of the listener.

Since we know today that such microphones do not pick up the same sound pressure differences in level and time delay as at the entrances to our outer ears,[2] it is understandable that these presentations sometimes sound strange.

It is typical that when one hears a pair of signals that are contrary to daily experience, our ears tend to localize the sound inside (or immediately next to) the head; such signals convey a senseless combination of sound pressure levels and delays to the central nervous system, which in turn creates a meaningless subjective acoustical impression. (Only the localization of one's own voice, with closed mouth, occurs more or less reasonably inside the head.)

Among the senseless signals are those sound pressure combinations with a time delay greater than 600 μs, which corresponds to the distance between the ears.

Unnatural sound pressure combinations can also be produced by loudspeakers. If a listener stands in front of two loudspeakers, symmetrically arranged with respect to his median plane, and both loudspeakers are driven in phase by the same signal, he gets the impression that there is only one source that lies midway between the two loudspeakers; this is called a 'phantom source'. If now the terminals of one loudspeaker are reversed ('contra-phased'), the ears tend to localize the source inside the head because the 180° phase-shift would correspond to a different direction for each frequency. Since a true broadband signal arriving from the side leads to phase differences that increase with frequency, the out-of-phase presentation is unnatural.[3]

The question of when the impression of something acoustically unusual occurs depends (as Plenge[4] has shown) on the immediately preceding impression to which the ear has adapted.

Imagine a person sitting in an anechoic room and listening to a dummy-head recording of a symphony orchestra in a large hall (see Section III. 2.7). If his earphones have pads made of foam rubber, he will also be able to hear sounds generated in the anechoic room. Under such

[2] Laws, P., Dissertation, TH Aachen, 1972.
[3] Schodder, G. R., *Acustica*, **6** (1956) 482.
[4] Plenge, G., *Acustica*, **26** (1972) 241.

circumstances, if 'dry speech' is presented by a loudspeaker in that room he will localize the added speech sounds inside (or at least at) his head. In the opposite case, where the listener is accustomed by the earphones to dry sound and then hears a reverberant signal presented by the loudspeaker in the anechoic room, he again localizes the second 'sound event' as a 'hearing event' inside (or at) his head.[5]

Indeed certain refinements can be observed with respect to in-head localization. The localization may be exactly in the middle of the head or it may be shifted a bit toward the side. One speaks in the latter case of 'lateralization'.

Much more important for a listener in a room is that he may have several impressions: he will surely have the impression of some sound sources in distinct directions; he may also have the impression of in-head localization for certain other sources; and finally he may feel that the sound comes to him from outside himself, but he cannot determine the particular direction; at most he can identify a directional region. Since Boerger's investigations with two Gauss-tones showed that the ear can distinguish no more than two directions in the horizontal plane simultaneously, and since this observation may also be valid for all directions of arrival, such an undefined directional perception is to be expected if the sound arrives simultaneously from many directions, as in the late phase of reverberant sound decay. The sound pressures at the left and right ears in this case no longer allow a sensible determination of one or two directions of arrival based on level and time differences. In spite of this, the ear does not interpret the impression as acoustically unusual and does not localize the sound inside the head.

Such vague localizations can be produced, for example, if we present a gradual transition from the extremes mentioned above where the two loudspeakers were excited once in phase and once in contra-phase. For this purpose Plenge[6] used the arrangement shown in Fig. 1.26. The listener is placed in an anechoic room so that the two loudspeakers are symmetrical about his median plane. Two noise generators, A and B, are provided, of which one (A) drives the loudspeakers in phase and the other (B) drives them out of phase. The outputs of these generators can be mixed in different ratios.

Following a procedure of Damaske,[7] each test subject was asked to

[5] These expressions for describing objective situations and subjective impressions were proposed by Blauert, J., Dissertation, TH Aachen, 1969, p. 2.

[6] Plenge, G., *Acustica*, **26** (1972) 241.

[7] Damaske, P., *Acustica*, **19** (1967/68) 199.

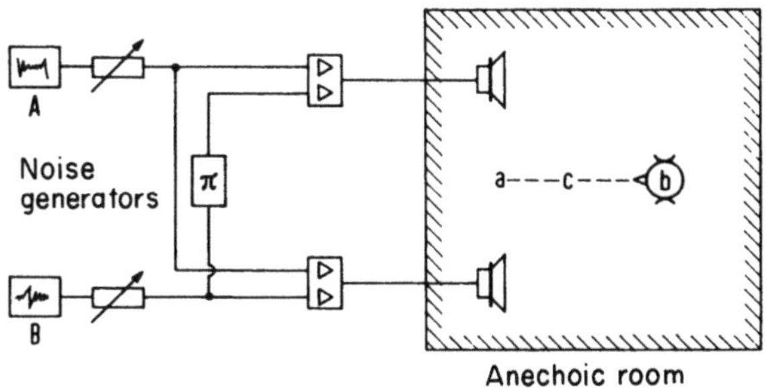

Fig. 1.26. Arrangement for the production of a transition between in-phase and out-of-phase excitation of two loudspeakers. (After Plenge.[6])

draw the area corresponding to the locus of the 'hearing event' of the source, on a sketch of the room showing the locations of the loudspeakers and the listener's head as guidepoints. The indicated areas for various subjects were overlaid on one another and the overlapping areas were then shaded to appear darker. Thus Fig. 1.27 (left) shows clearly a phantom source midway between the loudspeakers; for this observation the level L_A (from the in-phase generator) was predominant:

$$(L_A - L_B) > 25 \text{ dB}$$

The other extreme

$$(L_A - L_B) < -25 \text{ dB}$$

is represented in Fig. 1.27 (right). Here most of the subjects perceived the 'hearing event' inside their heads.

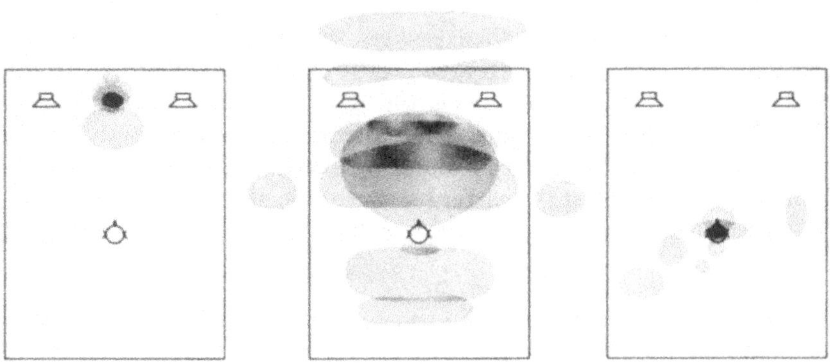

Fig. 1.27. 'Hearing events' drawn by ten test persons for different excitation of the loudspeakers in Fig. 1.26 by the noise generators A and B. (After Plenge.[6])

Between these two extremes lies the case:

$$(L_A - L_B) = 4 \text{ dB}$$

corresponding to Fig. 1.27 (centre). The localization of the 'hearing event' covered vague extended areas, rather different from one subject to another, but mostly in front of the listener's position. (At $(L_A - L_B) = -4$ dB, the areas appeared mostly behind the listeners.)

Using two incoherent[8] noise generators, Plenge measured sound pressures at the entrances to the ears that were only partially incoherent. From experiments with headphones, where the degree of coherence at the electrical input allows an immediate knowledge of the coherence of the respective sound pressures, it is known that the ears can discriminate between coherent, partially coherent, and incoherent signals in terms of small differences in lateralization in the head.[9] Coherent signals are localized in the middle while the localization of incoherent signals is split into two locations, shifted toward the outer ears.

Several sound waves arriving from different directions in a free field behave like waves that have been radiated from incoherent (or partially coherent) sources. The same situation occurs, as we shall explain in Chapter IV. 13 (Volume 2), in every reverberant room and thus is a well-known experience for the ears. For this reason there is no tendency toward localization inside the head. It was clear that the transition from the localization of Fig. 1.27 (left) to that of 1.27 (right) would take place outside the head, but it was not expected that localization would extend over such a wide area.

The possibility of directional localization diminishes as more and more incoherent sound sources are introduced at different directions around the listener. In this connection we refer to the room-acoustics experiment of Damaske,[10] with four loudspeakers in an anechoic room. Their location with respect to the test subject is shown by the arrows in Fig. 1.28 (top left). The circular areas in the figure represent the projection of the half-sphere around and above the listener, so that the points refer to the possible directions in the solid angle 2π from which the sound seems to arrive. (The point at the center represents sound coming from above.) The areas drawn by the test persons in the circles were cut out of black

---

[8] For a discussion of the concept of incoherence, see Cremer, L., *Acustica*, **35** (1976) 215.

[9] See, for instance, Gruber J., Dissertation, TU Berlin, 1967, where further references are given.

[10] Damaske, P., *Acustica*, **19** (1967/68) 199.

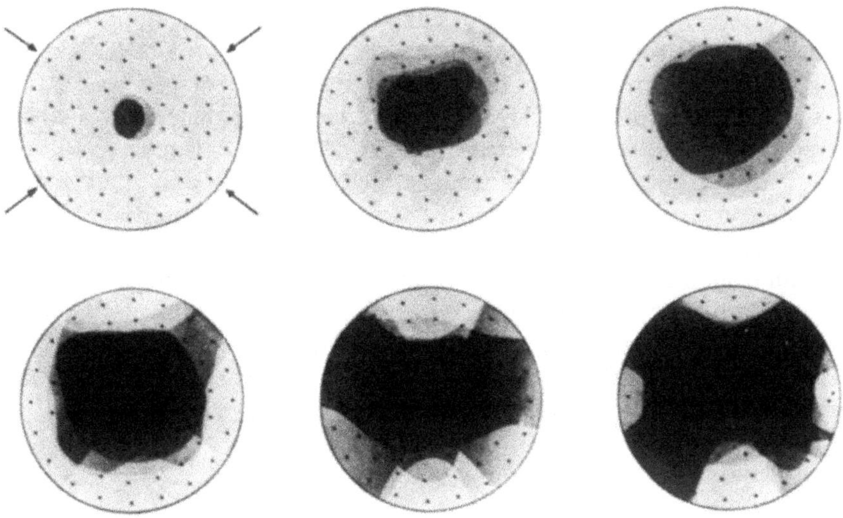

Fig. 1.28. Perceived spherical regions of direction, with sound radiation from four loudspeakers (located as shown by the arrows at the top left) in an anechoic room. The degree of coherence between the signals decreases corresponding to the increasing distances between the four microphones in the reverberation room that fed the four loudspeakers. (After Damaske.[10])

paper and photographed, all on the same negative and with the same exposure. Thus the shade of darkness on the positive is a measure of the relative frequency with which the different areas were designated by the test subjects.

The test signal was broadband random noise between 250 and 2000 Hz which was processed by radiation in a reverberation room. In this room four microphones were placed at the corners of a tetrahedron whose edge-length d could be varied. Each microphone was connected to a different loudspeaker.

The length d corresponding to the sequence of circles in Fig. 1.28 from left to right, top, and then left to right, bottom, was increased as follows:

3·3	10	18
24·5	34	59 cm

As we shall explain in Chapter IV. 13, this represents a transition from almost complete coherence to more and more incoherent signals.

In the case of coherence (top left) we get again a phantom source, which appears this time at the zenith above the listener's head. As the distance *d* is increased, thus decreasing the degree of coherence, the spherical area indicating the region of direction of incidence expands more and more and finally covers almost the entire half-sphere.

By this means Damaske demonstrated that with only four loudspeakers it is possible to produce the same 'subjective diffuseness'[11] in an anechoic room as we would experience in a reverberation room. (We refrain from mentioning the degrees of coherence that Damaske actually measured at the microphones in the reverberation room; in the first place, we would then have to go into greater detail about their meaning, and in the second place, we are not so much interested in the coherence at the microphones as in the coherence at the ears.)

[11] If we decide eventually to quantify this concept, it will perhaps be wise to give it another name, so that the overused word 'diffuseness' can at least be restricted to 'sound events'.

Chapter III. 2

Possibilities for Judging the Acoustical Qualities of Auditoriums

III. 2.1 The Difference Threshold for Reverberation Time

In the last chapter, we have discussed some elementary acoustical perceptions, such as loudness, timbre, the limit of echo perceptibility, and directional impressions; we considered either a single signal (the direct sound, so to speak) or a direct signal followed by a second signal that might represent a first reflection.

It would be possible now to increase the number of reflections, a step at a time, and again to vary the level-differences, time delays and directions, and thereby to approach more and more closely the condition in real rooms.

Indeed, such an approach has already been undertaken with the help of electroacoustically produced sound fields, as we shall see in Section III. 2.5. But this can be done in practice only with a very restricted number of combinations of reflections, in comparison with the enormous number of possible variations.

Such a restriction is actually justified on psychological grounds. As we have already stated in Section I. 2.2 in connection with the source images in a rectangular room with hard walls, the number of reflections ΔN in a given time interval Δt increases with the square of the elapsed time t:

$$\Delta N = \frac{4\pi c^3}{V} t^2 \Delta t \tag{2.1}$$

If we combine this result with the largest number of impulses in unit time (see eqn. (1.15a)) below which the ear is able to distinguish differences in impulse density:

$$\bar{m}_{\max} = \left(\frac{\Delta N}{\Delta t}\right)_{\max} = 2000\,\mathrm{s}^{-1} \tag{2.2}$$

we find the 'beginning time' for statistical reverberation:

$$t_{st} = \sqrt{\frac{V}{4\pi c^3}\left(\frac{\Delta N}{\Delta t}\right)_{max}} \tag{2.3}$$

This may be converted to the approximate rule-of-thumb formula already mentioned in Section II.7.2 as eqn. (7.2):

$$t_{st} = 2\sqrt{V} \text{ ms} \tag{2.3a}$$

with V in m^3.

It seems reasonable that the time interval (from $t = 0$ to $t = t_{st}$) that cannot be handled by statistical methods lasts for only the first 25 ms in a conference room of 160 m^3 and for the first 250 ms in a 16 000 m^3 concert hall.

Evidently a room with rough, diffuse-reflecting walls, such as we discussed in Section I.7.4, Fig. 7.6, will achieve 20 reflections in a 10 ms interval (see eqn. (1.15a)) much sooner; indeed, the reason that such 'diffuse' rooms are usually judged to be better may be that the 'beginning time' for statistical behavior occurs earlier. Thus, the actual value of t_{st} may be regarded as another criterion for good acoustical quality in rooms.

But in large rooms, the duration of 0 to t_{st} cannot be arbitrarily shortened. The time delays for the first reflections from the ceiling and side walls cannot be changed by surface treatment, such as coffers or niches. The most that can be achieved is to see that the first reflections not immediately related to the direct sound are so broken up in different directions that statistical field qualities occur soon after their first appearance.

But here the question arises as to when these small, rapidly successive reflections are to be considered as separated; this may depend on the shortness of the signal impulses.

Although we can no longer regard the reverberation time, which characterizes the reverberation process for $t > t_{st}$, as the only room-acoustics criterion, and perhaps not even the most important one, nevertheless it remains the criterion of interest for the late portion of the sound decay and it is still important that we choose the optimal range for it.

We start here, therefore, with the reverberation time and this, in fact, corresponds to the actual historical development. But it will be expedient to take advantage of our present state of knowledge and to order the results systematically, rather than historically, which means that we will sometimes introduce later findings earlier.

For instance, we begin by discussing just-perceptible differences in reverberation time before considering (in later sections) the optimal choice of reverberation time.

By means of electroacoustical apparatus it is possible to carry out fundamental tests based on comparisons of different reverberation times following immediately one after the other, without the test person being aware of the changes in the apparatus.

Again Békésy[1] was the first to attack the problem of comparisons between different exponentially decaying tones. But much more detailed studies of the difference limen for reverberation time were made by Seraphim,[2] based on a large number (500!) of test subjects who were asked to compare decaying band-pass noise signals.

For this study he defined the difference limen to be the relative change in reverberation time at which 75% of the test subjects, comparing a constant standard reverberation with a randomly varied reverberation, correctly identified the longer decay.

The reverberation events that were compared in Seraphim's study had a level–time dependence like that shown in Part II, Fig. 1.2 (bottom); that is, they consisted of a rapid build up of level, a steady-state interval, and a linear reverberant decay which ended in a constant masking noise level. After 1 s the event was repeated. When the pause between events was increased, the just-perceptible difference in reverberation time also increased, indicating a degraded capability for distinguishing differences in reverberation time.

Figure 2.1 (top) shows the relative difference limen ($\delta T/T$) for experienced subjects as a function of the standard reverberation time T; these results were based on an octave band of noise from 800 to 1600 Hz and a level difference of 30 dB between the beginning and the end of the reverberant decay. In the region of greatest sensitivity (T between 0·6 and 4·0 s), the value of $\delta T/T$ is between 3 and 4%. For untrained persons (whose abilities may be more important for practical applications), the value of $\delta T/T$ is a bit higher. Changes of frequency and reduction of the level difference to 20 dB made little difference in the results.

Thus we may keep in mind as a round figure for the relative difference limen of reverberation time the value:

$$\delta T/T = 4\% \tag{2.4}$$

[1] Békésy, G. v., *Ann. Phys. (Leipzig), V*, **16** (1933) 844.

[2] Seraphim, H. P., *Acustica*, **8** (1958) 280.

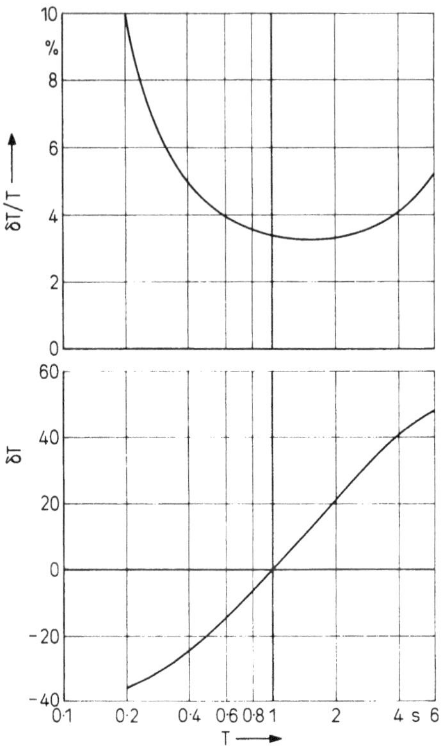

Fig. 2.1. Relative difference limen for reverberation time (top) (after Seraphim[2]), and the derived subjective scale (number of distinguishable δT steps, taking 1 s as the zero-point) (bottom).

For shorter reverberation times the difference limen increases, reaching about 12% for $T = 0{\cdot}2$ s. Therefore, below $T = 0{\cdot}6$ s the absolute, rather than the relative, change in T determines the difference limen; its value is approximately

$$T = 0{\cdot}024 \text{ s} \tag{2.4a}$$

It is possible that in this case it is the 'duration' of the reverberation time that is judged (see Section II. 1.1).

The relative difference limen also increases for longer reverberation times. But since it is nearly constant in the most important middle T-region, the sum of the numbers of distinguishable steps δT increases linearly with increasing log T (see Fig. 2.1, bottom). Therefore it seems subjectively appropriate, in plotting the frequency dependence of reverberation time, to use a logarithmic scale for T. In the range from 1 to 3 s, it is reasonable to round off the reverberation time to 0·1 s, or at the utmost 0·05 s. At least it is scarcely justified to state the reverberation time to three or more decimal places.

Since the dependence of reverberation time upon frequency is of special interest, Plenge[3] investigated the 'absolute perceptibility' of 'humps' and 'valleys' in the curve of reverberation time versus frequency. For this purpose he used real reverberation processes excited by pistol shots in rooms, in order to depart as little as possible from natural conditions in the tests. He analysed the reverberation in frequency groups (see Section III. 1.3). In two neighboring frequency groups, a variable gain amplifier was connected so that the decay coefficient for those frequency groups could be changed from δ_0 to $\delta_0 + \Delta\delta$; subsequently all the frequency groups were added together. The test subject (mainly, the author himself) changed the value of $\Delta\delta$ until he could perceive a difference between the two reverberation processes.

Figure 2.2 shows two examples of the absolute perceptibility threshold for relative change in reverberation time $\Delta T/T$, determined for the

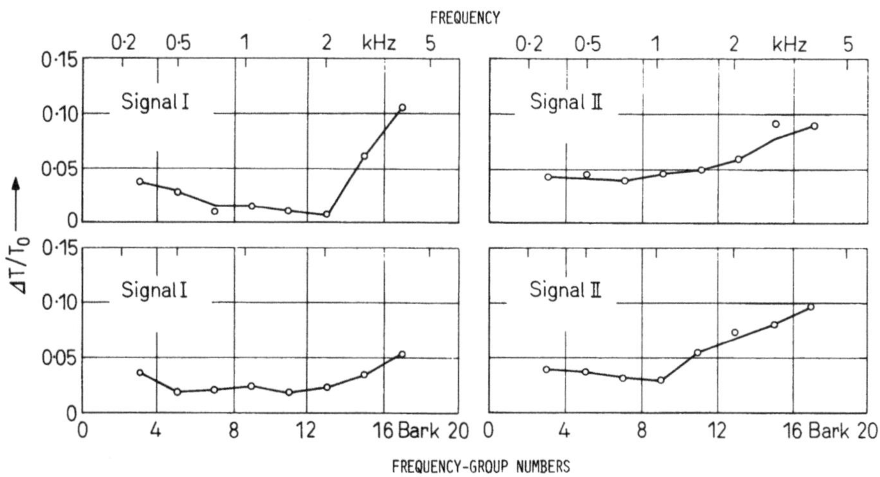

Fig. 2.2. Difference limens for valleys (top) and humps (bottom) having a breadth of two neighboring frequency groups, in reverberation time curves with a smooth dependence of reverberation time on frequency; the first 200 ms of all the decays was suppressed in the tests. Left: $T_0 = 1$ s. Right: $T_0 = 1{\cdot}5$ s. (After Plenge.[3])

[3] Plenge, G., *Acustica*, **16** (1965/66) 269.

different mean reverberation times of the comparison examples; they are plotted as a function of the mean number of the two neighboring changed frequency groups. In all these examples the first 200 ms of the decay was suppressed so that only the statistical portions of the decays were compared. The performance time was limited to 0·8 s. If the first 200 ms of the decays had been included, other, not necessarily higher, values of $\Delta T/T$ would have been observed.

In general, the 'humps' are more easily recognized than the 'valleys'.

The limen (for decays with the first 200 ms suppressed) decreases for frequencies below 1000 Hz. The results at lower frequencies approximately coincide with the values found by Seraphim for octave-band noise between 800 and 1600 Hz.

Finally we must consider reverberation processes composed of several exponential decays, especially those two-decay reverberations that we encountered in Chapter II. 3 in connection with coupled rooms. Here the initial part of the decay may differ considerably from the later part. These differences are made especially clear if the reverberation following a steady excitation is recorded according to Schroeder's method[4] of integrating an impulse response. This procedure is indispensable if the initial reverberation does not qualify for statistical treatment because of low reflection density.

Atal *et al.*[5] investigated whether the impression of reverberation for running speech and music is at all affected by the later part of the decay, which is mostly masked by the initial reverberation of new signals. The test signals, which included speech and music, were presented to the subjects by means of earphones; the signals were modified by a computer that simulated reverberation of different kinds. Exact exponential decays were compared with decays in which the slope changed during the decay. The reverberation times of the purely exponential decays were varied and the test subject was asked to decide which of the two decay processes seemed more reverberant.

The reverberation time of the pure exponential decay that split the subjects' judgments at 50% in both directions was defined as the 'subjective reverberation time' T_s. It corresponded to the slope of the initial part of the non-exponential decay (that is, to the first 160 ms or the first 15 dB decrease in level, see Section II. 7.1). The evaluation of

[4] Schroeder, M. R., *J. Acoust. Soc. Am.*, **37** (1965) 409.

[5] Atal, B. S., Schroeder, M. R. and Jessler, G. M. *Proc. 5th. ICA, Liège*, 7–14 *Sept.*, 1965, Paper G-32.

reverberation time from the standardized level region of the decay, from −5 to −35 dB below the steady level, gave significantly different results.

According to these results there is no point in attempting to increase the sensation of reverberation in a rather dry room by coupling it to a reverberant room (such as the attic) if the added reverberation appears only during the later part of the decay. This would not be noticeable in running music but only at sudden stops. On the other hand, this effect could be quite attractive in some circumstances.

Plenge also investigated the extent to which a reverberation decay with a definite 'knee' in the decay curve can be distinguished from a linear decay; he used impulsive excitation of the room. He found that reverberation times determined from only the first 10, 15 or 20 dB of such decays must differ much more from one another, in order to be distinguishable, than Seraphim found for decays with smooth slopes.

At least we can conclude from both results that there is not much sense in arguing over differences of tenths of seconds for reverberation times between 1·5 and 2·5 s.

III.2.2 Articulation Tests

Now knowing the difference limen for reverberation time, if we wish to go on to discuss the question of what is the optimal reverberation time, we must—as in all subjective acoustical evaluations—distinguish between the requirements for speech and those for music. This difference is important not only because we prefer higher reverberation times for music than for speech but even more because, with speech, we are interested almost exclusively in the intelligibility.

The intelligibility of speech is easy to determine by presenting test subjects with a series of words or sentences and asking them to write down what they understand. Preferably such intelligibility tests are carried out with a speaker who articulates clearly, presenting nonsense syllables, such as 'gul, peg, raf' (preferably in groups of three). Such a test is called an 'articulation test' and the percentage of correctly understood syllables is called the 'percentage syllable articulation' (PSA).

Knudsen[1] used this criterion in the 1920s, when the problem of establishing guidelines for the optimal reverberation time first arose. He determined, for step-wise changes in the reverberation time in rooms

---

[1] Knudsen, V. O., *Phys. Rev.*, **26** (1925) 287; *J. Acoust. Soc. Am.*, **1** (1929) 56.

with volumes between 6000 and 8000 m^3, how the PSA changed with increasing values of *T*. His results are presented in Fig. 2.3 (left): the PSA decreased from 90% at $T = 1$ s to 50% at $T = 8$ s.

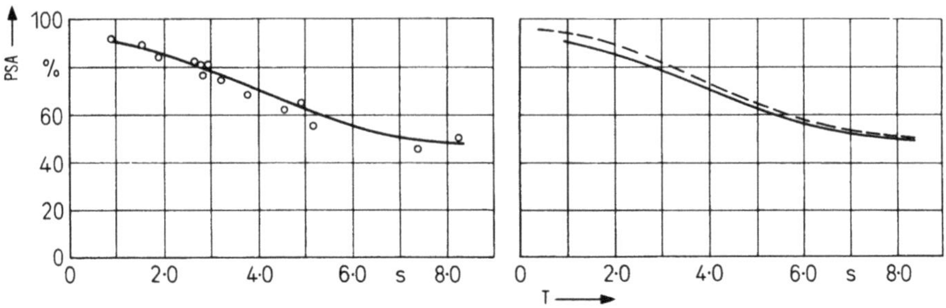

Fig. 2.3. Dependence of the percentage syllable articulation (PSA) on the reverberation time (*T*). Left: in rooms with volumes from 6000 to 8000 m^3. Right: the same curve (———) and a corresponding curve (– – –) corrected to equal loudness levels for steady excitation, according to Fig. 2.4. (After Knudsen.[1])

It was to be expected that longer reverberation times would be detrimental to speech intelligibility; therefore it was most surprising to find that, even for $T = 3$ s, the PSA was still as high as 80%, a value that readily allows clear understanding of coherent sentences. At any rate, we might suppose from Fig. 2.3 (left) that shorter reverberation time leads to better intelligibility of speech. This would indeed be correct, except that a decrease in reverberation time is always accompanied by a decrease in 'useful sound'. Thus the conclusion is justified only if this loss can be compensated by electroacoustical means.

If no artificial amplification of the natural speech is provided, we must take into account the dependence of the PSA on the loudness level of the speech—which we may call the 'speech level'—and also the possible dependence of the loudness level on the reverberation time.

For this purpose the results of an extensive investigation by Fletcher[2] are important, as shown by the dashed curve in Fig. 2.4. There is a broad maximum around 80 dB (implying that higher loudness levels can be detrimental) and a steep falling off below 50 dB. The solid curve presents similar results, based on the results of many subsequent tests under improved conditions, published by Fletcher[3] thirty years later.

[2] Fletcher, H., *J. Franklin Inst.*, **193** (1922). The abscissa in Fig. 2.4 has been adjusted to the present-day reference pressure of 20 μPa.

[3] Fletcher, H., *Speech and Hearing*, 2nd. edn., Van Nostrand, New York, 1953, Fig. 221.

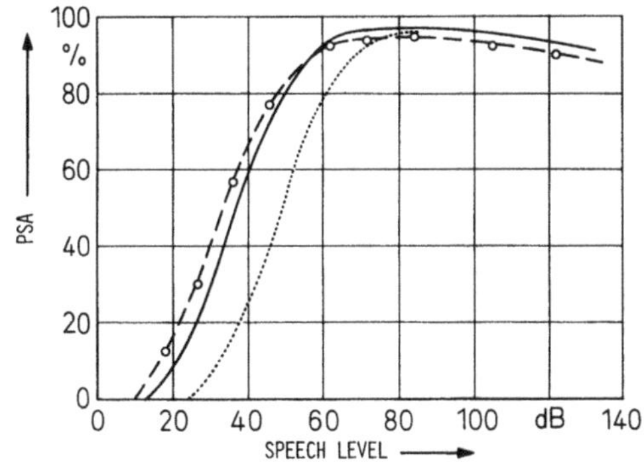

Fig. 2.4. Dependence of the percentage syllable articulation (PSA) on the loudness level of speech. ——— and - - - - without background noise; . . . . . with background noise level of 40 dB. (After Fletcher.[3])

Knudsen tried to eliminate this additional dependence on level by measuring the loudness level in one case and calculating the change to be expected for steady-state excitation according to eqn. (1.12) in Section II. 1.3:

$$L_1 - L_2 = 10 \log \frac{A_2}{A_1} = 10 \log \left(\frac{V_2}{V_1}\frac{T_1}{T_2}\right) \tag{2.5}$$

By this means he determined the dashed line in Fig. 2.3 (right) which he regarded as the correct dependence of PSA on reverberation time. The difference between this curve and the solid curve is small because all of his experimental conditions lay within the region of the broad flat maximum of Fig. 2.4. Knudsen extrapolated his dashed curve to $T = 0{\cdot}5$ s, divided all the PSA values by the value found for $T = 0{\cdot}5$ s, and thereby defined a 'degradation factor' k_r for characterizing the dependence of PSA on reverberation time only.

In a similar manner, by dividing all the values of the dashed curve in Fig. 2.4 by its maximum value of 96%, Knudsen defined another degradation factor k_l for the dependence of PSA on loudness level only.

Since k_r decreases monotonically with increasing T, and k_l increases monotonically with increasing T for given speech power, the product $k_r k_l$ reaches a maximum in a situation where, starting with a low value of T in a room, the reverberation time is gradually increased by removing absorptive materials. By this means, Knudsen sought to determine the optimal reverberation time for speech in that room.

Table 1 gives an example. The first column presents the reverberation times; the second column gives the loudness levels calculated for steady-state excitation with a speech power of 50 μW and a room volume of 11 000 m^3; the third and fourth columns give the values for the two degradation factors k_r and k_l; the last column gives their product. It can be seen that a maximum of intelligibility is reached at about $T = 1$ s. But this maximum is so flat that we may rather conclude that the reverberation time has practically no influence in this broad region, since in the range between $T = 0{\cdot}5$ s and $T = 2$ s the value of PSA changes by only 3%.

Table 1: Evaluation of the optimum reverberation time from the degradation factors k_r and k_l, giving the dependence of percentage syllable articulation (PSA) on reverberation time and loudness level. (After Knudsen.[1])

$T(s)$	$L(phon)$	k_l	k_r	$k_l k_r$
0·5	48	0·850	1·00	0·850
1·00	51	0·885	0·982	0·870
2·00	54	0·910	0·924	0·840
4·00	57	0·936	0·752	0·704
8·00	60	0·959	0·510	0·489

Knudsen also took into account the possibility that the speaker actually speaks louder in a large room than in a small one. (Nowadays, of course, no lecturer in a large hall would be obliged to speak without electroacoustical reinforcement (see below). But in that case differences in power would have an even greater influence on the value of PSA.)

By measuring the loudness during various lectures at the University of California at Los Angeles (UCLA), unknown to the lecturers, he determined that the mean speech power in a room of 770 m^3 was only 27 μW (see Table 2), but in a room of 6800 m^3 it was about 49 μW; and in a room with a volume of 175 m^3 at another institution, he found a mean speech power of 10 μW. Thus he seemed to have enough evidence to take this influence into account, at least approximately.

He summarized his results by stating that the optimal reverberation time for speech would be 0·8 s in a room of 770 m^3 and that this optimal value increases linearly with the logarithm of the room volume, so that it reaches 1 s at $V = 11\,300$ m^3. Analytically this means a dependence

Table 2: Differences of speech power for different speakers. (After Knudsen.[1])

Speaker no.	$A(m^2)$	$P(\mu W)$
4	54·9	8·5
6	66·5	30·0
5	67·9	4·5
2	70·6	66·2
3	82·8	23·0
1	108·9	32·2
Average value	—	27·4

according to:

$$T_{opt} = [0{\cdot}32 + 0{\cdot}17 \log V]\,\mathrm{s} \tag{2.6}$$

with V in m^3.

He found a more rapid increase for larger rooms but this has no practical importance here; in the first place, statistical considerations lose their validity in larger rooms and, in the second place, no one would speak in such large rooms without amplification.

In this first attempt to derive an acoustical rule-of-thumb for buildings from the results of articulation tests, Knudsen was well aware of the doubtfulness of his assumptions and the subsequent evaluations. The measured speech power, especially, showed an enormous scatter from speaker to speaker. Table 2 shows a series of results measured in a 770 m^3 lecture room; in addition to the speech power P, the values of equivalent absorption area A are also given, corresponding to the different numbers of students present. Since the observations are ordered according to increasing values of A, it is evident how little the lecturers took into account the increasing number of students with respect to increasing their vocal effort, and also how great the individual differences were between speakers. The highest and lowest speech powers of individual speakers differ by a factor of 15, the power levels differ by 12 dB!

Furthermore, the tabulated values of speech power are averages over time; the temporal fluctuations of speech power, the so-called 'dynamics' of the speaker, could account for another 20 dB of variation.

Equally important is the influence of the background noise level; in this context, also, Knudsen[4] made detailed measurements of PSA. We may

[4] Knudsen, V. O., *Phys. Rev.*, **26** (1925) 133.

surmise the influence of the background noise from the shift of the solid curve in Fig. 2.4 (determined without background noise) to the dotted curve which was found by Fletcher with a background noise level of 40 dB. For this situation a speech level of 50 dB produces a value for PSA of only 58%, which cannot be regarded as sufficient.

Such a noise level is by no means uncommonly high; a layman would not even notice it, so long as the speech level is higher. It corresponds to what is expected in large towns even with closed windows; only at night is the noise level lower. For this reason Knudsen carried out his studies which produced the results shown in Fig. 2.3 only in the quiet night hours.

The differences in PSA caused by different speakers and different background noise levels are, unfortunately, much greater than the changes that can be achieved with purely room-acoustical means—if we exclude serious mistakes. But this by no means relieves us of the responsibility to use these means to the best of our knowledge. Rather it points out, once more, how little weight can be given to individual lay judgments when the speaker's quality and the background level are unknown.

Since poor speakers and high noise levels (especially at sports and social events) are likely to occur fairly often, it is recommended that a public address system should always be available for rooms seating more than 1000 people.

If the system is used for speakers who do not need it, however, this is a regrettable misuse, especially since it will probably be operated by an unskilled person who, in order to be safe, will run the sound level as high as possible (i.e. near the feedback—or 'howl'—limit).

In fact, increasing the loudness level does not always produce an improvement in intelligibility, as Fletcher's tests have demonstrated (see Fig. 2.4), even if the sound system is not already overloaded. This holds especially for large rooms; small echoes that would otherwise not be noticeable may suddenly become annoying. In particular, an unnaturally high loudness disrupts the sense of personal intimacy because of the disproportion between what one sees and what one hears.

For this reason a loudspeaker system should never be operated at levels louder than just necessary; and this may require a different gain adjustment for different individual speakers on the same program. The first few words of the speaker may not be so important for a satisfactory adjustment, particularly if the audience is still noisy; the long-term gain adjustment for each speaker should be made in terms of his 'steady state'.

Finally it must be mentioned that if too much amplification is used for the speaker's voice, the audience will be tempted to become more noisy, for example by talking to their neighbors as is often done in cinemas. The fact that one must listen rather carefully in a drama theater may be one of the reasons that it is more exciting than the cinema.

We have presented here in some detail the results of the classic investigations of Knudsen because they were the first studies in which the audience reaction, in terms of the speech intelligibility, or PSA, was taken as the basis for a criterion of auditorium quality. At the time of Knudsen's studies, the distinction between useful (early) and detrimental (late) sound (which we discussed in Section III. 1.4) was not fully elaborated. From our present-day viewpoint, it appears that Knudsen was wrong to suppose that it does not matter whether a stationary speech level is achieved by the use of distortionless electronic amplification (as in Fletcher's investigations) or by room reflections that arrive with suitable time delays.

Knudsen thought it adequate to assume that the energy element $P\,dt$ of running speech is equally distributed throughout the room volume and that it should be integrated over the entire reverberation duration:

$$\int_0^\infty \frac{P}{V}\,10^{-6t/T}\,dt = \frac{PT}{13{\cdot}8\,V} = \frac{4P}{cA} \tag{2.7}$$

Today we would think it proper to integrate over, at most, the first 50 ms of the reverberation:

$$\int_0^{50\,\mathrm{ms}} \frac{P}{V}10^{-6t/T}dt = \frac{PT}{13{\cdot}8V}(1-10^{-0{\cdot}3/T}) \tag{2.8}$$

It is easy to see that this change would reduce the influence of the reverberation time in comparison with the loudness level: a six-fold change of T from 0·5 to 3·0 s changes the value of the integral in eqn. (2.8) from 0·375 to only 0·63, a ratio of only 1·0 to 1·7.

Nevertheless, it would be wrong to add the reverberant energy after 50 ms to the background noise level; instead one should retain Knudsen's degradation factor k_r since it is based on just that later part of the reverberation.

From the standpoint of our current knowledge, it seems more appropriate to derive the PSA from the loudness level and the background noise level only, adding the reverberant energy accumulated before a certain time limit to the former quantity and adding the energy after that time limit to the latter, as was proposed by Lochner and Burger.[5]

Those authors avoided a sharp limit between useful and detrimental sound by proposing a gradual transition from one to the other, based on earlier measurements of the PSA.[6] They added to the primary signal a second, delayed, signal and determined the amount by which the primary signal level would have to be raised to achieve the same PSA as with the addition of the second signal. Fig. 7.7, Part II, shows the results: the upper curve corresponds to primary and secondary signals with equal level; the lower curve corresponds to the case where the secondary signal level is 5 dB below that of the primary signal.

Based on these observations, Lochner and Burger introduced a weighting function $a_{(t)}$ for evaluating useful sound; it is constant and equal to unity for t up to 35 ms and then decreases linearly with increasing time, reaching zero at $t = 95$ ms:

$$E_n \sim \int_0^{95\,\mathrm{ms}} p_{(t)}^2 a_{(t)}\,\mathrm{d}t \tag{2.9}$$

All the subsequent reverberant sound energy is regarded as detrimental:

$$E_s \sim \int_{95\,\mathrm{ms}}^{\infty} p_{(t)}^2\,\mathrm{d}t \tag{2.10}$$

When they applied this concept to an exponential reverberation process, they found, surprisingly, a stronger dependence of the optimal reverberation time on room volume than did Knudsen. This may come from the other dependences of PSA on the signal level and the background noise level, as observed by Fletcher. It is curious, therefore, that they found no dependence on room volume when they applied their formulae to a non-continuous reverberation. For this analysis they undertook a detailed reckoning of the early reflections in a rectangular room with length 1·5 D, width D, and height 0·5 D, with the source at a distance 0·1 D from the front wall and the receiver at a distance 0·4 D

[5] Lochner, J. P. A. and Burger, J. F., *Acustica*, **10** (1960) 394.

[6] Lochner, J. P. A. and Burger, J. F., *Acustica*, **8** (1958) 1.

from the back wall; they used a constant, given background noise level and a variable but uniformly distributed and angle-independent sound absorption α.

The fact that PSA depends essentially on the direct sound and the early reflections implies that it is a criterion whose value depends on the location of the listener in the room. From this it follows that it would not be possible to derive from this criterion an optimal value for the reverberation time.

On the other hand, we can expect a close relation between PSA and those objective criteria that were developed as measures of 'distinctness' and 'clarity' (see Section II. 7.4) involving a comparison of the early and the late sound energies. Figure 2.5 shows the relation between PSA (here designated by the Gothic letter, $\mathfrak{v}$) and the 'center-time' which Kürer[7] evaluated at various locations in two very different rooms. The PSA ($\mathfrak{v}$) is divided here by its maximum value $\mathfrak{v}_{L0}$, which is constant below the center-time value $t_s = 70$ ms. Since, under statistical conditions $t_s = 70$ ms corresponds to $T = 1$ s, we may conclude that it is not worthwhile to reduce the reverberation time below 1 s in order to increase the distinctness.

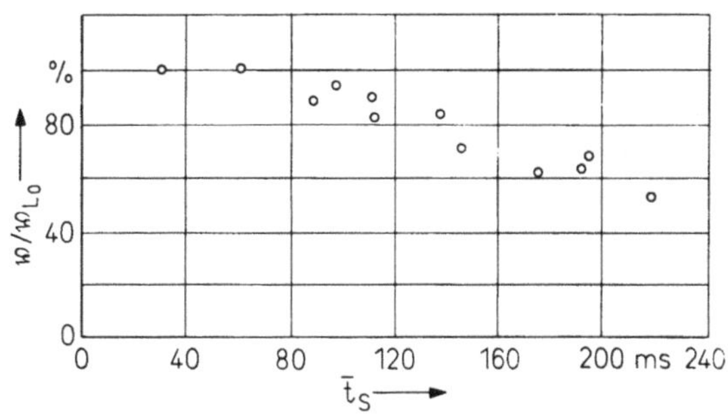

Fig. 2.5. Dependence of the percentage syllable articulation, PSA (*Silben-Verstandlichkeit* $\mathfrak{v}$), on the center-time t_s (see Section II. 7.4). (After Kürer.[8])

Above $t_s = 70$ ms there is a clear decrease of $\mathfrak{v}$, which Kürer approximated first by a straight line and later by a parabola. According to Kürer's comparisons,[8] the dependence of PSA on t_s was better correlated than its dependence on 'early decay time' or the 'distinctness coefficient'.

The possibility of such a correlation between a subjective criterion of

---

[7] Kürer, R., *Proc. 7th. ICA, Budapest, 1971*, Paper A5, p. 23.

[8] Kürer, R., *Akustik u. Schwingungstechnik*, Tagung, Berlin, 1970; VDI Verlag, Düsseldorf, 1971, p. 143.

room-acoustical quality and an objectively measurable quantity is important because the subjective evaluation takes much more time. To determine the PSA, many syllables must be heard and many test subjects are required if the results are to be statistically significant and of general value.

III.2.3 Acoustical Quality Information through Inquiry

All of the remaining sections in this chapter deal with the very complicated questions raised by the evaluation of room-acoustical quality for musical presentations. Here no such simple criterion as the PSA exists, as it does for speech.

Certainly the ability to recognize the different musical instruments and the notes played by them plays a role. Thus it might seem feasible to evaluate this ability in terms of a 'note intelligibility' rating. But quite apart from the fact that writing down the notes of a musical passage takes much more time than writing words, and, moreover, can be done only by a small circle of musically trained people, such a note intelligibility rating is not the only, and not even the most important, aspect of musical quality in an auditorium.

Just as a pianist sometimes uses the sustaining pedal to allow the struck strings to vibrate freely with their own long natural decay, so there are musical instruments and compositions for which it is desirable that the tone sequences should blend into one another. Therefore, the optimal reverberation times are higher for music than for speech.

Even on the dramatic stage, high intelligibility is not the only important requirement; some additional reverberation may be desirable for aesthetic reasons.[1]

The lack of an unequivocal criterion for music, such as percentage syllable articulation for speech, raises the question whether there can ever be general agreement about the acoustical quality of a hall for musical performances. It is possible that the musical tastes of different people are simply too different for such an agreement to be reached. We are not thinking here of the unavoidable scatter in people's judgments of loudness comparisons or of speech intelligibility. So long as such scattered judgments are normally (Gaussian) distributed about a mean value,

[1] Izenour, G. C., *Theater Design*, McGraw-Hill, New York 1977, Chapter 9, Section 2, particularly pp. 481–2. Lifschitz, S., *Vorlesungen über Bauakustik*, Moskau, 1923; German translation, Stuttgart, 1930, p. 48.

it is always possible to state this mean, along with the standard deviation as a measure of the scatter. But it may be possible that the scattered judgments do not define a single peak in the distribution of responses, but rather several or even none.

Indeed, this fundamental question cannot be answered with either a yes or a no, as we shall demonstrate at the end of this chapter. There are several aspects of acoustical quality on which the majority of test persons agree as to general tendencies; but there are others for which significant differences between groups of listeners either may be regarded as only a matter of taste, or may perhaps be based on physiological differences that we do not yet understand.

We shall discuss now the different procedures that have been used in past investigations, in an attempt to define criteria of musical acoustical quality in rooms. We discuss them in the order of their importance to the science of room acoustics. We begin with direct inquiries as to the listeners' opinions, then go on to judgments based on changes in the room(s), then to judgments based on changes in a synthetic sound field, and finally to the aural comparison of tape recordings made under different acoustical conditions in rooms.

Certainly those acoustical quality judgments about different auditoriums that are based on nothing but memory are the most doubtful. The person being questioned can say only that he remembers the degree of his satisfaction or his discontent, just as someone may declare that he relished a special dish in a certain restaurant some years ago to a greater or lesser extent than this particular one.

The weight of such an opinion is increased if the listener has made notes of his impression in a diary or a letter. For this reason written questionnaires are preferred, because they retain their validity even after a number of years.

But even such written statements do not exclude the possibility that the same person might revise his judgment if he could relive the same experience today, with an immediate comparison of then and now.

Furthermore it is practically impossible to separate the judgment of the acoustical quality of a hall from the judgment of the musical quality of the performance. Therefore concert halls in which outstanding artists play regularly (for example, halls in large cities or in festival buildings) stand a better chance of winning a reputation for good acoustics than halls in which important musical events seldom take place.

In spite of all these difficulties the method of direct inquiry will always be used. We content ourselves here with reference to three examples.

When Watson[2] first tried to develop a quantitative relation between the optimal reverberation time and the volume of concert halls, he calculated the reverberation times of six 'accepted' halls. As for the basis of the choice of those particular halls, we are told only that they are 'pronounced good by public opinion'. He plotted the calculated reverberation times against the values of $V^{1/3}$ (see Fig. 2.6), in order to see whether the increase in reverberation time is based only on the increase in mean free path; this was reasonable since at that time the shapes of the halls were not very different from one another and neither were the building materials. If one tries to draw straight lines that pass through both the origin and the calculated points for the empty and occupied halls, as shown by the dashed lines in Fig. 2.6, then, setting $l_m = 0{\cdot}62\,V^{1/3}$ as in Section II. 2.1, eqn. (2.8), we find mean absorption coefficients of 0·17 for the empty halls and 0·3 for the occupied halls.

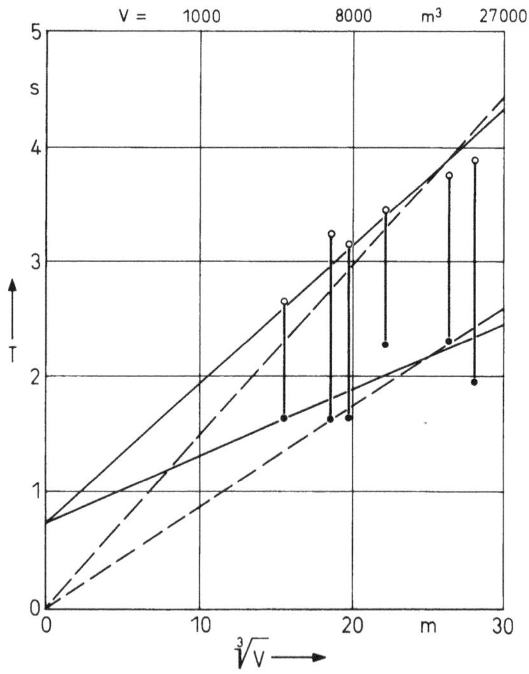

Fig. 2.6. Dependence of the calculated reverberation times for six 'accepted' concert halls on their volumes. Fully occupied, ●; empty, ○. (After Watson.[2])

[2] Watson, F. R., *Acoustics of Auditoriums*, New York, 1923; see also, *J. Franklin Inst.*, **198** (1924) 73.

The rather low value for the empty halls results from the fact that in those days the seats were made of wood or had only thin upholstery. The second value, however, offers some guidance for the desirable mean absorption coefficient in occupied rooms.

Even though the T-values calculated by Watson were probably too high, he was the first to state that the reverberation time increases less quickly than the mean free path. It is in accordance with the accuracy of his results that Watson endeavored to express this tendency by linear dependences on $V^{1/3}$, with the formula for the empty halls:

$$T_{opt} = 0{\cdot}75 + 0{\cdot}12\,\sqrt[3]{V}\ \text{s} \tag{2.11a}$$

and the formula for the occupied halls:

$$T_{opt} = 0{\cdot}75 + 0{\cdot}057\,\sqrt[3]{V}\ \text{s} \tag{2.11b}$$

with V in m^3.

No simple similarity law exists between T and V, because the height of the halls tends to increase less rapidly than the length and the width, for economical reasons. This means that the largest absorptive area (the floor covered with audience) always represents a higher percentage of the total surface area than the less absorptive wall surfaces, and this leads to increasingly higher values for the mean absorption coefficient as the volume increases. The reverberation times must therefore increase less rapidly than if they were proportional to the mean free paths.

Watson would probably have found the same result if he had made an arbitrary choice of the halls that he studied.

Lifschitz,[3] who tried to develop a formula for $T_{opt(V)}$ on the basis of physiological assumptions that are no longer acceptable today, also had to base his constants on a room with good acoustics. He chose the Säulensaal des Moskauer Verbandshauses (Pillar Hall of the Moscow Union House), which was well-known to him because the famous conductor Arthur Nikisch had declared it to be the best concert hall in Europe. (There are in fact a number of such 'best halls', several different ones sometimes being named as 'best' by the same conductor!)

The results of public inquiries may claim general validity only if they are based on the judgments of a sufficient number of musically competent persons.

Parkin *et al.*[4] addressed their written questionnaires concerning twelve older British concert halls to various musically competent listeners, such

---

[3] Lifschitz, S., *Phys. Rev.*, **25** (1925) 391; **27** (1926) 618.

[4] Parkin, P. H., Scholes, W. E. and Derbyshire, A. G., *Acustica*, **2** (1952) 97.

as music critics, professors of music, composers, etc., rather than only to the performing artists. They argued that the artists tend to evaluate the halls only from their experience on the stage; but, as we have already seen in Section III. 1.3, at that location the acoustical relations are quite different from those at the seats of the listeners, more distant from the source.

Only half of the 170 experts to whom the questionnaires were sent returned answers, which shows how few of the selected 'competent listeners' felt ready and able to respond to the questions. Even of the 75 questionnaires that were returned, only 42 could be evaluated, since only in those 42 responses had the respondent enough experience in at least three of the named halls to be able to judge them. Of those 42 persons, 14 were music critics, 10 were composers, 9 were professors of music; there were 9 'others'.

Table 3 shows the analysis of their judgments, based on a scale of acoustical quality with three steps: good, fair, and bad. Since the list did not include halls with obvious defects, the number of 'bad' judgments was small. The list is arranged in order of decreasing acoustical quality, based on the percentage of 'good' judgments. The last three halls have been identified in the published report only by the letters A, B and C, in order to avoid any embarrassment to those halls that might come from an inquiry whose results, despite all the care, are doubtful in principle.

Restricting the group of test persons to those who had extensive experience in six (instead of three) of the halls did not change the order of judged acoustical quality.

Table 3: Distribution of subjective evaluations of some concert halls.

Room	*No. of observers*	*'Good'*	*'Fair'*	*'Bad'*	*Percentage of 'good' responses*	$V(m^3)$	$T(s)$ (500 *Hz*)
Liverpool Philharmonic	25	22	2	1	90	13 500	1·6
Covent Garden	39	31	7	1	80	10 100	1·1
St. Andrew's Hall	17	12	5	0	70	16 100	2·2
Usher Hall	21	14	6	1	65	15 700	1·8
de Montfort Hall	16	10	5	1	65	12 500	1·6
People's Palace	29	17	11	1	60	8 900	1·4
Hastings Pavilion	18	11	7	0	60	7 900	1·3
A	33	14	13	6	40	—	—
B	26	10	15	1	40	14 400	1·8
C	18	7	8	3	40	16 900	2·1

The measured reverberation time (at 500 Hz) of the hall that was judged best was 1·6 s; with a volume of 13 500 m^3, this is surprisingly low. But even more astonishing is that the halls judged second and third best had reverberation times of 1·1 and 2·2 s, a very great difference!

In general, the increases in reverberation time in Table 3 correspond to increases in the hall volumes, but in a much stronger sense than would be predicted by any of the formulae for $T_{\text{opt}\,(V)}$ mentioned until now.

From this we can conclude that, if there *is* an optimal reverberation time, it comprehends a rather broad range. Now Covent Garden is a horseshoe-shaped opera house with many balconies; the reverberation time of 1·1 s is not unexpected since it has won its reputation from opera performances, although occasionally, as in all opera houses, orchestral concerts are also given there. But the differences in reverberation time between the Liverpool Philharmonic and the (no longer existing) St. Andrew's Hall in Glasgow are much larger than one would expect on the basis of their volumes; the ratio even exceeds the relation between their mean free paths.

If both of these halls are highly esteemed, it proves only that each of them is good in its own way, independent of the reverberation. According to the investigations of Kuhl, which we will discuss later in Section III. 2.6, we may expect the Liverpool hall to be more suitable for classical music and the Glasgow hall to be more suitable for romantic music. But this is not to say that knowledgeable people would regard the Liverpool hall as unfit for Bruckner and the Glasgow hall unsuitable for Mozart. It would be very boring if all concert halls were acoustically equal. To hear the same musical composition under different acoustical conditions is just as exciting as to hear it at the hands of different conductors, with different tempos and different dynamics. In this case the existing reverberation time will influence the choice of the tempo.

All of this demonstrates that room-acoustical qualities are influenced not only by technical conditions to which optimal values can be assigned, such as are adequate to specify the lighting and heating requirements, but they also enter into the realm of artistic performance and are subject to its laws. Just as there exist tempos that nearly every listener would regard as too fast or too slow, there also exist reverberation times that every listener would think too short or too long. But these reverberation limits are broad; they allow considerable latitude within which it is more important that the visual and aural events are well matched and that other acoustical criteria are respected.

The richest existing collection of room-acoustical data and descriptions of concert halls and opera houses was published by Beranek.[5] Although many of his colleagues contributed information to this compilation, Beranek himself personally visited the 54 halls to check the data against the plans and specifications; and where he found discrepancies he resolved them through correspondence with the persons who supplied the original reference data.

By comparison with his extreme care in handling the objective data about the halls, the corresponding subjective judgments of the acoustical qualities, which are the topic of this section, seem rather arbitrary. They presumably represent a small selection of acoustical judgments that Beranek solicited from 23 conductors and 21 music critics, as Winckel[6] had done before him. Based on these judgments and on his own evaluations of the halls, he ranked the 54 halls into five groups:

A+	excellent
A	very good to excellent
B+	good to very good
B	fair to good
C	fair

The reader learns the names of the famous conductors, from Sir John Barbirolli to Bruno Walter, and also the names of the (mostly American) music critics who responded. But unfortunately Beranek does not report, as did Parkin *et al.*, how many judgments were given for the individual halls and how the halls have been assigned to the five classes. At least he was apparently convinced that it is possible to make an unequivocal evaluation of the acoustical quality of a hall, and went on to develop quantitative rating schemes for the acoustical quality of concert halls and opera houses, based on objectively measurable quantities.

Starting from his original grouping of the halls into five classes of acoustical quality, he investigated the respects in which the excellent halls differed acoustically from the less-good halls. Although for many years the reverberation time had been regarded as the decisive criterion in room acoustics, Beranek pointed out, as had others before him, that at most it may be only one criterion among many. This was evident from the fact that the six halls in group A+ had reverberation times ranging from 1·7 to 2·05 s; those in group A ranged from 1·2 to 2·0 s, and so on.

[5] Beranek, L. L., *Music, Acoustics and Architecture*, John Wiley, New York, 1962 (reprinted Krieger Publishing Co., Huntington, New York, 1979); particularly Chapt. 6.

[6] Winckel, F., *Baukunst u. Werkform*, **8** (1955) 751.

He concluded that other desirable qualities may compensate for extreme values of reverberation time. Among those other qualities he found the most important to be the time delay between the direct sound and the first strong reflection (as observed in specially selected 'good' seats). He called this quantity the 'initial time delay gap' (see Section II. 7.3). He could show that the initial time delay gap was less than 20 ms in all the halls in group A+, less than 33 ms in the halls of group A, and less than 57 ms in the halls of group B; for the single hall in group C (the Royal Albert Hall in London) the value was 70 ms. This criterion can be determined (trusting to the validity and reliability of ray-diagramming) from the drawings of the hall. He gives it a weight of 40 points out of 100 in his rating scheme.

The reverberation time at mid-frequencies, given by $(T_2+T_3)/2$, was assigned a weighting of only 15 points.[7]

Another of his criteria characterizes the frequency dependence of T:

$$(T_0+T_1)/(T_2+T_3)$$

he called this quantity the 'bass ratio' and assigned it 15 points in his rating scheme. This quantity, based on objective measurements, is correlated in Beranek's scheme with the subjective quality of 'warmth'. Other criteria enter into the rating scheme with even smaller weightings; we refrain from mentioning them all because the assignment of quantitative values could only be based on his individual evaluations.

Beranek required quantitative criteria that could be objectively determined because it was his goal to develop a rating scheme that would allow him to classify each hall into one of the above-mentioned categories according to its total rating score. With 90–100 points it would belong to group A+; with 80–90 points, to group A, and so on. A hall with less than 50 points was regarded as unusable. This could happen if the hall exhibits serious defects such as pronounced echoes, excessively loud air-conditioning noise, etc. For such faults a prescribed number of points (up to as many as 50!) can be subtracted from the total score.

If we leave these 'penalty points' out of consideration, Beranek's rating scheme was the first with which to calculate the quality (*Güte*) g_k of a hall k as the sum of the products of quantifiable aspects of judgment f_{ik} and the corresponding weightings w_i:

$$g_k = w_{\mathrm{I}} f_{\mathrm{I}k} + w_{\mathrm{II}} f_{\mathrm{II}k} + \ldots w_m f_{mk} \tag{2.12}$$

[7] T_2 and T_3 are the reverberation times at 500 and 1000 Hz according to the definition of the frequency index, see eqn. (1.10).

If we have to consider N halls, we may summarize the N rating equations in a single matrix equation:

$$(g_1 \ldots g_N) = (w_1 \ldots w_m) \begin{pmatrix} f_{11} \cdots f_{1N} \\ \cdots\cdots\cdots\cdots \\ \cdots\cdots\cdots\cdots \\ f_{m1} \cdots f_{mN} \end{pmatrix} \tag{2.12a}$$

We introduce this manner of writing here in order to compare this first attempt of Beranek to calculate the quality of a concert hall with the method of factor analysis commonly used nowadays, which we will explain in Section III. 2.8 and apply in Sections III. 2.9 and III. 3.1. In that procedure also, the judgment of acoustical quality by one person is composed of a sum of products, as in eqn. (2.12). It is also assumed that the objects of judgment (particular seats in particular halls) can be characterized by different aspects of judgment (called 'factors'); but it is not supposed that all the test persons give the same weighting to these aspects. In this case the single-row matrices of eqn. (2.12a) expand to several rows, in which the row number corresponds to the number of the observers n, or at least to the number of groups that weight the aspects differently:

$$\begin{pmatrix} g_{11} \cdots g_{1N} \\ \cdots\cdots\cdots\cdots \\ \cdots\cdots\cdots\cdots \\ g_{n1} \cdots g_{nN} \end{pmatrix} = \begin{pmatrix} w_{11} \cdots w_{1m} \\ \cdots\cdots\cdots\cdots \\ \cdots\cdots\cdots\cdots \\ w_{n1} \cdots w_{nm} \end{pmatrix} \begin{pmatrix} f_{11} \cdots f_{1N} \\ \cdots\cdots\cdots\cdots \\ f_{m1} \cdots f_{mN} \end{pmatrix} \tag{2.12b}$$

The most important difference from Beranek's procedure is that it is not assumed that the number m of judgment aspects ('factors'), or the meaning of these aspects, or their respective weightings are known—if they were known, the qualities (on the left) could be immediately calculated. Instead the judgments of the test persons are the materials of observation from which one can first determine how many 'factors' are needed to explain those judgments and then how the weighting matrix is composed.

III. 2.4 Changes in the Room

The first person to investigate the question of whether there is any unanimity in the judgment of room-acoustical properties for musical events was W. C. Sabine himself.[1] For this purpose he introduced immediate changes in the test room. The reverberation time, which he recognized as an easily measurable quantity, could claim to be a meaningful criterion of room-acoustical quality only if the users of the room could evaluate changes in reverberation time unequivocally and reproducibly.

He was most interested in the judgments of musicians. Therefore complaints about excessively reverberant practice rooms ($V = 70$ to 200 m^3) in a local conservatory of music offered him a welcome opportunity to reduce the reverberation (by bringing seat cushions into the rooms) and to let a small circle of musicians decide whether or not the reverberation time was appropriate for playing and listening to piano music. He allowed this panel of five music professors to decide the conditions that seemed to them most favorable, as one of them played the piano while the others listened.

Table 4, taken from Sabine's publication, shows the number of seat cushions introduced in each case (at the left), the calculated reverberation time (at the center), and the acoustical judgment (at the right); this judgment was evidently the result of common discussion and not the mean of independent judgments.

Table 4: Test series for judging the optimal reverberation time. (After W. C. Sabine[1])

Number of cushions	*Reverberation time*	*Judgment of 5 musicians*
0	1·64	too reverberant
13	0·60	insufficiently reverberant
11	0·70	better
8	0·83	better still
6	0·95	satisfactory
4	1·22	too reverberant

Test room volume, 74 m^3; reverberation time (empty) = 2·43 s

The room in its existing state was thought to be too reverberant so Sabine brought in 13 seat cushions; then the reverberation was considered insufficient. Removing two cushions improved conditions, and

[1] Sabine, W. C., *Collected Papers on Acoustics*, No. 2, Harvard University Press, Cambridge, 1929.

removing three more cushions gave further improvement. But not until two more cushions were removed did the judges declare themselves satisfied; when two more cushions were removed the room was again considered too reverberant.

In the same manner Sabine determined the optimal reverberation time in four other rooms, with the following results (including the example in Table 4):

Room volume:	74	91	96	133	210	m^3
T_{opt}:	0·95	1·10	1·16	1·09	1·10	s

Since these experiments did not lead to an unequivocal dependence of optimal reverberation time on room volume, Sabine concluded that the deviations were random and calculated the mean value:

$$T = 1{\cdot}08\,\mathrm{s}$$

with a probable error of about 0·02 s. [Sabine's value of 0·02 for the probable error is correct only if the first room is omitted; if all five rooms are considered, the probable error is 0·05. T.J.S.]. He regarded this error as surprisingly small and took it as a justification of the 'accuracy of musical taste'. He even ascribed to this accuracy the fact that musicians are so seldom satisfied:

'This surprising accuracy of musical taste is perhaps the explanation of the rarity with which it is entirely satisfied, particularly when the architectural designs are left to change in this respect.'

At least this interpretation would permit the assumption that the musicians could detect a change of only one seat cushion, that is, a change in reverberation time of 0·05 s, which is in agreement with Seraphim's result (see Section III. 2.1).

Musical taste is not always so accurate, as Békésy[2] showed in a series of tests at the Hungarian Broadcasting Institute. He, too, systematically changed the equivalent absorption area in a studio (2000 m^3); but he tried to refine Sabine's test procedure inasmuch as, after determining the optimal reverberation time by approaching it from one side, he repeated the experiment approaching from the other side (see Table 5).

In the initial gradual increase of room absorption, a judgment of 'good' was reached for 24 absorption units, and at 28 units the room was judged 'much too highly damped, quite useless'. When the process was reversed and the absorption was gradually removed, however, even when

[2] Békésy, G. v., *Elektr. Nachr. Techn.*, **11** (1934) 369.

Table 5: Test series for judging the optimal reverberation time. (After Békésy[2])

Number of absorption units	*Acoustical judgments*
4	unusable, because too reverberant
9	a bit better
12	still too reverberant
14	beginning to be good, but still needs more damping
21	already rather good, but still needs more damping
24	good
28	much too strongly damped, unusable
24	better than before
20	still better than before
16	still a bit too much damped
14	already begins to blur in fast passages
10	much too reverberant

Test room: 2000 m^3.

there were only 16 units remaining in the room, the judgment was 'still a bit too much damped'.

This 'pulling effect', well-known from psychological testing, corresponds to a desire to change as much as possible a tendency originally recognized to be wrong. Another example of this effect is that auditoriums of all kinds that are entered by way of a reverberant lobby are often judged as rather 'dry' halls. For this reason Békésy recommended that such lobbies be heavily treated with sound-absorptive materials, a procedure which is also beneficial in helping to isolate the auditorium from intrusive noises and is standard practice in broadcasting studios.

As a result of this experiment, Békésy recommended that in determining optimal acoustical parameters, the optimal value should be approached step-wise from both sides, a practice that is good for all sensory–psychological tests.

Such a procedure of changing the equivalent absorption area is not only expensive but it also risks that the amount of time needed to make the changes between tests is so great as to make the comparisons unreliable.

We have already mentioned (in Section III. 1.2) another experiment of Békésy's in which he changed the listening room conditions. He introduced into the room absorptive treatments having different dependences of absorption coefficient on frequency, in order to find the optimal dependence of reverberation time on frequency.

Similar tests have been carried out by Jordan[3] in a studio of the Danish Broadcasting Corporation. He equipped this studio with variable wall-coverings which at one extrème led to a reverberation time increasing with frequency, and at the other extreme achieved a decrease with frequency. In contrast to Békésy's results Jordan found that some increase in reverberation time toward low frequencies was preferred.

Later it became customary to equip broadcasting studios with variable absorbing elements, especially after Kuhl demonstrated (see Section III. 2.6, below) that the optimal reverberation time depends on the style of the music.

Even in auditoriums the concept of variable reverberation time is used more and more often (see Section II. 6.6), at least in the so-called multipurpose halls which must accommodate all kinds of events from pure speech (congresses) to orchestral concerts, even including chorus. It is evident that the cost of such treatment increases rapidly with the room size and especially with the number of upholstered seats, against whose total equivalent absorption area the variable absorptive elements must work: the greater the basic absorption in the auditorium, the more variable absorption elements are required to effect a given change in reverberation time.

For this reason a number of expensive approaches have sometimes been used to vary the acoustics of a hall, such as raising and lowering the ceiling in order to change the room volume and thus to change the reverberation time. For example, Peutz[4] has provided both variable ceiling height and variable absorptive elements on the walls of the large studio (L'Espace de Projection) at IRCAM, in the Centre Pompidou, Paris. And Izenour has provided variable volume in a number of American concert halls.[5]

For studies of the optimal reverberation time, however, rooms with variable ceiling height are not very useful because they are mostly usable only in the two extreme ceiling positions.

In large rooms, changes in the height and inclination of sound reflectors above the orchestra can create remarkable acoustical changes on the stage and at certain locations in the audience.[6,7]

---

[3] Jordan, W., *J. Acoust. Soc. Am.*, **19** (1957) 972.

[4] Peutz, V.M.A., IRCAM: L'Espace de Projection, *L'Architecture d'aujourd'hui*, No. 199, October 1978, Paris, 'Les lieux du spectacle', pp. 52–63.

[5] Izenour, G. C., *Theater Design*, McGraw-Hill, New York, 1977; see particularly Chapt. 7.

[6] Cremer, L., *Schalltechnik*, **13** (1953). No. 5.

[7] Beranek, L. L. and Schultz, T. J., *Acustica*, **15** (1965) 307; particularly Sections 7 and 8. See also Schultz, T. J., *IEEE Spectrum*, June (1965); particularly pp. 63–4.

Unfortunately, acoustical consultants who provide such variable acoustical elements in the design of their halls, with the intention of providing a benefit for the musicians, are usually disappointed. It is astonishing how seldom the musicians in the orchestra, and even the conductors, take notice of these opportunities to vary the acoustics, not to mention using them routinely and correctly! The consultant even runs the risk that they will make changes that are totally contrary to his own tendencies and convictions. When they do attempt to make an acoustical adjustment it is usually with the aim of achieving the highest amount of reverberation in a hall with upholstered seats. It is to be observed that the performing artists generally prefer longer reverberation times than the listeners.

The reason for this is not entirely the 'sustaining pedal effect' of the reverberation, which smooths over small irregularities in the playing; good orchestras do not need that help. Rather it is the tendency of longer reverberation to enhance the 'fullness of tone' that excites the musicians. Above all, they know very well the notes they are playing and have less need than the listeners for information about what notes are played and how.

Even if there were more generally valid research material available on the possibilities for acoustical changes in rooms, there would remain the disadvantages that a significant amount of time must always elapse between the evaluations of different room conditions and that the test persons are always aware of the changes that have been made.

III. 2.5 Electroacoustical Simulation of Room-acoustical Conditions

If the various room-acoustical conditions to be evaluated are presented to the test person by means of loudspeakers or earphones, this guarantees that he has no clue as to what changes are made and thus the evaluation is based only on his acoustical perceptions.

We begin—again deviating from the historical sequence—with the possibility already mentioned at the beginning of this chapter, that the test person sits in an anechoic room surrounded with loudspeakers that simulate a room-acoustical condition. Figure 2.7 shows the first such 'loudspeaker firmament' as it was set up in the anechoic room of the Third Physics Institute in the University of Göttingen.[1] This arrangement made it possible to combine a direct sound with 13 reflections of

[1] Meyer, E., Burgtorf, W. and Damaske, P., *Acustica*, **15** (1965) 339.

Fig. 2.7. Loudspeaker 'firmament' in the anechoic room of the Third Physics Institute at Göttingen. (After Meyer *et al.*[1])

different intensities and time delays and from different directions. It was intended that this synthetic sound field would be used for planning, or at least improving, actual concert halls; but this hope was not fulfilled. Seraphim[2] mentions that the acoustics of the lecture room of the Institute were simulated in the 'firmament' and that the simulation was compared with the original. The comparison would have been satisfactory except that the reverberation produced by the reverberation plate had a 'metallic' character (which may have been a consequence of the dispersion of flexural waves on the plate).

In spite of the abundant information that has been developed in this Göttingen facility about the possibilities for reducing the number of reflections and number of directions to be simulated, it is still difficult for the operator to tell from the building drawings which reflections have to be taken into account in the simulation.

Also, although it is possible, in principle, in this facility to evaluate the impulse responses recorded at the ears of a dummy head, no program for doing this exists so far.

On the other hand, the synthesized sound field has been very useful for answering special scientific questions.

The simplest case, namely, the interaction between the direct sound and its reflections, has already been treated in the last chapter in connection with psychoacoustical matters of general interest, such as the

[2] Seraphim, H. P., *Acustica*, **13** (1963) 75.

inertia of the ear, the limit of perceptibility and the law of the first wavefront.

Here we leave open one room-acoustical question that is of special interest for the synthesis of sound fields with loudspeakers: namely, when can we omit a reflection from the synthesis without the test person's missing it? Certainly the question is answered if the addition of the reflection is not perceived at all.

Seraphim[3] investigated this absolute threshold of perception (aWS = *absolute Wahrnehmungsschwelle*) for running speech and Schubert[4] did the same for different examples of running music. Figure 2.8 shows their results for the case of the direct sound and the reflection arriving from the same direction. The steeply sloping long-and-short-dashed line concerns running speech; its slope corresponds to the relaxation time of the ear, discussed in Section III. 1.4. The dashed line is for pizzicato violin notes and the solid line is for a 'dry' recorded

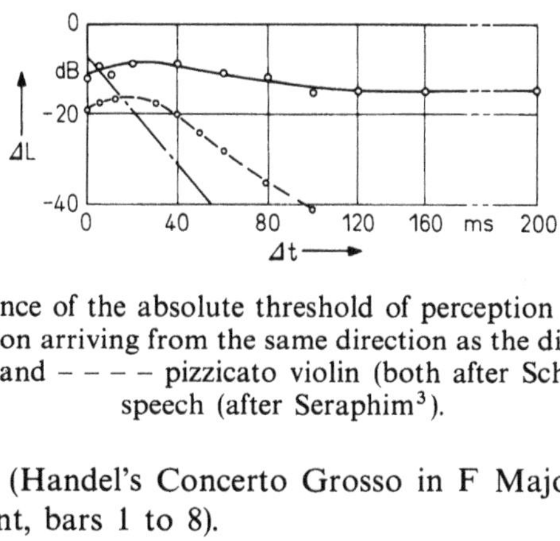

Fig. 2.8. Dependence of the absolute threshold of perception on the time delay, for a single reflection arriving from the same direction as the direct sound. ——— Orchestral music and – – – – pizzicato violin (both after Schubert[4]); —·—·— speech (after Seraphim[3]).

orchestral work (Handel's Concerto Grosso in F Major, Op. 6, No. 2, second movement, bars 1 to 8).

These comparisons show clearly the strong influence of the choice of signal for room-acoustical psychological research. Schubert explains the higher values of aWS for music, especially the orchestral example, by the perception of differences in timbre that are of no special interest in the case of speech and are hardly noticeable in pizzicato.

The tendency of the music curves to bend downward near $\Delta t = 0$ is explained by the increased loudness as the two signals coincide, which effect in this case takes the place of a change in timbre.

[3] Seraphim, H. P., *Acustica*, **11** (1961) 80.
[4] Schubert, P., *Hochfr. u. Elektroak.*, **78** (1969) 230.

One may be tempted to interpret this result, found with only one reflection, to mean that all of the reflections in an echogram that fall below the solid curve will be masked in orchestral music and can safely be omitted from the synthesis. But here we must remember Fig. 1.16, which showed that the ear combines three weaker reflections if they follow quickly one after the other, and the perception of the echo may be, for short delays, nearly as strong as that of a single echo with three times the energy.

The problems become more complicated with an increase in the number of reflections, as was shown in the tests of Seraphim.[5] After studying the effects of a single reflection, he added a second one and found an essential influence due to differences in the directions of arrival of the direct sound and the two reflections. Figure 2.9 (left) shows an

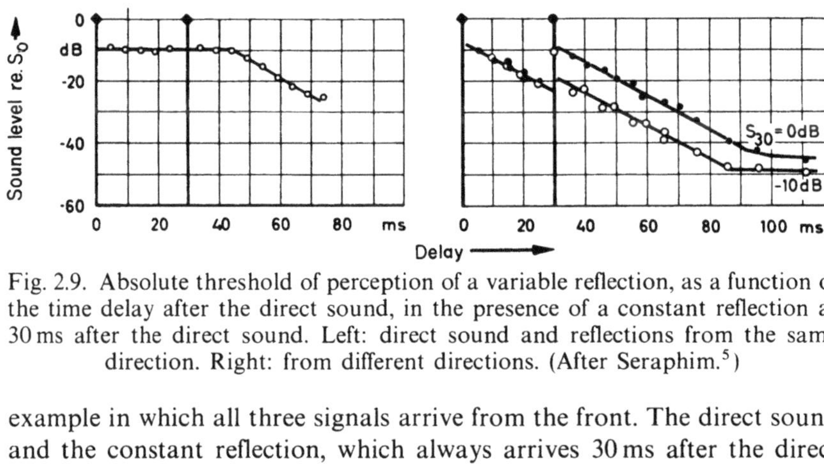

Fig. 2.9. Absolute threshold of perception of a variable reflection, as a function of the time delay after the direct sound, in the presence of a constant reflection at 30 ms after the direct sound. Left: direct sound and reflections from the same direction. Right: from different directions. (After Seraphim.[5])

example in which all three signals arrive from the front. The direct sound and the constant reflection, which always arrives 30 ms after the direct sound, have the same level. Another reflection occurs at different times with different levels. The level of the variable reflection when it is just imperceptible is plotted as a function of its time delay with respect to the direct sound. In contrast to Fig. 2.8, the absolute threshold for the variable reflection remains constant until it comes 15 ms after the 30 ms reflection and then slopes off according to the relaxation time of the ear. Figure 2.9 (right) shows another behavior, in which only the direct sound arrives frontally; the constant reflection arrives from 30° to the side and the variable reflection from 60° to the side. Here the level of the just-imperceptible variable reflection slopes off immediately after the direct

[5] Seraphim, H. P., *Acustica*, **13** (1963) 75.

sound and then again, from the same starting level, immediately after the constant reflection. It is as though the constant reflection replaces the direct sound in this range of delay. Evidently, the threshold level, the delay time and the direction of arrival are not independent of each other.

Burgtorf,[6] who carried out similar investigations with rectangular noise impulses and with the impulse-like syllable 'zack', could demonstrate furthermore that, with respect to the horizontal plane, there is no need for sources spaced more closely than about $\Delta\phi = 30°$. The loudspeakers in the arrangement shown in Fig. 2.7 correspond to about that number and disposition.

Sound fields with only one or two reflections occur mostly in open-air theaters but they do not give the impression of a closed room. For such an impression we have to add a non-localizable statistical reverberation. As we showed in Fig. 1.28, four loudspeakers in the horizontal plane are sufficient to create that impression provided that they are driven with incoherent signals, which can be produced by taking the reverberant sound from four widely spaced microphones in a reverberation room. But the reverberation alone does not give a room impression. Only when we add the direct sound do we achieve the simplest form of room representation.

In any case Reichardt and Schmidt[7] cleared up a fundamental room-acoustical question, using the arrangement shown in the upper part of Fig. 2.10. With this means they produced the direct sound pressure p_D by two loudspeakers, 1 and 2, symmetrically placed with respect to the median plane of the test person, thereby creating a phantom sound source in the middle. The sound pressure p_H (measured, like p_D, in the steady state) was produced with the help of a reverberation plate from which four practically incoherent signals were picked up and transmitted to the four diagonally placed loudspeakers (3, 4, 5 and 6). Figure 2.10 (bottom) shows the corresponding schematic $L_{(t)}$-echogram at the location of the test person; it consists of a 'column', representing the direct sound, and a reverberation process which appears as a straight line sloping off linearly. Like all the other authors who have synthesized a room sound field with statistical reverberation, Reichardt and Schmidt provided a time delay between the direct and the reverberant sounds. They chose a constant 50 ms time delay and a 2 s reverberation time but they varied the levels of the direct and reverberant sounds and thus

[6] Burgtorf, W., *Acustica*, **11** (1961) 97.

[7] Reichardt, W. and Schmidt, W., *Acustica*, **17** (1966) 175.

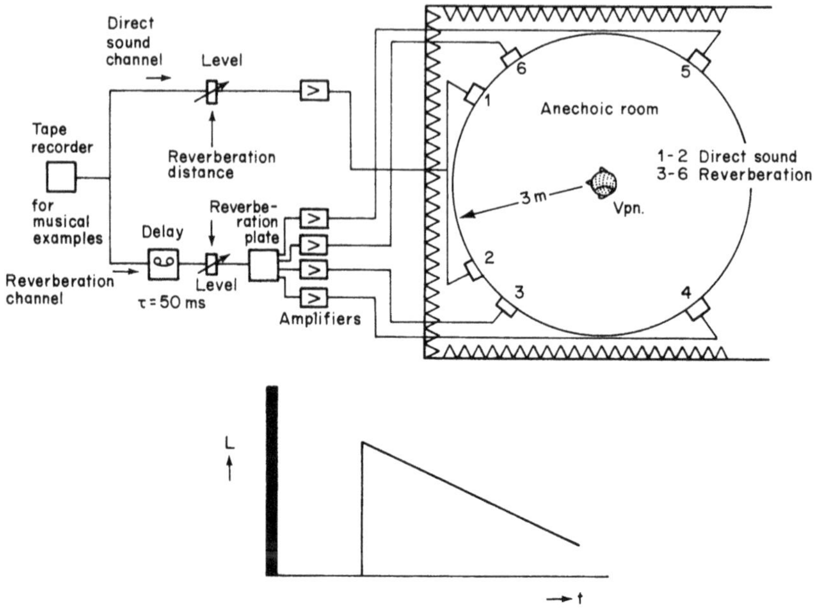

Fig. 2.10. Top: arrangement for producing a synthesized sound field consisting of the direct sound and statistical reverberation, produced by a reverberation plate. Bottom: corresponding schematic $L_{(t)}$-echogram. (After Reichardt and Schmidt.[7])

varied the ratio p_D/p_H; this is equivalent to changing the 'reverberation distance' (*Hallabstand*) or the negative 'liveness index' (see Section II. 7.3, eqn. (7.7a)):

$$H = 20 \log (p_D/p_H) \text{ dB} \tag{2.13a}$$

They invited 30 test persons to compare two such reverberation distances, H_1 and H_2, and to state which of them gives more 'room impression'. The values of $(H_2 - H_1)$ which brought to 50% the percentage of judgments of this difference with the right sign can be regarded as determining the difference threshold for the perceptible 'reverberation distance'. They established the existence of 14 distinguishable steps on the H-scale, seven for positive H ($p_D > p_H$) and seven for negative H ($p_D < p_H$). These are plotted in Fig. 2.11.

These steps are quite different for large and for small values of H; this is a consequence of taking the logarithm of the ratio p_D/p_H, a procedure that once more does not yield a scale corresponding to the subjective

Scale for predominance of reverberant sound component

Step	0	-1	-2	-3	-4	-5	-6	-7
H	0	-2	-4	-6·5	-9·5	-13	-17·5	<-23 dB

Scale for predominance of direct sound component

Step	0	1	2	3	4	5	6	7
H	0	2	4	6·5	9	12	16	>22 dB

Fig. 2.11. Difference limen for 'reverberation distances' H. (After Reichardt and Schmidt.[7])

impression. (The authors show that the ratio $(p_H/(p_H+p_D))$—in which the arithmetic summation of two signals delayed with respect to one other is problematic—results in a nearly constant difference limen of 0·06.)

It is of general interest that the number of distinguishable steps on the H-scale is twice the number of scale steps that would be found by asking the test persons to state their room impressions in categories between 'very dry' and 'very reverberant'. In such cases people can distinguish only about six or at most seven steps.

It may be assumed that the same double number of distinguishable differences holds for the logarithm of the ratio of late to early sound (see Section II. 7.4, eqn. (7.10a)):

$$R = 10 \log \frac{\int_{50\,\text{ms}}^{\infty} p^2 \, \mathrm{d}t}{\int_{0}^{50\,\text{ms}} p^2 \, \mathrm{d}t} \qquad (2.13\text{b})$$

which we proposed to call the 'late-to-early-sound index'. It has the advantage that it considers the reflections arriving in the first 50 ms as useful, thus giving them the same importance as is given to the direct sound in eqn. (2.13a).

The quantity R was, as we have mentioned in Section II. 7.4, first used by Beranek and Schultz[8] as an objective criterion for characterizing rooms as 'dry' or 'reverberant'; they also used a synthetic sound field with loudspeakers for determining an optimal value for R. They did not, however, use an anechoic room for their tests. Instead they put the

[8] Beranek, L. L. and Schultz, T. J., *Acustica*, **15** (1965) 307.

primary loudspeaker (which radiated dry recorded music) into a reflecting corner, in order to simulate the environment of a natural concert hall stage, therby adding some very early reflections to the direct sound. Four other loudspeakers brought the late sound, made electronically reverberant. (The fact that the four reverberant signals were not incoherent must be regarded as a shortcoming.) They asked the test persons to increase the level of the reverberant (late) sound, first to the point where the statistical reverberation was just perceptible, and then further to the point where the reverberation was 'excessive'. The range between these two levels was typically about 10 dB.

It should be mentioned again that the procedure in which the test person adjusts the signals is regarded in sensory-psychology as less reliable than the presentation of fixed stimuli to the test subjects for judgment. On the other hand, the test person undoubtedly feels much more involved with a phenomenon when he has the amplitude under his own control.

A synthetic sound field is much more room-like if some reflections from well-defined directions are introduced between the direct sound and the statistical reverberation. Even then it is important whether the reflection arrives from overhead (as from the ceiling) or from the side (as from a wall), a difference with considerable room-acoustical significance.

Reichardt and Schmidt[9] used a synthetic sound field to study this question. Two reflections were introduced between the direct sound and the two-second statistical reverberation starting after 50 ms. The first reflection arrived with a delay of 35 ms; its azimuth angle was $\phi = 60°$, its elevation angle was $\vartheta = 0°$, thus simulating a lateral wall reflection. The second reflection arrived with a somewhat greater delay; its azimuth angle was $\phi = -10°$, its elevation angle was $\vartheta = 60°$, thus simulating a slightly oblique ceiling reflection. The level of the lateral wall reflection had to be lowered to 10 dB below the direct sound for it to be imperceptible; for the ceiling reflection, 6 dB was sufficient. For equal initial levels, the difference limen was 1·5 dB for the first case, and 2·5 dB for the second. Finally, for the first reflection, an increase of 7 ms in the delay was perceptible; for the second, an increase of 12 ms was needed. From all this the authors conclude that reflections from the ceiling are not as important as lateral reflections in contributing to the room impression.

[9] Reichardt, W. and Schmidt, W. *Acustica*, **18** (1967) 274.

Lehmann and Wettschureck[10] reached the same conclusion. They introduced only one reflection with a delay of 27 ms between the direct sound and the reverberation, which began at 50 ms. But that reflection could arrive from in front ($\phi = 0°, \vartheta = 0°$), from the side ($\phi = 60°, \vartheta = 0°$), or from above ($\phi = 0°, \vartheta = 60°$). In the first case the loudspeaker for the direct sound also supplied the reflection. The transition between lateral wall reflection and oblique ceiling reflection could be made by a 90° rotation of the dummy head that was placed in the sound field. The sound pressures arriving at the ears of the dummy head were presented to the test subjects through earphones (see Section III. 2.7). As a subjective criterion they used the percentage syllable articulation (PSA). For this quantity they found, when the reflection was added, the following values:

Frontal reflection	75%
Lateral reflection	81%
Oblique, from above	74%

These different values seem at first to contradict the relation between percentage syllable articulation $\mathfrak{v}$ and the center-time t_s shown in Fig. 2.5, since one microphone, instead of the two in the dummy, would give the same impulse response in all three cases and thus the same center-time. But this equality changes if we record the two sound pressures at the dummy head, evaluate the center-times separately, and calculate and plot the percentage syllable articulation $\mathfrak{v}$ from their mean value $\bar{t}_s$. Here the influence of the sound pressure at the ear that is nearest the reflection outweighs that at the shadowed ear. At least the lateral and frontal cases in Fig. 2.12 agree with the dependence shown in Fig. 2.5. It remains unclear why the percentage syllable articulation $\mathfrak{v}$ for the reflection in the median plane lies below the downward sloping straight line through the data points of Fig. 2.12.

We have explained in Section I. 5.6 that one of the reasons for the good acoustical quality of the concert halls of the nineteenth century is their small width (for example, the Grosser Musikvereinssaal in Vienna is only 19·8 m (or 65 ft) wide; there are only 22 seats across, with side balconies of two rows each). As mentioned in Section III. 2.3, Beranek emphasized that the small width results in a short time delay between the direct sound and the first strong reflection, which he thought especially desirable for producing intimacy and clarity. Marshall[11], on the other

[10] Lehmann, P. and Wettschureck, R., *Proc. 7th ICA, Budapest*, 1971, Paper 24 S 17.

[11] Marshall, A. H., *J. Sound Vibr.*, **5** (1967) 100.

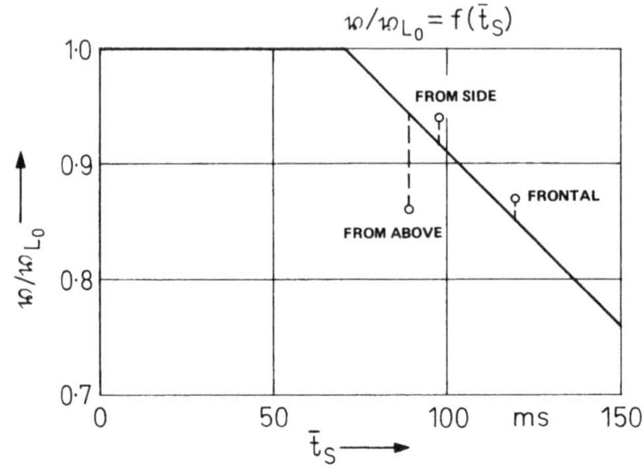

Fig. 2.12. Dependence of percentage syllable articulation (PSA) on the mean center-time for a 27 ms delayed reflection, arriving from above, from the side and from the front. (After Lehmann and Wettschureck.[10])

hand, has advanced the concept that only the lateral reflections are advantageous since only they contribute to produce what he called a 'spatial responsiveness'. Meyer and Kuhl[12] had already observed that one effect of lateral reflections is to expand the apparent width of a point source without destroying the ability to localize the source.

The expression 'apparent sound width' used by Keet[13] in this connection seems not fully adequate. Perhaps this limitation may be related to his experimental method. He used single loudspeakers (i.e. point sources) to radiate 'dry' recorded music in the concert halls under test. Microphones in stereophonic arrangement at different locations recorded the sound in the halls. The playback of these recordings was presented in an anechoic room by two loudspeakers hidden behind a scrim on which a distance scale was marked. By this means the test subjects could quantify their room-acoustical impression by estimating the 'apparent source width'. This expression is questionable. If listeners in a concert hall are asked to state the extent of the sound source with closed eyes they will place it between the last musicians on either side of the stage because of the law of the first wavefront. Nevertheless they would also be

---

[12] Meyer, E. and Kuhl, W., *Acustica*, **2** (1952) 77.

[13] Keet, W. de V., *Proc. 6th. ICA, Tokyo*, 1968, Paper E–2–4; see also his Dissertation, Johannesberg, 1969.

aware of strong lateral reflections that give the impression of receiving sound from the side. Kuhl[14] proposed to call this latter phenomenon *Raumlichkeit*, which we may translate by Marshall's expression 'spatial responsiveness'. Here we must remember that when Reichardt and Schmidt asked the test persons for their 'room impressions', they meant the balance between the direct sound and the reverberant sound without regard to the direction of incidence. Perhaps the use of the expression 'spatial responsiveness' or 'spatial impression' (see below) would avoid misunderstanding.

This impression is produced not by the statistical reverberation but by early lateral reflections, as was proved by Barron.[15] For his experiments he used a synthesis of direct sound and several delayed reflections presented by loudspeakers in an anechoic room. The test person had to adjust the level of the lateral ($\phi = 40°$) reflection presented with various time delays, against the direct sound level (dry recorded orchestral music) so that the 'spatial responsiveness' (which he called, more adequately to the subjective character, the 'spatial impression') was perceived to be the same for the different test signals. He determined that changes in the delay between 20 and 80 ms had only a slight influence on spatial impression. By contrast, the difference in level between the direct sound and the lateral reflections is very important.

We may conclude that narrow concert halls are good not only because of the short time delays for the early reflections, but also because of the greater intensity of the lateral reflections, compared to wide halls. By contrast, the newer multi-purpose halls tend to be ever wider, not only because economic reasons demand great seating capacity but also because their use for congresses mostly excludes balconies, particularly at the sides. Under such conditions it is practically impossible to provide early lateral reflections to the seats in the center. The early reflections can be provided only from above, either by a rather low ceiling or by overhead reflectors (see Section I. 5.3).

Marshall not only regarded the lateral reflections as more important than ceiling reflections, he was even fearful that ceiling reflections arriving before the lateral reflections could mask the latter—a misinterpretation of the results of Seraphim and Schubert shown here in Fig. 2.8. Barron's research found his fear not to be confirmed: according to his results, the ceiling reflection does not contribute to spatial impression but to loudness and distinctness, and therefore should not be eliminated.

---

[14] Kuhl, W., Deutsche Arbeitsgemeinschaft für Akustik (DAGA), Heidelberg, 1976.

[15] Barron, M., *J. Sound Vibr.*, **15** (1971) 475.

This agrees with the experience of the authors. Reflections from above (and below) are by no means unimportant. The best example is the excellent sound in seats in the center of the highest balcony in opera houses, immediately below the ceiling (provided that the ceiling is hard and not interrupted by beams or large steps). Also the importance of the reflections from the 'orchestra' in antique amphitheaters is well known. Since these reflections do not contribute to the spatial impression, it seems that this quality does not have such an exclusive importance for acoustical quality as is sometimes pretended nowadays.

Incidentally, no other room-acoustical characteristic is so dependent upon the absolute sound level as the spatial impression. This was found both by Keet[13] with loudspeakers as sources and (better adapted to musical practice) by Kuhl[16] in tests with dummy-head recordings of orchestral performances (see Section III.2.9). Burgtorf[17] had shown earlier that the perception of a reflection depends on the power of the source. Wettschureck[18] studied this question by means of a synthetic sound field, taking into account the fact that a higher sound level is usually accompanied by a change in spectrum. Figure 2.13 (left) shows his experimental arrangement. The loudspeakers D_1 and D_2 radiate the

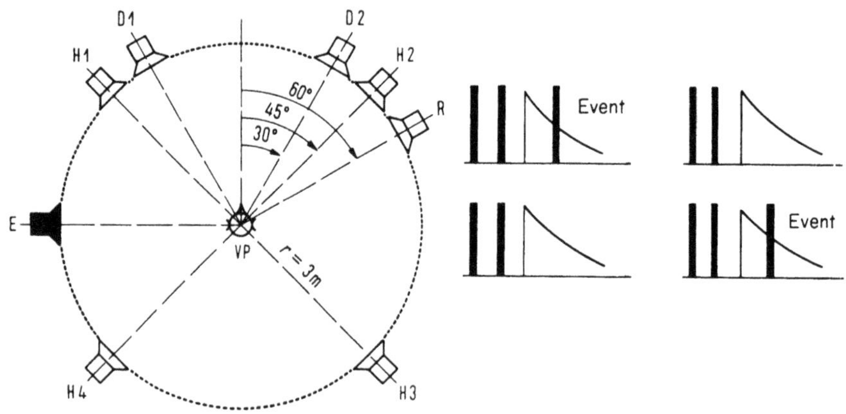

Fig. 2.13. Left: arrangement of equipment for studying the perceptibility of an added reflection (black loudspeaker) for different power levels of the source. Right: schematic echograms of alternative pairs of signals. (After Wettschureck.[18])

[16] Kuhl, W., *Acustica*, **40** (1978) 167.

[17] Burgtorf, W., *Acustica*, **11** (1961) 97.

[18] Wettschureck, R., *Acustica*, **32** (1975) 284; see also his Dissertation, TU Berlin, 1976.

direct sound, producing a phantom source in the middle. Loudspeakers H_1 to H_4 produce a statistical reverberation starting with a time delay of 50 ms. Loudspeaker R introduces an early reflection to make the synthesized field sound more natural. Loudspeaker E (in black) radiates a second reflection which comes from the left side during the statistical reverberation. Its level, with respect to that of the direct sound, was changed until 75% of the test persons could tell whether this additional reflection was presented with the first or the second of the test signals ('two-alternate forced-choice' method) (see Fig. 2.13, right). Wettschureck used running speech for his test signals taking advantage of the fact that its spectral distribution changes, depending on the vocal effort of the speaker. He could eliminate this latter additional effect by using different amplification for the loud and the soft speech.

Accordingly, we get four different level differences between the direct sound and the test reflection, L_D–L_E, the quantity which Wettschureck took as a measure for the absolute threshold of perceptibility.

These values of (L_D–L_E) are arranged in a quadrant scheme in Fig. 2.14. The differences between soft and loud amplifications are far greater

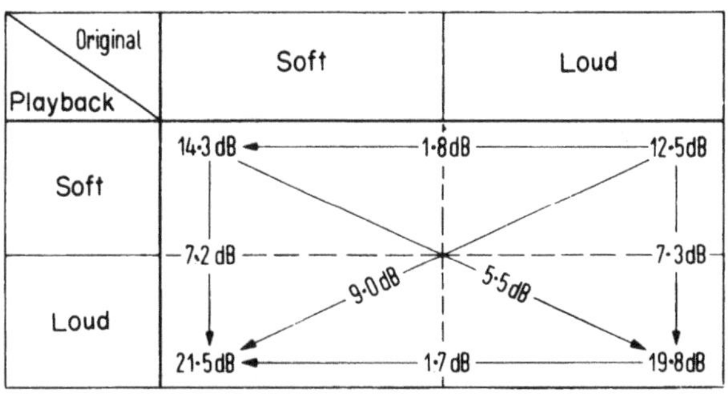

Fig. 2.14. Absolute threshold of perceptibility of the test reflection in Fig. 2.13. Horizontal: soft and loud speech. Vertical: small and large amplification. (After Wettschureck.[18])

than the differences between soft and loud original speech. The test reflection is easier to recognize for soft speech because of the prominent high frequencies in the sibillants. The reverse effect is to be expected for musical instruments where the energy content of the partial tones always increases with a change from piano to forte.

The non-spectral part of this effect, which (for equal level differences with respect to the direct sound) makes the 'weak' reflection more perceptible with increasing loudspeaker level, is explained by the shape of the curve of log N versus L (see Fig. 1.3). The perceptibility seems to be determined by the ratio of the loudness of the direct sound N_D to that of the test sound N_E; equal changes in N_D/N_E correspond to equal increments in log N. Since the log N versus L curve of Fig. 1.3 increases more rapidly for small L than for large L, equal changes in ΔL produce smaller changes in ΔN at high levels.

As a last example of the application of synthetic sound fields for the investigation of room-acoustical questions, we discuss the experimental attempts of Reichardt *et al.*[19] to find a musical analog to the speech articulation test: they suggest 'transparency' ('*Durchsichtigkeit*') as the simplest possible objectively measurable criterion. Their test subjects, who had both musical training and considerable room-acoustics experience, were asked to state for each test signal whether they regard the transparency as sufficient or insufficient. The judgments relating to a test example consisting of the first 36 bars of the Finale of Mozart's Jupiter Symphony, containing fast passages, were taken as pertaining to 'temporal transparency'. The judgments relating to the polyphonic music of bars 380 to 410 in the same movement were taken as ratings of the 'register-transparency'.

For the synthesized sound field they used two loudspeakers, driven stereophonically, to present the direct sound. The reflections, which came from both sides, were correspondingly distributed to both channels, such that first only one and then two reflections were presented. The single ceiling reflection was derived from a summation of both channels. Again, four diagonally placed loudspeakers provided delayed incoherent statistical reverberation with a reverberation time of 1·7 s. In most of the experiments the initial level of the reverberation was varied, while in others only the level of the first four reflections was changed. With this arsenal a multitude of variations was possible; but this was necessary if the question was to be answered for quite different room situations.

The selected echograms, schematically represented in Fig. 2.15 as $L_{(t)}$-dependences, give some insight. On the left, the statistical reverberation starts after 50 ms, on the right, after 120 ms. In the upper part of the figure the levels of the individual reflections are, as usual, decreasing with time; in the lower part they increase with time.

In view of the great variety of test signals, it is all the more astonishing that the authors were able to characterize the threshold of sufficient

[19] Reichardt, W., Abdel Alim, O. and Schmidt, W., *Acustica*, **32** (1975) 126.

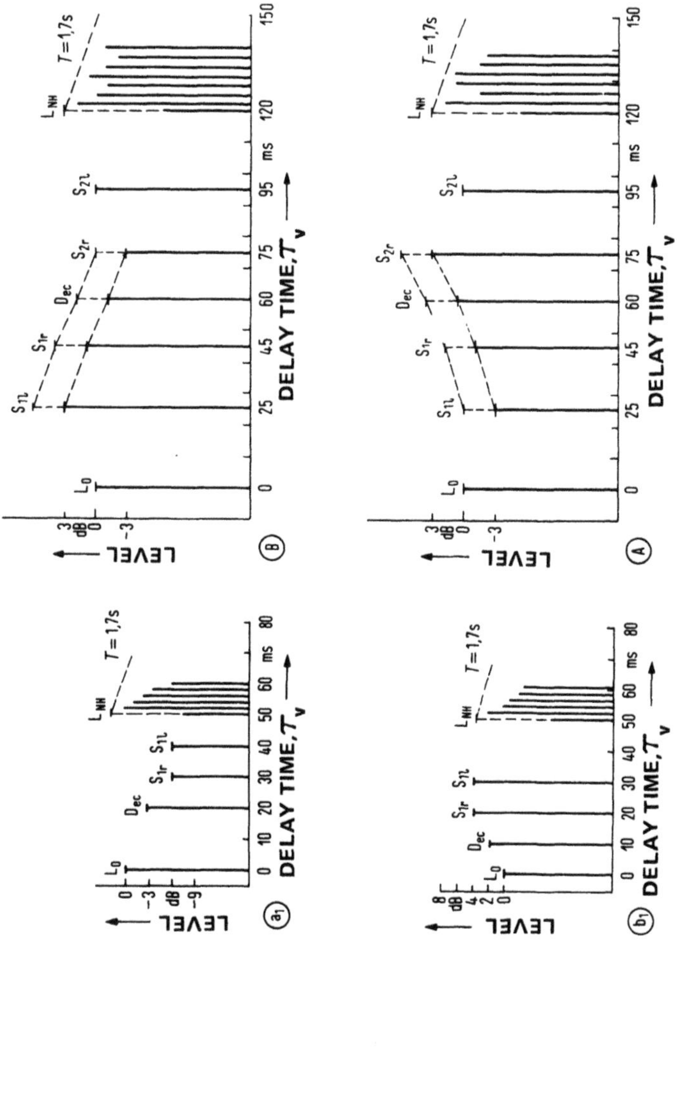

Fig. 2.15. Schematic $L_{(t)}$ echograms of the synthetic sound test fields used for the judgments of 'transparency'. (After Reichardt *et al.*[19])

transparency successfully with a single objective criterion: the clearness index C (see Section II.7.4, eqn. (7.11)). It is similar to the 'late-to-early-sound index' (see eqn. (2.13b), above), except that here the useful sound appears in the numerator of the energy ratio.

Reichardt *et al.* also considered splitting the integral

$$\int_0^\infty p^2 \mathrm{d}t$$

into useful and detrimental parts according to the time of arrival and the direction of arrival, but this presented severe experimental difficulties. Finally they found it sufficient only to shift the time limit from 50 to 80 ms. They assumed that this was justified by the fact that most transients of musical instruments last about 100 ms. Thus they proposed the clearness index as a useful criterion:

$$C = 10 \log \frac{\int_0^{80\,\mathrm{ms}} p^2 \mathrm{d}t}{\int_{80\,\mathrm{ms}}^\infty p^2 \mathrm{d}t} \ \mathrm{dB} \tag{2.14}$$

which has the same simplicity as the 'late-to-early-sound index' R. They even found an easily remembered value separating too much and too little transparency:

$$C_g = 0 \pm 1{\cdot}6 \ \mathrm{dB} \tag{2.15}$$

(The subscript g is from the German *Grenze* = boundary.) Even if this rule-of-thumb (that the clearness index should be about 0 dB) lacks a certain finesse, it nevertheless recommends itself by its simplicity, which was the chief intention.

At the end of this extended section we may conjecture that synthesized sound fields may soon lose their importance, since it has now become possible to store room-acoustical impressions and to reproduce them with only two channels, corresponding to the ears of a dummy head. The quality of reproduction is so convincing that the listener feels as though he is sitting in the concert hall. This also makes it possible to introduce interesting room-acoustical changes by electronic modification of the stored signals. For this purpose the use of computers is especially suitable.[20,21] We presented one rather simple example at the end of Section III.2.1.

Since each room-acoustical situation is fully defined by the impulse responses at the two ears recorded with a dummy head, and since each musical signal can be synthesized by a series of impulses, it is possible to simulate with a computer the performance of every musical example under the desired room-acoustical conditions.

III. 2.6 Comparisons of Single-channel Magnetic Recordings

It is, however, unlikely that synthetic sound fields will be developed much farther for the purpose of simulating real halls, since it is now possible to store original room-acoustical situations on magnetic tape with the help of two microphones mounted in dummy-heads and to reproduce the signals by two loudspeakers or earphones.

Comparisons of tape recordings are in fact even older than synthetic sound fields. The first such comparisons were made with monophonic recordings in a period when orchestra concerts were broadcast with a single monophonic channel. In those days it was reasonable to ask what reverberation time in the recording studio is desirable in order to please the listener in his 'dry' living room. Kuhl[1] undertook a round robin test in which more than 100 test persons compared tape recordings of three different musical compositions, recorded in 20 different rooms with volumes between 2000 and 14 000 m^3 and with mid-frequency (average of 500 and 1000 Hz) reverberation times between 1·3 and 2·7 s. The subjects were asked if they thought the reverberation 'too short', 'just right', or 'too long'.

In order that the judgments should be made primarily on the reverberation time, Kuhl insisted that the microphone for the original recording should be at least 5 m from the nearest instrument so that the room sound was predominantly heard. In spite of this restriction it is not certain that these judgments can be meaningfully evaluated at all, because different orchestras and different conductors participated in the recordings; since their artistic freedom naturally could not be restricted,

[20] Schroeder, M. R., *Proc. 6th. ICA, Tokyo, 1968*, Paper GP-6-1.
[21] Eysoldt, U., Dissertation, Göttingen, 1976.
[1] Kuhl, W., *Acustica*, **4** (1954) 618.

the recordings differed in respects other than the different room-acoustical conditions.

It was therefore a welcome reward for all the hard work that went into the experiment that the percentage of judgments 'reverberation too long' (to which the judgments 'reverberation just right' have been added with half-weight) show a monotonic tendency (see Fig. 2.16). If we draw

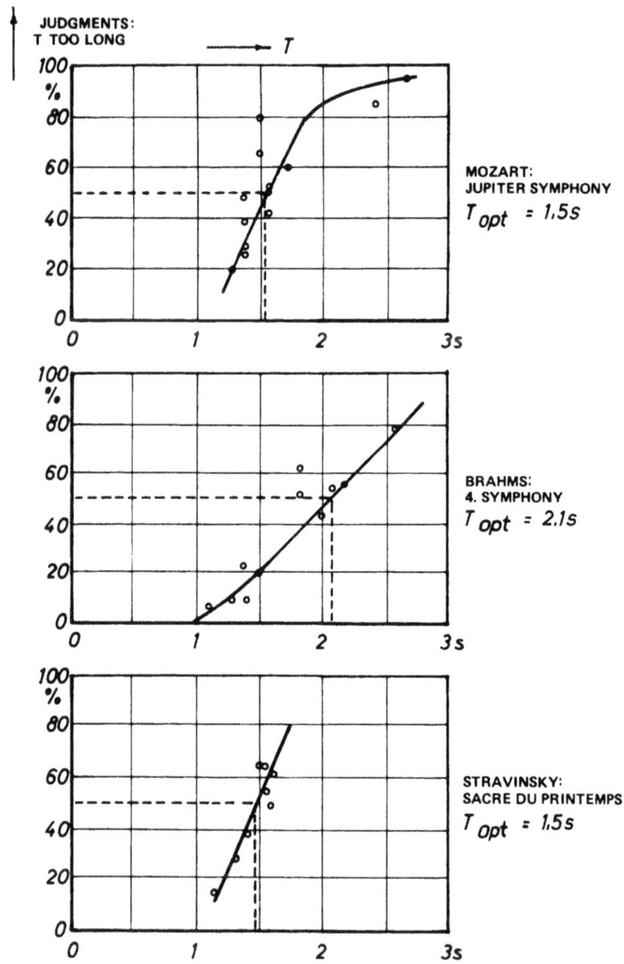

Fig. 2.16. Percentage of judgments 'reverberation too long' plotted against the reverberation time for monophonic tape recordings for music in three different styles. (After Kuhl.[1])

through the test points a curve like the cumulative distribution of a Gaussian function, we can adopt the 50% point as a mean value for the optimal reverberation time.

The large amount of scatter in the judgment data points for this experiment is especially interesting. In contrast to Sabine's statement about the 'accuracy of musical taste' (see Section III.2.4), it is evident here that not all the listeners approve or disapprove the same room-acoustical conditions. For instance, in the case of the Brahms symphony a reverberation time of 2·6 s was regarded as too long by 80% of the listeners, nevertheless 20% thought this too short; another 20% of the listeners found a reverberation time of 1·5 s to be too long. This experiment clearly demonstrates how impossible it is to satisfy every musical taste—or probably we should better say every mood.

On the other hand, the preference for different reverberation times for different kinds of music is quite evident despite the data scatter. In the first example of Fig. 2.16 (the first movement of Mozart's 'Jupiter' Symphony) the 50% point corresponded to $T = 1{\cdot}5$ s; for the second example (the fourth movement of Brahms' 4th Symphony) it was 2·1 s; and for the last example (Stravinsky's 'Rite of Spring') it was again 1·5 s. Kuhl suggested that these values not only characterize the preferences for these particular musical examples but also, in general, for the musical styles that they represent ('classical', 'romantic' and 'modern'). His view may be supported by the argument that the concert halls of the late 18th century were smaller than those of the late 19th century, that therefore the earlier halls had shorter reverberation times than the later ones, and that the music of the times is composed to fit the most familiar room-acoustical conditions. This argument fails to explain why the modern work demands a short reverberation time; but it may be because the 'Rite' was conceived as ballet music, which is typically performed in the rather dry room-acoustics of an opera house rather than a concert hall. It is doubtful that such a short reverberation time is suitable for all modern compositions. We must mention, moreover, that in Békésy's tests at the Hungarian Broadcasting Institution[2] he found no difference in the preferred reverberation times for Mozart's 'Marriage of Figaro' overture and for Wagner's 'Tannhauser' overture (see Section III. 2.4). Instead he found differences depending on the number of musicians.

All of these tests agree that the preferred reverberation time depends on

[2] Békésy, G. v., *Elektr. Nachr. Techn.*, **11** (1934) 369.

the kind of musical performance. If we consider, furthermore, that a large concert hall must accommodate all kinds of music, from a lone soloist to an oratorio, it becomes evident that the same room cannot be equally suitable for all these events, even if it is used only for orchestral concerts. The task of the acoustical engineer consists not so much in attaining optimal conditions for the particular musical purpose of most importance, but in guaranteeing adequate acoustical quality for all events.

It is astonishing that Kuhl's results did not show any systematic influence of the different room volumes in halls with the same reverberation time; in larger rooms the mean free paths are longer, typically leading to the longer reverberation times that we as listeners have come to expect. This may only prove, however, that there is a fundamental difference for the listener whether he hears music *in* a room or *from* a room.

III.2.7 Possibilities for Two-channel Storage of Room-acoustics Impressions

It cannot be expected that monophonic recording and playback of the sound processes at a particular room location would convey the acoustical impressions of a listener at that location, since such recordings fail to provide any sense of the direction of arrival of sounds.

Even the more recently developed two-channel reproduction, which has been called 'stereophonic' by analogy to stereoscopic vision (though the visual and aural processes are quite different), cannot provide a satisfactory solution to the problem, since stereophonic reproduction presents directional impressions only within the limited region between the two loudspeakers placed symmetrically in front of the listener.

As for the newer four-channel system called 'quadraphony', in which the two customary stereo channels are supplemented with an additional two channels feeding loudspeakers behind the listener, it is an open question whether it can reproduce the impression of sound coming from any direction. But since the process of recording with two microphones in a dummy head has been so greatly improved, it is no longer necessary to go to the added complexity of four-channel recording.

It was suggested in the first German edition of this book[1] that satisfactory directional realism might be achieved by recording with two

[1] Cremer, L., *Statistische Raumakustik*, Hirzel, Stuttgart, 1961, p. 218.

microphones spaced at ear-distance, and playing back through earphones; as explained here in Section III. 1.7 this would only lead to an in-head localization. Knowing now the importance of sound diffraction around the head, discussed in Section III. 1.5, and of the exact location and shape of the pinna of the ears, we can by means of dummy-head recordings reproduce sound events with great accuracy at the listener's ear canals (or eardrums).

Such dummy heads were developed independently and simultaneously in Göttingen[2] and in Berlin.[3] Figure 2.17 shows the Berlin dummy-head made of gypsum with rubber outer ears. The photo at the left shows the assembled head; at the right the front part of the head is removed to show the two condenser microphones and the connections to the outputs.

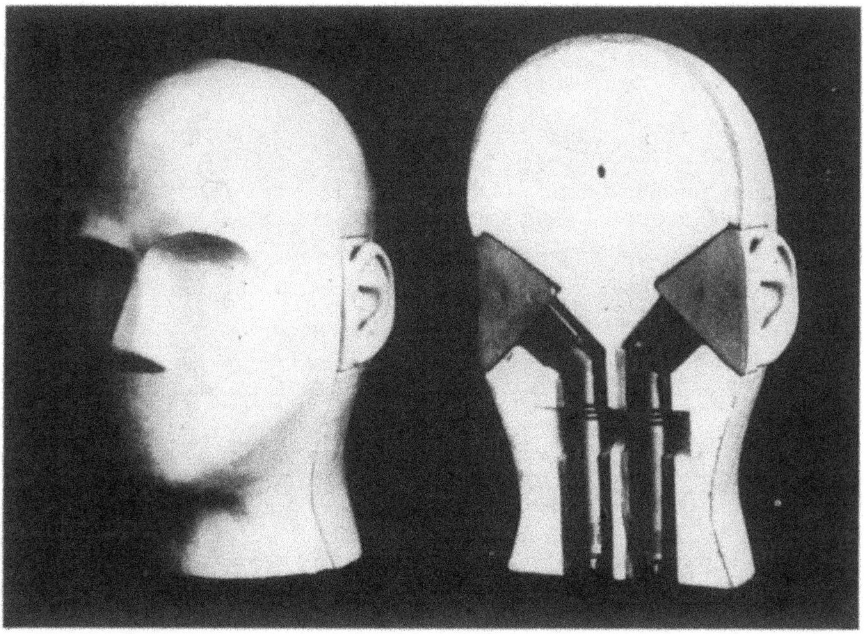

Fig. 2.17. Photographs of the Berlin dummy head. Left, assembled; right, with the front half removed. (After Kürer *et al.*[3])

[2] Damaske, P. and Wagener, B., *Acustica*, **21** (1969) 30; Mellert, V., *J. Acoust. Soc. Am.*, **51** (1972) 1359.

[3] Kürer, R., Plenge, G. and Wilkens, H. *37th AES convention, New York, 1969*, no. 666, H-3; Wilkens, H., *Acustica*, **26** (1972) 213; Wilkens, H., Dissertation, TU Berlin, 1975.

Figure 2.18 shows the directional characteristics measured by Wilkens for one ear at various frequencies, for the angular half-circle from ahead to behind on the side nearest the ear. The solid lines are for a male head; the dashed lines correspond to a dummy-head recording reproduced at the

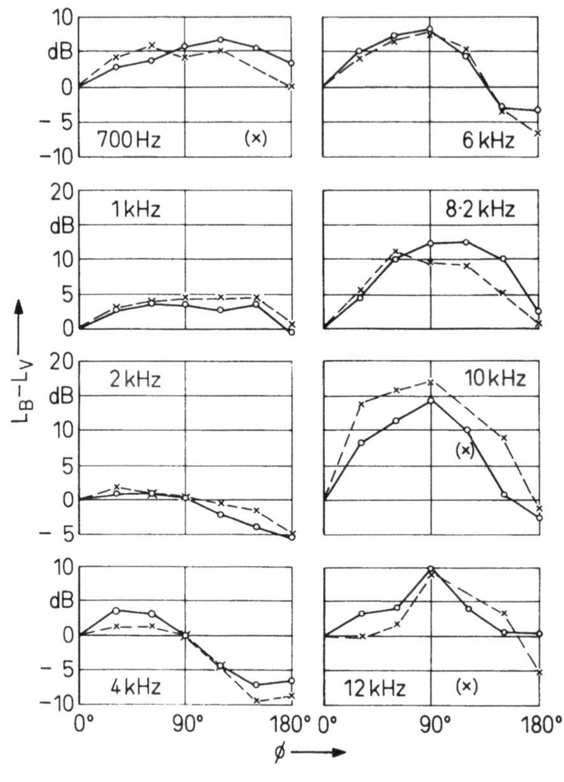

Fig. 2.18. Directional characteristics for one ear for the lateral half-circle on the near side. Male head: ○———○. Dummy-head recording and earphone reproduction: ×- - -×. (After Wilkens.[3])

ear entrance of a listener. In both cases these directional characteristics were determined by allowing the listener to compare two signals, one from directly ahead and one from the direction of interest, and to adjust the source levels for equal loudness. (This procedure has the advantage of taking into account any possible excitation of the ear drum, or even of the inner ear, that is not due to sound pressure at the ear entrance and may not even pass through the ear canals.)

Figure 2.19 compares the frequency dependence of sound pressure levels measured with probe microphones in the ear canal for frontal sound incidence, first (solid line) for directly received sound and then (dashed line) for sound recorded with a dummy head and reproduced (with suitable transformation) through earphones. The apparent good

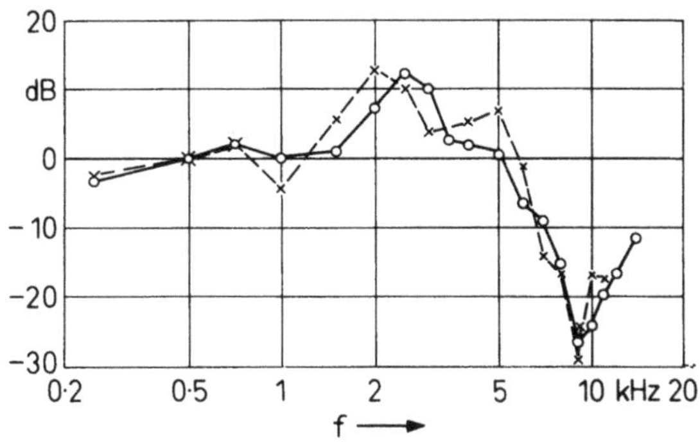

Fig. 2.19. Frequency dependence of the sound pressure in the ear canal for frontally incident sound. Directly received sound: ×– – –×. Dummy-head recording and earphone reproduction: ○———○. (After Wilkens.[3])

agreement between the objectively measured sound pressures encourages one to hope for correspondingly good agreement between the subjective impressions, live and recorded. Indeed we can already point to a number of acoustical impressions for which the recorded reproduced sound (via dummy head and earphones) corresponds to the sound in the original room. The correct dependence on the distance from the source to the head has been demonstrated,[4] as well as the dependence on reverberation[5] and on the direction of arrival (with the exception of a small frontal angular region where the judgments differed).

According to studies by Lehmann *et al.*,[6] front/rear interchanges occurred in about half of the tests; but it is worth mentioning that among those test persons who had no trouble in determining frontal direction was the person whose outer ears had been used for modelling the dummy head. The other test persons had first to accustom themselves to the pinna of the dummy head (see Section III. 1.5). Furthermore, it is

[4] Wettschureck, R., Plenge, G. and Lehringer, F., *Acustica*, **29** (1973) 260.

[5] Plenge, G. and Romahn, G., *J. Acoust. Soc. Am.*, **51** (1972) 421.

[6] Lehmann, U. and Abdel Alim, O., *Z. elektr. Inf. u. Energie–Technik*, **4** (1974) 169.

understandable that the ability to evaluate acoustical impressions differs from person to person and is better developed in people like recording engineers. Finally, the lack of a visual cue to the location of the sound source may tempt the test person to suppose that it is behind him.

There is, however, no reason to exclude those test persons, who found difficulty with frontal localization in purely directional tests, from participation in the comparisons described below where we will be concerned with problems such as: whether different people agree in their judgments of acoustical quality; how many acoustical 'factors' must be considered; whether these factors correspond to particular room-acoustical impressions; and whether they are significantly correlated with some of the objective criteria mentioned in Chapter II.7. Wilkens[7] was able to justify this decision by repeating test examples for test persons equipped with an apparatus developed by Boerger *et al.*[8] such that, as the head is rotated about the vertical axis, the time delays and levels between the ears are changed in the same way as in the original sound field. This arrangement immediately restores the impression of frontal incidence. By reversing the polarity, this 'frontal' impression can be changed to 'sound source behind'. But the judgments that the test persons made with respect to room-acoustical impressions remained the same, with 95% confidence limits.

The difficulties with frontal localization, mentioned above, created problems only with respect to the weights given to particular factors and to the rules derived from them. This was especially true for the perception discussed in Section III.2.5 under the name 'Spatial Impression', which may become exaggerated in the dummy-head earphone reproduction.[9]

Now it is possible that 'head-oriented stereophony' with earphones will be improved still further. But it is also possible that our front/back localization judgments depend on other information, still unknown, that is not transmitted by way of the two inner ear passages.[10]

In any case it is not surprising, in view of the residuum of listeners who are not satisfied with the spatial impression conveyed by earphones,

[7] Wilkens, H., Dissertation, TU Berlin, 1975, Section II, 1.2.6.

[8] Boerger, G. and Kaps, U., DAGA-Tagung, Aachen, 1973.

[9] Kuhl, W. and Plantz, R., *Rundfunkt. Mitt.*, **19** (1975) 120.

[10] Since the publication of the second German edition of this book, translated here, it has been found that the construction of an adequate dummy head can be significantly improved by closer adaptation to real heads. See the special issue of *Rundfunktechnische Mitteilungen*, September 1981, dealing with the theme, 'Artificial-Head Stereophony', with contributions from Plenge, Hudde-Schröter, Theile and Wollherr.

that a two-loudspeaker presentation was developed in Göttingen (after pilot tests with earphones) and that later, in Berlin, a two-channel presentation using four loudspeakers was also developed. These procedures, however, sacrifice an important advantage of earphone presentation, namely, that the same test can be given simultaneously to a number of test persons. The loudspeaker presentations are valid only if the listener occupies a particular narrowly restricted position. Specifically, the test can be done with only one person at a time in an anechoic chamber and with a fixed head position and direction.

The arrangement developed by Damaske and Mellert[11] in Göttingen was based on a principle proposed by Atal and Schroeder.[12] If one wishes the sound pressure at the left ear to depend only on the radiation from the left loudspeaker and that at the right ear to depend only on the radiation from the right loudspeaker, it is necessary to cancel the signal arriving at the right ear from the left loudspeaker and vice versa.[13]

Figure 2.20 shows how this can be done. The left loudspeaker radiates the impulse recorded by the left ear of the dummy head. This impulse appears first at the left ear of the listener (see sketch at bottom left) but it also appears at the right ear, a bit delayed and attenuated on account of shadowing by the listener's head (see sketch at bottom right). This unwanted impulse at the right ear can be cancelled by an out-of-phase impulse 2 from the right loudspeaker with suitable time delay and attenuation (see sketches at right, top and bottom). But since this second 'corrective' impulse also reaches the left ear, we need a third impulse from the left loudspeaker to cancel impulse 2 at the left ear, and so on.

For sufficiently oblique positions of the loudspeakers ($\alpha > 35°$), impulse 3 is already very small compared with impulse 1 on account of the shadowing of the head, so that it is effectively masked. Thus in practice only the corrective impulse 2 is required.

The delay times and the attenuations needed for this procedure were first produced by recording, one after the other, the sound from the single loudspeakers with an auxiliary dummy head in an anechoic room and combining the sound pressures in the outer ear canals to drive the

[11] Damaske, P. and Mellert, V., *Acustica*, **22** (1969/70) 153.

[12] Atal, B. S. and Schroeder, M. R., *Gravesaner Blätter*, **27/28** (1966) 124.

[13] Note: The arrangement described here as developed by Damaske and Mellert for research purposes, after a principle proposed by Atal and Schroeder in 1966, was earlier proposed by Benjamin Bauer at the CBS Laboratories in 1960. It is now available on the American market for a reasonable price, to connect to home hi-fi systems, with a special calibration procedure for speaker separation and distance to listener position. The effect of creating the impression of spatial surround is astonishing, but only for stereo signals. T. J. S.

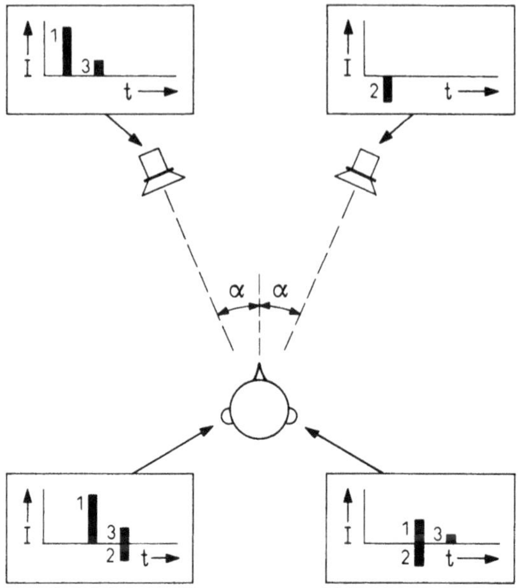

Fig. 2.20. Principle of cancelling an impulse 1 from the left loudspeaker heard at the right ear, by an opposite impulse from the right loudspeaker, and so on. (After Atal and Schroeder.[12])

loudspeakers to which the test persons listened. This rather expensive method was later replaced with a specially designed filter between the dummy head and the test loudspeakers.

The Göttingen group made use of this method to present recorded pairs of the same musical example in rapid succession for two seats in 25,[14] and later 30,[15] different concert halls.

This large number of examples was made possible by using as the sound source in all the various halls two loudspeakers, placed 3 m upstage of the apron, 1 m above the stage floor and 5 m apart. The loudspeakers were fed with 'dry' stereophonic recordings of orchestral music (recorded in a large anechoic room):[16] the reprise and coda of the Finale of Mozart's 'Jupiter' Symphony.

No doubt this procedure assured the reproducibility of the hall

[14] Siebrasse, K. F. and also Gottlob, D., Dissertations, Göttingen, 1973.

[15] Gottlob, D., Siebrasse, K. F. and Schroeder, M. R., DAGA, Braunschweig, 1975, p. 467, Physik Verlag, Weinheim.

[16] Burd, A. N., *Rundfunktechn. Mitt.*, **13** (1969) 200.

excitation in each case but, on the other hand, this means of excitation bears no resemblance to the excitation of the hall by a real orchestra. The most important difference is that with the loudspeakers we have only two point sources of sound, whereas the orchestra covers a large area. Reichardt and Ganev[17] have proposed that if one is interested in judging the balance between instrumental groups in the orchestra, one must distribute the orchestral sound among at least five loudspeakers. For this purpose they studied the 'mean spectra' and 'mean directional characteristics' of the strings, the woodwinds, the brass, and the percussion tuned for mid-frequencies, and for all the bass instruments separately.

But even though it may be possible in this way to simulate an orchestra with a reasonable compromise between affordable expense and desirable acoustical accuracy, there still remain the technical problems of recording. Of particular concern is the usual compression of the dynamic range (i.e. the difference in sound level between pianissimo and fortissimo) in the recording. It is probable in fact that the musicians themselves contribute to this dynamic compression, since otherwise they could not hear each other playing pianissimo in the anechoic room. Wilkens[18] reports differences in judgments with respect to the contrasts 'large/small' and 'reverberant/dry' (see Section III. 2.9, below) between recordings played back with reduced dynamic range, once at listening levels near the soft end of the true range of the composition and once near the loud end. This is not surprising since we have already encountered, in the last section, the dependence of room-acoustical impressions on the sound level. Still different judgments were obtained when a live orchestra played the same musical selections in a hall, a situation that undoubtedly introduced changes far beyond mere differences in listening level.

Finally we should not overlook the fact that it is very unusual for musicians to play in an anechoic room; indeed it is almost presumptuous to ask that they do so! They need the normal sound reflections of a room and may even adjust their way of playing to accommodate these reflections. In this sense the reaction of the musicians to a hall is one of the room-acoustical properties of the hall.

In reviewing the drawbacks of room-acoustical tests using anechoic orchestral recordings as the sound source, we must in fairness consider whether the alternative procedure, namely, successive performances of the same music by an orchestra, is sufficiently reproducible. The only

---

[17] Reichardt, W. and Ganev, S., *Z. elektr. Inf. u. Energietechn.*, **2** (1972) 249; **3** (1973) 137.
[18] Wilkens, H., Dissertation, TU Berlin, 1975, p. 35.

time when this question does *not* arise is when a number of dummy heads are used simultaneously to compare the sound at different seats in the same hall during a single musical performance. Wilkens[19] used this procedure during a rehearsal in the Berlin Philharmonie to determine the least distance between listening positions required to give distinguishably different room-acoustical impressions. Figure 2.21 shows the seats that he selected for his test in Block C; at the left, for seats in the same row; at the right, for seats behind each other. The matrix system in the lower

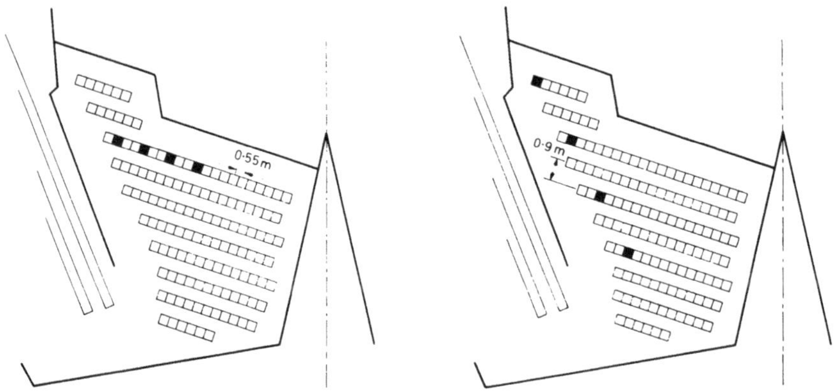

ROW 3 SEAT	2	5	8	11
2		43·9	66·1	41·5
5			36·9	36·0
8				13·1
11				

ROW/SEAT	1/1	3/2	5/3	7/3
1/1		57·0	42·6	42·6
3/2			58·2	72·4
5/3				22·3
7/3				

Fig. 2.21. Top: seats in Block C of the Berlin Philharmonie, used to determine just-noticeable differences in room-acoustics impressions. Bottom: percentage of perceived differences at seats in the same row and seats one behind another. (After Wilkens.[19])

part of the figure gives the percentages of test persons by whom the sounds in these seats (defined by the rows and columns of the matrix) could be distinguished. We may summarize these results by stating that at least 2 m in both directions are necessary, but not always sufficient, for 50% of the test persons to be aware of a difference. But more important,

[19] Wilkens, H., *Proc. 7th. ICA, Budapest, 1971*, Paper 24 S 5.

even for the greatest distances investigated in this study, no difference in room-acoustical 'goodness' could be determined. For this, greater distances between seats or different halls are necessary.

The Berlin group had the unique opportunity to accompany the Berlin Philharmonic Orchestra on a concert tour of the six halls whose floor plans are shown (drawn to the same scale) in Fig. 2.22. The scheduled musical program was different for each hall on the tour and thus was not suitable for making room-acoustical comparisons among the occupied

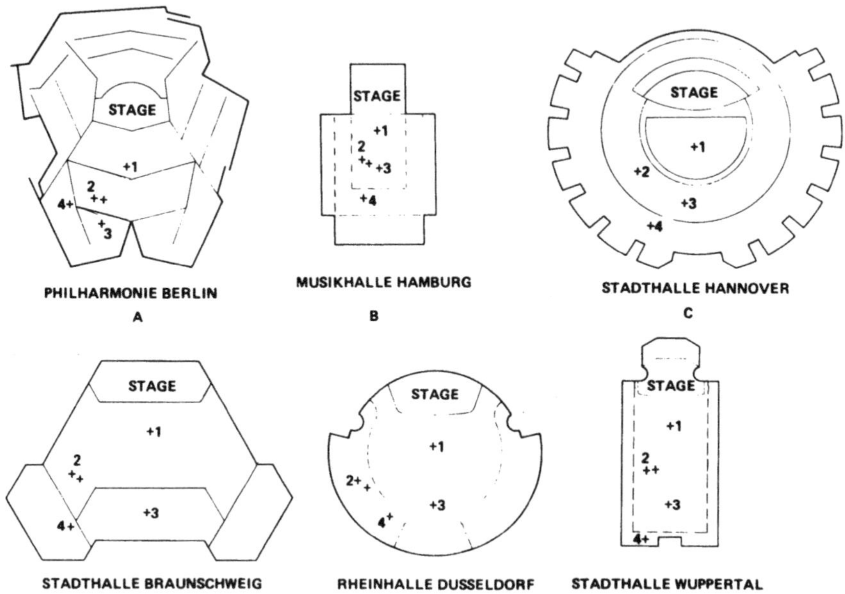

Fig. 2.22. Floor plans of six halls, with indicated recording locations, for a concert tour of the Berlin Philharmonic Orchestra. (After Wilkens.[19])

halls. Therefore before each concert the orchestra was requested to play, as a special contribution to the research, the same three musical examples, spanning three different musical styles:

Mozart: 'Jupiter' Symphony, 1st movement, bars 1–23.
Brahms: First Symphony, 4th movement, bars 61–118.
Bartok: Concerto for Orchestra, Introduction, bars 51–95.

In fact, in order to get some idea of the reproducibility of the musical 'signals', each example was played twice.[20]

Plenge[21] later showed that the differences in performances were smaller than the differences in room acoustics of the halls that were under study, thanks to H. Prim, who conducted the musical examples taking great care to make the performances as similar as possible.

Surely the Berlin Philharmonic Orchestra is the most precious 'sound source' ever used in an extensive comparison of the room acoustics of concert halls. The orchestra was available for this purpose only because the members of the orchestra were generous enough to undertake this additional chore without payment. It may be that they were so kind because they and their leader, Herbert von Karajan, had been favorably impressed by the dummy head/earphone reproductions that they had heard.

III. 2.8 The Purpose and the Limitations of Factor Analysis

The expense of equipment and travel and the generous support that were necessary to obtain the tape recordings made by the Göttingen and Berlin groups made it obligatory that the subsequent data analysis should yield the greatest amount of information with the greatest possible certainty. For these evaluations modern methods of psychometry were employed, the so-called multivariate analysis. Among these the most frequently used—and the best adapted to room-acoustical problems—is factor analysis. In this context the word 'factor' does not mean a multiplier, as is otherwise usual in mathematics and physics. Instead, as we have already mentioned in Section III. 2.3, the factors are aspects that influence judgments with respect to a number N of objects. (In Section III. 2.3 these were criteria for the acoustical quality of a room.) These data need not be quantified subjective perceptions; the results of objective measurements can also be investigated by factor analysis. We may therefore speak of the n initial attributes of N objects. The data to be analysed consist of a data matrix

$$X = \begin{pmatrix} x_{11} \cdots x_{1\mathrm{N}} \\ \cdots\cdots\cdots\cdots \\ \\ \cdots\cdots\cdots\cdots \\ x_{n1} \cdots x_{n\mathrm{N}} \end{pmatrix} \tag{2.16}$$

[20] G. Plenge, P. Lehmann, R. Wettschureck and H. Wilkens took part in this orchestra tour.
[21] Plenge, G., DAGA, Stuttgart, 1972, p. 154, VDI Verlag, Düsseldorf.

whose rows correspond to the attributes i and whose columns correspond to the objects j.

The goal is to transform this matrix into a product of two matrices of which the second contains, instead of the n initial attributes, a smaller number m of new ones called factors, while the first contains the weightings[1] with which the different factors appear in the initial attributes:

$$\begin{pmatrix} x_{11} \cdots x_{1N} \\ \cdots\cdots\cdots \\ \cdots\cdots\cdots \\ x_{n1} \cdots x_{nN} \end{pmatrix} \approx \begin{pmatrix} w_{11} \cdots w_{1m} \\ \cdots\cdots\cdots \\ \cdots\cdots\cdots \\ w_{n1} \cdots w_{nm} \end{pmatrix} \begin{pmatrix} f_{11} \cdots f_{1N} \\ \cdots\cdots\cdots \\ f_{m1} \cdots f_{mN} \end{pmatrix} \tag{2.16a}$$

The psychologists who developed factor analysis for their own purposes required for their choice of factors that they should be as few as possible and that they be self-explanatory, that is, that they be explainable by well-established concepts. For that reason they did not insist that the factors be unequivocal. Instead the proper choice of factors was determined in terms of their simplicity, their explainability, and the degree of approximation of equality of the left and right sides of eqn. (2.16a). (For this reason, we used the 'approximately equal' sign in that equation.)

By contrast we prefer in room-acoustical problems the important limiting case in general factor analysis where the W-matrix (weightings) follows unequivocally from the analysis.

In the literature of factor analysis this limiting case is often called 'component analysis'.[2] This analysis has, in principle, the disadvantage that the number of factors is as great as the number of initial attributes, if we wish to represent the data matrix exactly.

But since it is possible (and this is the proper aim of the analysis) to order the factors systematically according to their importance for the approximation to the ideal data matrix, the number of factors that must be considered depends on the degree of approximation that is intended.

As Siebrasse[3] has demonstrated, nearly all the characteristics of this limiting case of factor analysis can be explained in the simplest case, $n=2$, where the initial attributes and the factors determine only two planes.

---

[1] This is Siebrasse's term; the more commonly used expression, 'factor loads', is subject to misunderstanding.

[2] Harman, H. N., *Modern Factor Analysis*, University of Chicago Press, Chicago, 1968. This book is recommended for readers who wish to know more about factor analysis.

[3] Siebrasse, K. F., Dissertation, TU Berlin, 1973, p. 41.

For example, we choose as attributes the values of reverberation time and the values of 'center-time' (*Schwerpunktzeit*) measured by Lehmann[4] at comparable seat locations (identified by the number 2) in the six halls shown in Fig. 2.22:

		A	B	C	D	E	F	
T	=	2·15	2·05	1·90	1·75	2·50	2·40	s
t_s	=	131	144	135	97	157	152	ms

(The number of objects ought really to be larger in order to comply with the statistical conditions assumed in factor analysis; the number is restricted here for greater clarity.)

When two variables are involved, whether or not there is any dependence between them, it is customary to plot one as a function of the other. If it is possible to draw a monotonic curve through closely clustered points, then even rather complicated functions can be determined to describe the functional dependence of the two variables.

But if such a relation exists only for special cases, as would be true here for T and t_s only with exponential decays, the data points generally describe a 'galaxy' rather than a curve. Then it seems reasonable to inquire only to what extent a linear relationship exists, since this is what is assumed in the factor analysis.

It is always possible to lay a linear regression through the data points:

$$(\hat{x}_{2j}-\bar{x}_2)=\frac{s_{12}}{s_{11}}(x_{1j}-\bar{x}_1) \tag{2.17a}$$

(the bar characterizes the mean value, the circumflex identifies the expected value); but to do so supposes that the variable on the right, corresponding to the abscissa (here the reverberation time x_1), is regarded as an independent variable. Such a distinction is not justified in general for the rows of a data matrix and it is not valid here for T and t_s. But if we take the other variable, the center-time x_2, as independent, the corresponding equation reads:

$$(\hat{x}_{1j}-\bar{x}_1)=\frac{s_{12}}{s_{22}}(x_{2j}-\bar{x}_2) \tag{2.17b}$$

Both straight lines (see Fig. 2.23, where the abscissa corresponds to

[4] Lehmann, P., Dissertation, TU Berlin, 1976.

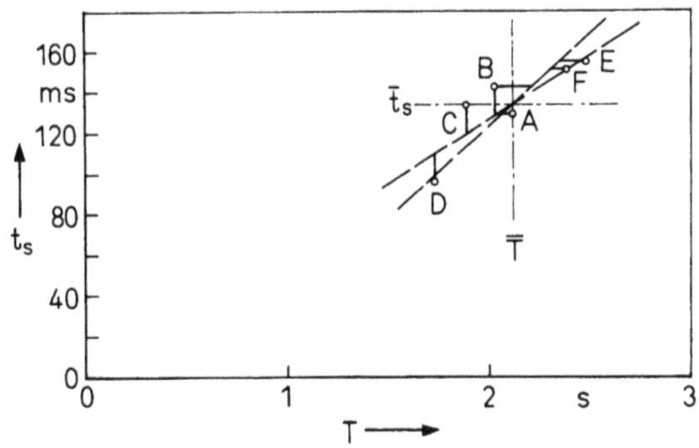

Fig. 2.23. Two-dimensional example of factor analysis, in which the object points lie in the plane of the two initial attributes.

$x_1 = T$ and the ordinate to $x_2 = t_s$) go through the middle-point of the data cluster, characterized by $\bar{x}_1(=\bar{T})$ and $\bar{x}_2(=\bar{t}_s)$, but they exhibit different slopes given by the variances:

$$s_{11} = \frac{1}{N-1} \sum_{j=1}^{N} (x_{1j} - \bar{x}_1)^2 \qquad (2.18a)^5$$

$$s_{22} = \frac{1}{N-1} \sum_{j=1}^{N} (x_{2j} - \bar{x}_2)^2 \qquad (2.18b)$$

and the so-called covariance:

$$s_{12} = \frac{1}{N-1} \sum_{j=1}^{N} (x_{1j} - \bar{x}_1)(x_{2j} - \bar{x}_2) \qquad (2.18c)$$

The slopes become equal only if $s_{12}/s_{11} = s_{22}/s_{12}$, i.e. only if s_{12} is the geometric mean of s_{11} and s_{22}.

But now the question arises as to whether there exists one exceptional straight line for which both variables are of equal significance. This may be easily answered by analogy to the mechanics of a rigid body. If we regard the data points as equal mass-elements connected by massless rigid rods, we have a two-dimensional rigid body. The quantities s_{11} and s_{22} are (if we neglect a common multiplier) the moments of inertia for axies

[5] Under special conditions, it is more convenient to define these quantities with N instead of $N-1$ in the denominator. For the conclusions that follow, this difference, which diminishes with increasing N, is of no importance.

parallel to the ordinate and abscissa through the center of gravity of the body, and s_{12} is the centrifugal moment or product of inertia. These three quantities form a tensor of second order which can be expressed by the symmetrical matrix:

$$\begin{pmatrix} s_{11} & s_{12} \\ s_{12} & s_{22} \end{pmatrix}$$

Now it is known that a rigid body possesses three axes, perpendicular to one another, that have special properties; in this case two of them lie in the plane $x_1 x_2$. The moments of inertia about these axes, s_{I} and s_{II}, represent extreme values. In particular, the principal axis *I*, about which the moment of inertia s_{II} is the least, is the exceptional line that we are seeking. Even more important for factor analysis is that the other moment of inertia s_{I} represents a maximum of variance.

Moreover, the orientation of the ordinates to the principal axes leads to another important result, for here the centrifugal moment, i.e. the covariance, vanishes. Thus the variance–covariance tensor may be transformed by the new coordinates into a diagonal matrix:

$$\begin{pmatrix} s_{\mathrm{I}} & 0 \\ 0 & s_{\mathrm{II}} \end{pmatrix}$$

This analogy shows that only the deviations of the attributes from their mean value enter into the determination of the orientation of the principal axes. But these deviations may represent quite different numbers, depending on the definitions and the units chosen for the quantities plotted. Factor analysis regards all attributes as equally important, since it divides all of the differences $(x_{ji}-\bar{x}_i)$ by the square root of the corresponding variance, the so-called standard deviation:

$$\sigma_i=\sqrt{s_{ii}}$$

Thus a new set of initial attributes is defined:

$$z_{ij}=\frac{x_{ij}-\bar{x}_i}{\sigma_i} \tag{2.19}$$

with the corresponding data-matrix:

$$Z=\begin{pmatrix} z_{11} & \cdots & z_{1N} \\ \cdots & \cdots & \cdots \\ z_{n1} & \cdots & z_{nN} \end{pmatrix} \tag{2.19a}$$

With this normalization the variances for these new variables become:

$$s_{11}=s_{22}=1 \tag{2.19b}$$

and the covariance becomes the so-called correlation coefficient:

$$r_{12}=\frac{1}{N-1}\sum_{j=1}^{N} z_{1j}z_{2j} \tag{2.19c}$$

For our example, it is $r_{12}=0{\cdot}854$. And the normalized initial-tensor takes the simple form:

$$R=\begin{pmatrix} 1 & r_{12} \\ r_{12} & 1 \end{pmatrix} \tag{2.20a}$$

But equal variances (and therefore equal initial moments of inertia) mean that the principal axes are rotated by 45° with respect to the original coordinates (see Fig. 2.24(a)). Every textbook on the mechanics of a rigid body shows that, for the present two-dimensional case of equal moments of inertia about orthogonal axes, in that plane the principal moments of inertia are equal to the sum of the initial moment of inertia plus or minus the centrifugal moment. The tensor takes the form:

$$\begin{pmatrix} 1+r_{12} & 0 \\ 0 & 1-r_{12} \end{pmatrix}=\begin{pmatrix} \lambda_{\text{I}} & 0 \\ 0 & \lambda_{\text{II}} \end{pmatrix}=\Lambda \tag{2.20b}$$

The symbols λ_{I} and λ_{II}, which represent the variances in the directions of the principal axes, are chosen in accordance with principles discussed below. (Although both R and Λ designate identically the same tensor, it is convenient for our present discussion to use different letters for the two forms.) It may be stated here only that

$$\lambda_{\text{I}}+\lambda_{\text{II}}=2 \tag{2.20c}$$

i.e. equal to the sum of the variances of the normalized initial-attributes, and that $\lambda_{\text{I}}/2$ and $\lambda_{\text{II}}/2$ indicate which relative part of the total variance the new 'principal coordinates', y_{I} and y_{II}, represent.

In our example, λ_{I} 'explains' 92·7%, so that λ_{II} explains only the remaining 7·3%.

Although the principal coordinates, which are related to the original coordinates by the transformation equation

$$\begin{pmatrix} x_{1j} \\ x_{2j} \end{pmatrix}=\begin{pmatrix} \cos 45^\circ & -\sin 45^\circ \\ \sin 45^\circ & \cos 45^\circ \end{pmatrix}\begin{pmatrix} y_{\text{I}j} \\ y_{\text{II}j} \end{pmatrix}=\frac{1}{\sqrt{2}}\begin{pmatrix} 1 & -1 \\ 1 & 1 \end{pmatrix}\begin{pmatrix} y_{\text{I}j} \\ y_{\text{II}j} \end{pmatrix} \tag{2.21}$$

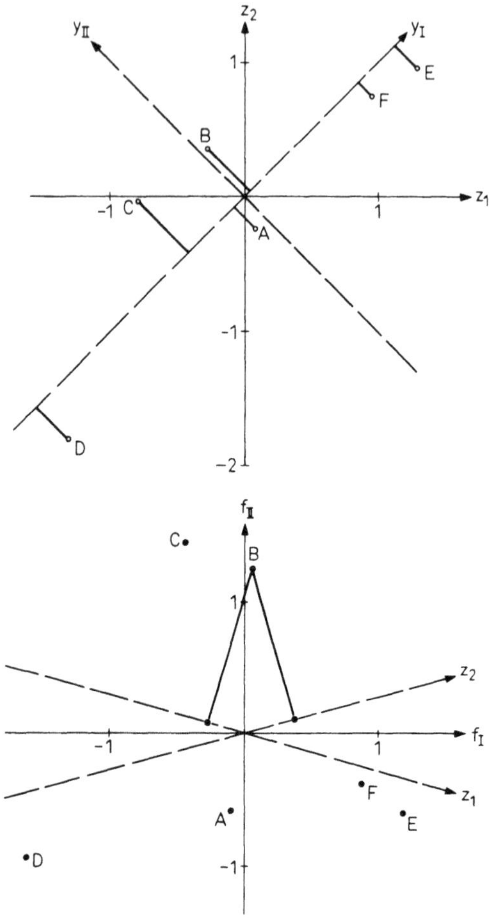

Fig. 2.24. Two-dimensional example of factor analysis. Top, as in Fig. 2.23, but after normalization; bottom, as above, but after rotation and extension.

exhibit the extreme values of variances:

$$\frac{1}{N-1}\sum_{j=1}^{N} y_{Ij}^2 = \lambda_I \tag{2.22a}$$

$$\frac{1}{N-1}\sum_{j=1}^{N} y_{IIj}^2 = \lambda_{II} \tag{2.22b}$$

as well as the vanishing of the correlation coefficient

$$\frac{1}{N-1}\sum_{j=1}^{N} y_{\mathrm{I}j}y_{\mathrm{II}j}=0 \tag{2.22c}$$

they still are not the factors f_I and f_II that we are looking for. Not only is it expected that these factors are independent, that is, that their correlation coefficient also vanishes:

$$\frac{1}{N-1}\sum_{j=1}^{N} f_{\mathrm{I}j}f_{\mathrm{II}j}=0 \tag{2.23a}$$

but also the variances are both expected to equal 1, as was the case with the initial attributes:

$$\frac{1}{N-1}\sum_{j=1}^{N} f_{\mathrm{I}j}{}^2=1 \tag{2.23b}$$

$$\frac{1}{N-1}\sum_{j=1}^{N} f_{\mathrm{II}j}{}^2=1 \tag{2.23c}$$

This requires an 'extension' of the principal coordinates by the multipliers $(1/\lambda_\mathrm{I})^{1/2}$ and $(1/\lambda_\mathrm{II})^{1/2}$, i.e. for $\lambda_\mathrm{I}>1$, a shortening. (The general expression 'extension' may remind us that every test of tensile strength involves both an extension and a contraction of the sample in two orthogonal directions.)

In the transition from Fig. 2.24 (top) to 2.24 (bottom), rotation and extension are combined, since the pure rotation would be easy to execute in Fig. 2.24 (top). In contrast to eqn. (2.21), the combination of rotation and extension corresponds to the transformation equation:

$$\begin{pmatrix} x_{1j} \\ x_{2j} \end{pmatrix} = \frac{1}{\sqrt{2}}\begin{pmatrix} \sqrt{\lambda_\mathrm{I}} & -\sqrt{\lambda_\mathrm{II}} \\ \sqrt{\lambda_\mathrm{I}} & \sqrt{\lambda_\mathrm{II}} \end{pmatrix}\begin{pmatrix} f_{\mathrm{I}j} \\ f_{\mathrm{II}j} \end{pmatrix} \tag{2.24}$$

The most important advantage of this transformation is that the axes of the initial attributes, whose orthogonality does not correspond to the correlation between the x_{1j} and x_{2j}, now form an acute angle.

If we give eqn. (2.24) the form of (2.16a) and introduce the row vectors $\mathfrak{w}_1$ and $\mathfrak{w}_2$, we get:

$$\begin{pmatrix} x_{1j} \\ x_{2j} \end{pmatrix} = \begin{pmatrix} w_{1\mathrm{I}}w_{1\mathrm{II}} \\ w_{2\mathrm{I}}w_{2\mathrm{II}} \end{pmatrix}\begin{pmatrix} f_{\mathrm{I}j} \\ f_{\mathrm{II}j} \end{pmatrix} = \begin{pmatrix} \mathfrak{w}_1 \\ \mathfrak{w}_2 \end{pmatrix}\begin{pmatrix} f_{\mathrm{I}j} \\ f_{\mathrm{II}j} \end{pmatrix} \tag{2.24a}$$

This representation shows that the directions of the vectors $\mathfrak{w}_1$ and $\mathfrak{w}_2$, which are associated with the initial attributes in the plane of the factors by the angles:

$$\vartheta_{11} = \operatorname{arctg} \frac{w_{1\,\text{II}}}{w_{1\,\text{I}}} \tag{2.25a}$$

$$\vartheta_{21} = \operatorname{arctg} \frac{w_{2\,\text{II}}}{w_{2\,\text{I}}} \tag{2.25b}$$

determine axes on which the object points j with their coordinates $f_{\text{I}j}$ and $f_{\text{II}j}$ projected, give the initial coordinates x_{1j} and x_{2j}. In Fig. 2.24 (bottom) this is demonstrated for the object B.

In the example of Fig. 2.24, the vectors $\mathfrak{w}_1$ and $\mathfrak{w}_2$ (indicated in Fig. 2.24 (bottom) by z_1 and z_2) depart from the principal axis I by equal but opposite angles. The smaller this angle the higher the correlation between the initial attribute 1 and the factor I, and the greater the chance of representing the data matrix by this single factor.

The correlation coefficient r_{12} is also presented in this figure; it is easy to calculate as the cosine of the angle between the vectors $\mathfrak{w}_1$ and $\mathfrak{w}_2$:

$$r_{12} = \cos\,(\vartheta_{11} - \vartheta_{21}) \tag{2.25c}$$

The angular relation of the vectors $\mathfrak{w}_1$ and $\mathfrak{w}_2$ remains unchanged if we rotate the orthogonal axes of the factors. Here it is important that the distribution of the 'factor scores' in Fig. 2.24 (bottom) does not exhibit a centrifugal moment, i.e. a correlation coefficient $r > 0$, for any rotated position of the factor axes. All of the factor scores that are referenced to such rotated axes are in this sense independent of one another.

In factor analysis this is regarded as an advantage, since it allows us to adapt one factor to a well-known initial attribute. But only the factors defined by the directions of the principal axes, which are called for this reason 'principal factors', are ordered according to extremes of explained variances.

Furthermore, the transition from Fig. 2.24 (top) to 2.24 (bottom) makes it clear that the rotation must be executed before the extension. Therefore the exact path from the initial attributes to the factors, even those that are the result of a later rotation, cannot avoid the use of the transformation to the principal axes.

The analogy between variances and moments of inertia can be applied to the three-dimensional case, $n = 3$, as well. But the concept of principal axes in the theory of matrices was extended long ago to any arbitrary value of n, without mention of the original mechanical significance at all. The same holds for factor analysis. Therefore we may now extend the rules derived for $n = 2$ to arbitrary values of n, without detailed proof. The condition (2.23) for the factor scores f_{kj} reads in general matrix form:

$$\frac{1}{N-1} FF' = I \tag{2.26}$$

Here F is the matrix of the f_{kj} already introduced at the right in eqn. (2.16a), F' is the transformed matrix with rows and columns exchanged, and I is the unit matrix, i.e. a matrix with diagonal elements $=1$ only.

If we define, for arbitrary n, the general correlation matrix

$$R = \frac{1}{N-1} ZZ' \tag{2.27}$$

corresponding to the data matrix Z (according to eqn. (2.19a)), and again regard our goal to be to express Z by the product

$$Z = WF \tag{2.28}$$

we get, on account of eqn. (2.26), the fundamental equation of factor analysis:

$$R = WW' \tag{2.29}$$

It shows that the correlation matrix R is the proper goal of the analysis and that we must synthesize R from the elements of the weighting matrix. (This may be the reason that these elements of W are also called factor loads, although they depend not only on the choice of the factors but also on the choice of the initial attributes.)

We need to know the matrix W first, before we can calculate the matrix of the factor scores by finding the inverse of eqn. (2.28):

$$F = W^{-1} Z \tag{2.30}$$

The conditions expressed by eqns. (2.26) to (2.30) do not themselves require an orientation to the principal axes. So long as this condition is not required in addition, we have an infinite ensemble of W-matrices, all of which fulfil the principal condition (2.29). If we require that the factors are so chosen as to explain extreme values of the variances of the initial attributes and the factors are ordered according to their contributions to explanation of the variances, then eqn. (2.29) has an unequivocal solution.

In order to show how, under these conditions, the elements of W can be found for an arbitrary value for n, we now refer to an analogy between factor analysis and the analysis of a vibrating system, which may be more meaningful to the acoustician than the methods for calculation of variations, used in the literature of factor analysis. Here again the case for $n = 2$ shows all the essential aspects.

If we have two equal masses connected to fixed points by springs with the stiffnesses s_{10} and s_{20}, and also coupled together by a spring with stiffness s_{12}, the application of Newton's Laws to the two masses leads to the 'system equations':

$$m\ddot{x}_1 + (s_{10} + s_{12})x_1 + s_{12}x_2 = 0$$
$$s_{12}x_1 + m\ddot{x}_2 + (s_{20} + s_{12})\,x_2 = 0 \qquad (2.31)$$

In order to describe the motion of the two masses we may think first of using the displacements x_1 and x_2 (in eqn. (2.31) they are directed opposite to each other). But the displacements are not independent of one another; instead they are coupled by the terms $s_{12}x_1$ and $s_{12}x_2$ in (2.31). Since we must expect linear differential equations of the second order in all vibration problems, we may set:

$$x_1 = \mathrm{Re}\{\underline{\hat{x}}_1\, e^{i\omega t}\}, \qquad x_2 = \mathrm{Re}\{\underline{\hat{x}}_2\, e^{i\omega t}\} \qquad (2.32)$$ [6]

that is, we assume that the system is able to vibrate with a specific eigenfrequency as a simple oscillator with one kinematic degree of freedom. Then the differential equations (2.31) become linear equations in the complex amplitudes, which we may transform to the letters used in factor analysis by introducing:

$$s = s_{10} + s_{12} = s_{20} + s_{12} \qquad (2.33a)$$

and the abbreviations:

$$\frac{s_{12}}{s} = r_{12}, \qquad \frac{\omega^2 m}{s} = \lambda \qquad (2.33b)$$

Thus we get:

$$(1-\lambda)\underline{\hat{x}}_1 + r_{12}\underline{\hat{x}}_2 = 0$$
$$r_{12}\underline{\hat{x}}_1 + (1-\lambda)\underline{\hat{x}}_2 = 0 \qquad (2.34)$$

It is evident here that the stiffnesses correspond to the elements of the correlation matrix R.

[6] With respect to representing sinusoidal vibrations by the real part of complex quantities (underlined here), see Section IV. 1.7 (Volume 2).

These two equations have non-vanishing solutions only if the determinant of the coefficients is zero:

$$\begin{vmatrix} 1-\lambda & r_{12} \\ r_{12} & 1-\lambda \end{vmatrix} = 0 \tag{2.35}$$

Thus the condition on λ is:

$$(1-\lambda)^2 - {r_{12}}^2 = 0 \tag{2.35a}$$

which is called the characteristic equation.[7] Its roots

$$\begin{aligned} \lambda_{\mathrm{I}} &= 1 + r_{12} \\ \lambda_{\mathrm{II}} &= 1 - r_{12} \end{aligned} \tag{2.35b}$$

are called eigen-values.[8]

This method of evaluating the extremes of the variances explained by the factors (which we first compared with extremes of moments of inertia) is obviously applicable for arbitrary values of *n*.[9]

But the same holds for locating the directions of the principal axes. The two eigen-values correspond to the squares of the two eigen-frequencies ω_{I} and ω_{II}. The general solution for the initial coordinates x_1 and x_2 is composed of two eigen-vibrations. If we write these in the form:

$$\begin{aligned} x_1 &= \mathrm{Re}\{w_{1\mathrm{I}}\underline{\hat{x}}_{\mathrm{I}}e^{i\omega_{\mathrm{I}}t} + w_{1\mathrm{II}}\underline{\hat{x}}_{\mathrm{II}}e^{i\omega_{\mathrm{II}}t}\} \\ x_2 &= \mathrm{Re}\{w_{2\mathrm{I}}\underline{\hat{x}}_{\mathrm{I}}e^{i\omega_{\mathrm{I}}t} + w_{2\mathrm{II}}\underline{\hat{x}}_{\mathrm{II}}e^{i\omega_{\mathrm{II}}t}\} \end{aligned} \tag{2.36}$$

it becomes evident that $\hat{x}_{\mathrm{I}}$ and $\hat{x}_{\mathrm{II}}$ are the (complex) amplitudes of coordinates that correspond in factor analysis to principal coordinates. With respect to vibration systems they are also called in the English literature normal coordinates.[10] (In Germany the designation principal (*Haupt-*) coordinates was introduced into the analysis of vibrating systems in the last century.)

---

[7] We learned this term in Section II. 3.3 in discussing the theory of coupled rooms.

[8] In mathematics the λ are called characteristic numbers and their reciprocals are called eigen-values.

[9] For the conditions which an analogous vibrating system must fulfil, see Cremer, L., *Acustica*, **42** (1979) 1.

[10] Lord Rayleigh, *The Theory of Sound*, Vol. I, Dover Publications, New York, 1945, p. 108. Rayleigh cites in this same connection Thomson and Tait's *Natural Philosophy*, 1st. edn., 1867, Section 337.

But the dependence on time in eqn. (2.36) lies outside the analogy with factor analysis, though it holds with respect to the weighting-multipliers w that appear in (2.36). They, too, exhibit an eigen-configuration. They can be calculated by introducing the data for a particular eigen-vibration into the system eqns. (2.34). For example, we may choose those characterized by the index I in the first equation of (2.36) and those with the index II in the second and get:

$$\begin{aligned}(1-\lambda_{\mathrm{I}})w_{1\,\mathrm{I}}+r_{12}w_{2\,\mathrm{I}}&=0\\ r_{12}w_{1\,\mathrm{II}}+(1-\lambda_{\mathrm{II}})w_{2\,\mathrm{II}}&=0\end{aligned} \tag{2.37}$$

The results here are independent of r_{12}:

$$\frac{w_{2\,\mathrm{I}}}{w_{1\,\mathrm{I}}}=1; \qquad \frac{w_{1\,\mathrm{II}}}{w_{2\,\mathrm{II}}}=-1 \tag{2.38}$$

This determines two configurations, one in which the masses vibrate in the same direction and the other in which they vibrate in opposite directions. It is characteristic in problems of vibrating systems that normalizing the column vectors of the W-matrix, which would read for each n:

$$W=(\mathfrak{w}_{\mathrm{I}},\ \mathfrak{w}_{\mathrm{II}},\ldots) \tag{2.39}$$

is possible but not necessary, and thus is arbitrary. This marks a significant difference compared with factor analysis problems and their analogy to normalized moments of inertia. In all three cases, the column-vectors, $\mathfrak{w}_1$ and $\mathfrak{w}_2$, are called 'eigen-vectors', even when their lengths remain undefined. (They are not to be confused with the row-vectors of the W-matrix introduced in eqn. (2.24a), which are associated with the initial coordinates and the initial attributes.)

Extended to arbitrary n, this analogy between initial attributes and initial coordinates, and between principal factors and principal (normal) coordinates, leads to the following procedure.

Suppose we are given a matrix:

$$R=\begin{pmatrix}1 & r_{12} & \cdots & r_{1n}\\ r_{12} & 1 & \cdots & r_{2n}\\ \cdots & \cdots & \cdots & \cdots\\ r_{1n} & r_{2n} & \cdots & 1\end{pmatrix} \tag{2.40a}$$

We first look for the corresponding diagonal matrix:

$$\Lambda = \begin{pmatrix} \lambda_{I} & 0 & .. & 0 \\ 0 & \lambda_{II} & .. & 0 \\ \cdots & \cdots & \cdots & \cdots \\ 0 & 0 & .. & \lambda_n \end{pmatrix} \tag{2.40b}$$

We find its elements by solving the characteristic equation, which may be written:

$$\text{Det}\,(R-\lambda I)=0 \tag{2.41a}$$

where the coefficients are known, or by

$$\text{Det}\,(\Lambda-\lambda I)=0 \tag{2.41b}$$

which comes from the characteristic equation broken into the linear factors:

$$(\lambda_{I}-\lambda)(\lambda_{II}-\lambda)\,\ldots\,(\lambda_n-\lambda)=0 \tag{2.41c}$$

It follows from this form that the eigen-values which are the solutions of the characteristic equation are the elements of the diagonal matrix for which we are looking.

We get the column vectors $\mathfrak{w}_k$ of the W-matrix—apart from their lengths—from the n equations:

$$(R-\lambda_k I)\mathfrak{w}_k=0 \tag{2.42a}$$

which impose only $(n-1)$ conditions, since they depend on one another on account of the characteristic equation.

The additional normalization required by factor analysis with respect to their lengths

$$|\mathfrak{w}_k|=\sqrt{\lambda_k} \tag{2.42b}$$

is already included in eqn. (2.29), if we assume that it applies to the principal factors. Since the following equalities hold:

$$\text{Det}\,R=\text{Det}\,(WW')=\text{Det}\,(W'W) \tag{2.43}$$

the characteristic equation may also be written:

$$\text{Det}\,(W'W-\lambda I)=0 \tag{2.44}$$

Now, the column vectors of the W-matrix are orthogonal, i.e.

$$\sum_{i=1}^{n} w_{ik} w_{il} = 0, \qquad \text{for } k \neq l \tag{2.45}$$

because this condition is necessary to fulfil eqn. (2.26). Therefore, $W'W$ is a diagonal matrix:

$$W'W = \begin{pmatrix} \sum w_{iI}^2 & 0 & \dots & 0 \\ 0 & \sum w_{iII}^2 & \dots & 0 \\ \dots & \dots & \dots & \dots \\ 0 & 0 & \dots & \sum w_{in}^2 \end{pmatrix} \tag{2.46}$$

A comparison of eqn. (2.44) with (2.41b) shows that the elements of this diagonal matrix are the λ_k, which proves eqn. (2.42b) generally.

Furthermore, the fact that they represent the part of the total variance that is explained by the specific factors becomes obvious if we split the total variance into the matrix-like two-dimensional form:

$$\begin{array}{l} w_{11}^2 + w_{12}^2 + \dots + w_{1n}^2 \\ + w_{21}^2 + w_{22}^2 + \dots + w_{2n}^2 \\ + \dots\dots\dots\dots\dots\dots \\ + w_{n1}^2 + w_{n2}^2 + \dots + w_{nn}^2 \end{array} \tag{2.47}$$

Here the sums of the rows represent the variances of the different initial attributes, which are (according to the normalization (2.19)) all equal to 1; the sums of the columns represent the variances λ_k, corresponding to the factors. Their sum must be n:

$$\lambda_I + \lambda_{II} + \dots \lambda_n = n \tag{2.48}$$

We now extend our example from $n = 2$ to $n = 3$ by adding the data on 'strength coefficients' (see Section II. 7.3, eqn. (7.8)), measured by Lehmann for the same objects:

$$\begin{array}{ccccccc} & \text{A} & \text{B} & \text{C} & \text{D} & \text{E} & \text{F} \\ \gamma = (& 1{\cdot}6 & 4{\cdot}5 & 1{\cdot}1 & 1{\cdot}1 & 2{\cdot}5 & 2{\cdot}7) \times 10^{-3} \end{array}$$

From this we get the correlation matrix:

$$R = \begin{pmatrix} 1 & 0{\cdot}854 & 0{\cdot}388 \\ 0{\cdot}854 & 1 & 0{\cdot}572 \\ 0{\cdot}388 & 0{\cdot}572 & 1 \end{pmatrix}$$

The rather high correlation ($r_{12} = 0{\cdot}859$) between reverberation time and center-time is not changed by the introduction of the new variable, the strength coefficient, into the data matrix. The fact that the latter is less well correlated ($r_{13} = 0{\cdot}388$) with reverberation time may be the result of the great differences in hall volumes. The correlation between t_s and γ ($r_{23} = 0{\cdot}572$) is intermediate between the other two correlations.

The characteristic equation yields the following eigen-values:

$$\lambda_{\mathrm{I}} = 2{\cdot}228; \qquad \lambda_{\mathrm{II}} = 0{\cdot}653; \qquad \lambda_{\mathrm{III}} = 0{\cdot}119$$

This time we must divide the eigen-values by 3 to get the explained percentages of variances:

$$74{\cdot}3\%; \quad 21{\cdot}8\%; \quad 3{\cdot}9\%$$

Again it appears reasonable to neglect the last factor.

But first we evaluate the elements of the W-matrix according to eqns. (2.42a) and (2.42b):

$$W = \begin{pmatrix} 0{\cdot}891 & 0{\cdot}398 & 0{\cdot}218 \\ 0{\cdot}954 & 0{\cdot}149 & -0{\cdot}257 \\ 0{\cdot}723 & -0{\cdot}687 & 0{\cdot}071 \end{pmatrix}$$

Figure 2.25 presents the row vectors, identified with the corresponding initial attributes. Because of the small value of λ_{III} these vectors have small components in the direction of f_{III}.

If we neglect this third factor-dimension we get projected vectors $\mathfrak{w}_i$ in the f_{I}–f_{II}-plane that are shorter than 1. We conclude from this foreshortening that the respective attributes cannot be entirely explained by the factors in that plane.

After dropping the third factor-dimension there remains the possibility of rotating the orthogonal axes of the remaining factors so that they coincide as much as possible with the directions for T and γ.

Finally the factor scores (coordinates in the factor-space) for the object A have also been evaluated by eqn. (2.30); and the vector OA, from the origin to A, has been shown as an example in Fig. 2.25.

We mentioned at the beginning of this section that we would treat only a special case of factor analysis. It was characterized not only by the mutual independence of the factors and their orientation to the principal axes, but also by the assumption that each factor is correlated with at least two initial attributes.

In contrast, the creators of factor analysis made a distinction, from the beginning, between 'common factors' and 'unique factors' which are correlated with one attribute only. Mathematically this means that the

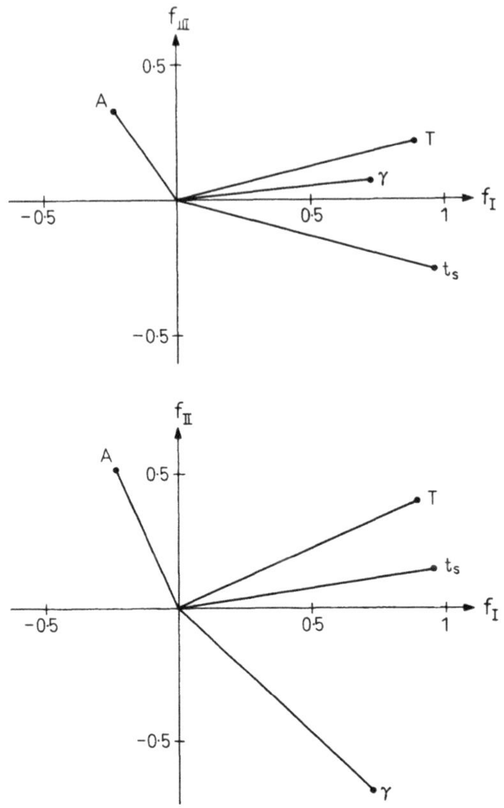

Fig. 2.25. Example of factor analysis as in Fig. 2.24 (bottom), but extended to three dimensions.

correlation matrix is composed of a sum of one square matrix and one diagonal matrix:

$$R = \begin{pmatrix} r_{11} r_{12} \ldots r_{1n} \\ r_{12} r_{22} \ldots r_{2n} \\ \ldots\ldots\ldots\ldots\ldots \\ r_{1n}\; r_{2n} \ldots r_{nn} \end{pmatrix} + \begin{pmatrix} 1-r_{11} & 0 \ldots\ldots\ldots 0 \\ 0 & 1-r_{22} \ldots\ldots 0 \\ \ldots\ldots\ldots & \ldots\ldots\ldots\ldots\ldots \\ 0 & 0 \ldots\ldots 1-r_{nn} \end{pmatrix} \qquad (2.49)$$

Although this appears at first to be more complicated it need not be, because the required number m of common factors for the first part may be much smaller.

This part of the correlation matrix may be treated according to the same rules that we learned for the 'component analysis'. The elements r_{ii} in the principal diagonal, which now appear instead of 1, namely:

$$r_{ii} \leq 1 \tag{2.50}$$

are called 'communalities' because they approach their maximum value 1 more closely the more the respective attribute is 'explained' by the common factors.

This diminution exists in principle; it is not caused, as in the case of the correlation coefficients r_{ki} ($k \neq i$), by unavoidable scatter in the measured results.

More crucial is the question as to whether the separation of individual factors is as fully adequate to the principles of factor analysis as is so often claimed. If this were true then the communalities should be able to be unequivocally evaluated from the correlations between the initial attributes. But this is not the case. In fact various recommendations are offered as to how to estimate their values, and the recommendations from different sources differ among themselves. Furthermore the reader is often not informed about the chosen method and the results.

Thus there is a basic defect of ambiguity in factor analysis that remains even when the common factors are treated as principal factors belonging to the first matrix in eqn. (2.49), with the elements r_{ii} instead of 1. Only the principal factors calculated with $r_{ii} = 1$ (also called the principal components) comply with the basic principle of the natural sciences that the results of a method must be independent of the researcher and of any more-or-less arbitrary decisions that he may make.

Furthermore it is very doubtful that the unique factors and their contributions will be mentioned at all in future discussions of explained variances.

This procedure may have found favor initially because it permitted a reduction in the number of 'common factors', i.e. factors for which the calculation of eigen-values would have been necessary—a matter of some importance before the advent of the computer.

But the principal-component method also permits a reduction in the number of factors to be considered and does so as a methodical approximation.

On the other hand, we note that the n differences between r_{ii} and 1 in the diagonal have less influence on the result, the larger the total number n^2 of all the elements.

At least the choice of the communalities is less important for room-acoustical consequences than the arbitrariness in the choice of attributes and objects. Factor analysis cannot find relations that are not contained in the data matrix.

Nevertheless, the use of factor analysis marks a great step forward in comparison with the earlier unavoidable and arbitrary assumptions about the dependence or independence of the selected attributes and about their weightings.

III.2.9 Multivariate Evaluation of Room-acoustical Observations

Multivariate methods, which were developed to determine the minimum number for the aspects of acoustical judgment and the possibility of the existence of individual differences, were first applied not to the comparison of tape recordings in room-acoustical problems but to immediate subjective observations in concert halls. Yoshida[1] asked test persons to scale their perceptions of 'pleasingness of reverberation', 'length of reverberation' (not necessarily the same as reverberation time), 'softness', 'richness', and 'goodness from all points of view', for seven different seats in the Sendai Public Hall, where he presented reproductions of operatic arias, orchestral music and speech, by means of loudspeakers.

He evaluated mean values for the scaling data given by the various test persons and thus derived for each seat the scores of the attributes. Factor analysis showed that they could all be explained by only two factors, which appeared in the different initial attributes with different weightings.

Hawkes and Douglas[2] evaluated significantly more factors by means of questionnaires in which the test persons were asked to rate their acoustical impressions along one-dimensional scales whose end-points were given names with opposite extremes:[3]

1. Disliked very much	—	Enjoyed very much
2. Live	—	Dead
3. Cold	—	Warm

[1] Yoshida, T., *Proc. 5th ICA, Liège, 7–14 Sept. 1965* Paper B–18.

[2] Hawkes, R. J. and Douglas, H., *Acustica*, **24** (1971) 235.

[3] Most of these pairs were taken over by L. L. Beranek in *Music, Acoustics and Architecture*, John Wiley, New York, 1962 (reprinted Krieger Publishing Co., Huntington, New York, 1979), p. 64; but he also took other aspects into account.

4. Clear	—	Muddy
5. Dull	—	Brilliant
6. Even	—	Boomy
7. Distant	—	Close
8. Dry	—	Resonant
9. Blended	—	Unblended
10. Balanced	—	Unbalanced
11. Public	—	Intimate
12. Reverberant	—	Unreverberant
13. Unresponsive	—	Responsive
14. Large dynamic range	—	Small dynamic range
15. Poor definition	—	Good definition
16. Good	—	Bad

Those pairs that are based upon the same word (for example, 'Blended—Unblended') not only have the advantage of making the scaling easier because the concept under evaluation is sharpened, but they lead to a quantitative definition of that concept.

Each of these pairs expresses a subjective perception without assuming that there is an objective quantity that is highly correlated with it. In fact test subjects who are familiar with the concepts of room acoustics must guard against associating the word 'reverberant' with 'the mid-frequency reverberation time', for example. Whether or not such a correlation exists must be proved by factor analysis.

Hawkes and Douglas also determined mean values for the scalings of the 16 attributes given by the test persons in their questionnaires, first for different seats in the Royal Festival Hall (London) at the same concert, and later in three other concert halls with quite different musical compositions ranging from Bach to Stravinsky.

Regarding the different seats as the 'objects', they investigated how many factors are needed to explain the variances. Beranek had already distinguished between dependent and independent aspects (i.e. factors); but still he found it necessary to consider nine different factors whereas Hawkes and Douglas required only six, taking all their halls into account, and only four for the Royal Festival Hall concert. These were:

Definition
Reverberance
Balance and blend
Proximity

While the authors that we have mentioned so far were interested only in evaluating the least number of judgmental aspects and assumed that all the test persons gave them the same weighting, Yamaguchi[4] was the first to investigate whether or not there are systematic differences between the judgments of different people. He was also the first to use recordings in order to prevent the subject's judgment from being biased by the sight of the auditorium or the knowledge of the seat location. (Who would dare to rate Vienna's Grosser Musikvereinssaal as anything but excellent, in any seat whatever?)

He recorded at several seats in the Yamaha Music Hall with two microphones separated by 50 cm; the signal was stereophonic 'dry' music radiated by loudspeakers with omnidirectional characteristics. The two channels were later reproduced for the test persons, either by two loudspeakers or two earphones. The subjects were not asked for their preference with respect to two test examples following immediately one after the other, but rather whether they could distinguish any difference at all between the two examples.

These discrimination tests were evaluated according to a method called 'multi-dimensional scaling', which is similar to factor analysis insofar as the objects appear in a multi-dimensional space; the number of dimensions of this space determines how many independent aspects ('factors') must be considered in order to explain the variances of the discrimination tests with sufficiently good approximation.

According to our discussion in Section III. 2.7 of the difficulties of accurately reproducing acoustical impressions from a hall, it is doubtful that the arrangement and the apparatus used by Yamaguchi could succeed in conveying adequate room-acoustics perceptions. Nevertheless he was able to show that such perceptions depend very much on the loudness level of the reproduction. It is also interesting that he required three 'factors' to explain the discrimination tests carried out by different people; this implies that different people may not give the same weighting to the different attributes.

The comparison tests made with dummy-head recordings in Göttingen and Berlin had not only the advantages of recordings that simulated the room impressions more accurately and of a greater number of objects (more seats in more halls), but they fulfilled the condition that the 'signals' within a comparison-series remained the same. This avoids influences that have nothing to do with the hall and reduces the comparisons to purely acoustical impressions.

---

[4] Yamaguchi, K., *J. Acoust. Soc. Am.*, **52** (1972) 1271.

The papers[5, 6, 7, 8] written about these tests provide such clear information about the test procedures that they are especially suitable for demonstrating the advantages of factor analysis.

We begin with the question, raised by Siebrasse, of people's preferences for objects (halls, seats), represented by pairs of recordings played one immediately after the other; this procedure allows us to relate the results to Beranek's calculations of acoustical quality (see Section III. 2.3). Siebrasse requested the test subjects to write down after each test pair a rating number, $p_{ij} = 1, -1$, or 0, depending on whether they preferred i or j or had no preference at all. The quality judgment g_{ki} of the kth test person with respect to the ith object was given, for all the pair-comparisons involving object i, by:

$$g_{ki} = \sum_{j=1}^{N} (p^k_{ij} - p^k_{ji}) \tag{2.51}$$

Thus even those cases could be included in which the test person contradicts himself, for example, by first preferring A over B, then B over C, then C over A. Such occurrences are not necessarily signs of unreliable judgment, because the test person may have based his judgment in the first two comparison-pairs on other acoustical aspects than in the last pair.

The numbers g_{ki} can then be transformed into normalized initial attributes according to the principles of factor analysis:

$$z_{ki} = \frac{g_{ki} - \bar{g}_k}{\left(\sum_{i=1}^{N} (g_{ki} - \bar{g}_k)^2 \right)^{1/2}} \tag{2.52}$$

(Siebrasse uses a slightly different normalization here than we discussed in the last section but this is without influence on the conclusions.)

Having at his disposal 13 test persons and 25 concert halls (some with several different seats), Siebrasse found from his factor analysis that at least four factors are worthy of consideration. Figure 2.26 shows his results for four halls (E, P, Q, T), with sometimes two and sometimes three different seats. Since he takes four factors into account we have the problem of representing a four-dimensional preference space. But in our

[5] Siebrasse, K. F., Dissertation, Göttingen, 1973.
[6] Gottlob. D., Dissertation, Göttingen, 1973.
[7] Wilkens, H., Dissertation, TU Berlin, 1975.
[8] Lehmann, P., Dissertation, TU Berlin, 1976.

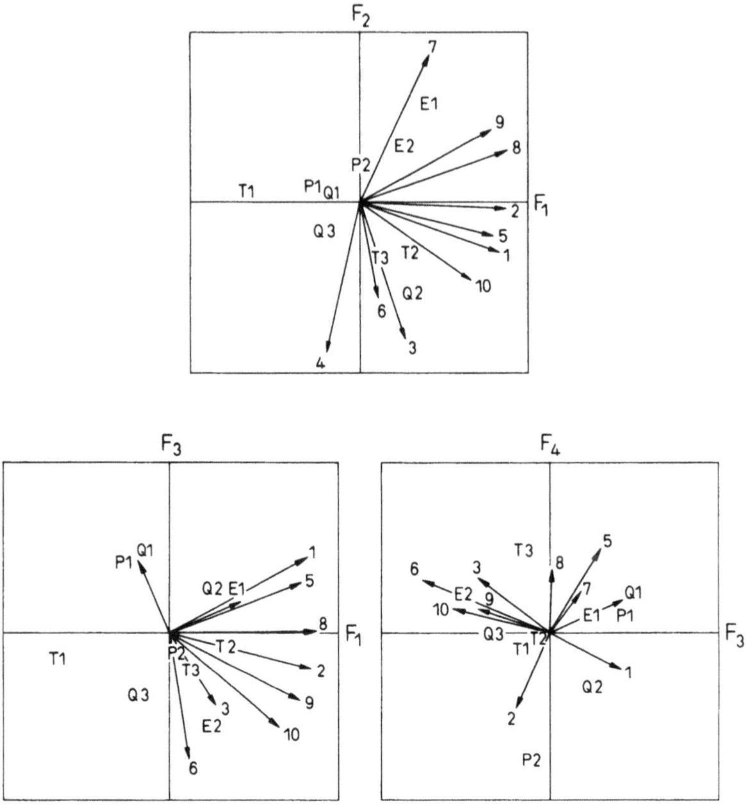

Fig. 2.26. Representation of a four-dimensional factor space by three factor-planes. The letters denote different halls, their indices denote different seats in them, and the straight lines with numbers correspond to different test persons. (After Siebrasse.[5])

three-dimensional example (see Fig. 2.25) we have already made use of projection onto planes. We can proceed here in the same way, by using, for instance, the planes (see Fig. 2.24):

$$(F_1, F_2), (F_1, F_3) \text{ and } (F_3, F_4)$$

But we cannot combine such projections into a three-dimensional space in which we are accustomed to move.

On the basis of Fig. 2.26 we can conclude that different halls do not differ any more from one another than different seats in the same hall; we do not even consider here extreme locations such as the conductor's podium or seats deep underneath a balcony. This conclusion already makes the notion of rank-ordering concert halls very questionable. It would be possible (but tedious) to average the preference judgments over the different seats in a hall; but this would not adequately take into account the different weighting of the factors by the different test persons.

For one of the factors, F_1 (which is the one that explains the greatest percentage of the total variance), all but one of the arrows form an acute angle with its axis; that is, they point in the same direction. With respect to this factor there seems to be some 'consensus'.[9] But the agreement is not good enough to justify neglecting the other factors. For factor F_2 we might at most speak of 'groups of people with different tastes'. For factors F_3 and F_4 not even groupings can be detected.

Figure 2.27 shows by the heights of the columns the percentages of explained portions of the total variance, plotted against the numbers of the factors, ordered in this respect. The circles connected by straight lines

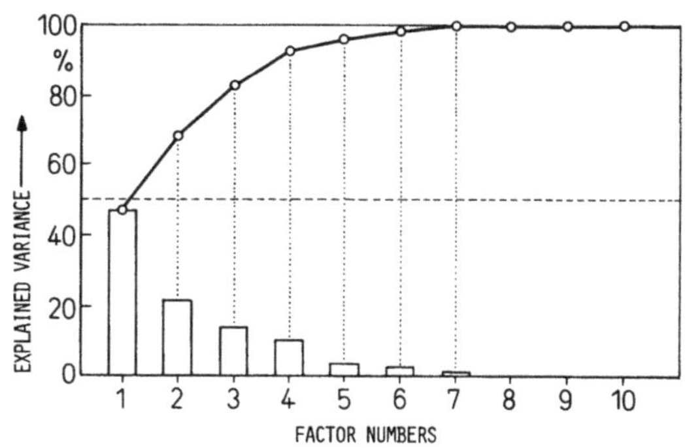

Fig. 2.27. Percentages of the explained portions of the total variance corresponding to the respective factors (columns) and their sum (circles), see text. (After Siebrasse.[5])

give the sums of those percentages up to the respective factor number. We see that factor 1 explains less than 50%; factors 1 and 2 together explain no more than 67·4%; the first three factors explain 81·9% and the first four (which are the basis for Fig. 2.26) explain 91%. One may wonder whether, for room-acoustics problems, it might not be sufficient to consider only the first three factors.

[9]The term 'consensus' was introduced by J. D. Caroll. See Sheppard, R. N., Romney, A. K. and Nerlove, S. B., *Multidimensional Scaling: Theory and Applications in the Behavioral Sciences*, Vol. 1. *Theory*, Vol. 2 *Applications*, Academic Press, New York, 1972, p. 106.

Siebrasse also shows how much the portions of explained variance, and thus the number of factors that need to be retained in the analysis, depend on the choice of the objects.

At the beginning of his tests he included two halls with reverberation times less than 1·2 s. He found that the good agreement among the test subjects in disliking these two halls increased the explained variances for the first two factors so greatly as to reduce the number of factors that had to be considered. On the other hand, we may ask whether factor analysis should not include all the kinds of halls that are used in practice: the more nearly alike the objects are, the less certain are the distinctions between them.

Eysoldt,[10] therefore, in his later factor analysis tried to assess the certainty of the judgments, taking as a measure of certainty the ratio of the used variance to the maximum possible variance of the specific attribute.

Still another limitation of the Göttingen tests must be mentioned: in the playback of the recorded examples the level was always adjusted to give equal loudness at the listener's position, rather than equal source power. This was because it was found, as stated earlier by Yamaguchi, that the loudness significantly influences room-acoustical judgments and thus must be eliminated. With this procedure Siebrasse actually considered five factors as significant, not four. But what (we may ask) is the meaning of 'equal loudness' for running music in halls with different reverberation times? Certainly not that of the direct sound only; but also surely not that of the direct sound and all the reflections!

This indecision appears as a further reason for adjusting the playback levels to equal sound power as was done in the Berlin tests. It is hardly likely that the orchestra, playing in halls of different size, adjusted to these differences by playing with more or less power than they are accustomed to produce in the Berlin Philharmonic Hall.

Wilkens[11] also asked the test persons for their preferences between objects at first; but he concluded that this does not lead to unambiguous ranking of the halls because of individual differences in taste.

His investigations differ in principle from those of Siebrasse; he asked, in addition, that the test persons rate their impressions on scales whose endpoints were named with opposite extremes, as Hawkes and Douglas did earlier.

---

[10] Eysoldt, U., Dissertation, Göttingen, 1976.

[11] Wilkens, H.;[7] also *Berichte z. 10 Tonmeistertagung*, (1975) 166; also *Acustica*, **38** (1977) 10.

Recognizing that the result of factor analysis depends on the choice of these pairs, where either some important aspect of hearing may be overlooked, or not be taken sufficiently into account, or another factor may be exaggerated by too many interrelated attributes, Wilkens gives exact information concerning the principles that governed his choices. He first collected all the descriptive terms that had been used in the literature and also those used in daily life by musicians and recording engineers. The number of opposing pairs collected in this way exceeded 90!

If we anticipate here that he finally found only three factors to be necessary, it simply demonstrates the difficulty and uncertainty in describing specific room-acoustical perceptions with everyday words. He tried to avoid terms that suggest a physical state, for instance, 'resonance'. Furthermore he excluded from the preliminary tests word-pairs that are either irrelevant to room-acoustical quality, unclear or even ambiguous; and finally he eliminated those that were synonymous with word-pairs already chosen. The following pairs were finally used for the main body of his research:[12]

ENGLISH

		1	2	3	4	5	6	
1	small	---	--	–	–	--	---	large
2	*pleasant*	---	--	–	–	--	---	*unpleasant*
3	unclear	---	--	–	–	--	---	clear
4	soft	---	--	–	–	--	---	hard
5	brilliant	---	--	–	–	--	---	dull
6	rounded	---	--	–	–	--	---	pointed
7	vigorous	---	--	–	–	--	---	muted
8	*appealing*	---	--	–	–	--	---	*unappealing*
9	blunt	---	--	–	–	--	---	sharp
10	diffuse	---	--	–	–	--	---	concentrated
11	overbearing	---	--	–	–	--	---	reticent
12	light	---	--	–	–	--	---	dark

[12] The English translations of the German word-pairs given here were proposed by Wilkens himself, with one exception: where he used the word 'clear' twice (to translate both *deutlich* and *klar*) we have changed one of them (the latter) to 'transparent'. See Wilkens, H. and Plenge, G., in *Auditorium Acoustics*, Chapt. 18, Applied Science Publishers Ltd, London, 1975.

13 muddy	--- -- - - -- ---	transparent
14 dry	--- -- - - -- ---	reverberant
15 weak	--- -- - - -- ---	strong
16 emphasized treble	--- -- - - -- ---	treble not emphasized
17 emphasized bass	--- -- - - -- ---	bass not emphasized
18 *beautiful*	--- -- - - -- ---	*ugly*
19 soft	--- -- - - -- ---	loud

GERMAN

1 klein	groß
2 *angenehm*	*unangenehm*
3 undeutlich	deutlich
4 weich	hart
5 brillant	matt
6 rund	spitz
7 kräftig	gedämpft
8 *gefällt*	*gefällt nicht*
9 stumpf	scharf
10 diffus	konzentriert
11 aufdringlich	zurückhaltend
12 hell	dunkel
13 verschwommen	klar
14 trocken	hallig
15 schwach	stark
16 höhenbetont	nicht höhenbetont
17 tiefenbetont	nicht tiefenbetont
18 *schön*	*häßlich*
19 leise	laut

Each such choice is open to discussion no matter how much care is taken. For example, those acousticians who have recently given early lateral reflections such great importance (see Section III. 2.5) may feel that these have not been adequately accounted for in word-pairs 1 and 10. Also Beranek's terms 'balance' (between different instrumental groups) and 'blend' (the fusion of sound from different instrumental groups) do not appear here explicitly. But these concepts depend strongly on non-room-acoustical conditions, such as the kind of musical composition, the desires of the conductor, the placement of the musicians, etc.

Each word-pair implies a scale with numbers from 1 to 6 on which the test person completing the questionnaire is to rate his room-acoustical impressions. The order of the word-pairs (left-to-right and vice versa) was deliberately mixed up in order to avoid any suggestion that 'favorable' attributes always tend to occur on one side.

The terms printed in italics represent unspecific word-pair judgments concerning general quality. The responses to these items could have been used, like the preferences for a factor analysis, to show the differences between persons. For these word-pairs, in contrast to all the others, an average over all the test persons would have eliminated important differences in taste. Therefore Wilkens investigated how the rating numbers 1–6 were distributed for certain special attributes. Fig. 2.28 shows as an example the results for the word-pair 'pleasant–unpleasant'. This distribution, like some of the others, shows two clear peaks. He concluded that it would be adequate to split the test persons into two groups with apparently different tastes; that is, with respect to quality categories he considered in effect two factors fewer than Siebrasse.

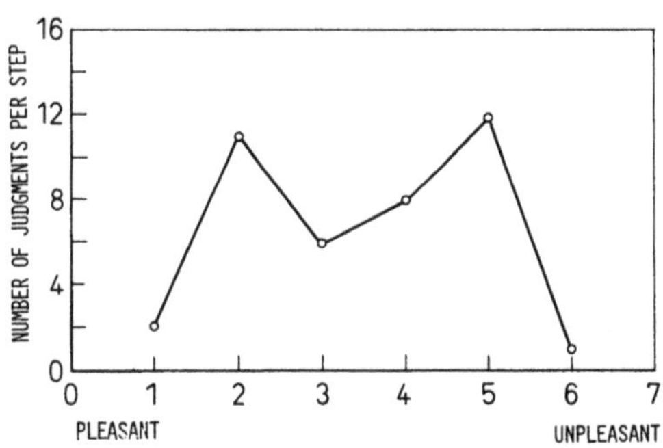

Fig. 2.28. Number of judgments plotted against the rating numbers 1–6 for the opposing word-pair, 'pleasant–unpleasant' for one object: Bartok, Hamburg, Seat no. 2. (After Wilkens.[11])

He also found that these two nearly equal groups (15 and 17 persons) correlated the 'pleasant–unpleasant' pair in an opposite sense to the 'large–small' pair, see Fig. 2.29. The test persons for these comparisons were psychologists, recording engineers and acousticians. But no relation could be found between the splitting of the acoustical tastes into two

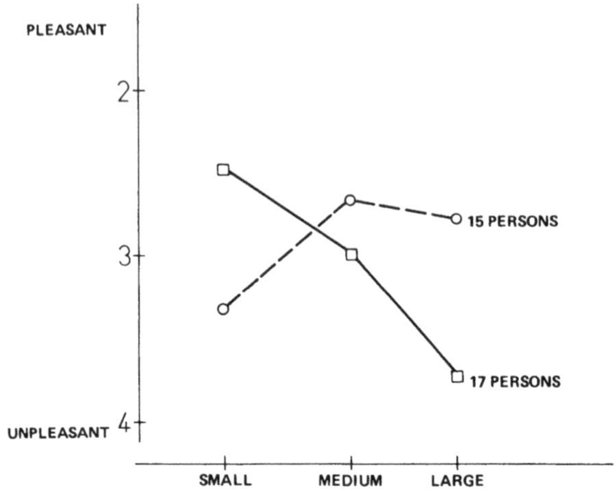

Fig. 2.29. Mean values of the rating numbers for two groups with different tastes with respect to the opposing word-pair 'pleasant–unpleasant', plotted against those for the word-pair 'large–small'. (After Wilkens.[11])

groups and the kinds of education and listening experiences of the subjects. (Siebrasse had come to the same conclusion.)

Wilkens evaluated, separately for the two groups, a three-dimensional factor-space for the 19 attributes that could explain in both cases nearly 90% of the variances. He could, through subsequent rotation of the axes of one of the spaces, bring them into such agreement that the vectors of all the attributes (except the pure quality categories 2, 8 and 18) practically coincided. This supports the assumption that all people are in agreement, whether they find an object (a hall, a seat, or a musical example) to be 'strong', 'clear', etc. They differ only in the weightings that they assign to these judgments.

Figure 2.30 shows the factor-space for the opposing word-pairs listed on pp. 585–6 and for the two taste groups mentioned above. Since the space is three-dimensional, we show the planes $(F_1 F_3)$ and $(F_1 F_2)$ one above the other; the upper one can be regarded as a vertical section (x, z) and the lower one as a horizontal section (x, y). For greater clarity only one extreme of each word-pair is indicated. The pure quality categories appear twice, once for each taste group. (The positive directions for F_1 and F_2 have been reversed with respect to Wilkens' representation, so that the judgments 'pleasant', 'appealing', and 'beautiful' appear on the positive side.)

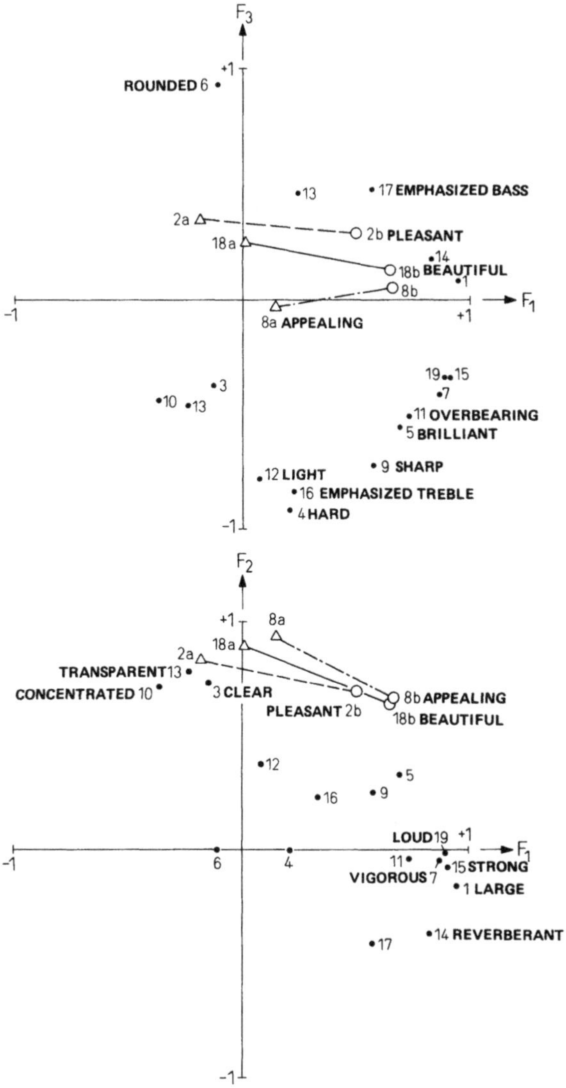

Fig. 2.30. Three-dimensional factor-space for the opposing word-pairs listed on pp. 585–6, of which only one extreme is marked. For the quality-pairs 2, 8 and 18, two points are given, corresponding to two groups with differing taste. (After Wilkens.[11])

Wilkens concludes from the relative orientation of the attribute vectors that factor 1 is correlated with the perception of strength and extension of the source, factor 2 with clearness, and factor 3 with timbre.

The two different evaluations of the pure quality attributes are mainly shifted only in the direction of factor 1. There exist, as we have already discussed in connection with Fig. 2.29 about the word-pair 'small–large', two different tastes with respect to strength and extension of the source.

Eysoldt[13] also found such different preferences with respect to the reproduction levels that they, too, could be split into two groups. Since we expect that stronger reinforcement of the direct sound by late reflections results in some loss of clarity, we may perhaps say that one group is prepared to accept this, the other not.

But both groups are interested in clearness, because all of the attributes that include this aspect are highly correlated with factor 2, and in this direction the two groups do not lie far apart. This holds also for factor 3 concerning 'timbre'.

Finally, Wilkens carried out a 'multi-factor variance-analysis' of his results. Figure 2.31 (left) shows the rating numbers for the attribute

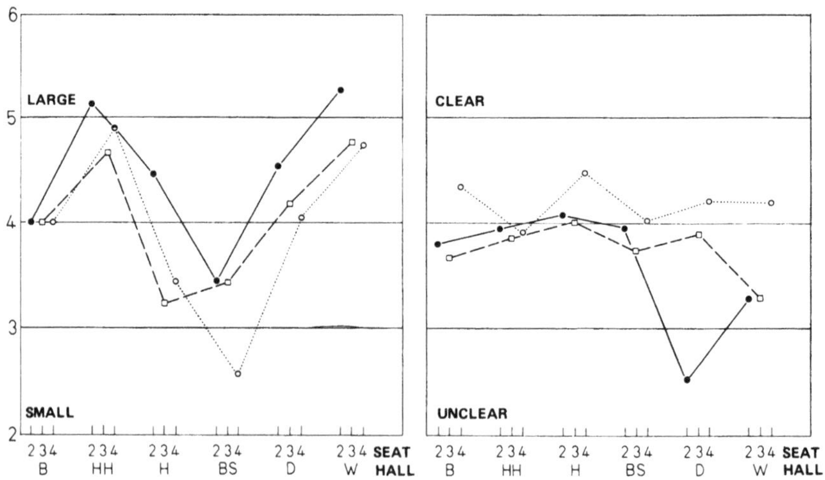

Fig. 2.31. Left: rating numbers of the attribute 'small–large' averaged over all test persons and music examples. Right: the same for the attribute 'unclear–clear' plotted against the different halls and the different seats in the halls. (After Wilkens.[11])

[13] Eysoldt, U., Dissertation, Göttingen, 1976.

'small–large', averaged over all the test persons and music examples and plotted against the different halls and the different seats. The values belonging to rather similar seat locations are connected by straight lines; they depart from one another, in general, less than the different halls. It is seen (see Fig. 2.31, left) that the attribute 'small–large' (and thus factor 1) is strongly dependent on the hall but less so on the seat location in the hall. The opposite is true for the attribute 'unclear–clear', see Fig. 2.31 (right). Here the differences within the same hall are larger than the differences of the mean values of the seats from hall to hall, although here again we have left out of consideration extreme seat locations. This proves that clearness depends on the fine structure of the reverberation and on the existence of strong early reflections, as we assumed in our discussion of objective criteria in Section II. 7.4.

Chapter III.3

The Consequences for the Design of Auditoriums

III.3.1 From Subjective 'Factors' to Objective Criteria

Although, no doubt, the subjective judgments about the acoustical quality of a hall are ultimately decisive, the method of subjective testing falls down in room-acoustical practice because we usually lack the required number of test subjects. Furthermore, such tests can be carried out only in the finished hall and therefore they can help only in the last-minute adjustments of the variable acoustical elements in that hall.

Model techniques today are far from being able to provide accurate room-acoustical impressions of a hall, whether by loudspeakers or by earphones; and even if the electroacoustical problems could be solved it would require such a detailed investigation into the characteristics of the hall under test that not only would the models become very expensive, but it must be doubted that the time needed for construction of the models and for adequate tests is available during the planning and construction of a hall, after the necessary funds have been provided.

Thus it is understandable that the practice of room acoustics, from the beginning, has sought objective and measurable, or, if possible, calculable criteria. The latter hope applies so far only to the Sabine reverberation time, which explains its still predominant place in room acoustics today. Although nowadays we regard the reverberation time as only one among many acoustical criteria, the possibility of estimating its value rather easily and early in the design process, based on the building plans and known or measurable absorption coefficients, assures its continuing importance in the design of halls.

It is more difficult—if it is possible at all—to calculate or even form a rough estimate of the other criteria mentioned in Chapter II.7. These criteria are derived from the fine structure of an impulse response: for example, from the first 10, 15 or 20 dB of the decay following steady-state excitation (so as to take special account of that portion of the decay that is not masked by subsequent signals); or from a distinction between useful (i.e. early) and detrimental (i.e. late) reflections, in order to get a measure for the distinctness of speech or the clearness of music.

We have already seen in Chapter II.7 that the multitude of such criteria demands that we select only the best, particularly since they differ from one another only slightly. Such a selection can be based only on the degree of their correlation with subjective attributes. The further question as to how far these criteria are independent of one another can again be answered best by factor analysis.

Such an analysis was first published by Lehmann,[1] followed a bit later by Gottlob.[2] Their results do not agree in all details but this is not surprising because, at least in part, their criteria were different. In both cases the criteria were based on echograms but certainly some differences were introduced because of differences in their test objects.

In spite of these differences the two studies agree in their essential results. Both authors needed only two factors to explain 90% (for Lehmann) or 80% (for Gottlob) of the total variance. Also in both representations the vectors in the factor-plane, which are associated with objective criteria, call to mind the vectors of Fig. 2.26 which Siebrasse associated with the individual preferences of the different test persons. This relation is understandable if we imagine test persons who orient their judgments only according to the initial reverberation time and others who do it only according to the distinctness coefficient, and so on. Seen in this light we may regard highly correlated criteria as defining a single 'taste group'.

It is not surprising that in Gottlob's factor plane (see Fig. 3.1) the four different versions of reverberation time, which differ only in the different level ranges over which they are defined, should form such a group. They are designated here (as also in Section II. 7.1) as:

T reverberation time evaluated from -5 to -35 dB (thus over 30 dB);

T_A *Anfangs* reverberation time, evaluated from 0 to -20 dB;

T_I Initial reverberation time, evaluated from 0 to -15 dB;

T_E Early decay time, evaluated from 0 to -10 dB.

[1] Lehmann, P., DAGA, Stuttgart, 1972, p. 162, VDI Verlag.

[2] Gottlob, D., Dissertation, Göttingen, 1973.

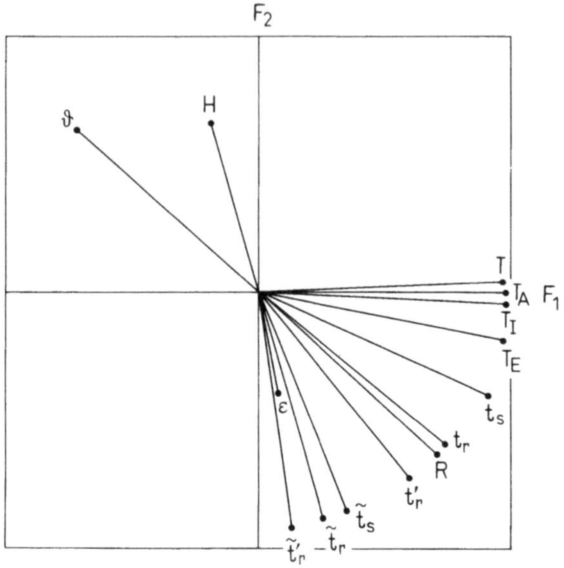

Fig. 3.1. Results of the factor analysis of objective criteria. (After Gottlob.[2])

It is understandable that the last of these quantities differs the most from the first quantity. Therefore if it is felt that a further measure of reverberation time is needed, beyond T (i.e. Sabine's reverberation time, with experimental restrictions), the choice should be for T_E.

In addition, those criteria that refer to distinctness (see Section II. 7.4) essentially form such a group:

ϑ	distinctness coefficient;
R	late-to-early sound index;
t_r	rise-time, according to Jordan;
t'_r	rise-time, according to Reichardt;
t_s	center-time.

Here it is irrelevant that in the definition of the distinctness coefficient the sign was chosen so that larger ϑ corresponds to more distinctness while for all the other criteria this holds for smaller values.

Since all the criteria in Fig. 3.1 are normalized to their mean values this difference leads only to a change in sign: to judge the relation of ϑ to the other criteria it would be better to plot the vector for $-\vartheta$. Here the vectors for $-\vartheta$ and R practically coincide, which is not surprising since these two quantities are unequivocally dependent on each other. The fact that the $-\vartheta$ and R vectors are not identical follows only from their non-linear relation, which always results in a correlation coefficient less than 1.

Since all the distinctness criteria are partly correlated with the reverberation time—in fact for an ideal exponential decay they would be fully dependent on one another—their group of representative vectors cannot lie perpendicular to that of the reverberation time (the F_1-axis); they are directed more or less at 45° to that axis.

If we eliminate this correlation by dividing t_s, t_r and t'_r by their expected values for an exponential decay (see Section II. 7.4, concluding table), as Lehmann did, we get the normalized (dimensionless) criteria $\tilde{t}_s$, $\tilde{t}_r$ and $\tilde{t}'_r$. The vectors associated with these normalized criteria lie nearly orthogonal to that of the reverberation time in Fig. 3.1. (See Section II. 7.4.)

The same is to be expected for the last two criteria represented in Fig. 3.1:

H reverberation 'distance' (see Section II. 7.3, eqn. (7.7a));
ε echo coefficient (see Section II. 7.5, eqn. (7.16)).

But these last two quantities are not highly correlated with the normalized criteria for distinctness; they are hardly related to them at all and also not to one another. Each of them introduces other influences. The reverberation 'distance' H depends, in contrast to R, on the volume of the room and on the distance between source and receiver. As for the echo coefficient, the decisive matter is the integrated exceedance of the mean reverberation decay.

This lack of relation to the other criteria is shown in Fig. 3.1 by the fact that the corresponding vectors are significantly shorter.

Now it is generally true that the vector characterizing one attribute i and having the components $w_{i1}, w_{i2}, \ldots w_{im}$, appears in a factor plane (F_1, F_2) with the length:

$$w_{i12} = (w_{i1}^2 + w_{i2}^2)^{1/2} < 1 \tag{3.1}$$

The length could be 1 only if w_i is explainable by factors 1 and 2 alone and if the so-called communalities $r_{ii} > 1$ (see Section III. 2.8) are not used. In fact it is surprising that the other vector lengths in Fig. 3.1 are all so nearly unity. This makes it possible to estimate their correlations from their differences in direction.

The lengths of w_{H12} and $w_{\varepsilon 12}$ are significantly shorter than the others, proving that these attributes can be only partly explained by the factors

F_1 and F_2. But this does not allow us to conclude that those criteria are unimportant. That the factors F_1 and F_2 do not explain H and ε, but do explain 90% of the total variance, can be attributed to the fact that the other criteria are highly intercorrelated—proving once again how the choice of the attributes influences the results of factor analysis. One may also conclude that too many criteria in the literature take account of more or less the same aspects of the subjective impression!

The same holds true for the choice of objects. Gottlob demonstrated this by plotting the distinctness coefficient against the reverberation time for all the objects (halls and seats) for which he had measured both quantities (see Fig. 3.2). If a certain combination of data occurred several times, the size of the corresponding data point was enlarged. Taking into consideration all of the objects, Gottlob found a negative correlation coefficient of −0·66, corresponding to the cosine of the angle between the ϑ and T vectors in Fig. 3.1. But if he considered only the objects in the favorable region around $T = 2$ s, marked by the two vertical dashed lines in Fig. 3.2, the correlation coefficient decreased to −0·35 and the vectors would have formed a correspondingly different angle in the factor-plane.

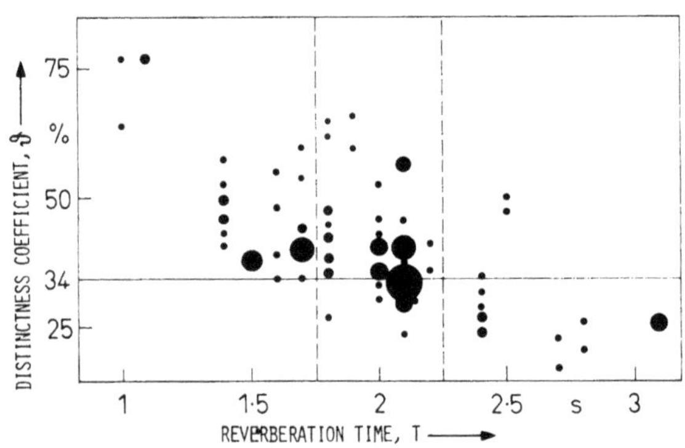

Fig. 3.2. Distinctness coefficients plotted against reverberation times, measured for many objects; when a certain combination appears several times, the size of the data point is made correspondingly larger. (After Gottlob.[2])

The exclusion of the higher reverberation times is justified here because the measurements were carried out in unoccupied halls with rather thinly upholstered seats, so that the reverberation times were unusually long compared to those for the occupied halls.

The exclusion of the shorter reverberation times is more doubtful. They appear typically in opera houses, where symphonic concerts are also sometimes performed and are very much enjoyed despite the short reverberation time.

One point was missed by Gottlob himself in the choice of criteria for his first investigation: none of them takes into account the distribution of the directions of the reflections. But his main goal was the interpretation of the subjective factors found by him and Siebrasse (see Fig. 2.24) by correlation with objective criteria. For this purpose he considers, in addition, the 'maximal interaural correlation coefficient' κ (see Section II. 7.6) that results from correlating the sound pressures at the left and right ears of a dummy head during an integration time of 50–100 ms with a maximal time delay of $\tau = 1$ ms. The smaller the value of κ, the more probable it is that the reflections come from outside the median plane, i.e. that they are more-or-less lateral.

The objective criteria considered so far, which can be nearly represented with only two factors, could not be sufficient to explain the four factors derived from the preferences (even for equal loudness). Therefore Gottlob dropped from consideration the two factors for objective criteria in Fig. 3.1 in his further studies of the relations between the subjective factors and the objective criteria.

We may mention here that Yamaguchi (see p. 580) had already tried to correlate his purely subjective judgments of acoustical quality with 44 (!) objective criteria. The large number results from his attempt to include the frequency dependences of the criteria at eight frequencies, as far as such a dependence was to be expected. We do not discuss his results in detail here because of the doubtful electroacoustical methods of storing and reproducing the signals, and because his data pertain to only one hall. Nevertheless it should be mentioned that Yamaguchi evaluated the correlations while executing a common factor-analysis of his subjective factors and the objective criteria; this led to new factors and reduced his purely subjective factors to only three, at the same time demonstrating which of the objective criteria are most strongly correlated with the subjective factors.

In contrast to this procedure Gottlob adopted a four-dimensional factor-space as reproduced in Fig. 2.26; but he considered only the plane (F_1, F_2) for the two most important factors. Here F_1 represents the 'consensus-factor'.

In order to represent adequately the objective criteria in a plane of factors that was not determined by them, Gottlob chose for the directions of the corresponding vectors those where the correlation coefficient ρ for the measured data projected on that axis becomes maximal, and he took this coefficient itself for the length of the vector (see Fig. 3.3).

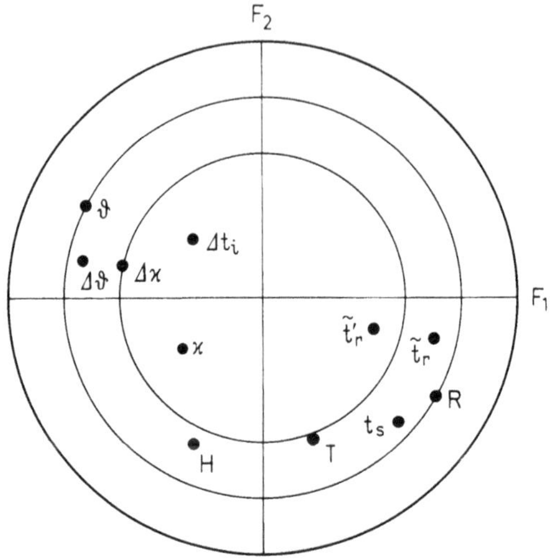

Fig. 3.3. Representation of objective criteria in the factor-plane F_1, F_2 of Fig. 2.26. (After Gottlob.[2])

This choice brings to mind factor analysis, insofar as the length of the vector which there represents the attribute under consideration (here, the objective criterion) is shorter the less well it is correlated with the factors that define the plane. Even less certain here is the calculation of the maximal correlation coefficient ρ. Gottlob makes this clear with the use of three circles, as shown in Fig. 3.3. The outer limiting circle corresponds to $\rho = 1$; no criterion reaches this value. The middle circle corresponds, for the number of objects under consideration, to a level of significance with a probability of error of 1%, the inner circle to 10%. Criteria with corresponding correlations inside the latter circle must be regarded as rather uncertain.

This means that the indicated criteria of distinctness are significant and furthermore are highly correlated with the consensus factor F_1. These aspects of distinctness and clarity clearly play an important role. Thus it is all the more surprising that the distinctness coefficients and the consensus factor are correlated with a negative sign—signifying that the test persons preferred less distinctness. This is in contrast to the results of Reichardt (Section III. 2.5), with respect to the desirable minimum for his clarity index, and with the results of Wilkens' factor analysis. It also contradicts the experience of the authors.

Gottlob himself thinks it is at least possible that very small values for ϑ, such as may be expected in very reverberant rooms, are unfavorable and that it may be better to introduce, instead of ϑ, the absolute deviation from an optimal value of ϑ:

$$|\Delta\vartheta| = |\vartheta - \vartheta_0| \tag{3.2}$$

He refers here to the investigations of Beranek and Schultz,[3] who found for the 'late-to-early-sound index' (which can be unequivocally derived from ϑ) an optimal value which would correspond to an optimal value of $\vartheta_0 = 0{\cdot}34$. Since $\Delta\vartheta = 0$ corresponds to the optimum, $|\Delta\vartheta|$ can be correlated to the consensus factor, assumed to increase with increasing acoustical quality, but with negative sign since $|\Delta\vartheta|$ increases as we move away from the optimum.

We have discussed this problem in some detail because it shows a further shortcoming of factor analysis; namely, it assumes linear relations between all the attributes and the factors. Thus it can lead only to those quality aspects that increase or decrease monotonically with a quantified sense of 'desirability'. Criteria that are characterized by optimal values must first be transformed into functions that correspond to this condition; but here the choice of function is at the author's own discretion. Surely the absolute deviation from the optimal value, as in eqn. (3.2) with its sharp break rather than a rounded maximum at the optimal value, seems an inadequate description of the actual psychoacoustical conditions.

We are sure that reverberation times that are too short as well as those that are too long are perceived as unfavorable, and that between these extremes there is a rather flat optimal region. Therefore it is not surprising that T appears at the inner circle in Fig. 3.3. Gottlob's introduction of the function $|\ln (T/T_{opt})|$ does not eliminate the principal objection to such a procedure and we refrain from further discussion of the question here.

But it merits interest that for the criterion κ, the 'maximal interaural correlation coefficient' that Gottlob used to characterize the directional

[3] Beranek, L. L. and Schultz, T. J., *Acustica*, **15** (1965) 307.

distribution of the reflections (see Section II. 7.6), the introduction of the absolute deviation from an optimal value

$$|\Delta\kappa| = |\kappa - \kappa_0| \tag{3.3}$$

appears to be successful, as shown in Fig. 3.3. Gottlob chose for the optimum value $\kappa_0 = 0{\cdot}23$. At least it shifts the corresponding point from well inside the inner circle for κ to the inner circle itself for $|\Delta\kappa|$. Gottlob regards this as proof of a good correlation of $|\Delta\kappa|$ with the consensus factor. Again it is negative because $|\Delta\kappa|$ increases with increasing deviation from the optimum. One consequence of this correlation would be that both a lack and a preponderance of lateral reflections must be regarded as detrimental.

On the other hand, both Keet and Barron found that the 'short-time correlation coefficient' (designated $\kappa_{(0)}$ and defined slightly differently from κ, see Section II. 7.6, eqn. (7.25)) is suitable as an objective criterion that may replace the 'apparent source width' of Keet, the 'spatial impression' of Barron, or the *Raumlichkeit* of Kuhl. They even accepted a simple linear relation $(1 - \kappa_{(0)})$ as a suitable representation of this room-acoustical impression.

But since this spatial impression depends on the level, the linear relation must also be level-dependent.

Kuhl[4] investigated this matter by personal observation, listening to the Berlin recordings described above with a four-loudspeaker arrangement.[5] He found that the scale steps from 0 to 4 that he chose for the sensation of *Raumlichkeit* increased linearly with the reproduction level: one step corresponded to a level change of 5 dB. A hall, or a seat in a hall, where a given source power on the stage leads to a higher observed sound level leads also to a stronger 'spatial impression' for a given value of $\kappa_{(0)}$. Nevertheless the 'spatial impression' must be regarded as a further independent room-acoustical criterion because the same step on the spatial impression scale may be achieved in one hall with soft music, but in another hall only with loud music.

A most exciting possibility is that the 'spatial impression' is confined to the stage in piano passages, but as the music becomes louder more and more reflections from different directions become evident and the music 'fills the hall'. It is still not yet known how much weight should be

[4] Kuhl, W., *Acustica*, **40** (1978) 167.
[5] Kürer, R., Plenge, G. and Wilkens, H., *Radio Mentor Elektronik*, (1973) 512.

attached to this 'dynamics of the hall' or whether there exist differences in musical taste in this respect.

So far we are altogether lacking systematic studies of the influence of the spectrum on this criterion. There is no doubt that both κ_{max} and $\kappa_{(0)}$ are spectrum-dependent, and it is not yet experimentally proved to what extent the sensation 'spatial impression' is correlated with this criterion.

Eysoldt,[6] who was able to change κ by electronic means for his comparison tests, felt confronted with the difficulty that such changes are always accompanied by changes in timbre; therefore he could not decide whether the expressed preferences were more influenced by the timbre or by the 'spatial impression'.

Inside the inner circle of Fig. 3.3 we also find the 'initial time delay gap' Δt_i between the direct sound and the first strong reflection, a quantity to which Beranek gave an especially high weighting (see Section II. 7.3). Gottlob has even made this concept applicable to echograms by requiring that the 'first reflection' peak should not be more than 10 dB below that of the direct sound. The rather small weighting given in Fig. 3.3 to this criterion, which was at first widely accepted because of its simplicity, may be explained by the fact that it does not take into account the integration ability of the ear and the direction of the first strong reflection.

Finally Gottlob also introduced, as further objective criteria, certain geometric data of the halls such as the volume, the source–receiver distance (which was always chosen proportional to the length), the width and the height. These are not represented in Fig. 3.3, partly because they would have appeared inside the inner circle, but mainly because the width and height cannot be defined unequivocally in all halls: these dimensions are practically restricted to simple rectangular rooms. Furthermore each geometrical configuration represents only one physical possibility of realizing special acoustical requirements; the dimensions are not mandatory criteria for those conditions.

In contrast to Yamaguchi and Gottlob, who related their objective criteria immediately to the subjective quality-factors found in the preference tests, Lehmann tried to correlate the subjective factors found by Wilkens by the use of his 'contrast pairs', see pp. 585–6, with adequate objective criteria.

Beranek had already interposed between the quality categories and the objective criteria certain specific subjective impressions, such as intimacy

[6] Eysoldt, U., Dissertation, Göttingen, 1976.

or warmth, largely because musicians find it difficult to accommodate the quantitative objective criteria to their own qualitative terms. But he had no other choice than to present such combinations according to his own personal discretion. As clear and useful as these proposals have been for acousticians, only correlation tests with extensive and tedious comparisons of different halls could provide them with a firm scientific basis.

With respect to Lehmann's results we may begin with those factors where no differences between the 'taste-groups' appeared. Here the factor that explains the most variance concerns the sensation of 'distinctness'. Therefore it is not surprising that Lehmann found a high correlation between the scores for this factor and the measured center-times for the different objects (see Fig. 2.22), as shown in Fig. 3.4(a). (The special choice of t_s was the result of his comparison of the correlations of factor 2 with other objective criteria, which all showed smaller correlation coefficients.) The straight line corresponds to the least-squares deviations in the direction of the ordinate, a so-called linear regression.

In the same way, both groups agreed with respect to the third factor concerning the influence of the hall on the timbre. Clearly, here the frequency-dependence of an objective criterion was in question and at least an analysis in octave intervals seemed to be sufficient. The reverberation times T_0 to T_3, which had already been used in defining Beranek's 'warmth' (see Section III. 2.3), could have been used. Instead Lehmann used the 'early decay times', for lack of sufficient level-differences at low-frequencies to allow accurate determination of the T's. But this change may actually be subjectively preferable, because we are more interested in the changes of timbre during running music than in the decays after the rather infrequent full stops. Furthermore he took as his mathematically general criterion the slope, T'_E in s/oct., of the best-fit (least squares) straight line through the data points, including T_4 for a frequency of 2000 Hz. For this slope the following formula holds:

$$\bar{T}'_{E,04} = \frac{1}{10} \sum_{i=0}^{4} (i+1) T_{E,i} - \frac{3}{10} \sum_{i=0}^{4} T_{E,i} \tag{3.4}$$

In Fig. 3.4(b) the timbre-factor 3 is plotted against this quantity.

The fact that the factor values scatter so far from one another—more than in Fig. 3.4(a)—does not imply a large spread or uncertainty in the measurements but is rather a consequence of the normalization of the factors, which magnifies the differences in those judgments that make use of only a small region of the scale. On the other hand, it should be mentioned that the slope of the $T'_{E(\log f)}$ line represents a criterion with

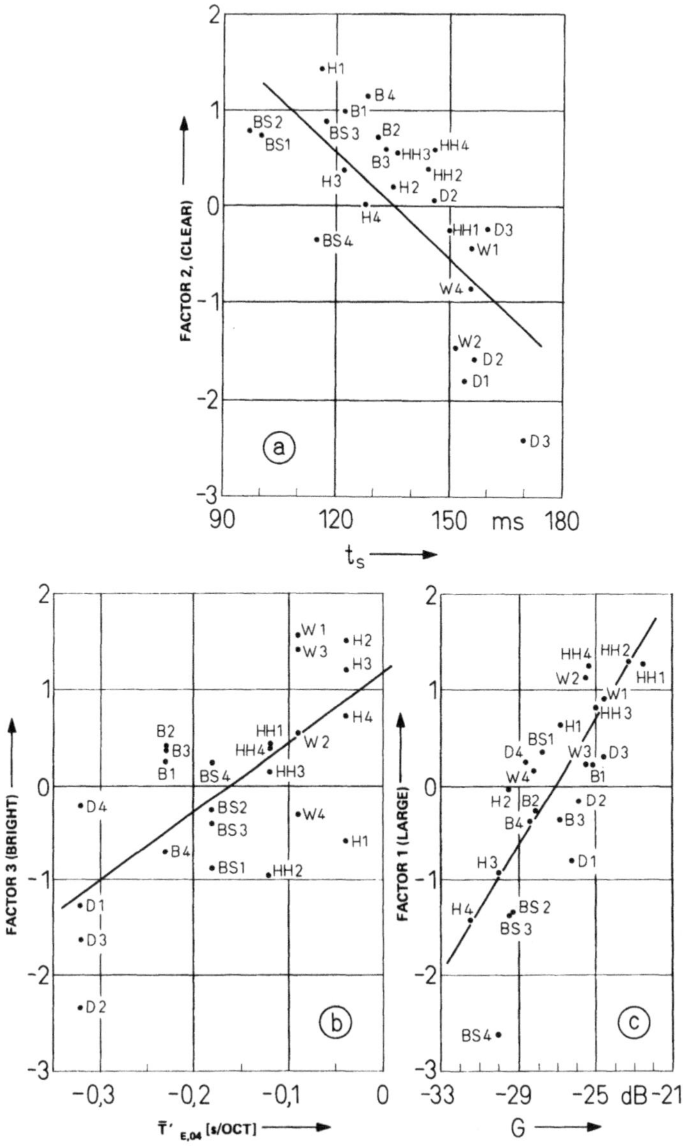

Fig. 3.4. Correlation of the subjective factors of Fig. 2.30 (ordinate) with objective criteria. (a) Distinctness-factor plotted against center-time. (b) Timbre-factor plotted against $\bar{T}'_{E,04}$, according to eqn. (3.4). (c) Strength-factor plotted against G, according to eqn. (3.5). The data points in this figure are designated by a combination of the city-name abbreviations (see p. 562 and Fig. 2.22) and the seat number in the hall.

an optimal value and is therefore not immediately suitable for correlation with a quality factor.

For the last factor—which is the one that explains the greatest portion of the variance, according to Wilkens—we consider that one for which two 'taste groups' could be defined and which Wilkens characterized as 'strength and extension of the source'. If we take into account only the 'strength' aspect, which to some extent (according to Wettschureck[7]) also implies extension of the source, it is obvious that the relative levels of an echogram recorded at a receiver position are not adequate, but that the absolute values of the sound pressures must be taken into account. And these must be compared with a sound pressure that is proportional to the square root of the radiated sound power. For the latter, Lehmann chose—for his spark source 1·5 m above the floor of the stage—the pressure of the direct sound arriving between $t=0$ and $t=\Delta t$, measured at a distance x_0 of 5 m; he normalized this measured pressure to a distance where the area of a sphere around the source would be 1 m^2. In this way, he determined the strength coefficient that we have already mentioned in Section II. 7.3, as well as its logarithmic form:

$$G=10\log\left[\int_0^{\infty} p^2(x,t)\,dt \bigg/ \left(\int_0^{\Delta t} p^2(x_0,t)\,dt\right)(4\pi x_0^2)\right]\text{dB} \qquad (3.5)$$

with x_0 in m.

Lehmann found these quantities adequate for factor 1, because (he stated) the impression of 'strength' is not produced only by the direct sound, nor even by the 'useful' sound, but rather by *all* of the reflections arriving at the listener's position. Figure 3.4(c) shows the good correlation between the 'strength-factor' and the strength-index G. Here again the strength-index was found to have the highest correlation coefficient of all the criteria that he considered, even higher than the strength coefficient, i.e. the argument of the logarithm in eqn. (3.5).

Since in general the center-time increases with increasing G, we may assume that the group that prefers greater strength is willing to accept less distinctness. (The additional diminution of distinctness at higher sound levels, to be expected from Wettschureck's investigations, is not taken into account here.) The proof that group I is more interested in G while group II is more interested in t_s was provided by Lehmann's calculations of the correlation coefficients between the attribute 'pleasing'

[7] Wettschureck, R., Dissertation, TU Berlin, 1976.

and the criteria G and t_s for both groups. He got:

	G	t_s
Group I	0·66	0·46
Group II	−0·55	−0·73

The greater interest of group I in G and group II in t_s makes it reasonable to plot the attribute 'pleasing' against G in the first case and against t_s in the second (see Fig. 3.5). The object points do not depart very much from the linear regression lines.

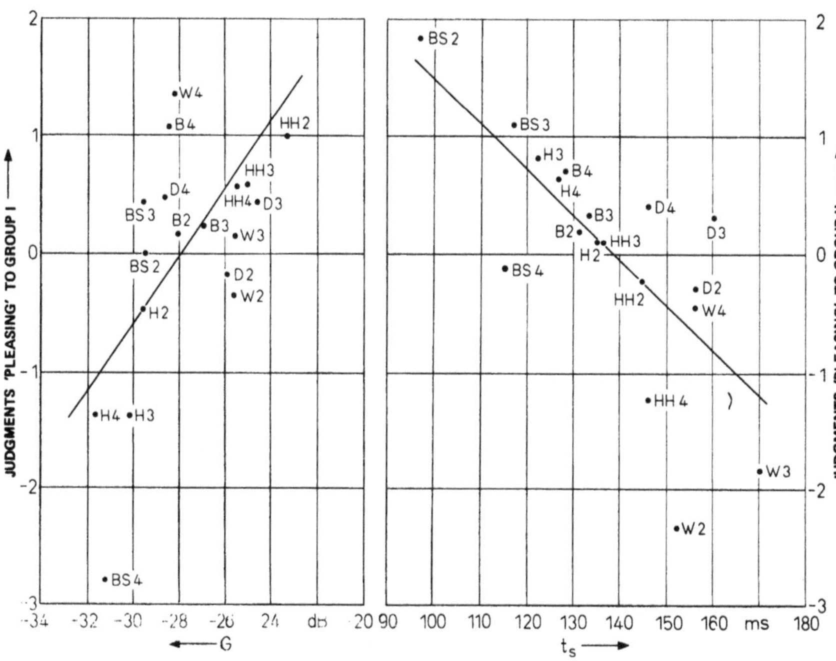

Fig. 3.5. Relation between the attribute 'pleasing' with the objective criteria. Left: for Group I with G, according to eqn. (3.5). Right: for Group II with the center-time t_s. (After Lehmann.[1])

III. 3.2 Tasks and Options in Acoustical Consulting

There is no doubt that the investigations based on binaural recordings of room-acoustical situations, judgments of these examples by numerous test persons, and evaluation of the judgments by factor (or other correlation) analyses have led to progress from the scientific standpoint, because the results are free from non-acoustical influences and from arbitrary interpretation. Nevertheless the hope has not been fulfilled that this effort and expense would result in generally accepted guidelines and in more reliable predictions of the acoustical quality of concert halls.

There are still differences in the ranking of the subjective aspects and the corresponding objective criteria. In Göttingen the distinctness and the distinctness coefficient occupy first place, while 'spatial impression' and the binaural correlation coefficient take second place. In principle the same rank-order is recognized in Dresden, although there another criterion, the 'clarity index', is preferred and other methods for evaluating the directional distribution of early reflections are under discussion.

In Berlin the strength factor is regarded as the most important (i.e. it explains the greatest portion of the variances); but this factor is subject to differences in the tastes of two groups of test subjects. The distinctness factor, best correlated with the center-time, takes second place; the timbre factor takes third place. Since the last two factors exhibit no differences in taste among the test subjects they could be combined into one consensus factor.

More serious is the discrepancy that shows up in the opposing recommendations with respect to distinctness. Here Berlin and Dresden agree that greater distinctness means higher acoustical quality and that at least a minimal value of the clarity index should be exceeded (or a maximal value of the center-time should not be exceeded). In Göttingen lower values of distinctness were preferred in general, although Gottlob now concedes that there may be an optimal value of the distinctness coefficient and that his first conclusion may have resulted from the fact that most of the test objects (halls and seats) presented large values of ϑ, the distinctness coefficient. Thus the discrepancy mentioned above might be a consequence of a choice of different test objects in the different investigations. But it may also be true that the use of loudspeakers as sound sources (in the Göttingen studies) presented room-acoustical impressions that approached normal orchestral sound only for lower values of distinctness.[1]

[1] In a recent paper (*Acustica*, **45** (1980) 256, Fig. 9), Lehmann and Wilkens compared, for the same musical example, the mean spectra of dry-recorded loudspeaker-reproduced sound with the sound of an orchestra performance in the Berlin Philharmonie in a seat near the stage. They found that the first spectrum contains fewer low-frequency components. They conclude that this reduces the apparent source-width and that there is no need for more distinctness or for preserving it.

This could also explain the desire for more lateral reflections. On the other hand, we cannot dismiss the possibility that the reproduction of the sound by earphones in Berlin at that time, which did not always give a very satisfying frontal localization, may have resulted in a preponderance of the perception of lateral reflections and therefore a requirement for increased distinctness.

Here, obviously, a number of questions remain to be clarified. We must recognize that dummy heads and correlation analyses by computers have been in use for only a short time in investigating the problems of room-acoustical quality; the procedures and conclusions are subject to change. On the other hand, a book about the principles of a science must deal with the most recent trends, and therefore it runs the risk of becoming quickly out of date. Indeed part of its task is to hasten this process!

Despite the fact that at the moment acoustical experts do not agree on all details about what is most important and desirable, it would be altogether wrong to conclude that it is pointless to apply acoustical considerations to the design of concert halls. The disagreements that we have been discussing are exceptions. For most of the questions that the acoustical consultant must answer there is clarity and unity.

Moreover, there is one essential point on which the investigations evaluated by factor analysis are in agreement: namely, an unequivocal rank-ordering does not exist. There may be differences in taste with respect to the *n* factors that are needed to explain a high percentage of the variances. On this point Yamaguchi, Siebrasse and Wilkens agree, in spite of the fact that the first considers three, the second four and the last two preference-factors as necessary.

This means also that there is no optimal shape for a concert hall, despite the fact that the acoustical consultant is always expected to recommend the 'best configuration' at the start of design of a new hall and that such recommendations can be found in the literature. If such an optimal shape did exist, and if there were unequivocally optimal choices of the surface materials, then it would be possible to 'standardize' the concert hall—which would at least have the advantage that orchestras on tour would find the same acoustical conditions everywhere. Such a possibility of perfect exchangeability of halls would also make the job of the recording engineers easier, who nowadays are accustomed to exchangeability in their standardized apparatus. As a matter of fact we will become acquainted with an example of such standardization in the next section; but it is restricted to conference and lecture rooms with volumes less than $1000\,m^3$ where only unamplified speech is of interest, rooms that turn up very often.

For theaters and concert halls, however, such a standardization would be the end of architecture. Variability in these facilities is attractive not only for the eye but also for the ear. This variability is also justified by the reasonable assumption that the acoustical optima, if they exist at all, are at least rather broad so that it becomes most important simply to avoid exceeding certain limits. Even the undoubted existence of different tastes supports the principle of variation in design.

Certainly there exist particular solutions that have proved themselves. One of the authors has therefore been asked why such examples should not be copied. The fact is that economic considerations immediately make this impossible because of changed construction techniques and changed building codes, not to speak of the effrontery of asking today's architects to mimic, for example, the stucco decorations of the 19th century.

We know, today, how far we may deviate from such successful prototypes without risking significant acoustical differences; but this will not prevent certain listeners from complaining that the success of the prototype was not achieved. Such convictions are especially to be expected if the prototype no longer exists. It is, then, hardly by chance that concert halls that were destroyed in World War II, such as the Leipzig Neues Gewandhaus, the old Berlin Philharmonie, the Munich Odeon or the London Queen's Hall, nowadays enjoy a posthumous reputation for acoustical quality that is unlikely to be achieved again.

But the effort to create an 'optimal hall' in spite of the differences in acoustical tastes overlooks the much more modest task that confronts the acoustical consultant in general. It is not he who creates the hall, but the architect—who is much more interested in the visual aspects of the building where he can express both his particular gift for shaping space and his flair for decoration. Usually he has already settled upon a design concept for the building before he contacts the acoustical consultant. Often he is committed to carry out the design scheme that, by a jury decision, won him the job. If the acoustical consultant is invited to attend the deliberations of such a jury he does so only as a technical assistant, but he does not play a decisive role.

On the other hand, the architect expects that his consultant will endeavor to realize his visual concept as far as possible, and that he would refuse only obvious and serious mistakes, such as concave curved surfaces, excessive hall width and seats deep under balconies without early reflections.

The wishes of the acoustical consultant are often in conflict with the architect's preferences.

The same applies to the recommendations of the other experts on whom the architect must rely, and finally also to the restrictions imposed by the building costs; for generally much more money is spent to please the eyes than the ears. Here again the acoustical consultant must vigorously defend his standpoint, if he is sure that otherwise serious mistakes will be made. With respect to these matters there is no disagreement among the acoustical experts.

As a result there is hardly any hall whose acoustical consultant would not have done things a bit differently if he had been in full control.

But his task is not only defensive in nature. He may suggest changes or even alternative approaches to the shaping of the hall that will improve the acoustical conditions in accordance with acknowledged rules. If he succeeds in those purposes and in the avoidance of mistakes, he will have done all that is required and all that is possible for him.

Insofar as his work includes guarantees to meet acoustical criteria he must propose those himself. (We exclude here measures of noise control, where maximum acceptable noise levels are more or less standardized.) Sometimes he is asked to guarantee particular reverberation times within a tolerance of 0·1 s; such guarantees are common in other fields such as lighting, heating, etc. But what is really requested in the acoustical consultant's field is good hearing conditions, and he himself must decide upon the appropriate criteria for each special case.

If he is asked to guarantee particular reverberation times—and this term has come to be more and more well-known to clients, architects and musicians—he must regard the proposed figures very critically. Not only are reverberation times sometimes requested that are quite impossible to achieve on account of the small room volume or the sound attenuation in the air; but he must even question whether the requested values for T are consistent with the expectations of the client or the proposed purposes of the room. Here he may also have to resolve conflicting wishes. For example, the performing artists often prefer longer reverberation times than the listeners (if they were asked). For the artists there is no risk that the extra reverberation will spoil the transparency of the music, because they are in the neighborhood of the sound sources; also they are quite familiar with the work being performed. On the other hand, they usually like to hear some room response.

In short, the acoustician must attempt to translate the wishes of the client and the future users into objective criteria. He may even be able to guarantee certain figures within reasonable tolerances, but only under the condition that all of his requirements are fulfilled. Since this is almost never the case, all claims for liabilities and for compensation by insurance companies are doubtful. The acoustician can be made responsible only for carelessly or willfully ignoring the present state of science and technology, the proof of which could only be given by another acoustical expert. In no case can the acoustical consultant guarantee that all subsequent users of the hall will be satisfied!

III. 3.3 Recommended Values for the Reverberation Time

After W. C. Sabine's pioneering work, the reverberation time that he introduced was regarded until the 1940s as the only measurable criterion for acoustical quality. But even when it became evident that this could not be the only criterion of quality and other criteria were offered again and again as supplements, the reverberation time remained the only criterion for which guideline values were presented in textbooks and tables. As we have already mentioned in Section III. 3.1, this resulted from the fact that the reverberation time is still today the only acoustical criterion that can be calculated or even estimated in advance, based on the hall drawings and known sound absorption coefficients. No responsible acoustical consultant would fail to estimate the values of reverberation time for the important frequencies between 125 and 2000 Hz and perhaps even at 4000 Hz. (The value at 63 Hz would also be very interesting for musical purposes but there exist almost no reliable data on sound absorption coefficients at that frequency, because most reverberation rooms are too small for accurate measurements.)

We will first discuss recommended values for reverberation time based on experience, rather than on subjective tests evaluated according to factor analysis (see the discussion in Section III. 3.1, following eqn. (3.2), on some shortcomings of factor analysis).

It is obvious that there may be values of reverberation time that are too high with respect to the distinctness of speech and the clarity of music; also the acceptable upper limit of reverberation time depends on the kind of performance. We found the same difference in the limit of perceptibility for sound reflections: 50 ms for speech, 80 ms for music.

It follows therefore that recommendations for suitable reverberation time will depend on the intended purposes of the room: whether it is to serve for speech only, as in conference or lecture halls or in drama theaters; or whether it will be especially (or additionally) used for musical performances, as in concert halls, opera houses, churches, or multi-purpose auditoriums.

Since the laws of sound propagation, including reflection and absorption, are independent of the purpose of the hall, we took no account of such distinctions in Parts I and II of this book. But they become essential now in the discussion of guideline values for acoustical criteria. Furthermore, in speech rooms it is extremely important whether electroacoustical speech reinforcement equipment will be used or not.

In principle this would also be true for music rooms. It is even surprising, given today's standards of electroacoustical technique, that these aids are usually operated in such a way that they are conspicuous, and even near the limit of electroacoustical feedback; such systems are sometimes even operated at sound levels that are dangerously high with respect to potential hearing damage. But perhaps it is this very misuse that dissuades artists, who are able to 'fill the hall' without electroacoustical reinforcement, from taking advantage of a bit of 'acoustical make-up' that would be imperceptible to the listeners. This need not remain so.

The following recommendations for rooms used for music or for speech concern only those cases where natural sound sources are in use.

Reverberation times that are too low can be detrimental for two reasons. The first assumes that, both for distinctness and clarity, there are upper and lower acceptable limits for reverberation time, with a broad optimal region in between. Even speech offered without reverberation may sound unattractive; but especially music without reverberation is perceived as too 'dry'. Particularly the Gregorian chant, developed in large cathedrals, not only tolerates a long reverberation time but actually demands it. In this case as in all *a capella* singing the long reverberation may help to maintain the proper pitch. But there also exists an upper limit, and both limits are independent of the volume of the hall.

It is different with respect to the second reason for the lower limit on reverberation time, which seems to be the more important according to the subjective investigations discussed in Section III.3.1. This regards the contribution of sound reflections to the loudness.

We became acquainted with this question in Section III.2.2 in connection with the problem of articulation tests. There it was problematic that Knudsen included not only the early but also all the late reflections. For music, however, this seems to be justified, according to Lehmann's correlation test between the subjective impression of strength and the strength-index, which takes account of all reflections. This means that the energy density in the steady state is decisive; for a given sound power, in statistical room acoustics, the energy density is proportional to the ratio of reverberation time to the room volume (see Part II, eqn. (1.12)). Maintaining a minimal value for E would thus lead to a minimum acceptable value of reverberation time which increases linearly with the room volume.

Such a strong increase of T with V would not be possible to realize with room-acoustical means alone, quite apart from the fact that it would quickly come into conflict with the upper limit on reverberation time. As we have explained in Section III. 2.3 according to the law of similarity, for equal room shape and equal mean absorption coefficient, the reverberation time can increase only with $V^{1/3}$. The formulae that Watson developed on the basis of calculations for real concert halls showed, by the addition of a constant term, an even smaller increase which could be explained by deviations in room shape caused by the relative restrictions in height in larger building construction. But it is certain that the requirement for sufficient strength of sound sets a lower limit on the reverberation time which depends on the room volume.

Therefore we will take over the dependence of the lower limit of reverberation time on the room volume and will do this separately for the different purposes of the hall. But we will have to accept such a dependence for the upper limit also, because it is physically unavoidable. For both limits, however, we will adopt smaller slopes in the corresponding log T–log V diagrams than would correspond to the law of similarity.

If we assume that for most hall volumes we have already exceeded the upper limit of T and are therefore interested in decreasing T, but that we would like, on the other hand, also to increase E, then we must look for a proper balance between the relative changes $-\mathrm{d}T/T$ and $\mathrm{d}E/E$. Since the difference-limen for the first is 0·04 (see Section III. 2.1) and that for the level is about 1 dB (i.e. $\mathrm{d}E/E=0{\cdot}26$), we find an adequate balance given by

$$-\mathrm{d}E/E=6\ \mathrm{d}T/T \tag{3.6}$$

On the other hand, since E is proportional to T/V we have

$$\mathrm{d}E/E=\mathrm{d}T/T-\mathrm{d}V/V \tag{3.7}$$

Equations (3.6) and (3.7) together result in:

$$0=7\ \mathrm{d}T/T-\mathrm{d}V/V \tag{3.8}$$

that is, in:

$$\log (T/T_1) = (1/7) \log (V/V_1) \tag{3.9}$$

The straight lines in Figs. 3.6, 3.8 and 3.10 correspond approximately to this slope.

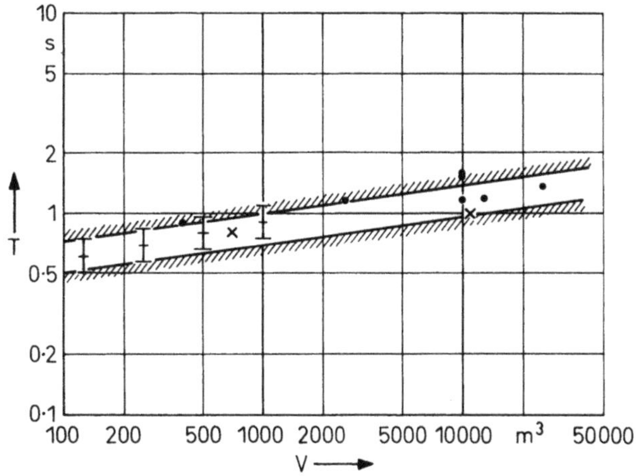

Fig. 3.6. Region of recommended reverberation times for speech rooms. + according to DIN 18041; × according to Knudsen; ● according to measurements of $T_{2,3}$ presented in Table 6.

They mark lower and upper limits, instead of an 'optimal' reverberation time; not only because all calculated optima are rather flat, so that a rather wide tolerance is always permitted, but in principle there are only upper and lower limits with some 'room for play' between them that allows a certain freedom of choice (*variatio delectat*).

The fact that the plotted measurement points for mean values of the reverberation times T_2 and T_3 (for 500 and 1000 Hz) in good concert halls actually fill up this 'room for play' does not exclude the possibility of an optimal value. According to Beranek's rating scheme (see Section III. 2.3), a deviation from the optimal value can be compensated by better values for the other criteria. But in the view of the authors these data points justify the location of the limiting straight lines.

Figure 3.6 applies to small rooms used for unamplified speech only. To the extent that these are lecture rooms, conference rooms, courtrooms and the like, these are covered by a German standard[1] which prescribes

[1] DIN 18041, Beuth-Vertrieb, Berlin/Köln.

for the room volumes:

$$V = \quad 125 \qquad 250 \qquad 500 \qquad 1000\,\text{m}^3$$

the corresponding reverberation times:

$$T = \quad 0{\cdot}6 \qquad 0{\cdot}7 \qquad 0{\cdot}8 \qquad 0{\cdot}9\,\text{s}$$

with a tolerance of $\pm 20\%$. These proposals are included in Fig. 3.6.

The value of 0·8 s for a volume of 700 m^3, proposed by Knudsen on the basis of his articulation tests, lies within the recommended region, a bit below the middle; and even his value of 1 s for a volume of 11 000 m^3 is still inside the limits, although at the lower boundary. The other plotted points refer to the measured values given in Table 6, which also include much larger rooms.

Table 6: Reverberation time in occupied halls for speech.

Hall	*V(m³)*	*f(Hz)* 125	250	500	1 000	2 000	4 000	*Data source*
Council Chamber, empty, but with curtain closed	390	1·2	1·15	1·0	0·8	0·9	0·7	ITA
Large Physics Lecture Hall, ETH, Zürich	2 600	1·25	1·2	1·2	1·1	0·9	0·6	Furrer
Kresge Auditorium, Cambridge, Mass.	9 750	1·65	1·5	1·5	1·45	1·35	1·3	Beranek
Audit. Max. Univ., Hamburg	9 800	1·2	1·1	1·1	1·2	1·1	0·95	ITA
Benjamin-Franklin Congress Hall, Berlin:								
empty	13 000	1·9	1·85	1·4	1·2	1·2	1·15	ITA
occupied, calculated		1·6	1·6	1·25	1·15	1·15	1·1	
Aula Magna, Caracas	24 800	2·0	1·9	1·4	1·3	1·2	1·0	Beranek

We may raise the question here whether it makes sense to try to include rooms with such different volumes as 125 m^3 and 25 000 m^3 in the same guidelines, especially since, for the larger rooms, some electro-acoustical reinforcement will always be used and thus the reinforcement by natural room reflections becomes irrelevant.

The question of when such reinforcement becomes necessary depends on the quality of the speaker and on the background noise level. It is astonishing how well speech is understood in drama theaters with a capacity of 2000 seats or more (in Epidauros, 14000!) without any amplification, because the artists speak with clear articulation and the listeners are careful to be as quiet as possible. By contrast, in congresses we have to take into account not only bad speakers—in the purely acoustical sense—but also a constant coming and going of the delegates and the not-so-*sotto voce* conversations among neighbors; in short, a rather high background noise level. It will be even higher if there is entertainment for the attendees. And an unnecessarily loud public address system can further increase the noise level.

As we have already emphasized in Section III. 2.2, loud speech does not necessarily lead to better understanding, even when the sound reinforcement system is not overloaded. But often we must observe with regret that lecture rooms for only 300 persons are senselessly burdened with loudspeakers. The layman often assumes in such cases that the lack of intelligibility is the fault of the room.

Now it is true, independent of the room size, that amplified speech sounds better and more natural, the more the room is designed so that a good speaker without amplification can be understood in quiet conditions. In this respect the reverberation time limits shown in Fig. 3.6 can be applied whether or not electroacoustical sound reinforcement is provided.

The same holds for cinema theaters. It was earlier believed that they should have short reverberation times so as not to influence the reverberation times associated with the sound from the film (see Section II. 3.5). But it was soon found that the sound reproduced in over-dry rooms had an unnatural effect.

Theaters also, if they are used exclusively for drama, can tolerate reverberation times 10–20% higher than recommended in Fig. 3.6; on the one hand, the speech of actors does not lose too much in intelligibility, and on the other hand, dramatic speech in classical plays benefits from a certain 'ring' in the hall.

A much more important difference between a drama theater and a lecture room, with respect to the reverberation time, is that in a theater as in a dwelling the reverberation depends on the furnishings, specifically, the scenery and the draperies around the stage. We will come back to this matter when we discuss opera houses. The reverberation times recommended in Fig. 3.6 refer to conditions with the fire curtain closed and with a full audience.

Now the influence of an audience on the reverberation depends strongly on the kind of auditorium seats (see Section II. 6.6), even if, as is usual nowadays, they are upholstered. Moreover, the reverberation times of the empty rooms can be estimated with greater confidence and they can be measured with more care and time, particularly since, in the occupied hall, one must either rely on suddenly stopped fortissimo chords in the music for the reverberation signals, as in Fig. 1.1, Part II, or bother the audience with pistol shots, etc.

Therefore it is recommended that the hall be planned so that, in the empty room, the upper limit of reverberation time will be attained or even slightly exceeded, in which case the occupied room will still have a reverberation time above the lower limit. The reverberation time values in Table 6 all refer to the occupied halls; but since some of these values were measured in the occupied halls, and some were measured in empty halls and corrected for the occupied condition, and since different signals and methods for evaluation were used, Table 6 contains some uncertainties.

Such values should by no means be used to compare the acoustical qualities of the various rooms. They serve only as characteristic examples for the special kinds and volumes of the rooms.

Except for a few in large cities, most theaters must serve for both drama and opera. In opera the acoustics for the music takes priority over that for the text, which is usually not understood because of the orchestra anyway. But music not only allows a higher reverberation time than speech, it demands it; thus, partially for musical–aesthetic reasons and partially in order to achieve a greater strength-sensation, it is recommended that the values of $T_{2,3}$ (500–1000 Hz) should be greater than in pure speech rooms of the same volume. A reverberation time in the range from 1·4 to 1·8 s is suitable for opera.

Nevertheless, we find in Table 7 highly esteemed opera houses with values of $T_{2,3}$ in the range of 1·0 to 1·2 s, which would lie, according to their volumes, within the recommended limits for speech rooms in Fig. 3.6. If one of the measured sets of values for La Scala lies about 0·3 s higher than the other set, this is not a sign of measurement uncertainty but is the effect of the stage condition. For the opening concert, the stagehouse was closed off by an orchestra enclosure, while in the second case the stagehouse was coupled to the auditorium. This second case, in general, is quite undefined, as indicated in Fig. 3.7(a); it presents the reverberation times measured in the auditorium of the Munich National Theater with the proscenium curtains drawn back to show the open stage: once (the upper curve) with a reverberant stagehouse ($T_{2,3} = 4·5$ s measured

Table 7: Reverberation times for occupied opera houses.

Opera house	$V(m^3)$	*f(Hz)* 125	250	500	1000	2000	4000	*Data source*
Staatsoper, Berlin: before reconstruction	7500	1·4	1·2	1·0	0·9	—	—	Meyer Jordan
after reconstruction		1·3	1·0	1·0	1·0	0·9	0·9	Reichardt
La Scala, Milan: opening concert	10 000	1·2	1·0	0·9	0·9	0·8	0·8	Furrer
stage condition unknown		1·5	1·5	1·25	1·0	0·8	0·6	Reichardt
Covent Garden, London	10 100	1·2	—	1·1	—	1·1	1·1	Parkin *et al.*
Staatsoper, Vienna, stage condition unknown	11 600	1·5	1·5	1·4	1·2	1·2	1·15	Reichardt
Deutsche Oper, Berlin	10 800	1·65	1·6	1·6	1·35	1·2	1·1	ITA
Festspielhaus, Bayreuth: stage condition unknown	11 000	1·7	1·6	1·5	1·4	1·4	1·3	Reichardt
empty, fabric over the seats	11 000	1·65	1·9	1·85	1·9	1·8	1·7	Venzke
National Theater, Munich, main curtain closed	11 000	2·1	1·7	1·7	1·7	1·6	1·2	Müller

in the stagehouse), and once (the lower curve) with the stagehouse highly absorptive. Often these extreme values of reverberation contradict the kind of acoustics suggested by the scenery: the '*teure Halle*' that Elizabeth praises in 'Tannhäuser' usually is quite free of reverberation, whereas for the open-air festival in 'Die Meistersinger', the rather reverberant stagehouse makes it clear that this scene is not played outdoors—which is no disadvantage, musically!

Figure 3.7(b) shows some dependence of the reverberation time with normal variations in stage scenery. For the grand finale of the second act of 'Aida', the stage director used not only heavy draperies at the sides of the stage but also an extremely thick foam carpet on the floor, to allow

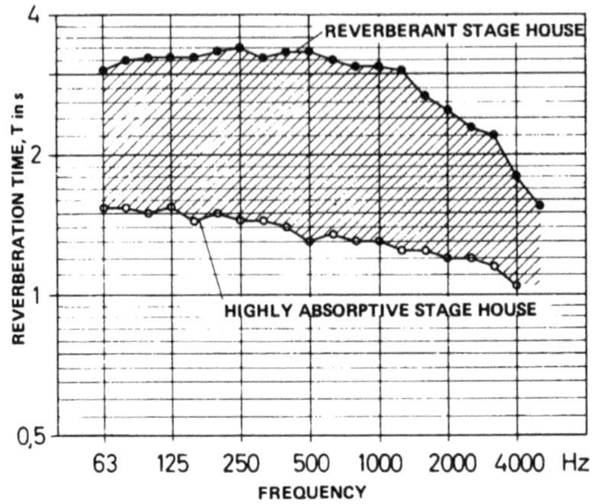

Fig. 3.7(a). Reverberation times in the Munich National Theater with open stage. (After Müller.)

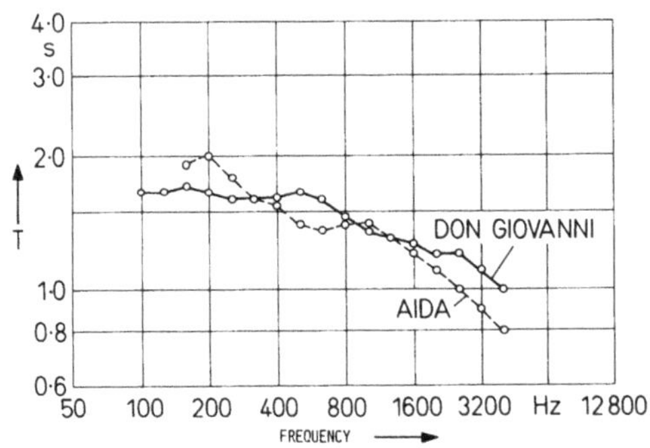

Fig. 3.7(b). Reverberation times for the occupied Deutsche Oper, West Berlin, with different stage scenery. (After Cremer, Nutsch and Zemke.)

the Egyptian warriors to storm wildly onto the stage without disturbing the music with their footsteps—an acoustical use of the scenery, no doubt, but without any consultation with the acoustical consultant. The lack of reflections from the sides and from the floor gave a deplorable sound.

But even if the voice of the acoustical consultant is heard, be it only in the sense of a 'veto', this influence is usually effective only for the opening performances in the house, after which the acoustics are at the mercy of the scenery and the production. The only remedy would be to require a minimum of acoustical knowledge from the stage manager and the producer.

For orchestral concerts in opera houses, in which the musicians occupy the stage, it is possible to make careful use of the longer reverberation in the stagehouse in order to approach an acoustical situation typical of a concert hall (see Section II. 3.4).

The low reverberation times of the opera houses of the 18th and 19th centuries result in part from the lavish use of plush furnishings and in part from the many balconies with deep sound-absorbent boxes. At first the amphitheater-shaped Bayreuth Festival Theater had longer reverberation times, as shown in Table 7 for two different conditions. This change in reverberation time corresponded to a transition from the Italian to the High Romantic style of opera. Since in other opera houses both styles must sound well, it was intended during the restoration or rebuilding of the houses that were destroyed during the war to provide longer reverberation times through the use of less heavily upholstered seats and the avoidance of absorptive walls and draperies. The result is shown in the lower half of Table 7 for the opera houses in West Berlin, Munich and Vienna. In Fig. 3.8 the limiting straight lines are therefore drawn so as to include only those examples with higher reverberation times.

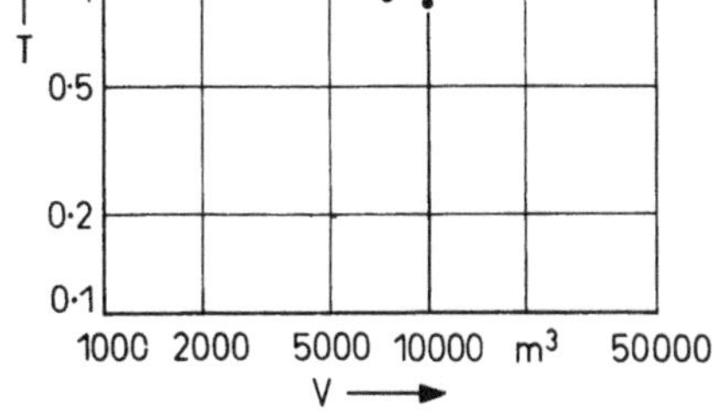

Fig. 3.8. Range of recommended reverberation times for opera houses.

In contrast to the opera houses with their many horseshoe-shaped balconies, the oblong concert halls with one, or at most two, balconies and a horizontal floor could more easily achieve a greater room volume per seat and thus a higher reverberation time—more easily even than in modern halls of the same seating capacity, because nowadays the sloping of the audience to provide good sightlines exposes the individual listener more fully to the sound field and therefore makes him more sound-absorptive.

Since it is usually difficult to attain the upper limit recommended in Fig. 3.9, the lower limit is the more critical in practice. Therefore it is often considered desirable to make the reverberation time as high as possible, and the hall with the higher value for $T_{2,3}$ is regarded as better. (In fact, this tendency is not entirely without its influence on the calculated and measured values of reverberation time!) It is clear, however, to anyone who has had to cope with empty halls with un-upholstered seats, that there are upper limits for acceptable reverberation time.

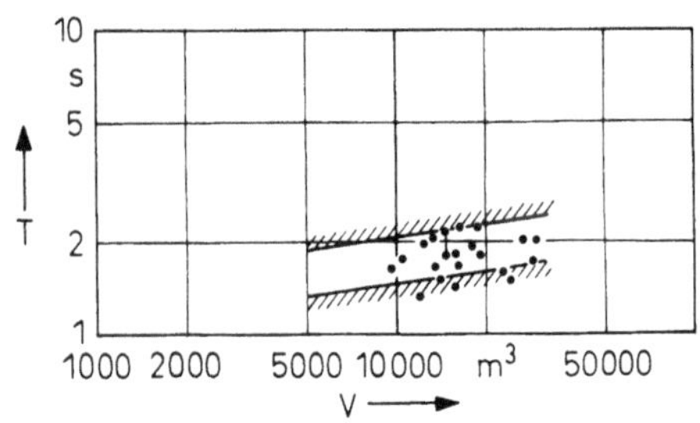

Fig. 3.9. Range of recommended reverberation times for concert halls.

Nevertheless, Table 8 includes several concert halls with excellent reputations, such as the earlier Queen's Hall in London and the Academy of Music in Philadelphia, whose low reverberation times bring to mind those of opera houses that are also sometimes used for concerts. The lower limit in Fig. 3.9 is therefore drawn only a bit above the corresponding data points for those halls.

The dependence of reverberation time on room volume has some special significance here. The smaller rooms are used preferably for solo recitals and for chamber music, while the larger rooms are used for full orchestra concerts. The question as to whether oratorios or other works with very many musicians can be performed depends not so much on the volume of the hall as on the size of the stage. This dependence is in agreement with the fact that solo recitals and chamber music require greater clarity and transparency than performances with full orchestra.

In practice the choice of a hall for a particular event is often more a question of selling tickets than of the acoustical qualities. Again and

Table 8: Reverberation times for occupied concert halls.

Hall	$V(m^3)$	*f(Hz)* 125	250	500	1000	2000	4000	*Data Source*
Musikhochschule, Berlin	9600	1·6	1·8	1·7	1·6	1·6	1·3	ITA
Konzertsaal, Turku	9600	1·9	1·75	1·6	1·6	1·55	1·3	Beranek
Stadtkasino, Basel	10500	2·2	2·0	1·8	1·6	1·5	1·4	Furrer
(Old) Queen's Hall, London	12000	1·3	—	1·3	—	1·3	1·2	Parkin *et al.*
Saal d. Senders Freies, Berlin	12500	2·3	2·2	2·0	1·9	1·8	1·4	ITA
Herkulessaal, Munich	13400	2·1	1·8	2·0	2·0	1·9	1·6	Müller
Philharmonic, Liverpool	13500	1·3	—	1·6	—	1·5	1·3	Parkin *et al.*
Neues Festspielhaus, Salzburg	14000	1·7	1·6	1·5	1·5	1·4	1·3	Schwaiger
Musikvereinssaal, Vienna	14600	2·4	2·4	2·2	2·1	2·0	1·3	Ravag, Vienna
Musikvereinssaal, Vienna	14600	2·8	2·0	1·9	1·7	1·4	1·3	Bruckmayer
Academy of Music, Philadelphia	15600	1·4	1·7	1·45	1·35	1·25	1·15	Beranek
Usher Hall, Edinburgh	15700	1·9	—	1·8	—	1·6	1·5	Parkin
Liederhalle, Stuttgart	16000	—	1.9	1·6	1·7	1·7	1·4	ITA
Beethovenhalle, Bonn	16000	2·0	1·7	1·7	1·7	1·7	1·6	Meyer, Kuttruff
St. Andrew's Hall, Glasgow	16100	2·1	—	2·2	—	2·1	2·0	Parkin *et al.*
(Old) Philharmonie, Berlin	18000	2·6	2·5	1·9	1·9	1·9	1·7	Meyer, Jordan
Symphony Hall, Boston	18800	2·2	2·0	1·8	1·8	1·7	1·5	Beranek
Concertgebouw, Amsterdam	19000	2·25	2·25	2·2	2·2	1·9	1·7	Geluk
Mann-Auditorium, Tel-Aviv	21250	1·55	1·5	1·55	1·55	1·5	1·3	Beranek
Royal Festival Hall, London[a]	22000	1·4	1·5	1·5	1·5	1·4	1·3	Parkin
Meistersingerhalle, Nürnberg	23000	2·2	2·05	2·05	1·9	1·7	1·5	Müller
Carnegie Hall, New York	24300	1·8	1·8	1·8	1·6	1·6	1·4	Beranek
New Philharmonie, Berlin	24500	2·4	2·0	1·9	2·0	1·9	1·7	ITA
De Doelen, Rotterdam	27000	2·3	2·2	2·1	2·2	2·2	1·9	de Lange

[a] Without 'assisted resonance'

again it happens that worthwhile musical performances are presented, not in halls that are available and appropriate to them, but in sports or exposition halls that are neither intended nor suitable for such performances, but that allow many more tickets to be sold and thus support higher salaries. The managers of such halls have a natural interest in praising them as being extraordinarily good!

If in opera houses the stage furnishings have a great influence on the reverberation time, in concert halls it is the number of musicians seated on the stage. This effect becomes most evident in a comparison of the empty hall with the hall occupied only by the musicians, a situation that is important for rehearsals and especially for recordings. We call this the 'studio state'.[2] In a performance of Beethoven's Ninth Symphony with which the Berlin Philharmonie was opened, the reverberation times for this studio state lay just in the middle between those for the completely empty hall and those for the hall fully occupied by audience and musicians.[3] Thus in Table 8 the two measurements for the Vienna Musikvereinssaal are not necessarily contradictory; it may be assumed that the larger values correspond to a small classical orchestra, and the smaller values to a large orchestra with chorus. With respect to the data for the Liederhalle, Stuttgart, they correspond to a performance of Orff's 'Carmina Burana' with a fully occupied stage. Thus, again as in Table 6, the reverberation times of Table 8 are not always comparable. (It is a regrettable paradox that chamber music, which needs shorter reverberation times, is performed on a nearly bare stage and thus with a longer reverberation time; on the other hand, oratorios, which were composed for the long reverberation times of large churches, fill the stage with singers in addition to the orchestra and thus lower the reverberation time.)

As the last group of performance spaces, we must mention churches; they comprehend the longest reverberation times. But this condition arises only from historical development out of a respected tradition. From the purely acoustical standpoint, the very long reverberation times (such as Lottermoser[4] measured in the Cloister Churches of Ottobeuren and Weingarten, see Table 9—and even these are not the longest!), are

[2] International Standard ISO 3382. This standard prescribes conditions for the measurement of reverberation time in concert halls; unfortunately, these were not known for most of the measurements in Table 8.

[3] Cremer, L., *Schalltechnik*, **14** (1964) No. 57.

[4] Lottermoser, W., *Acustica*, **2** (1952) 109.

Table 9: Reverberation times for churches.

Church	$V(m^3)$	*F (Hz)* 125	250	500	1 000	2 000	4 000	*Data source*
Kirche am Hüttenweg, Berlin, empty	2 500	3·0	2·4	2·4	2·2	2·0	1·7	ITA
Ev. Kirche am Lietzensee, Berlin, empty	3 200	2·1	2·5	2·6	2·7	2·0	2·1	ITA
Thomas-Kirche Leipzig:	18 000							
Before reconstruction		2·45	3·5	3·5	3·1	2·9	2·1	Meyer
After reconstruction								
empty		2·65	3·65	4·0	3·9	3·3	2·0	Kuhl
occupied		1·9	2·0	2·0	1·85	1·6	1·5	Kuhl
Klosterkirche, Weingarten								
empty	*c.* 90 000	7·0	8·5	8·0	6·5	5·5	4·1	Lottermoser
occupied		6·0	6·3	6·0	5·4	4·1	3·0	Lottermoser
Klosterkirche Ottobeuren, empty	*c.* 130 000	6·0	7·0	6·0	7·0	6·0	4·0	Lottermoser

suitable only for Gregorian Chant. They are too long for the polyphonic music for choir and organ which was composed at the same time.

Even if one had to take into account only the aesthetic requirements of church music, surely those conditions are preferable which Johann Sebastian Bach found in the Thomas-Kirche in Leipzig, the acoustical conditions of which have scarcely changed since his day.[5,6] Its reverberation time in the occupied state is less than the upper limit for occupied concert halls (see Fig. 3.9); but it would be wrong to see this as profaning the church music. On the contrary, conditions that have been found to be best on purely artistic musical grounds should be adopted for church services. It should not be overlooked that in Roman Catholic churches, from the Middle Ages to the times of the Renaissance and the Baroque, only the clerics assembled in the Choir of the church to follow

[5] Bagenal, H. and Bursar, G. *J. Roy. Brit. Architects*, (1930) 154.
[6] Keibs, L. and Kuhl, W., *Acustica*, **9** (1959) 365.

their Latin texts, possibly separated from the congregation in the nave by a rood screen. The sermon had only minor importance.

After the Reformation, however, the sermon delivered in the vernacular assumed primary importance in the church service. But the sermons would have been unintelligible had not pews and balconies been introduced at the same time, which reduced the reverberation times to values more suitable for speech. The same holds today for most Roman Catholic services as well.

It is clear, therefore, that the preferences expressed for church music by the choir director, the organist and (above all) the organ builder represent only one consideration in the choice of an appropriate reverberation time for churches; and this preference generally runs contrary to the requirement for clear intelligibility of the sermon and the prayers.

It is often argued today that the latter requirement can be fulfilled by the use of electroacoustical speech reinforcement systems. In large reverberant churches this is possible only with a distributed loudspeaker system; according to the law of the first wavefront, such a system destroys the localization of the sound at the position of the live speaker since the sound from the nearest loudspeaker determines the localization. Thus a congregation gathered together in a large church is acoustically separated from the preacher and from each other; with closed eyes they may even have the impression of listening to a radio broadcast. Although it is possible to avoid this effect through the use of delay systems (see Section I. 5.2), and lately such systems have become less expensive, nevertheless such systems require careful installation and adjustment; therefore, at least in Europe today, they are very seldom used in churches. The discrepancy between the visual and the aural impressions is mostly regarded as unavoidable.

Again, as with concert halls, the conditions for loudspeaker installations in churches are better, the more the reverberation time is adapted to natural speech. Moreover, the loudspeakers may be installed high above the pulpit or high in the nave, so that the natural time delay guarantees the unity of the visual and aural orientation (see Section I. 5.2, Fig. 5.5). But with reduced reverberation times, the natural (i.e. room-acoustical) means of speech reinforcement are often sufficient.

In spite of the concern for intelligibility of the sermon and prayers, the reverberation times in (occupied) churches must be higher than for pure speech rooms in order to accommodate the music, which was written, after all, *ad majorem gloriam Dei.* Here we are confronted with the fact that the number of people in the congregation varies greatly from service to service. Sometimes the attendance is very poor. Thus it is very desirable that the pews and the kneeling benches (prie-dieux) be upholstered. (See also the arrangement in Section II. 6.6, Fig. 6.16.)

In order to take into account the case of small attendance, Fig. 3.10 shows two upper limits for reverberation time in churches, one for the fully occupied condition, the other for nearly empty churches. Furthermore the measured values of $T_{2,3}$ from Table 9 are included.

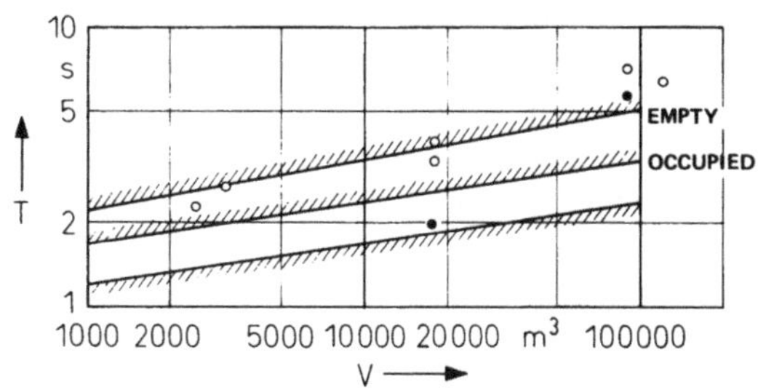

Fig. 3.10. Range of recommended reverberation times for churches. Two upper limits are given: one for the fully occupied church, the other for the nearly empty church. Measurements in empty (○) and occupied (●) churches.

Whereas in churches, speech and music always alternate in the same service, there are other halls that are used for quite different purposes on different occasions; we call them 'multi-purpose halls'. It would be best in such halls if the reverberation time could be changed to be appropriate for each of the several uses. Since this is not always possible, or at least is possible only at great expense, some compromise may be unavoidable, and this must take into account the relative frequency of occurrence of the various kinds of use. Here the decisions depend on the special needs and wishes of the individual client.

Since in order to get a nearly exponential decay, reverberation times are always measured in octave or third-octave bands of frequency, the dependence of reverberation time on frequency is a further available criterion of acoustical quality, and it can be taken into account even in the planning of the hall. For a single-number criterion, this dependence can be characterized by the mean slope of the (T–$\log f$) curve according to Lehmann's formula (eqn. (3.4)), which he uses for the practically incalculable and position-dependent early decay time. Although the slope for the first 10-dB drop may be most important for evaluating the impression of timbre in running music, the application of eqn. (3.4) to the values of T_0–T_4 (125–2000 Hz) will also yield a useful criterion, particularly since Wilkens found that the subjective timbre-factor is nearly independent of location in a hall. Thus it seems reasonable always to check whether the so-calculated $\bar{T}_{04}'$ lies in the range $-0{\cdot}1$ to $-0{\cdot}2$ s/oct., which Lehmann found to be preferred. Carrying out this calculation for the reverberation time frequency dependences shown in Table 8, we find that the Berlin Philharmonie, the Berlin Broadcasting Studio (Sender Freies, Berlin) and the concert halls in Basel, Boston and Vienna lie within these limits. For the last case (where two conditions are given), only the first falls within the limits; the second gives a slope of $-0{\cdot}31$, much steeper than Lehmann's upper limit of $-0{\cdot}2$ s/oct.

It is very surprising that we cannot determine any systematic dependence of the reverberation time slope on the room volume or on the kind of room, as we might have expected from the different dependence of loudness judgments on level in different frequency regions (see Section III. 1.2). This does not call Lehmann's recommendations into doubt, but merely suggests that they should be restricted to room volumes greater than 11 000 m^3, which were used in Wilkens' comparison tests. For small volumes it may be more suitable to follow Békésy's recommendation and make the reverberation time independent of frequency, as his test persons did (see Section III. 1.2). In churches, on the other hand, it may even be desirable, with regard to the lowest register of the organ, to let T increase more rapidly toward the low frequencies below 125 Hz.

Characterizing the frequency dependence of reverberation time by the mean slope of the curve is, naturally, not always sufficient. Also measurements with continuous noise excitation in octave-bands lead to

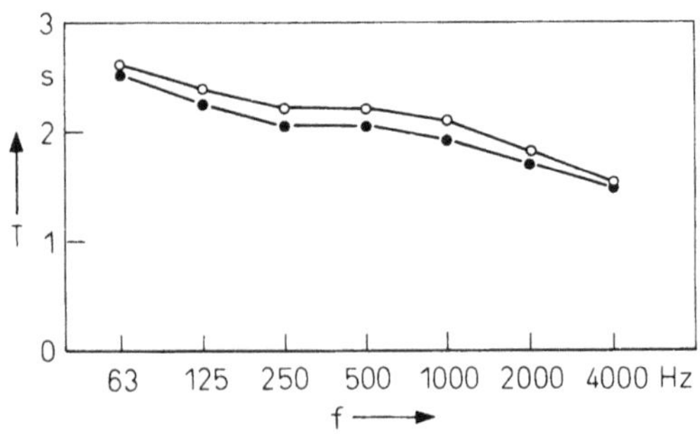

Fig. 3.11. Dependence of the reverberation time on frequency in the Meistersinger Hall in Nürnberg, measured in adjacent octave bands. Upper curve, empty hall; lower curve, occupied by 80 musicians and 2000 listeners. (After H. A. Müller.)

smoothed frequency dependence, of which Fig. 3.11 gives typical examples. If overlapping third-octave filters are used, at steps of 1/3 octave, we may get several small peaks and valleys in the reverberation time curve, as shown in Fig. 3.12; it would have to be checked whether (or how much) these would change for different source and receiver locations. So long as they are no more pronounced than those in Fig. 3.12, it may be assumed, according to Plenge's investigations (see Section III. 2.1), that they are perhaps perceptible but not critical. Certainly in planning a hall one should try to achieve a smooth curve of reverberation time against frequency. This is the reason that acousticians always try to avoid a large number of identically tuned absorptive elements.

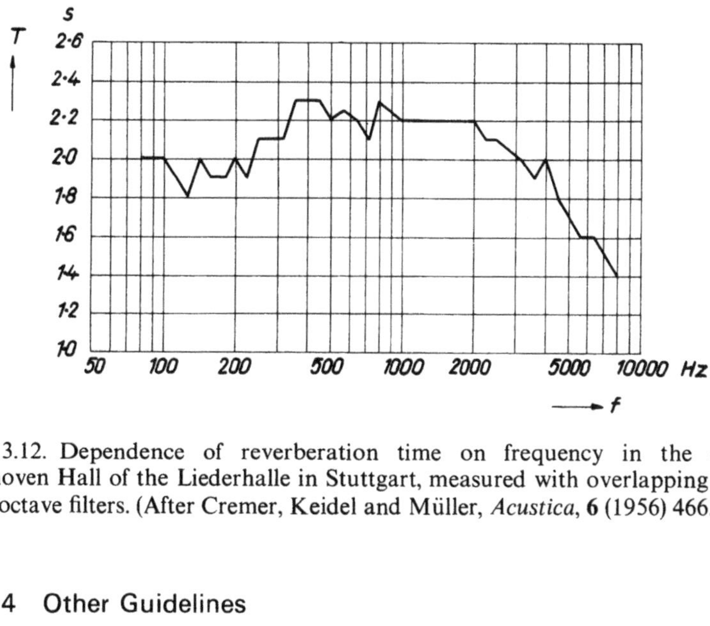

Fig. 3.12. Dependence of reverberation time on frequency in the empty Beethoven Hall of the Liederhalle in Stuttgart, measured with overlapping third-octave filters. (After Cremer, Keidel and Müller, *Acustica*, **6** (1956) 466.)

III. 3.4 Other Guidelines

We come now to the criteria that attempt to account for the subjective factor of distinctness, or clarity. It is not possible to check compliance with these criteria during planning by a simple calculation based on the building drawings, but they may be checked to some degree through the use of model tests. For this we have several options:

(a) the requirement of Reichardt *et al.* for a clarity index (see Section III. 2.5) of at least 0 dB;

(b) the assumption of Gottlob that the distinctness coefficient should approach the optimal value of 0·34, based on the optimal value found for the late-to-early-sound index by Beranek and Schultz (see Section II. 7.4);

(c) the recommendation by Lehmann of a maximum value of center-time of 140 ms, which he infers from Fig. 3.5.

If the sound field obeys the laws of statistical room acoustics, these recommendations would correspond to reverberation times of 1·6, 1·7 and 1·9 s, respectively. That these three values do not quite agree is less serious than the conclusion that higher reverberation times should generally be regarded as unfavorable, independent of the volume. At least for higher reverberation times we are always in the situation of compromise, which was the basis for eqn. (3.9).

Now the first reflections do not obey the statistical laws as do the late reflections. Therefore for the above-stated limiting reverberation times, it is possible that the first reflections follow each other at shorter time intervals and are stronger than would correspond to the exponential decay of a statistical impulse response. This means a faster buildup to the steady state and a faster initial decay in the reverberation. The 'sagging' curves of $L_{(t)}$, which are always to be avoided in the laboratory reverberation room, may very well be desirable in a concert hall. In addition, coupling the auditorium to more reverberant adjacent spaces, for example, to the stagehouse when orchestral concerts are performed in opera houses, could also (to some extent) be desirable.

All of the criteria that we have considered here so far could be determined from the impulse response of a single microphone at a single location. They are, so to speak, 'monaural' and give no information as to direction of arrival. The importance of direction for the room-acoustical impression is easy to demonstrate by closing one ear with a finger during a concert. On the other hand, it is so difficult to carry out the purely physical evaluation of directional distribution and its representation by the 'hedgehog' of Meyer and Thiele[1] (see Section II. 7.6), or even the dependence on direction in different time intervals,[2] that these procedures are quite out of the question for planning a hall with the use of acoustical models. It is therefore understandable if one hopes, by a comparison of the sound pressures at the left and right ears of a dummy

[1] Meyer, E. and Thiele, R., *Acustica*, **6** (1956) 425.

[2] Junius, W., *Acustica*, **9** (1959) 289.

head, to develop an easier method to get information about the directional distribution of sound and to do so in a manner that corresponds to the ear's impression of directions. Among the criteria proposed for this purpose, the 'interaural correlation coefficient' (see Section II. 7.6) is the easiest to determine electronically. But as Fig. 7.11, Part II, shows, it does not always allow unequivocal conclusions with respect to the directional distribution, even if the later diffuse distribution is suppressed by restricting the integration time to 50 ms. Also its application in models with correspondingly small dummy heads has not yet been tested. Thus this criterion is not ready to be used for planning of a hall; it can serve only to provide a supplementary evaluation of the acoustical quality of a finished hall, or at most to guide the improvement of a hall with suitable changes.

Nevertheless it should be recalled that Gottlob has recommended an optimal value of 0·23 for this dimensionless quantity; in other places it is recommended that this quantity should be as small as possible. It remains open for discussion whether it is more the lateral incidence or the diffuse sound field (see Section IV. 13.5, Volume 2) that is desirable. There are still many problems to be solved, particularly since our subjective spatial impressions depend on the sound level (see Section III. 3.1).

To the acoustical consultant it can be recommended, even without a quantitative criterion, that it is desirable to provide as many early reflections as possible, especially lateral reflections, and to take care that the 'late sound field' is homogeneous and isotropic.

Apart from such room-acoustical attributes as strength, distinctness, clarity, and directional impression, which are all positively correlated with good acoustical quality, there are also attributes that correspond to undesirable qualities. One of these is the non-uniform sound distribution, discussed in Chapter I. 4, which is caused by large concave reflecting surfaces and is sometimes combined with the echo impression.

On the other hand, the authors wish to emphasize, on the basis of their own experience, that echoes are not always to be regarded as negative attributes. The reflection from the rear wall of a hall, if it is not too strong, may be for the musicians on stage a useful clue to the acoustical size of the hall and a reassurance that their sound is reaching the most distant seats. Certainly it is possible to overdo this 'little echo', but the acoustician is sometimes surprised at how seldom this happens. When one of the authors asked the casting director of the Munich Opera (which at that time still gave its performances in the Prinzregenten

Theater) whether the focused cueball rear-wall echo (see Fig. 4.4, Part I) caused serious problems, he was told that that echo was well-known but it was not nearly so important as the 'echo in the Press'.

It was certainly helpful that Niese tried to develop an objective single-number criterion for the annoyance of echoes, and it is reasonable to regard only those reflections as echoes that arrive after a critical time and stand out above the general reverberation. But he established his quantitative criterion only with respect to speech and in this case it is best to make each echo as small as possible. An optimal value for musical purposes has never been investigated. It may not even be definable for all situations and all tastes: it may be like salt, without which a dish will be insipid but a bit too much of which will make the dish inedible.

A special kind of echo arises in the periodic sequence of reflections that occurs when a sound ray repeats the same path several times. We referred to this phenomenon as a 'flutter echo' in Section I. 4.3 and showed that it is not easy to detect in an echogram if it is mixed with other, more or less random reflections. But we pointed out (in Section II. 7.5) that such periodic echo-sequences become immediately evident in an auto-correlation function.

Kuttruff[3] proposed to use the ratio of the primary maximum amplitude in the auto-correlation function (at $\tau = 0$) to the amplitude of the largest secondary maximum ($\tau \neq 0$) as a characteristic criterion for a hall; he called it 'temporal diffusion'. The fact that this criterion has not, so far, entered into room-acoustical practice does not exclude its use in the future. At least such a criterion, probably somewhat differently defined, would take into account qualities in echograms that are not accounted for by the criteria introduced in factor analysis. Also there would be no difficulty in recording auto-correlation functions in models.

Every acoustician, along with the authors quoted above, assumes that flutter echoes are to be scrupulously avoided. Hence, another anecdote. One of the authors asked a famous violinist to play at two different locations on the bare stage of the Herkules Hall in Munich. At one of these locations a flutter echo could be heard from the parallel side-walls (it was not noticeable when the stage was occupied) and at the other location it did not occur because of sawtooth-shaped side-walls. At first the artist could not detect any difference in the sound. After having heard the flutter echo following hand-claps, he played some staccato passages and

[3]Kuttruff, H., *Acustica*, **16** (1955/56) 166.

[4]Städtler, H. J., Studienarbeit ITA, TU Berlin, 1969.

found the flutter effect 'interesting' (the German word is *apart*). Similar judgments were also given for comparisons of tape recordings in which violin passages were reproduced first with periodic repetitions and then with random repetitions.[4]

Nevertheless acoustical consultants would be wise to avoid flutter echoes; there surely are signals where flutters would lead to significant annoyance. Certainly a recording engineer would object if he found his microphone in such a self-repeating sound path!

One aspect of room acoustics can always be regarded as detrimental, and that is audible background noise. Just as calligraphy and drawing presuppose a spotless paper, so speech and music demand a nearly silent background. Here the rule for evaluation is always unambiguous: the less noise the better. Although this book cannot enter into all the problems of noise control in performance spaces, at least the criteria of acceptability must be mentioned. The acoustical consultant must demand that these criteria be respected even if he is not responsible for achieving the desired results.

These noise control measures are the only acoustical considerations that are likely to increase the building cost significantly, and subsequent corrective measures for noise control, if they are possible at all, would be very much more expensive.

The choice of the building location surely has a decisive influence on the cost of noise control. If, for instance, a congress hall must be located so that it is easily accessible, it is likely to be placed in the midst of noisy transportation systems and it will require costly building construction to keep out the noise.

In addition the distribution of the rooms inside the building can reduce the noise control costs if it is done with a care to sound isolation, particularly in buildings where several different independent events must occur simultaneously.

Even single performances bring with them their own instrusive noise sources. The performers and the audience need parking areas and toilets, they like to be free to talk and laugh in the lobbies, staircases, cloak rooms and bars. The sound absorbing elements treated in Chapters II.6 and IV.8, IV.9 and IV.10 help to decrease the unavoidable noise levels. They are of special importance in the 'sound locks' through which the auditorium is entered from the lobbies, if it is planned for serious musical or dramatic performances. These 'sound locks' consist of small rooms just outside the auditorium, closed by doors at both sides, only one of which is opened at a time. These sound locks require very heavy damping. It should be mentioned that locks for sound are also locks for light, all around the theater.

Finally there are noises whose sources are inside the auditorium. Here the air-conditioning equipment must be mentioned, even if the fans and other equipment are located in a distant part of the building. The decisive question is how much sound power may be permitted to enter through the ventilation openings (inlets and outlets) in the auditorium, and also whether the air flow velocities are so high that secondary noises are produced at the slits and grilles provided for flow control at these openings. Often the acoustician is assured that the planned layout for the ventilation system will make it 'totally inaudible'. Such a statement only proves that the system noise was previously unnoticeable because the equipment was applied only in rooms where other noises masked it, for example, hotel rooms for banquets and dancing. When lectures are presented in these same rooms, it is astonishing to note what high noise levels are accepted as inevitable by the organizers, not knowing that there exist standardized maximum acceptable noise levels for such events. The remedy most frequently used in such cases is to introduce loudspeaker equipment, which only produces a *circulus vitiosus*, because now the audience feels encouraged to make additional noise by talking with their neighbors, etc.

On the other hand, the costs must be restricted as much as possible for all noise control measures. For this reason maximum acceptable sound pressure levels have been standardized for various kinds of rooms. But since, as shown in Fig. 1.1, the sound pressure levels for the same loudness-sensation have quite different frequency dependences for different loudness levels, an evaluation adequate to the ear's sensitivity would have to analyse the noise with filters in narrow bands and evaluate each band separately. Here, also, acoustical practice is interested in characterizing noises with single-number criteria, which furthermore may be more easily measured and evaluated. For this purpose a frequency-weighted sound pressure has been defined:

$$L_A = 20 \log [a_{(f)} p/p_0] \text{ dB(A)} \tag{3.10}$$

in which the dimensionless weighting factor $a_{(f)}$ is so chosen that the dependence on frequency corresponds to that of the ear for low loudness levels, as is desired. Such a sound level is called 'A-weighted' and is designated by L_A; sometimes, though it is not necessary, the letter A is added behind dB for greater clarity: dB(A).

The maximum acceptable value of L_A depends on the purpose of the room. In lecture or conference rooms or in cinemas, noise levels of

35 dB(A) are admissible. In drama theaters, opera houses and concert halls, a maximum limit of 25 dB(A) should be demanded, for in such rooms there occur critical moments when the audience literally holds its breath.

It is certainly better, because it corresponds to the behavior of the ear to filter the noise into octave bands and to compare each partial band level with a 'grade curve' (see Fig. 3.13). In this case the higher curve corresponds to equally loud perceived levels of the octave bands. If this curve matches the octave band spectrum of the measured noise, then the A-weighted sound level is just 25 dB(A). Since the octave band spectra for air-conditioning systems usually exhibit curves that do not depart much from the shape of the upper curve in Fig. 3.13, the A-weighted sound level, for the measurement of which special instruments exist, may be regarded as a suitable criterion.

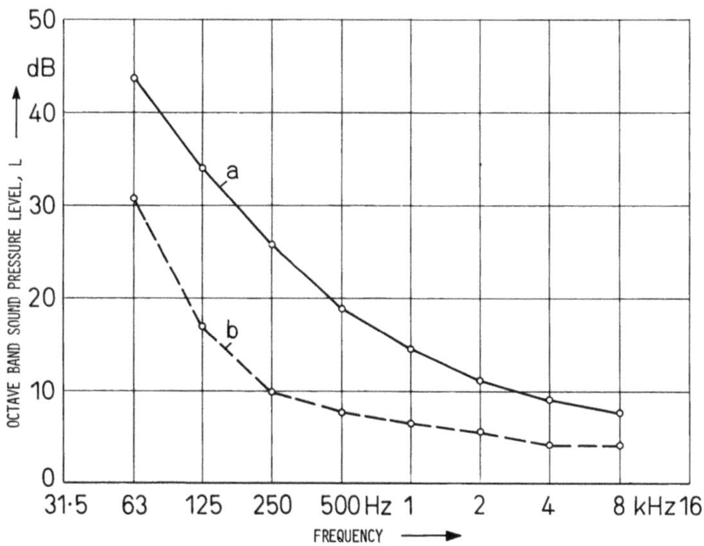

Fig. 3.13. Grade curves for octave band spectra of noises: (a) for drama theaters, opera houses and concert halls; (b) for recording studios. (After Kuhl.[5])

The same holds true for the special fans that cool the spotlights in a theater; these may sometimes be more annoying than the noise of the entire air-conditioning system.

But if the octave band spectrum of a noise has a very different shape and especially if it is concentrated in a single octave band, or even in a narrower band, then it is recommended to compare the measured curves with the grade curves directly. The requirement that the grade curve must not be exceeded in any octave band is certainly much more strict. The corresponding value of L_A may then be very far below 25 dB(A). But from the standpoint of the audibility of the noise, such a strict requirement is justified. Especially annoying at musical events are pure tones, such as are produced by electrical apparatus (e.g. transformers and fluorescent light ballasts) that are connected to a frequency-stable power supply system. Sometimes a single light ballast may annoy the entire audience.

Figure 3.13 also gives a lower grade curve, recommended by Kuhl for recording studios.[5] The still more stringent requirement is based on the fact that a listener in a hall finds it easier to discriminate between the noise and the sounds he is interested in, on account of differences in direction. If he listens to a reproduction of the sound in his own home he gets no such help. For recordings stored on disks or tapes, the repetition of background noises is especially annoying. Furthermore it is often unavoidable in recordings that the level-differences must be electronically compressed and thereby the softer passages are increased in level. In such cases the background noise level is also increased.

On the other hand, it is surprising how seldom musicians object to the background noise. It is possible that such noises degrade the perceived acoustical quality without their being conscious of the reason; the same holds for the listeners. But the disturbance may be quite sufficient to leave the impression that the 'acoustics of the room' are bad!

[5] Kuhl, W., *Acustica*, **14** (1964) 355.

Index

www.ingramcontent.com/pod-product-compliance
Ingram Content Group UK Ltd.
Pitfield, Milton Keynes, MK11 3LW, UK
UKHW020256051026
15470UKWH00020B/344

9 780932 146298